By David McCullough

The Johnstown Flood
The Great Bridge
The Path Between the Seas

DAVID McCULLOUGH

The Path Between the Seas

The Creation of the Panama Canal 1870–1914

A TOUCHSTONE BOOK
PUBLISHED BY SIMON AND SCHUSTER
NEW YORK LONDON TORONTO SYDNEY TOKYO SINGAPORE

A Touchstone Book
Published by Simon & Schuster
Rockefeller Center
1230 Avenue of the Americas
New York, New York 10020
TOUCHSTONE and colophon are registered trademarks
of Simon & Schuster

Designed by Edith Fowler
Manufactured in the United States of America

1 2 3 4 5 6 7 8 9 10
30 29 28 27 26 25 24 pbk.

Library of Congress Cataloging in Publication Data

McCullough, David G.
The path between the seas.

Bibliography: p.
Includes index.
1. Panama Canal—History. I. Title.
F1369.C2M33 972.875'04 76-57967
ISBN 0-671-22563-4
ISBN 0-671-24409-4 Pbk.

For Rosalee Barnes McCullough

Contents

Preface

The creation of the Panama Canal was far more than a vast, unprecedented feat of engineering. It was a profoundly important historic event and a sweeping human drama not unlike that of war. Apart from wars, it represented the largest, most costly single effort ever before mounted anywhere on earth. It held the world's attention over a span of forty years. It affected the lives of tens of thousands of people at every level of society and of virtually every race and nationality. Great reputations were made and destroyed. For numbers of men and women it was the adventure of a lifetime.

Because of it one nation, France, was rocked to its foundations. Another, Colombia, lost its most prized possession, the Isthmus of Panama. Nicaragua, on the verge of becoming a world crossroads, was left to wait for some future chance. The Republic of Panama was born. The United States was embarked on a role of global involvement.

In the history of finance capitalism, in the history of medicine, it was an event of signal consequence. It marked a score of advances in engineering, government planning, labor relations. It was a response to Sedan, a response to the idea of sea power. It was both the crowning constructive effort, "The Great Enterprise," of the Victorian Era and the first grandiose and assertive show of American power at the dawn of the new century. And yet the passage of the first ship through the canal in the summer of 1914—the first voyage through the American land mass—marked the resolution of a dream as old as the voyages of Columbus.

So this book is an attempt to give fitting scope to the subject, to see it whole. I have tried to discover underlying causes for what happened, to measure forces of national pride and ambition, to grasp the still untarnished ideal of progress.

What was the nature of that day and age now gone to dust? What moved people?

Primarily my interest has been in the participants themselves. Of great importance, I felt, was the need to show the enormous variety of people involved, and the skills and strengths called upon by such an undertaking, quite apart from technical competence. I wanted to see these people for what they were, as living, fallible, often highly courageous men and women caught up in a common struggle far bigger than themselves, caught up frequently by forces beyond their control or even their reckoning. I have tried to present the problems they faced as they saw them, to perceive what they did *not* know as well as what they did know at any given time, and to keep constantly in mind that like all mortals in every age, they had no sure way of telling how it would all come out. The book is their story.

Pure chance, fate if you prefer, played a major part, as it always does. Popular misconceptions, self-deceptions large and small, were determining factors all along the line from the time Ferdinand de Lesseps first set things in motion. One is struck, too, by what a moving, potent force personality can be—de Lesseps and Theodore Roosevelt being the outstanding examples. But no less impressive to me are the numbers of instances in which large events turned on the actions of individuals who had little notion that they were playing a part in history.

A good deal of what follows is new. It has been drawn from interviews, from unpublished sources, from published documents hitherto ignored. Several leading characters emerge as quite different from previous portrayals, and major portions of the book are set far indeed from the jungles of Panama.

Much of the French side of the story will, I expect, come as a surprise to many readers. To many readers, also, the Panama revolution and the bizarre chain of events surrounding it may seem more like the creations of fiction. But let me stress that nothing in the book has been invented. Documentation will be found in the Notes at the end.

I feel I should add a word of explanation concerning the current controversy over the canal.

My work was begun years before the canal leaped back into the headlines, and my purpose throughout has remained what it was at the start: to tell a large and important story beginning in 1870 and ending in 1914, because that was where the story belonged—back on the other side of the Great War. That was the world that built the canal.

The root causes of the present controversy are all here, however; they too are part of the story, as the reader will discover.

DAVID MCCULLOUGH

West Tisbury, Massachusetts
October 1976

Far better it is to dare mighty things, to win glorious triumphs, even though checkered by failure, than to take rank with those poor spirits who neither enjoy much nor suffer much, because they live in the gray twilight that knows not victory or defeat.

—THEODORE ROOSEVELT

BOOK ONE

The Vision 1870-1894

1
Threshold

There is a charm of adventure about this new quest...
—*The New York Times*

I

The letter, several pages in length and signed by Secretary of the Navy George M. Robeson, was addressed to Commander Thomas O. Selfridge. It was an eminently clear, altogether formal document, as expected, and had a certain majesty of tone that Commander Selfridge thought quite fitting. That he and the Secretary were personally acquainted, that they had in fact become pleasantly drunk together on one past occasion and vowed eternal friendship as their carriage rolled through the dark capital, were in no way implied. Nor is it important, except that Selfridge, a serious and sober man on the whole, was to wonder for the rest of his days what influence the evening may have had on the way things turned out for him.

His own planning and preparations had already occupied several extremely busy months. The letter was but the final official directive:

Navy Department
Washington, January 10, 1870

Sir: You are appointed to the command of an expedition to make a survey of the Isthmus of Darien, to ascertain the point at which to cut a canal

> from the Atlantic to the Pacific Ocean. The steam-sloop *Nipsic* and the store-ship *Guard* will be under your command . . .
>
> The Department has entrusted to you a duty connected with the greatest enterprise of the present age; and upon your enterprise and your zeal will depend whether your name is honorably identified with one of the facts of the future. . . .
>
> No matter how many surveys have been made, or how accurate they may have been, the people of this country will never be satisfied until every point of the Isthmus is surveyed by some responsible authority, and by properly equipped parties, such as will be under your command, working on properly matured plans. . . .

So on January 22, 1870, a clear, bright abnormally mild Saturday, the *Nipsic* cast off at Brooklyn Navy Yard and commenced solemnly down the East River. The *Guard*, under Commander Edward P. Lull, followed four days later.

In all, the expedition comprised nearly a hundred regular officers and men, two Navy doctors, five civilians from the Coast Survey (surveyors and draftsmen), two civilian geologists, three telegraphers from the Signal Corps, and a photographer, Timothy H. O'Sullivan, who had been Mathew Brady's assistant during the war.

Stowed below on the *Guard* was the finest array of modern instruments yet assembled for such an undertaking—engineers' transits, spirit levels, gradienters, surveyors' compasses and chains, delicate pocket aneroid barometers, mercurial mountain barometers, current meters—all "for prosecuting the work vigorously and scientifically." (The Stackpole transits, made by the New York firm of Stackpole & Sons, had their telescope axis mounted in double cone bearings, for example, which gave the instrument greater rigidity than older models, and the introduction of a simplified horizontal graduation reading allowed for faster readings and less chance of error.) There were rubber blankets and breech-loading rifles for every man, whiskey, quinine, an extra 600 pairs of shoes, and 100 miles of telegraph wire. Stores "in such shape as to be little liable to injury by exposure to rains" were sufficient for four months: 7,000 pounds of bacon, 10,000 pounds of bread, 6,000 pounds of tomato soup, 30 gallons of beans, 2,500 pounds of coffee, 100 bottles of pepper, 600 pounds of canned butter.

The destination was the Darien wilderness on the Isthmus of Panama, more than two thousand miles from Brooklyn, within ten degrees of the equator, and, contrary to the mental picture most people had, east of the 80th meridian—that is, east of Florida. They would land at Caledonia Bay, about 150 miles east of the Panama

Railroad. It was the same point from which Balboa had begun his crossing in 1513, and where, at the end of the seventeenth century, William Paterson, founder of the Bank of England, had established the disastrous Scottish colony of New Edinburgh, because Caledonia Bay (as he named it) was to be the future "door of the seas." Harassed by the Spanish, decimated by disease, the little settlement had lasted scarcely more than a year. Every trace of it had long since vanished.

Darien was known to be the narrowest point anywhere on the Central American isthmus, by which was meant the entire land bridge from lower Mexico to the continent of South America and which included the Isthmus of Tehuantepec, Guatemala, Honduras, British Honduras, El Salvador, Nicaragua, Costa Rica, and Panama, the last of which was still a province—indeed a most prized province—of Colombia. From Tehuantepec to the Atrato River in Colombia, the natural, easternmost boundary of Central America, was a distance of 1,350 miles as the crow flies, as far as from New York to Dallas, and there were not simply a few, but many points along that zigzagging land mass where, on the map at least, it appeared a canal could be cut. A few years before, Admiral Charles H. Davis had informed Congress that there were no fewer than nineteen possible locations for a Central American ship canal. But at Darien the distance from tidewater to tidewater on a straight line was known to be less than forty miles.

Because of the particular configuration of the Isthmus of Panama—with the land barrier running nearly horizontal between the oceans—the expedition would be crossing *down* the map. The men would make their way from the Caribbean on the north to the Pacific on the south, just as Balboa had. (Hence Balboa's designation of the Pacific as the Sea of the South had been perfectly logical.) The Panama Railroad, the nearest sign of civilization on the map, also ran from north to south. Its faint, spidery red line looked like something added by a left-handed cartographer, with the starting point at Colón, on Limon Bay, actually somewhat farther west than the finish point at Panama City, on the Bay of Panama.

They were to measure the heights of mountains and the depths of rivers and harbors. They were to gather botanical and geological specimens. They were to take astronomical observations, report on the climate, and observe the character of the Indians encountered. And they were to lose as little time as possible, since the rainy season—the sickly season, Secretary Robeson called it—would soon be upon them.

Six other expeditions were to follow. A Presidential commission, the first Interoceanic Canal Commission, would be established to appraise all resulting surveys and reports and to declare which was the chosen path. The commission would include the chief of the Army Engineers, the head of the Coast Survey, and the chief of the Bureau of Navigation. Nothing even remotely so systematic, so elaborate or sensible, had ever been attempted before.

But the Darien Expedition was the first, and the fact that it was to Darien, one of the wildest, least-known corners of the entire world, was a matter of extreme concern at the Navy Department. Sixteen years earlier, in 1854, well within the memory of most Americans, an expedition to Caledonia Bay had ended in a disaster that had the whole country talking and left the Navy with a profound respect for the terrors of a tropical wilderness. What had happened was this.

In 1850, Dr. Edward Cullen, an Irish physician and member of the Royal Geographical Society, had announced the discovery of a way across Darien by which he had walked from the Atlantic to the Pacific several times and quite effortlessly. He had been careful to mark the trail, Cullen said, and at no place had he found the elevation more than 150 feet above sea level. It was the miracle route everyone had been searching for and the story caused a sensation. A joint expedition to Darien was organized by England, France, Colombia (then known as New Granada), and the United States. But when the American ship, *Cyane*, reached Caledonia Bay ahead of the others, Navy Lieutenant Isaac Strain and a party of twenty-seven men started into the jungle without waiting, taking provisions enough for only a few days and fully expecting to pick up Cullen's trail. Balboa, when he started into this same jungle, had gone with a force of 190 heavily armed Spaniards and several hundred Indians, some of whom knew the way.

Strain was not seen again for forty-nine days. His troubles had begun from the moment he set foot on shore. The Indians, impressed by the guns of the *Cyane*, agreed to let his party pass, but refused to serve as guides. Cullen's trail was nowhere to be found. Within days the expedition was hopelessly lost. Food ran out; rifles became so rusted as to be useless. Strain picked up a large river—the Chucunaque—which he thought would take him to the Pacific but which, in reality, was leading him on an endless looping course eastward, through the very center of the Isthmus. When a band of Indians warned him that it was the wrong way, he decided they were deliberately trying to mislead him.

Verging on starvation, his men devoured anything they could lay hands on, including live toads and a variety of palm nut that burned the enamel from their teeth and caused excruciating stomach cramps. The smothering heat, the rains, the forbidding jungle twilight day after day, were unlike anything any of them had ever experienced. Seven men died; one other went temporarily out of his mind. That any survived was due mainly to the discipline enforced by Strain and Strain's own extraordinary fortitude. Leaving the others behind, he and three of the strongest men had pushed on in search of help. When they at last staggered into an Indian village near the Pacific side, Strain, who was torn and bleeding and virtually naked, turned around and led a rescue mission back to the others. A British doctor who examined the survivors described them as the most "wretched set of human beings" he had ever seen. "In nearly all, the intellect was in a slight degree affected, as evinced by childish and silly remarks, although their memory, and the recollection of their sufferings, were unimpaired. . . . They were literally living skeletons, covered with foul ulcers. . . ." Strain's weight was seventy-five pounds. A few years later, at Colón, having never fully recovered, Strain died at age thirty-six.

Strain had found the mountains at Darien not less than one thousand feet. From what he had seen, Darien was "utterly impracticable" as the route for a canal. Just the same, others were not quite willing to abandon the idea. While Strain's ordeal was taken as a fearful object lesson at the Navy Department, there were some who were still willing to accept the possibility that Edward Cullen had been telling the truth after all.

Cullen, who had come out with one of the British ships but then made a hasty retreat to Colón (and from there to New York) the moment it appeared something was amiss, turned up later as a surgeon with the British Army in the Crimean War. He also kept persistently to his story. The expedition had been deplorably misled, he argued. Strain had had no business proceeding without him or without his map, which by itself would have made all the difference.

Admiral Davis, Commander Selfridge, and, most importantly, Admiral Daniel Ammen, chief of the Bureau of Navigation, were among those who considered the case still very much open. "It is to the isthmus of Darien that we are first to look for the solution to the great problem," Davis had informed Congress. "The statements of Dr. Cullen had been so severely criticized," Selfridge was to explain, "and so persistently advocated by him, that I was inclined to put some faith in his representations." To Admiral Ammen, who had pored over every recorded detail of the episode, the critical clue was in Strain's own report. Days after he had started inland, at a time when he should have been well beyond earshot of Caledonia Bay, Strain had written in his journal of hearing the evening gun on the *Cyane*, and this, Ammen believed, was evidence of a low-lying valley running inland from the bay; otherwise the sound would have been blocked by intervening hills.

Interest in the new expedition was considerable in numerous quarters. The very times themselves seemed so immensely, so historically favorable. If there was one word to characterize the spirit of the moment, it was Confidence. Age-old blank spaces and mysteries were being supplanted on all sides. The summer before, the one-armed John Wesley Powell, in the interests of science, had led an expedition down the Colorado River into the Grand Canyon. The great geological and geographical surveys of the West had begun under the brilliant Clarence King. Poking about in godforsaken corners of the western desert, Othniel C. Marsh, of Yale, who was not yet forty and the country's first and only professor of paleontology, had unearthed the fossils needed to present the full evolution of the horse, the most dramatic demonstration yet of Darwin's theory.

People were reading Jules Verne's *Twenty Thousand Leagues under the Sea.* The Roeblings had begun their Brooklyn Bridge. Harvard had installed a chemist as its president. In Pittsburgh, experiments were being made with a new process developed by the English metallurgist Bessemer. And within the preceding nine months alone two of the most celebrated events of the century had occurred: the completion of the Union Pacific Railroad and the opening of the Suez Canal. All at once the planet had grown very much smaller. With the canal, the railroad, the new iron-screw ocean steamers, it was possible—in theory anyway—to travel around the world in a tenth of the time it would have taken a decade earlier, as Jules Verne would illustrate in his next *voyage extraordinaire.*

The feeling was that the revealed powers of science, "the vast strides made in engineering and mechanical knowledge," as Commander Selfridge would say, had brought mankind to a threshold. It was said that the power generated by one steamship during a single Atlantic crossing would be sufficient to raise from the Nile and set in place every stone of the Great Pyramid. Men talked confidently of future systems of transport that would bring all peoples into contact with one another, spread knowledge, break down national divisions, and make a unified whole of humanity. "The barrier is down!" a French prelate proclaimed on the beaches of Port Said when Suez was opened. "One of the most formidable enemies of mankind and of civilization, which is distance, loses in a moment two thousand leagues of his empire. The two sides of the world approach to greet one another . . . The history of the world has reached one of its most glorious stages."

There really seemed no limit to what man might do. While an official report of the kind Commander Selfridge was to submit might contain the expression "under Providence" (in conjunction with certain accomplishments), such terms seemed perfunctory. Man, modern man—the scientist, the explorer, the builder of bridges and waterways and steam engines, the visionary entrepreneur—had become the central creative force. In the summer of 1870, the summer Selfridge returned from Darien, thirty, perhaps forty, thousand people would fill London's Crystal Palace for a public reception that only a Nelson might have been accorded in an earlier day. Thousands of rockets would hurtle into the night and two hundred boys from the Lambeth Industrial Schools would wave four hundred colored flares in an "Egyptian Salute," all to honor the Frenchman Ferdinand de Lesseps, builder of the Suez Canal. This was jubilation of a kind not known before and that future generations would have some trouble comprehending. De Lesseps' desert passage of 105 miles had brought Europe 5,800 miles closer to India. The Near East had been restored to its ancient position as a world crossroads. Africa had been made an island at a stroke. And the fact that the project had been denounced by men reputedly far wiser than de Lesseps—most especially by Britain's own Robert Stephenson—made the ultimate triumph all the more thrilling.

Victoria, who was to give a name to the era, its elegance, its sense of purpose, its heavy, varnished furniture, its small and large hypocrisies, was very much in her prime at age fifty-one. Samuel Smiles, that most eminent Victorian, had published his *Lives of the Engineers*, wherein good and useful giants—Brindley of the English canals, Rennie of the Waterloo Bridge, the genius Telford—did good and useful work for the betterment of all. Paris was newly transformed by the brilliant Georges Haussmann, prefect of the Seine, and the picture-book troops of Napoleon III, in their kepis and *pantalons rouges*, were thought to be the most formidable on earth, the Franco-Prussian War being still over the horizon.

Among the American tourists to be found strolling Baron Haussmann's magnificent boulevards as the *Nipsic* and the *Guard* sailed for

Darien was an undersized eleven-year-old in the company of his parents, Theodore Roosevelt, whose ambition at the moment was to be a naturalist.

The President of the United States at this juncture was Ulysses S. Grant and it was he, the year before, who had instructed Admiral Ammen to organize the series of expeditions—"practical investigations," he called them. Grant, despite his subsequent reputation as a President of little vision or initiative, was more keenly interested in an isthmian canal than any of his predecessors had been. He was indeed the first President to address himself seriously to the subject. If there was to be a water corridor, he wanted it in the proper place—as determined by civil engineers and naval authorities—and he wanted it under American control. "To Europeans the benefits of and advantages of the proposed canal are great," he was to write, "to Americans they are incalculable."

Grant's blind faith in old friends was to prove his greatest failing as time wore on, but in Admiral Ammen, a friend since boyhood, he had made an excellent choice. Ammen had been reassigned from sea duty and put at the head of the Bureau of Navigation almost the moment Grant became President. A picture of authority, Ammen was whiskered, grizzled, like Grant himself, but with a large, imposing nose and a permanent scowl. Once, while in command of a training cruise to Panama, he had settled a mutiny on the instant by calmly shooting the two leaders. He also had an agile and resourceful mind.

The Navy was to provide the ships and most of the personnel. Ammen selected the officers. Thomas Oliver Selfridge, irrespective of any impression he may have made on Secretary Robeson, had been first in his class at Annapolis and distinguished himself as a commander of gunboats at Vicksburg and on the Red River. En route to Darien he would celebrate his thirty-fourth birthday. Captain Robert Shufeldt, who would lead the Tehuantepec Expedition in the fall of 1870, had had thirty years at sea. He was a physical giant who appeared equal to any wilderness and he had, besides, considerable tact. (Though it had been more than twenty years since the Mexican War, there was much apprehension over the reception an American expedition might receive in Tehuantepec.) And the studious, likable Edward P. Lull, who was in charge of the *Guard*, and who was later to command both the Nicaragua and Panama expeditions, was as able a young officer as was to be found in the Navy.

These particular officers, moreover, had been imbued with a star-spangled sense of American destiny in the Pacific Ocean. As a young lieutenant, Daniel Ammen had sailed on Commodore James Biddle's voyage to China and Japan, the voyage that resulted in 1846 in the first treaty between China and the United States. Selfridge also had begun his career with a South Pacific cruise and Shufeldt had been in command of the *Wachusett* in the Orient only the year before the Tehuantepec Expedition.

"Sufficient is it to add that advantageous as an interoceanic canal would be to the commercial welfare of the whole world, it is doubly so for the necessities of American interests," Selfridge was to write. "The Pacific is naturally our domain."

"It may be the future of our country lies hidden in this problem," Shufeldt would address his crew when the *Kansas* sailed for Tehuantepec. And from the rail of a battered little river steamer laboring against the brown current of the San Juan, his eyes squinting against the hard glare of a Nicaragua morning, Edward Lull would envisage American ships of the line riding the same path to the Pacific.

These were professional sailors, not remarkable men, or so they undoubtedly would have said. They were experienced in command, meticulous about details, physically very tough; but without airs or pretense. In the field, with their sun hats and field glasses, their blue northern eyes, they would look much like other English-speaking harbingers of civilization in other so-called "dark" corners of the world. But there was no overflowing ego among them, no Burton or Speke or Stanley possessed by visions of personal destiny. Nor were they great men in the way a Powell or a King was, intellectually and in orginality of purpose. Had they been asked, they undoubtedly would have said they were doing their job.

II

The seven Grant expeditions to Central America between 1870 and 1875 can be seen as a sharp, clean line through the whole long history of canal plans and proposals reaching back to an obscure reference concerning an obscure Spaniard, Alvaro de Saavedra, a kinsman of Cortez', who supposedly "meant to have opened the land of Castilla del Oro . . . from sea to sea." There had never been any serious possibility of a canal during Spanish times. "There are mountains, but there are also hands" was the lovely declaration of a Spanish priest of the sixteenth century, "and for a king of Castile, few things are impossible." The priest, Francisco López de Gómara, was the first to raise the issue of location, naming Panama, Nicaragua, Darien, and Tehuantepec as the best choices, in a book published in 1552. But he was sadly deceiving himself. Not for another three hundred years, not until the nineteenth century, would a canal, even a very small canal, become a reasonable possibility. It required certain advances in hydraulic engineering, among other things; and it required the steam engine.

The place most nineteenth-century North Americans expected to see the canal built, including the President, was Nicaragua. If not Darien, it would be through Lake Nicaragua; if not there, then probably it would have to be Panama. Tehuantepec had the virtue of being so much closer to the United States, but that was about all that could be said for Tehuantepec. The great overriding problem, however, was the extremely low level of reliable geographical information on Central America, and this despite more than fifty years of debate over where a canal ought to go, despite volumes of so-called geographical research, engineering surveys, perhaps a hundred articles in popular magazines and learned journals, promotional pamphlets, travel books, and the fact that Panama, Nicaragua, and Tehuantepec had all been heavily traveled shortcuts to the Pacific since the time of the California gold craze. As Admiral Davis had quite accurately stated, there were not in the libraries of the world the means to determine even approximately the most practicable route.

The earliest authoritative study of the problem, or rather the first to be taken as authoritative, appeared in 1811 and designated Nicaragua as the route posing the fewest difficulties. The author of this rather tentative benediction was Alexander von Humboldt, the adventurous German-born naturalist and explorer, and Nicaragua thereafter had been "Humboldt's route." Humboldt, as it happens, had never set foot in Nicaragua, or in any of the four alternatives he named. He had built his theories wholly from hearsay, from old books and manuscripts, and the few pitiful maps then available, all of which he plainly acknowledged. The precise location of the City of Panama was not even known, he warned. Nor had anyone determined the elevation of the mountains at Panama, or at any other point along the spine of Central America.

Panama he judged to be the worst possible choice, primarily because of the mountains, which he took to be three times as high as they actually are. Tehuantepec appeared to be too broad, as well as mountainous, and he feared the "sinuosity" of the rivers. About the best that could be done at either Panama or Tehuantepec would be to build some good roads for camels.

Humboldt was still comparatively unknown when he wrote his *Political Essay on the Kingdom of New Spain*, the book containing his long canal essay; his renown was limited still to scientific circles. No Peruvian current or glacier or river had been named for him; Humboldt, Kansas, and Humboldt, Iowa, were still unbroken prairie grass. His views, nonetheless, were to have more influence on the canal issue than everything that had been written previously taken together, for by mid-century he was to tower above all others as the beloved high priest of modern science, a university unto himself, as Goethe would say.

Humboldt's *Political Essay* was the result of a five-year journey through Spanish America, the likes of which would never be equaled. He had been up the Orinoco and the Magdalena; he had been over the Andes on foot. In Ecuador he had climbed Chimborazo, then believed to be the highest mountain on earth, and though he failed to reach the top, he had gone to nineteen thousand feet, which was higher—considerably higher—than any human being had ever been before, even in a balloon. If he had not been in Nicaragua or Panama or Tehuantepec or anywhere along the drenched, green valley of the Atrato River, the location of his two other possible pathways to the Pacific, he had been almost everywhere else and no one was assumed to have more firsthand knowledge of the American jungle. The rather vital fact that his canal theories were almost wholly conjecture was generally ignored. Moreover, those who used his name to substantiate their own pet notions, those who would quote and misquote him endlessly, would find it convenient to forget that it was he who insisted that no canal should be considered until the comparative advantages and disadvantages of *all* possible routes were examined firsthand by experienced people and according to uniform standards.

The Nicaragua canal he visualized was much along the lines of Thomas Telford's Caledonian Canal in Scotland, then the most ambitious thing of its kind. Lake Nicaragua, besides being navigable, would, like Telford's Scottish lakes, provide a natural and limitless source of water for the canal—a vast "basin"—at the very summit of the canal.

Should Nicaragua be found unsatisfactory, then perhaps one of the two routes on the Atrato would serve best. The Napipi-Cupica route, as he named it and as it is still known, would follow the Napipi River, a tributary of the sprawling Atrato, to its headwaters, then continue down to the Pacific at Cupica Bay.

The other Atrato scheme, the so-called "Lost Canal of the Raspadura," appealed mainly to his imagination. Years before, he had heard, a Spanish monk "of great activity" had induced some Indians to build a secret passage betweeen the Atrato and the Pacific, a passage large enough only for small boats, but one that followed a near-perfect path for a canal of larger size, somewhere off the Raspadura River, another distant tributary. All one had to do was find it.

How much of all this he may have discussed with Thomas Jefferson in the spring of 1804, at the end of the Spanish-American odyssey, is not known. But probably his stay at the White House marks the start of Presidential interest in the canal. It is known that Jefferson had shown prior curiosity on the subject while he was minister to France. Furthermore, the visit coincided with the departure of Lewis and Clark from St. Louis to seek, on Jefferson's orders, a northwest water passage to the Pacific. And Humboldt, a lean, deeply tanned, explosively energetic young man, had so enthralled Jefferson with accounts of his travels that Jefferson kept him on as a guest for two weeks. So it is difficult to imagine them *not* discussing a Central American corridor as they strolled the White House grounds or sat conversing, hours on end, at the big table in Jefferson's first-floor office, maps and charts all over one wall and Jefferson's pet mockingbird swinging in a cage overhead.

Humboldt's Spanish-American travels had been the result of an unprecedented grant from the Spanish Crown to investigate wherever he wished in the cause of scientific progress. Until then explorations of any kind by foreigners within Spain's New World realm had been strenuously discouraged. But once Spanish rule began to dissolve in the 1820's the way was open to almost anyone. And almost anyone was what turned up. Engineers, naval officers, French, English, Dutch, Americans, promoters, journalists, many of whom expressed grand visions of a canal, in the event political permission could be obtained, in the event the necessary capital could be assembled. A few of these were able people, but very few had any technical competence. Many of them were also perfectly genuine in their aspirations and sincerely believed in their rainbow-hued promises, however inept or naïve they may have been. Others, quite a good many others, were petty adventurers or outright crackpots.*

* One outstanding example was Charles de Thierry, or Baron de Thierry, as he preferred. An Englishman and graduate of Cambridge, he had so impressed some Maori chiefs who were visiting London that they asked him to come to New Zealand and rule as their king, or so he reported. His idea was to build a canal across Panama to further European trade with New Zealand and he thought the

The canals they had in mind, regardless of specified location, were invariably feasible technically, within range financially, and destined to be bonanzas for all investors and for whichever impoverished little Central American republic was to be involved. Emissaries from Bogotá and Managua and Mexico City were dispatched to the capitals of Europe and to Washington to enlist support. Even the pope was approached. Special agreements and franchises were signed and sealed with appropriate formality. The future was rich with possibilities.

With the opening of Telford's canal and the Erie Canal, both in the 1820's, reasonable men also felt justified in projecting comparable works across the map of Central America. "Neptune's Staircase," the spectacular system of locks on the Caledonian Canal, could lift seagoing ships—could lift a thirty-two-gun frigate, for example—a hundred feet up from the level of the sea. The Erie Canal, though built for shallow-draft canal barges, was nonetheless the longest canal in the world, and its locks overcame an elevation en route of nearly seven hundred feet. So on paper a canal at Panama or Nicaragua or any other place in favor at the moment did not seem unrealistic. Telford in his last years was considering "a grand scheme" for Darien. DeWitt Clinton, "father" of the Erie Canal, had joined with Horatio Allen, builder of the Croton Aqueduct, to plan a water passage through Nicaragua.

A skeptical or cautionary voice was the rare exception. The view of someone such as Colonel Charles Biddle, sent by President Andrew Jackson to appraise Panama and Nicaragua, stands in solitary contrast to almost everything else being written or said. Having made his way up the Chagres River by canoe, then overland to Panama City, a trek of four days, Biddle concluded that any talk of a Panama canal was utter foolishness and that this ought to be clear to all men, "whether of common or uncommon sense." (He did not bother to go see Nicaragua.)

Far more representative were the views of John Lloyd Stephens, which appeared about the time John O'Sullivan, editor of the *Democratic Review*, was writing that "our manifest destiny is to overspread the continent allotted by Providence for the free development of our yearly multiplying millions."

The task, declared Stephens, posed no major problems and ought not cost more than $25,000,000, a figure most people took to be absurdly high.

---

complete project could be finished without difficulty in three years. A railroad over the same route was quite out of the question, however, he said, since the ground was so uneven and covered with so many leaves.

Stephens was "the American traveler," an engaging, romantic, red-bearded lawyer and author of popular travel books who passed through Nicaragua on his way to the Mexican provinces of Chiapas and Yucatán in 1840. He was looking for the "lost" cities of the Maya, which he found, and the book describing those discoveries, *Incidents of Travel in Central America*, went through edition after edition. It was a classic, thrilling piece of work and can be seen now as the beginning of American archaeology. But Stephens had no more business issuing pronouncements on the feasibility of a Nicaragua canal from the little he had seen than had the engineer Horatio Allen from the comforts of his Manhattan office.

A Nicaragua canal posed no major problems, Stephens declared. Here was an enchanting land of blue lakes and trade winds, towering volcanic mountains, rolling green savannas and grazing cattle. Nicaragua could become one of the finest resorts on earth were a canal to be built. Like Humboldt he had scaled a volcano—Masaya—then, to the horror of his guide, descended bravely into its silent crater. "At home, this volcano would be a fortune, with a good hotel on top, a railing to keep the children from falling in, a zigzagging staircase down the sides, and a glass of iced lemonade at the bottom." The mountain, he noted, could probably be purchased for ten dollars.

The truth is that all the canal projects proposed, every cost estimated, irrespective of the individual or individuals responsible, were hopelessly unrealistic if not preposterous. Every supposed canal survey made by mid-century was patently flawed by bad assumptions or absurdly inadequate data. Assertions that the task would be simple were written by fools or by men who either had no appropriate competence or who, if they did, had never laid eyes on a rain forest.

The one important step taken prior to the California gold rush was of another kind, but very little was made of it.

On December 12, 1846, at Bogotá, a new American chargé d'affaires, Benjamin Alden Bidlack, of Wilkes-Barre, Pennsylvania, acting entirely on his own initiative, signed a treaty with the government of President Tomás Cipriano de Mosquera. The critical agreement was contained in Article XXXV. New Granada guaranteed to the United States the exclusive right of transit across the Isthmus of Panama, "upon any modes of communication that now exist, or that may be, hereafter, constructed." In exchange the United States guaranteed "positively and efficaciously" both the "perfect neutrality" of the

Isthmus and New Granada's rights of sovereignty there. (It was this agreement by which the Panama Railroad was to be made possible.)

In Washington the news was greeted with only moderate interest since Bidlack had acted without instruction and since there was much old, deep-seated distrust of "entangling" alliances. Not for another year and a half did the Senate act on confirmation and not until the government of New Granada had sent a special envoy to Washington, the very able Pedro Alcántara Herrán, to lobby for the agreement.

The Bidlack Treaty, as it was commonly called, was Bidlack's only diplomatic triumph. A small-town lawyer and newspaper editor, a congressman briefly before going to Bogotá, he died seven months after the treaty was ratified.

For three centuries the gold in the stream beds of the Sierra Nevada had gone undetected and for all the commotion over Central American canals in the first half of the new world-shaking nineteenth century, Central America remained a backwater. No canals, no railroads were built. There was not a single wagon road anywhere across the entire Isthmus. But in January of 1848 a carpenter from New Jersey saw something shining at the bottom of a millrace at Coloma, California, and within a year Central America re-emerged from the shadows. Again, as in Spanish times, gold was the catalyst.

There were three routes to the new El Dorado—"the Plains across, the Horn around, or the Isthmus over"—and for those thousands who chose "the Isthmus over," it was to be one of life's unforgettable experiences. The onslaught began first at Panama, early on the morning of January 7, 1849, when the little steamer *Falcon* anchored off the marshy lowlands at the mouth of the Chagres River and some two hundred North Americans—mostly unshaven young men in red flannel shirts loaded down with rifles, pistols, bowie knives, bedrolls, pots and pans, picks, shovels—came swarming ashore in one great noisy wave. To the scattering of native Panamanians who stood gaping, it must have seemed as if the buccaneer Morgan had returned after two hundred years to storm the Spanish bastion of San Lorenzo, the frowning brown walls of which still commanded the entire scene. The invaders shouted and gestured, trying to make themselves understood. Nobody seemed to have the least idea which way the Pacific lay and all were in an enormous hurry to get started.

Amazingly, all of this first group survived the crossing. They came dragging into Panama City, rain-soaked, caked with mud, hollow-eyed from lack of sleep, and ravenously hungry. They had gone up the Chagres by native canoe, then overland on mule and on foot, as Charles Biddle had and as thousands more like them would, year after year, until the Panama Railroad was in service. Old letters and little leather-bound journals mention the broiling heat and sudden blinding rains. They speak of heavy green slime on the Chagres, of nights spent in vermin-infested native huts, epidemics of dysentery, mules struggling up to their haunches in the impossible blue-black Panama muck. A man from Troy, New York, counted forty dead mules along the Cruces Trail, the twisting jungle path, barely three feet wide, over which they all came from the river to Panama City. Others wrote of human companions dropping in their tracks with cholera or the dreaded Chagres fever.

"I have no time to give reasons," a Massachusetts man wrote home after crossing Panama, "but in saying it I utter the united sentiment of every passenger whom I have heard speak, it is this, and I say it in fear of God and the love of man, to one and all, for no consideration come this route. I have nothing to say for the other routes but do not take this one."

Yet the gain in time and distance was phenomenal. From New York to San Francisco around the Horn was a months-long voyage of thirteen thousand miles. From New York to San Francisco by way of Panama was five thousand miles, or a saving of eight thousand miles. From New Orleans to San Francisco by Panama, instead of around the Horn, the saving was more than nine thousand miles.

Besides, how one responded to Panama depended often on the season of the year and one's own particular make-up. Many were thrilled by the lush, primeval spectacle of the jungle—"overwhelmed with the thought that all these wonders have been from the beginning," as one man wrote. For wives and parents left behind they described as best they could those moments when magnificent multicolored birds burst into the sky; the swarms of blue butterflies—"like blossoms blown away"; the brilliant green mountains, mountains to put Vermont to shame said a young man from Bennington who was having a splendid time traveling up the Chagres. "The weather was warm but we had a roof to our boat . . . and what was of more consequence still we had on board a box of claret wine, a bacon, bread, and a piece of ICE!"*

* The ice was supplied by the Boston and Panama Ice Company and it sold for as much as fifty cents a pound when first introduced on the Isthmus. One ship from Boston carried seven hundred tons of ice packed in sawdust all the way around the Horn to Panama City, with a loss from melting of only one hundred

The little railroad was begun in 1850, with the idea that it could be finished in two years. It was finished five years later, and at a cost of $8,000,000, six times beyond anyone's estimate. For a generation of Americans there was something especially appealing about the picture of this line across Panama, of a steam locomotive highballing through the jungle, pulling a train of bright passenger cars, a steam whistle scattering monkeys to the treetops—"ocean to ocean" in something over three hours. It was also the world's first transcontinental railroad—one track, five-foot (or broad) gauge, exactly forty-seven and one-half miles long—and the most expensive line on earth on a dollar-per-mile basis, expensive to build and expensive to travel. A one-way ticket was $25 in gold.

To its owners the railroad was the tiny but critical land link in the first all-steam overseas system to span the new continental United States. The Pacific Mail Steamship Company, with offices in New York, had been established just before the news of California gold reached the East, or when such an idea had looked dangerously, if not insanely, speculative. The ships operated to and from Panama on both oceans, providing regular passenger service and mail delivery to California. (A generous subsidy from the federal government to carry the mail had made it considerably less speculative.) William Henry Aspinwall, a wealthy New York merchant, was the founder and guiding spirit of the steamship line, and in the railroad venture he was joined by a banker named Henry Chauncey and by John Lloyd Stephens, who, in the time since his Nicaragua travels, had concluded that Panama was where the future lay. Stephens was the first president of the Panama Railroad Company and its driving force until his death at age forty-six. He was the one member of the threesome to stay with the actual construction effort in the jungle, and the result was an attack of fever, a recurrence of which was fatal in the fall of 1852.

Having, as it did, a monopoly on the Panama transit, the railroad was a bonanza. Profits in the first six years after it was finished were in excess of $7,000,000. Dividends were 15 percent on the average and went as high as 44 percent. Once, standing at $295 a share, Panama Railroad was the highest-priced stock listed on the New York Exchange.

---

tons. But in the process of getting the ice from ship to land to the Panama icehouse, a distance of two miles, another four hundred tons melted. Yet such was the demand that the sale of the remaining two hundred tons paid for the voyage. Within a few years, ice on the Pacific side was being supplied by ships from Sitka, from what was then known as Russian America.

So dazzling a demonstration of the cash value of an ocean connection at Panama, even one so paltry as a little one-track railroad, was bound to draw attention. Matthew Fontaine Maury, the pioneer oceanographer, had told a Senate committee as early as 1849 that a Panama railroad would lead directly to a Panama canal "by showing to the world how immense this business is," but nobody had been prepared for success on such a scale. The volume of human traffic alone—upward of 400,000 people between 1856 and 1866—gave Panama a kind of most-beaten-path status unmatched by any of the other canal routes talked of.

Surveys for the railroad had also produced two pertinent pieces of information. The engineers had discovered a gap in the mountains twelve miles from Panama City, at a point called Culebra, where the elevation above sea level was only 275 feet. This was 200 feet less than what had been considered the lowest gap. Then, toward the close of their work, they had determined once and for all that there was no difference between the levels of the two oceans. The level of the Pacific was not twenty feet higher than that of the Atlantic, as had been the accepted view for centuries. Sea level was sea level, the same on both sides. The difference was in the size of their tides.

(The tides on the Pacific are tremendous, eighteen to twenty feet, while on the Caribbean there is little or no tide, barely more than a foot. When Balboa stood at last on the Pacific shore, he had seen no rush of lordly breakers, but an ugly brown mud flat reaching away for a mile and more, because he had arrived when the tide was out.)

Yet, ironically, it was the experience of the railroad builders that argued most forcibly for some different path, almost any other location, for the canal. If humane considerations were to be entered in the balance, then Panama was the worst possible place to send men to build anything.

Panama had been known as a pesthole since the earliest Spanish settlement. But the horror stories to come out of Panama as the railroad was being pushed ahead mile by mile quite surpassed anything. The cost paid in human life for the minuscule bit of track was of the kind people associated with dark, barbaric times, before the age of steam and iron and the upward march of Progress. The common story, the one repeated up and down the California gold fields, the one carried home on the New York steamer, the claim that turns up time and again in the dim pages of old letters, is that there was a dead man for every railroad tie between Colón and Panama City. In some versions it was a dead Irishman; in others, a dead Chinese. The story was nonsense—there were some seventy-four thousand ties along the Panama line—but that had not kept it from spreading, and from what many thousands of people had seen with their own eyes, it seemed believable enough.

How many did actually die is not known. The company kept no systematic records, no body count, except for its white workers, who represented only a fraction of the total force employed over the five years of construction. (In 1853, for example, of some 1,590 men on the payroll, 1,200 were black.) However, the company's repeated assertion that in fact fewer than a thousand had died was patently absurd. A more reasonable estimate is six thousand, but it could very well have been twice that. No one will ever know, and the statistic is not so important as the ways in which they died—of cholera, dysentery, fever, smallpox, all the scourges against which there was no known protection or any known cure.

Laborers had been brought in by the boatload from every part of the world. White men, mostly Irish "navvies" who had built canals and railroads across England, "withered as cut plants in the sun." But of a thousand Chinese coolies, hundreds fell no less rapidly or died any less miserably of disease, and scores of Chinese workers were so stricken by "melancholia," an aftereffect of malaria, that they had committed suicide by hanging, drowning, or impaling themselves on sharpened bamboo poles.

Simply disposing of dead bodies had been a problem the first year, before the line reached beyond the swamps and a regular cemetery could be established on high ground. And so many of those who died were without identity, other than a first name, without known address or next of kin, that a rather ghoulish but thriving trade developed in the shipping of cadavers, pickled in large barrels, to medical schools and hospitals all over the world. For years the Panama Railroad Company was a steady supplier of such merchandise, and the proceeds were enough to pay for the company's own small hospital at Colón.

A reporter who visited this hospital in 1855, the year the railroad was finished, wrote of seeing "the melancholy rows" of sick and dying men, then of being escorted by the head physician to an adjoining piazza, "where, in conscious pride, he displayed to me his collection of well-picked skeletons and bones, bleaching and drying in the hot sun." It was the physician's intention, for the purposes of science, to assemble a complete "museum" representing all the racial types to be found among the railroad dead.

The worst year had been 1852, the year of Stephens' death, when cholera swept across the Isthmus, starting at Colón with the arrival of a steamer from New Orleans. Of the American technicians then employed—some fifty engineers, surveyors, draftsmen—all but two died. When a large military detachment, several hundred men of the American Fourth Infantry and their dependents, made the crossing in July en route to garrison duty in California, the tragic consequence was 150 dead—men, women, and children. "The horrors of the road in the rainy season are beyond description," wrote the young officer in charge, Captain Ulysses S. Grant, whose memory of the experience was to be no less vivid years later when he sat in the White House.

Nicaragua was different.

The United States and Great Britain had come close to war over Nicaragua, in fact, at the beginning of the gold rush, so seriously was Nicaragua's importance as a canal site regarded on both sides of the Atlantic. The Caribbean entrance to a Nicaragua canal would be San Juan del Norte, at the mouth of the San Juan River, and a British gunboat had seized San Juan del Norte in 1848 and renamed it Greytown. A crisis was averted by a treaty specifically binding the United States and Great Britain to *joint* control of any canal at Nicaragua, or, by implication, any canal anywhere in Central America. This was the Clayton-Bulwer Treaty of 1850—after John Clayton, the American Secretary of State, and Sir Henry Lytton Bulwer, the special British envoy involved—and it seemed a very good thing in Washington, in that it blocked a foothold for the British Empire in Central America and precluded any chance of a wholly British-owned and -operated canal in the Western Hemisphere. So important a document signed by the two powers had also put the Nicaragua canal in a class by itself.

Nicaragua and Tehuantepec both competed with Panama for the California trade, and though the Tehuantepec transit never really amounted to much, the one at Nicaragua did and far more so than is generally appreciated. In 1853, for example, traffic in both directions across Panama was in the neighborhood of twenty-seven thousand people; that same year probably twenty thousand others took the Nicaragua route, going from ocean to ocean on an improvised hop-skip-and-jump system of shallow-draft steamers on the San Juan, large lake steamers, and sky-blue stagecoaches between the lake and the Pacific. The actual overland crossing at Panama was shorter and faster, but Nicaragua, being closer to the United States, was the shorter, faster route *over all*—five hundred miles shorter and two days faster. A through ticket by way of Nicaragua also cost less and, perhaps as important as everything else, Nicaragua was not known as a deathtrap.

The Nicaragua system was the creation of Cornelius Vanderbilt, who became seriously enough interested in a Nicaragua canal to hire Orville Childs, a highly qualified engineer, to survey the narrow neck of land between Lake Nicaragua and the Pacific. And in 1851 Orville Childs had the good fortune to hike into a pass that was only 153 feet above sea level. He had found a place, in other words, that was a full 122 feet lower than the summit of the Panama Railroad, and by 1870 no lower point had been discovered anywhere else.

The impetus to resolve the canal question grew steadily as the steam engine transformed ocean travel on a global scale. In 1854 Commodore Matthew Perry with his "black ships" had forced Japan to open her ports to Western commerce. Seven years later the first Japanese delegation to the United States, eighteen lords wearing the swords and robes of samurai, passed through Panama on its way to Washington.

A Wall Street man named Frederick Kelley calculated that a canal through Central America could mean an annual saving to American trade as a whole of no less than $36,000,000—in reduced insurance, interest on cargoes, wear and tear on ships, wages, provisions, crews—and a total saving of all maritime nations of $48,000,000. This alone, he asserted, would be enough, irrespective of tolls, to pay for the entire canal in a few years, even if it were to cost as much as $100,000,000, a possibility almost no one foresaw.

Darien had been tried several times again since Lieutenant Strain's tragedy, as had the Atrato headwaters, all without luck. Small French exploring parties had begun turning up in both areas in the 1860's, and Frederick Kelley, who became the most ingenuous canal booster of the day, expended a fortune backing several disappointing expeditions, including one in search of Humboldt's "Lost Canal of the Raspadura." The leader of that particular Kelley venture was a hard-bitten old jungle hand, John C. Trautwine, who had worked on the Panama Railroad surveys. There was no lost canal, he reported, at the conclusion of a search across hundreds of miles of Atrato wilderness. Perhaps a Spanish priest had induced his flock to make a "canoe slide," but it was never anything more than that. "I have crossed it [the Isthmus] both at the site of the Panama Railroad and at three other points more to the south," Trautwine wrote in a prominent scientific journal. "From all I could see, combined with all I have read on the subject, I cannot entertain the slightest hope that a ship canal will ever be found practicable across any part of it."

But whose word was to be trusted? Which data were reliable?

The information available had been gathered in such extremely different fashions by such a disparate assortment of individuals, even the best of whom found it impossible to remain objective about his own piece of work. The more difficult it was to obtain the data, the higher their cost in physical hardship, time, or one's own cash, the harder it was to appraise them dispassionately. The conditions under which the field work had to be conducted were not only difficult in the extreme, but even the best-intentioned, most experienced men could be gravely misled if they allowed themselves to be influenced by the "feel" of the terrain, as nearly all of them had at one time or another.

The French explorers and engineers had little faith in American surveys; the Americans had still less regard for any data attributed to a French source. The only surveys of consequence were that of the Panama Railroad and the Nicaragua survey by Childs. Only one of these had been made with a canal in mind and it was really far from adequate. The organized approach Humboldt insisted on had never once been tried, for all the talk and energies expended. Nor, it must be added, had any serious body or institution–American, European, scientific, military–addressed itself to the critical question of the *kind* of canal to be built; whether in the interests of commerce and of future generations, it ought to be a canal cut through at sea level, such as the Suez Canal, or whether one that would lift ships up and over the land barrier with a system of locks.

III

Late in the afternoon of February 21, 1870, having made a stop at Colón to pick up a Colombian commissioner, Señor Don Blas Arosemena, who was to accompany the expedition, as well as a force of *macheteros*, the steam sloop *Nipsic* arrived at Caledonia Bay. The weather was delightful. The dry season at Darien is the time when trade winds blow fresh from the north and a heavy blue sea breaks all along the coast. Little rain falls, except in the mountains. Temperatures range in the low eighties; the sky is spotlessly clear day and night.

The line of march was to be over the abrupt green mountains that rise only a few miles in from shore. Selfridge would head for the Caledonia gap, which, from the bay, appears lower than it is and might well be taken even by an experienced observer as the perfect place for a canal. It was what had attracted Dr. Edward Cullen originally.

Selfridge issued strict orders concerning the Indians. Their property was to be "perfectly respected," no villages were to be entered without their consent. Any "outrage of their women" would be answered with the most severe punishment.

Operations commenced February 22, the morning after arrival. Selfridge met on the beach with the chief of the Caledonia tribe. "When you give an order to one of your young men, do you expect him to obey?" Selfridge asked. "I am sent here by my great chief," he continued, "with orders to pass through the country and I must obey. I shall cross to the Pacific, peaceably if possible, but if not I have ample force at my command." The Indian said the white men could go at will but he professed no knowledge of the interior. Like the other Indians to be seen on the beach, he was quite small in stature, but muscular and quick, with bright, intelligent eyes. "I was not able to discover their ancient form of worship," Selfridge wrote. "They believe in evil spirits, and . . . they believe that God made the country as it is, and that He would be angry with them and kill them if they assisted in any work constructed by white men."

Four days later, leaving a small party behind to organize a telegraph station and an astronomical observatory, Selfridge and a force of about eighty men, including Marines and *macheteros*, started inland to make a reconnaissance. In a week they were back, dirty, exhausted, and full of stories. They had found the Sucubti River, which flows to the Pacific, the river Strain should have followed. Once they had reached the mountains it had rained nearly the whole time, and in some places the trail had run along ridges only a few feet wide, with great gorges dropping off on both sides ("in the depths of which was heard the roaring of wild animals"). Some of the older men, veterans of the Civil War, said they had never experienced anything to equal the march. But they had crossed the divide.

On March 8, a full-scale surveying party got under way, stringing a telegraph wire as it went in order to report its progress back to the base camp. Two weeks later, on March 22, the chief telegrapher with the party, W. H. Clarke, sent the following message:

> I am at the front. We are progressing finely through the worst country I ever saw, on our way to the Pacific; impossible to write; everybody is well and in good spirits.

On March 30 came another message from Chief Telegrapher Clarke, this one to Commander Lull and the crew of the *Guard*.

The entire column of the Surveying and Telegraphic Corps unite in sending you and all friends on board, a greeting from the summit of the dividing ridge. Looking to the westward we see the long-looked-for slope of the Pacific stretching far away, seemingly all an impenetrable forest; to the northeast Caledonia Bay and the *Guard* is plainly visible; immediately around me I see Lieutenant Schulze, Mr. J. A. Sullivan, Ensigns Collins and Eaton of the *Guard*, Messrs. H. L. Merinden, J. P. Carson, T. H. O'Sullivan and Calvin McDowell, and as I telegraph this message they are singing "Jordan is a hard road to travel."

What was not reported, but already known by then, was that the lowest pass on the Sucubti was 553 feet above sea level, and that the mountains were indeed a thousand feet or more in elevation, just as Strain had reported. So Edward Cullen was a fraud after all.

Still the expedition continued and under considerable hardship. The terrain was often such that it was impossible to do the chaining and leveling for the survey, detours had to be made, progress on the survey slowed to not more than a thousand feet a day. The cameras of Timothy O'Sullivan, the heavy glass plates and the dark tent he had brought along, were just about useless because of the heat and humidity and the vegetation that shut out nearly all daylight.

The standard attire was a big straw hat, blue flannel shirt, duck trousers, shoes with canvas leggings. The flannel shirt was to be worn next to the skin, and the day began with a tablespoon of whiskey and two grains of quinine per man. To such precautions—"under Providence"—Selfridge attributed the "wonderful good health" of the command the entire time in the jungle.

Perhaps because of his preliminary orders, perhaps because of the conspicuous Marine guard, there were no troubles with the Indians of the interior, many more of whom were encountered than expected and none of whom had ever before seen a white man. Once on the Sucubti several Indians armed with poisoned arrows volunteered to serve as guides, then led the party along the most tortuous course possible. The Americans saw what was going on and said nothing, as "it was thought better not to offend them."

A few entries from the field diary kept by Selfridge give an indication of their days:

Thursday, April 7.—Took up our March at 6:30 A.M., the Indian Jim and others with us. . . . One of the Marines shot another private by accident in the arm, and he was left behind in camp. The Indians were very much surprised that the affair was taken so coolly, and two or three ran off to tell their chief. About 9 A.M. we struck the river again, and the Indians left us. . . . At half past 2 o'clock we forded the La Paz;

this was the deepest river we met, the water coming up to our armpits, and obliging us to carry our ammunition and provisions on our heads. Several bungo-trees full of monkeys were seen, as many as twenty or thirty in a tree; some were shot, and provided a pleasant and much-needed repast. . . .

Friday, April 8.– . . . Eugenio, the machetero, was bitten during the night by a scorpion or tarantula, and his leg and foot became so swelled that we were forced to leave him behind. . . . Passed a miserable night, tormented by mosquitoes and sand-flies.

Saturday, April 9.–Started down the right bank of the river. Left behind nine men who were shoeless. Cut through 5,000 feet, a dense mangrove [swamp]. . . .

Sunday, April 10.–Another sleepless night, on account of insects. . . .

"We were to find," he later wrote, "that in spite of the most careful preparations, the success of the expedition also depended upon extraordinary persistence and willingness to endure hardships." The torment inflicted by the sand flies and mosquitoes was indescribable–"mosquitoes so thick I have seen them put out a lighted candle with their burnt bodies." There was no longer any mystery, he mused, why the secrets of the Isthmus had remained locked up for so many hundreds of years.

At Caledonia Bay, a week later, Selfridge concluded that he had seen as much as needed of Cullen's route. So on April 20 the expedition packed up and steamed out of the bay for the Gulf of San Blas, another magnificent harbor on the Darien coast, approximately a hundred miles west, toward the Panama Railroad. Here again the mountains gave the appearance of a low pass, and from one of Frederick Kelley's expeditions it was known that the distance from tidewater to tidewater at this point was less than thirty miles. San Blas was a mere knife edge, where the two oceans came nearer to touching each other than at any other point in Darien or all of Central America.

Selfridge took his men ashore to search the Mandinga, the one large river on the Atlantic slope between the Atrato and the Chagres. By now the rainy season had returned and the bottom lands were a vast pulsing swamp. Frequently the men were "obliged to pass the night in trees, the water rising so rapidly as to drive them from their beds." In a week of relentless effort they were able to survey a bare two miles, and it was a full month later still by the time they had measured the mountain gap that from the sea had seemed so near. The elevation was a disappointing three hundred feet.

With provisions now runing low, his men worn out (". . . and no longer kept up by the charm of novelty"), with their entire stock of shoes used up—all six hundred pairs!—Selfridge thought perhaps he ought to pull back and sail for home. But ". . . could we carry our levels over the divide, we should be able to decide upon the practicability of this route." So on he went with a picked crew, moving their cumbersome, delicate equipment from point to point, putting down stakes, filling notebooks with pages of computations, observations on plants and animals, and geological notations. On June 7, at the top of the ridge, at an altitude of 1,142 feet by the barometer, they hammered in stake No. 96,000.

On the Pacific slope, the climate, the whole character of the country, changed. "Trees, soil, all different," Selfridge noted, "and the weather beautiful." They took their line to the point where it coincided with the one Kelley's people had mapped. Then, having followed the Kelley line far enough to be satisfied with its accuracy, they turned back, without going the whole way to the Pacific.

The San Blas route, Selfridge now could report, was no more practicable than the one at Caledonia Bay. A tunnel would be required, and even if enough locks could be built to lift ships over the mountains—to the preposterous altitude of a thousand feet—there were no rivers at that level to supply water for the canal.

Selfridge would return to Darien with a second expedition before the year ended, to search that section fronting on the Gulf of Urabá where the Isthmus joins South America. He would, on this second expedition, sail far up the Atrato to the Napipi to explore the route Humboldt had thought so promising. Later, in 1873, he would command a third expedition, this one to the Atrato headwaters. But none would compare to the Darien Expedition of 1870. It was the proudest accomplishment of his life. Nothing done before or after was so difficult or gave such personal satisfaction. It did not matter that they had failed to find the proper path, he would write near the end of a long life; they had led the way.

In the official report he filed with Secretary Robeson, Selfridge said merely that the effort had served to simplify matters—"the field of research is reduced and the problem narrowed." He was convinced that the determining factor must be the canal to be built. The canal "should partake of the nature of a strait, with no locks or impediments to prolong the passage . . ." It must be a "through-cut," at the level of the sea, he wrote, a canal like the canal at Suez, and, from what was known of Central America, the only feasible point for such a passage was Panama.

2
The Hero

How dull it is to pause, to make an end,
To rust unburnish'd not to shine in use!
—ALFRED, LORD TENNYSON, *Ulysses*

I

Independence, his vital source of strength, he often remarked, had come late in life to Vicomte Ferdinand de Lesseps. The charm, the pervasive, indomitable, world-famous de Lesseps charm that had carried him so very far, had been there right along, born in him, a family streak, it was said, like the zest for adventure and the good looks. From the very start of his career at Lisbon he had made a strong impression. Older observers likened him to his father and to his celebrated uncle, Barthélemy de Lesseps. Friends of both sexes were gathered effortlessly. "Ferdinand encounters friends everywhere," his first wife had written from the post at Málaga. "He is loved with true affection. . . . It is wonderful to have a husband so liked by everyone." And a little later on: "Ferdinand is so good, so amiable, he spreads life and gaiety everywhere."

He was gifted, passionate; he loved books, music, horses, his work, his children, his graceful, witty first wife, his stunning second wife, and occasionally, if we are to believe one admiring French biographer, the wives of others. But independence had not come until he was past forty, thrust upon him unexpectedly by forces not of his own making.

In the summer of 1870, when he stood on the flower-banked platform within the great Crystal Palace, beaming as the boys from the Lambeth Industrial Schools waved their "Egyptian Salute," Ferdinand de Lesseps was sixty-four years old, very nearly as old as the century. He had been born on November 19, 1805, the year of Austerlitz, in a beige-colored stone house with white shutters that still stands in the town of Versailles. Less than fifty yards from the house, through an iron gate at the end of the Rue de la Paroisse, were the gardens of the Versailles Palace, the great Neptune Basin with its spectacular fountains, and just beyond that, within a mile or so, the Grand Canal of Versailles, which once, in the time of Louis XIV, had been alive with brightly painted gondolas and had been the setting for mock naval battles staged by actual ships of the line.

His family was long distinguished in the French diplomatic service. The men were esteemed as "lovers of progress and movement"; they were cultivated, athletic, fond of extravagant living, and immensely attractive to women. A great-uncle, Dominique de Lesseps, had been ennobled for his services to the state a hundred years before Ferdinand's birth. Grandfather Martin de Lesseps had been French consul general to the court of Catherine the Great, and Ferdinand's father, Comte Mathieu de Lesseps, had been an accomplished Napoleonic diplomat, a friend of Talleyrand's. In Egypt, at the time of the British occupation, or shortly before Ferdinand was born, the vivacious Mathieu de Lesseps had worked miracles for Franco-Egyptian relations, and in 1818, when young Ferdinand was entering the Lycée Napoleon, Mathieu had been posted to the United States. Some sixty years later, at the unveiling of the Statue of Liberty in New York, Ferdinand would tell how his father had negotiated the first commercial treaty between France and the United States.

Barthélemy de Lesseps, the famous uncle, had been able to speak three languages by the time he was ten. While still in his twenties, he had sailed on the final expedition of the navigator La Pérouse, around Cape Horn to California and, at length, to Petropavlovsk, in Kamchatka. From there, in 1787, on orders from La Pérouse, all alone and with winter approaching, he had set out to find his way home to France. A year later, dressed as a Kamchatkan, he was presented to Louis XVI at Versailles, having traveled the entire distance across Siberia to St. Petersburg, mostly by dog sled, then on to Paris. He was a national hero overnight and in his subsequent diplomatic career—first under the Monarchy, then under the Empire, finally under the Restoration—he distinguished himself repeatedly, surviving three years of imprisonment in Turkey and the retreat of the *Grande Armée* from Moscow. So throughout his boyhood Ferdinand had been nourished on tales of valiant endurance, of heroic quests and heroic triumphs at the far ends of the world.

His mother was Catherine de Grivignée, whose French father had settled in Spain, prospered in the wine business, and married a Spanish girl of good family. His mother had lived her entire life in Spain until her marriage; Spanish was her first language and she was very Spanish in temperament, as Ferdinand would recall. He had grown up speaking Spanish as well as he did French, all of which would be offered later in explanation for the special allure of Panama, "a country made to seduce him."

There was never an overabundance of money in the family, appearances to the contrary. His mother's jewels had been pawned privately at least once to meet family expenses and his father had died all but bankrupt. Nor did Ferdinand attain great wealth. Like his father, he married well; like his father, he always lived in grand style. But the reputed de Lesseps fortune was a fiction.

Whether as a youth he ever envisioned a life other than the diplomatic service is impossible to say. But at age nineteen, having studied a little law, he was appointed *élève-consul* to his uncle, then the French ambassador to Lisbon. He served in Tunis afterward, with his father, until 1832, the year of his father's death; then came a Biblical seven years in Egypt, where being the son of Mathieu de Lesseps was a decided advantage. Later came Rotterdam, Málaga, and Barcelona. In 1848, at age forty-three, he was made minister to Madrid.

It was work he naturally enjoyed and he did it well. He was efficient; he was gallant. He sat a horse beautifully. He was a crack shot and a great favorite among sportsmen. ("These healthful occupations," wrote one high-Victorian biographer, "contributed largely to the promotion of that robust health and that iron constitution, thanks to which he was able to bear, without even feeling them, the innumerable fatigues, labors, and voyages in all parts of the world.")

Though of less than average height, he was handsomely formed. He had a fine head of thick black hair, a good chin, a flashing smile that people would remember. The eyes were dark and active. The mustache had still to make its appearance.

His wife, the former Agathe Delamalle, bore him five sons, only two of whom would live to maturity, and she appears to have been another important asset to his career. A French officer described her as "this young woman with the clear gaze, witty, decided . . ." "Diamonds glittered everywhere," reads another account from the time, a description of a ball she gave at Barcelona. "Madame de Lesseps received the guests with perfect grace. Her toilette was ravishing, and she wore it with that marvelous air of which only *Parisiennes* have the secret. Let us add that the affection which everyone bears her did not a little to increase the charm of this magnificent soirée, which lasted until dawn."

His interest in canal building began supposedly in Egypt in the early 1830's with the arrival of the Saint-Simonians, about twenty Frenchmen, many of them civil engineers, who were led by an improbable figure named Prosper Enfantin. They had come, they announced, to dig a Suez canal, a work of profound religious meaning.

Their messiah was the late Claude Henri de Rouvroy, the Comte de Saint-Simon, who had fought under Lafayette at Yorktown, then, back in France, founded his own radical philosophy aimed toward a new global order. It was he who wrote, "From each according to his ability, to each according to his work." Private property and nationalism were to be things of the past. The leadership of mankind was to be entrusted to an elite class of artists, scientists, and industrialists. Mainly the good society was to be attained through ennobling, regenerative work. The world was to be saved—from poverty, from war—through immense public improvements, networks of highways, railroads, and two great ship canals through the Isthmus of Suez and the Isthmus of Panama.

Prosper Enfantin had taken up the banner after the death of the Master, calling himself *Le Père,* "one half of the Couple of Revelation." The other half, he said, was a divine female who had still to make herself recognized. A "church" was established on the Rue Monsigny in Paris; lavish receptions were staged to welcome the female messiah, candidates for the honor being received in Father Enfantin's ornate bedchamber. Further, at a private estate near Paris, he founded an all-male colony for the faithful, where the prescribed habit, an outfit designed by the artist Raymond Bonheur, was a long, flowing tunic, blue-violet in color, tight-fitting white trousers, scarlet vest, and an enormous sash of richly embroidered silk. Enfantin, a big, bearded man, had the words "*Le Père*" embroidered across the front of his blouse. When he was taken to court for his advocacy of free love, he appeared in Hessian boots and a velvet cloak trimmed with ermine.

Asked to defend his behavior, he stood motionless and silent, then explained that he wished the court to have a quiet moment to reflect on his beauty.

But for all this he had a decisive intelligence. He had been an excellent student at the École Polytechnique, the ultimate in French scientific training. He was a financier of importance and converts to the creed included eminent financiers, respected business people, journalists, many of the ablest civil engineers in France.

Enfantin had judged Suez to be an easier undertaking than Panama. He was further inspired by a premonition that his female counterpart waited for him somewhere in the ancient cradle of civilization. So after serving a brief prison term, he had sailed for Egypt, and it was de Lesseps who persuaded the ruling viceroy of Egypt, Mohammed Ali, not to throw him out of the country. De Lesseps may also have provided Enfantin with financial assistance. At any rate, Enfantin and his engineers went into the Suez desert.

After four years, more than half of them had died of cholera and little of practical value had been accomplished. Nonetheless, the prospect of a Suez canal was being talked about in Europe with seriousness at last, as a result of Enfantin's proselytizing, and young de Lesseps, if not exactly a complete convert to Saint-Simonianism, had been uplifted by ideas that were to last a lifetime. "Do not forget that to accomplish great things you must have enthusiasm," Enfantin had said, repeating the deathbed exhortation of the Master.

There was, however, to be no immediate deviation from the progress of a model career, and by any reasonable standard of evaluation, nobody could possibly have prophesied the future the young diplomat had in store. What heights he personally aspired to can only be guessed at. Probably they were of the predictable kind.

Viewed in retrospect, de Lesseps' life stands out as one of the most extraordinary of the nineteenth century, even without the Panama venture. That he of all men of his time should have been the one to make "the miracle" happen at Suez is in itself miraculous. Suddenly there he was. Known after 1869 as "The Great Engineer," he was no such thing. He had no technical background, no experience in finance. His skills as an administrator were modest. Routine of any kind bored him quickly.

The great turning point, the traumatic personal watershed from which so much history was to flow, came in 1849. That it happened that particular year, the year of the gold rush, when Panama emerged from the shadows once again, seems a play of fate that not even a novelist of his day might have risked.

A French expeditionary force sent to subdue Mazzini's newborn Roman republic and restore papal rule had been unexpectedly thrown back at Rome by Garibaldi. De Lesseps, then in Paris, was told he was to leave at once to resolve the crisis. "Guided by circumstances," he was to please all parties and achieve a peaceful accommodation. With all eyes on him he had shown the incredible stamina and single-mindedness he could summon—and especially if all eyes were on him. Convinced that he could succeed, he very nearly had, and apparently quite blind to the fact that he was being used by his own government merely as a means to gain time. A temporary cease-fire was agreed to. But then French reinforcements arrived; Louis Napoleon, the new "Prince-President" of France, gave the order and the French army attacked.

Summarily recalled, de Lesseps was publicly reprimanded before the Assembly for exceeding his instructions. When Rome fell to the French army, he was left with no choice but to resign. The gossip was that the strain of the mission had been too much, that he had temporarily departed from his senses.

So at age forty-three he was without the career his background and natural gifts had so ideally suited him for, and to which he had given himself so wholeheartedly. The future was a blank page. He was in debt. Public disgrace was something he had never experienced. Yet outwardly he remained the man he had always been, jaunty, confident, up at dawn, busy all day. With his wife and three young sons he moved into a flat on the Rue Richepanse and for the next five years divided his time between Paris and a country estate in central France, an ancient, towered château in the province of Berri that had once belonged to Agnès Sorel, mistress of Charles VII. Known as La Chesnaye, it had been purchased at de Lesseps' urging by his mother-in-law, Madame Delamalle, who had recently come into a sizable inheritance. The estate was located near the little village of Vatan on an open plain, mostly wheat country and extremely good land, with a great belt of forest a few miles to the south. His ambition was to create a model farm and he plunged into the role of country gentleman.

To occupy his mind he returned to the old interest in an Egyptian canal, reading everything he could lay his hands on. He was in touch again with Prosper Enfantin, for whom the Egyptian dream still burned. Enfantin generously supplied studies and papers from his files in the belief that he and de Lesseps could join forces. De Lesseps, however, had no such intention. His destiny henceforth, he had decided, would be in his own hands. Once, years before in Egypt, Mohammed Ali had advised, "My dear Lesseps . . . when you have something important to do, if there are two of you, you have one too many."

France, meantime, had been wrenched by still another bloody political turn. The improbable Prince-President sprang a coup d'état, made himself dictator, and proclaimed the birth of the Second Empire. As Emperor Napoleon III, he would take France into a new age of progress, he said. "We have immense territories to cultivate, roads to open, harbors to deepen, canals to dig, rivers to make navigable, railroads to complete." The Saint-Simonians were among his strongest supporters.

He established a brilliant court at the Tuileries, and on a bright winter morning at Notre Dame, he married the spectacular Eugénie de Montijo, who was half Spanish, half Scottish, something of an adventuress, and a distant cousin of Ferdinand de Lesseps'. (His mother and her grandmother were sisters.) Young enough to be the daughter of her cousin Fernando, as she called him in Spanish, she had always looked to him for advice. Especially in her new responsibilities would she welcome his views, she wrote the week before the wedding.

A few months afterward, in the spring of 1853, Agathe Delamalle de Lesseps died of scarlet fever and a son, his father's namesake, died of the same cause. De Lesseps took refuge at La Chesnaye, pouring himself into routine projects and his canal studies. Life, he wrote to his oldest son, Charles, demanded courage, resignation, and trust in Providence. Charles, a bright, attentive boy of twelve, a student in Paris, had become a particular source of pride.

Then quite out of the blue came the news that Egypt's ruling viceroy had been murdered by two slaves. De Lesseps was on a scaffold working with some stonemasons on the old house when the postman appeared in the courtyard with the Paris mail. "The workmen passed my letters and papers from hand to hand. Imagine my astonishment when I read of the death of Abbas-Pasha . . . I hurried down, and at once wrote to the new Viceroy to congratulate him. . . ." The new viceroy was Mohammed Said, whom de Lesseps had befriended years before when Mohammed Said was a fat, unattractive, and friendless little boy.

Mohammed Said, for whom de Lesseps was to name Port Said, had since become a walleyed mountain of a man, a great eater and drinker and jovial teller of "French stories," a ruler who liked to have his pashas wade through gunpowder carrying lighted candles to test their nerve. More important, he was known for his generous impulses and so de Lesseps wasted no time in getting to Egypt. By way of welcome, Said arranged to go on maneuvers in the Western Desert with an army of ten thousand men. They were joined by Bedouin tribesmen and a military band. It was the sort of show de Lesseps adored. He traveled in style—his own private tent, mahogany furniture, quilted silk bedding, ice for his drinking water.

In the pages of his journal one senses a sudden exhilaration, a tremendous feeling of release and adventure.

He joined Said at his desert command post outside Alexandria on November 13, 1854. Both were in top spirits. Said expressed a singular desire to commence his regime with some great enterprise. Did Ferdinand have any ideas? But de Lesseps said nothing of the canal; he was waiting for a sign, as he explained later.

At night he searched the desert sky. Before dawn he was up and out of doors and the day was spent galloping miles over the desert on a magnificent Arabian steed. But the following morning, he knew the moment had come. He was standing at the opening to his tent, wrapped in a red dressing gown, looking and feeling for all the world like an Arab sheik. The description that follows is from his journal:

> The sun's rays were already lighting up the eastern horizon; in the west it was still dark and cloudy. Suddenly I saw a vivid-colored rainbow stretching across the sky from east to west. I must admit that I felt my heart beat violently, for . . . this token of a covenant . . . seemed to presage that the moment had come for the consummation of the Union between East and West. . . .

Before breakfast, but with everyone watching, he mounted his horse and went sailing over a high wall, a bit of imprudence, he calls it in the journal, but one "which afterward caused the Viceroy's entourage to give the necessary approval to my scheme. The generals with whom I shared breakfast congratulated me and remarked that my boldness had greatly increased their opinion of me."

And thus was launched the great Suez Canal. He broached the subject to Said at the close of day. Said asked a few questions, then declared the matter settled. His staff was summoned to hear the news.

Nothing had been said about cost. That de Lesseps had no experience faintly related to such an undertaking, that he represented no powerful organization, no combination of interests, that he had neither rank nor office nor any entrée to financial sources, seems not to have concerned either of them.

For the next fifteen years he was everywhere at once—Egypt, London, Constantinople, Paris—coaxing, flattering, convincing monarchs and newspaper editors, issuing endless reports, driving the work forward in the desert, watching over every detail, frequently overruling his technical advisers, defying the European bankers, and facing the scorn of the English prime minister, Palmerston, who called him a swindler and a fool and who saw the canal as nothing more than a cheap French grab for power in the Mediterranean.

The engineer Stephenson, builder of the Britannia Bridge, member of Parliament, rose from a bench in Commons to pronounce the scheme preposterous. De Lesseps, whose English was terrible and whose experience as a builder had begun and ended with the restoration work at La Chesnaye, hung a French flag from his hotel window on Piccadilly, and went traveling across England giving more than eighty speeches in a month. "They never achieve anything who do not believe in success," he loved to say.

When the Rothschilds wanted 5 percent for handling the initial stock subscription, he said he would hire an office and raise the money himself. "You will not succeed," said Baron de Rothschild, an old friend. "We shall see," de Lesseps had answered.

Approximately half the money had come from France (from twenty-five thousand small investors), the rest from Mohammed Said. When Said died, in 1863, his replacement, Khedive Ismail, was even more beneficent, so much so that by 1869 he had nearly put Egypt into bankruptcy. In the final stages it had been the colossal steam dredges designed by French engineers that made the difference. Nor can the repeated influence of the empress, her faith in her brilliant cousin, be discounted. Yet de Lesseps remained the driving spirit, and in truth he was something new under the sun; he had no historical counterpart. What he was—what he became—was the *entrepreneur extraordinaire*, with all the requisite traits for the role: nerve, persistence, dynamic energy, a talent for propaganda, a capacity for deception, imagination. He was a bit of an actor and as shrewd and silky a diplomat as anyone of his time.

He had no interest in making money, as he professed. "I am going to accomplish something without expediency, without personal gain," he once wrote in his quick, sure, upward-sloping hand. "That, thank God, is what has up to now kept my sight clear and my course away from the rocks." At any time he could have sold his precious concession and realized a fortune, but this he never did; his driving ambition throughout was to build the canal, "*pour le bien de l'humanité*."

"He persevered, you see," a grandson would recall. "He was a very stubborn man." Jules Verne called it "the genius of will." But de Lesseps spoke of patience. "I wait with patience," he wrote to a correspondent in the final year of the work, "patience which I assure you requires more force of character than does action."

On the morning of the Grand Opening, November 17, 1869, tens of thousands of people lining both banks of the canal saw him ride by. Radiant with health, his hair turned nearly white by now, he stood beside the empress on the deck of the imperial yacht, *Aigle*. She was wearing a big straw hat and waving a white handkerchief.

Khedive Ismail had spared no expense on the inaugural ceremonies. Six thousand invitations were sent, offering to pay all travel and hotel expenses. A Cairo opera house had been built for the occasion and Verdi had been commissioned to write a spectacular new work, *Aïda*.* Five hundred cooks and a thousand waiters were imported from Europe. At Lake Timsah, halfway down the canal, a whole town, Ismailia, had been created, trees planted, hotels put up, a palace built.

Behind *Aigle* steamed an Austrian frigate carrying Emperor Franz Josef, who was turned out in scarlet trousers, white tunic, and a cocked hat with a green feather. There were two Austrian corvettes, five British ironclads, a Russian sloop of war, several French steamers—fifty ships in all. "There was a real Egyptian sky," Eugénie would remember, "a light of enchantment, a dreamlike resplendence. . . ."

For the next eight months, until the outbreak of the Franco-Prussian War, he was Europe's reigning hero. The empress presented the Grand Cross of the Legion of Honor. The emperor hailed his perseverance and genius. He was cause for dozens of banquets in Paris. His name was constantly in the papers, his face in the illustrated magazines. And the fact that he had also become a bridegroom added immeasurably to his hold on the public imagination.

A small, private ceremony had been performed at Ismailia a few days after the opening of the canal. The bride was a stunning French girl of twenty, with large, dark eyes and great spirit, Louise Hélène Autard de Bragard, the daughter of an old and wealthy friend of de

* The opera was not ready in time, so the performance was put off until 1871.

Lesseps' and of a magnificent mother who, in her own youth, had been the inspiration for a sonnet by Baudelaire. She had been raised on the island of Mauritius, in the Indian Ocean, where her family, Huguenots, owned large plantations. According to the traditional story, it was love at first sight when she and de Lesseps met at one of Eugénie's "Mondays." By this second marriage he was to produce no fewer than twelve children—six sons, six daughters—which in some circles was considered a more notable accomplishment than the canal.

Palmerston was in his grave. In London, a few days after the great Crystal Palace reception, Prime Minister Gladstone informed the hero of Suez that Her Majesty had bestowed upon him the Grand Cross of the Star of India.

Few men had ever been so vindicated or extolled while they lived.

II

The first skirmish of the war, "*La Débâcle*" that overcame France with such appalling fury in 1870, was fought on August 4, the day Ferdinand de Lesseps returned from London, and the outcome, despite French heroism, was plain almost immediately. Napoleon III was suddenly aged and so ill he could barely sit a horse; yet he insisted on commanding an army in the field. An American observer, General Sheridan, wrote of the "marvelous mind" of Moltke and called the German infantry "as fine as I ever saw." The steel guns from the Krupp Works had twice the range of the French bronze pieces.

Within two weeks the main French army was penned in at Metz. On September 2, at Sedan, Napoleon III and 100,000 of his troops surrendered. It was the most stunning, humiliating defeat in French history. The Second Empire collapsed instantly. Sunday, September 4, Léon Gambetta climbed out onto a window sill at the Hôtel de Ville to proclaim to a Paris mob the birth of the Third French Republic. The empress, with the help of Ferdinand de Lesseps, escaped from the Tuileries and rushed to the home of her American dentist, a Dr. Evans, who got her to the Normandy coast and arranged for a yacht that carried her to asylum in England.

The war ended with the capitulation of Paris in January, after a siege of four months, during which the beleaguered citizens ate pet cats and elephants from the Paris zoo. The French dead were three times the German casualties, and by the peace terms France lost the rich, industrial provinces of Alsace and Lorraine. Further, Bismarck demanded an indemnity of 5,000,000,000 francs—$1,000,000,000—enough, he thought, to keep France crippled and subservient for another generation. And as a final humiliation, the despised German troops with their spiked helmets were to be permitted to parade down the Champs Élysées.

Then, with the return of spring, the tragedy was compounded. While a German army of occupation stood idly by, a vicious civil war raged; the savage days of the Commune became a bloodier time even than the infamous Terror.

Yet the Third Republic survived and the sudden resurgence of France after the war was as astonishing as her defeat. It was as if Sedan had released a vital inner resource. Everywhere people doubled their efforts, fired by a spirit of *revanche*. It was to be a revenge won on battlefields of "peace and progress"—for the while, anyway. In Paris the rubble was carted off and the new government carried on with the grandiose construction programs of Napoleon III and Baron Haussmann. Coal and iron production increased even without Alsace and Lorraine. Money was plentiful, furthermore, for capitalizing new enterprises, for foreign investments. Amazingly, the German indemnity was paid off in full by 1873, two years ahead of schedule. The days of *grandeur* were not past; France would be herself again.

For his own part Ferdinand de Lesseps was no more interested in retirement than he had been twenty years earlier. Inspirited by his new marriage and constant public attention, he was openly casting about for new worlds to conquer. He had been untarnished by the war; he was among the few. People spoke of him as the living embodiment of French vitality and the century's "splendid optimism." "We have had a lot of other men who have done things perhaps more remarkable and who have been less popular," a grandson would remember, "but that's the way he was." Once, on Bastille Day, when he was on his way to the station to take the train to his country place, a cheering crowd stopped his carriage, unhitched the horses, and pulled the carriage the rest of the way to the station. Gambetta called him *Le Grand Français*—The Great Frenchman, The Great Patriot—and the name was picked up by everyone.

He kept in excellent physical condition. He exercised regularly—fencing, riding—and with the zest of a man half his age. He looked at least ten to fifteen years younger than he was. An admiring American of the day described him as "a small man, French in detail, with . . . what is called a magnetic presence." A reporter for the New York *Herald* provided this description:

> He bears his years with ease and grace, showing no sign of age in his movements, which are quick and frequent, though never jerky. . . . His hair is almost white. His eyes are black, large, restless, and fringed by heavy lashes over which are shaggy eyebrows. His face is tanned . . . and ruddy with the evidence of perfect health. A mustache is the only one hirsute adornment on his face. It is small, iron-gray, bristling and has an aggressive look. In stature he is a little below medium height. His bearing is erect, his manner suave, courteous and polished.

Come winter he was usually off to Egypt with his wife and children, and wherever they went she attracted still more attention for him. "Her form is the admiration of every dressmaker in the French capital," reported the Paris correspondent of the Chicago *News*, "and a tight fitting dress sets off her elegant figure to the greatest advantage." They were seen riding in the Bois, at balls at the Élysée, where the stately Marshal MacMahon, president of the Republic, and his lady led "the decorous waltz" past flower-wreathed panels that still bore the imperial initials of "N" and "E."

They entertained often and grandly at a new apartment on the Rue Saint Florentin, a home with "every elegance"—Persian rugs, walls of family photographs and paintings in heavy gilt frames, a pair of tremendous elephant tusks in one antechamber, in another a display of his decorations. Presently, with her money, a larger, more impressive residence was purchased, a five-story private mansion, or *hôtel particulier*, on the chic new Avenue Montaigne, where, as at La Chesnaye, the custom was never-ending hospitality. There were always ten to twelve people at dinner, always some old Suez comrade or distant kinsman or other stopping over for the night and staying a week or six months.

As chairman and president of the Suez Canal Company, he remained for thousands of shareholders the charmed guardian of their fortunes, which kept gaining steadily as the value of the stock grew ever greater. He thrived on the public role expected of him, rising to all occasions—banquets, newspaper interviews—with exuberant renditions of his adventures in the desert, or, increasingly, with talk of some vast new scheme in the wind. He talked of building a railroad to join Paris with Moscow, Peking, and Bombay. He had an astonishing plan to create an inland sea in the Sahara by breaking through a low-lying ridge on Tunisia's Gulf of Gabès and flooding a depression the size of Spain.

Interestingly, when a special commission of engineers was appointed to appraise this particular scheme, his absolute faith in it was not enough. Among the members was Sadi Carnot, a future president of France, who would recall de Lesseps' performance years later. "We had no difficulty in showing him that the whole thing was a pure chimera. He seemed very much astonished, and we saw that we had not convinced him. Take it from me that as a certainty he would have spent millions upon millions to create his sea, and that with the best faith in the world."

It was said that he could command money as no one else alive, and encouragement came from every quarter. Victor Hugo urged that he "astonish the world by the great deeds that can be won without a war!"

A forum for de Lesseps' interests, now a favorite gathering place for those most intrigued by his future plans, was the Société de Géographie de Paris, where Humboldt had once been a reigning light. Geography, since the war, had become something of a national cause. Among men of position it had also become extremely fashionable. It was said that ignorance of the world beyond her borders had put France in an inferior position commercially, that it had contributed to her disgraceful performance in the late war. Geographical societies sprouted in the provinces. Geography was made mandatory in the schools. Membership in the Paris society increased four times, and Vice Admiral Clement Baron de La Roncière-Le Noury, president of the society, wrote of "this ardor for geography" as one of the characteristics of the epoch. When the first serialized chapters of *Around the World in Eighty Days* appeared in *Le Temps* in 1872, Paris correspondents for foreign papers cabled their contents to home offices as though filing major news stories. Nothing else had ever made the geographical arrangement of the planet quite so clear or so interesting in human terms. An extravagant stage production of the novel opened in Paris, complete with live snakes and elephants, and between acts audiences jammed the theater lobby to watch an attendant mark Phileas Fogg's progress on a huge world map.

Jules Verne, strictly an armchair adventurer, worked in a tower study in his home at Amiens, but came often to Paris to attend meetings of the Société de Géographie and to do his research in its library. When he was made Chevalier of the Legion of Honor, it was on the nomination of Ferdinand de Lesseps, and the sight of two such men at Société functions, talking, shaking hands with admirers, was in itself a measure of the organization's standing.

It was at an international congress held under the auspices of the Société the summer of 1875 that de Lesseps made his first public declaration of interest in an interoceanic canal. The meetings opened at the

Louvre, in conjunction with a huge geographical exhibition, the first of its kind, that took Paris by storm. Crowds ranged from ten to twelve thousand people a day.

Two issues must be resolved, he said. First was the best route; second was the type of canal to be built, whether at sea level (*à niveau* was the French expression) or whether a canal with locks. Several French explorers who had been to Darien spoke on their experiences and presented proposals. Joseph E. Nourse, of the United States Naval Observatory, reported on the recent American expeditions to Nicaragua and Panama. But de Lesseps was the center of attention, and when he declared that the canal through the American Isthmus must be *à niveau* and *sans écluses* (without locks), it seemed that side of the problem had been settled.

Events began to gather momentum. Aided by the Rothschilds, England suddenly acquired financial control of the Suez Canal, and de Lesseps, while still head of the company, with offices in Paris, found his influence substantially undercut. The beloved enterprise, the pride of France, had become the lifeline of the *British* Empire.

Then before winter was out came the decision of President Grant's Interoceanic Canal Commission. Having weighed the results of its surveys in Central America, the commission had decided in favor of Nicaragua. The decision was unanimous. Panama received little more than passing mention.

Within weeks it was announced that the Société de Géographie would sponsor a great international congress for the purpose of evaluating the scientific considerations at stake in building a Central American canal. The American efforts had been insufficient, it was stated.

III

Whether Ferdinand de Lesseps was merely an adornment for the Türr Syndicate or a willing confederate or its guiding spirit were to become questions of much debate. In some accounts he would be portrayed as the victim of forces beyond his control. "Inevitably the whirlpool began to draw Ferdinand nearer and nearer its vortex," reads one interpretation of events surrounding the origins of the Panama venture, and he is pictured struggling valiantly against the current. To a great many contemporary American observers he would appear more the innocent dupe of furtive schemers—"insidious influences," as one of the New York papers said—who were placing the old hero out in front of the French people like a goat before sheep.

Had things turned out differently, however, it is unlikely that the galvanizing leadership of the effort would ever have been attributed to anybody other than Ferdinand de Lesseps, which, from the available evidence, not to mention the man's very nature, appears to have been the truth of the matter. As he himself once remarked to an American reporter, "Either I am the head or I refuse to act at all."

The newly formed Türr Syndicate was quite small but made up of such "well-selected" figures as to command immediate attention and confidence. Its better-known stockholders included Senator Émile Littré, author of the great French dictionary, and Octave Feuillet, the novelist. (Littré declared that the five thousand francs he put in represented the first financial investment of his life.) There were General Claude Davout, Charles Cousin, of the Chemin de Fer du Nord, the Saint-Simonian financier Isaac Periere, and Jules Bourdon, who was curator of the Opéra. Dr. Henri Bionne, an official of the Société de Géographie, was an authority on international finance, a former lieutenant commander in the French Navy, who had degrees in both medicine and law. Dr. Cornelius Herz, a newcomer to Paris and an American, was a physician and entrepreneur who claimed a personal friendship with Thomas Edison.

The syndicate's formal title was the Société Civile Internationale du Canal Interocéanique de Darien. It had a capital of 300,000 francs represented by some sixty shares and de Lesseps was neither a shareholder nor an officer. The leadership and the bulk of the stock were in the hands of three directors. The most conspicuous of these was General Istvan Türr, a Hungarian who had covered himself with glory in Sicily as Garibaldi's second in command and who for a time had been employed by King Victor Emmanuel II for diplomatic missions. With his long, elegant figure, his long, handsome face and spectacular Victor Emmanuel mustache–it must have been the largest mustache in all Paris in the 1870's–Istvan Türr had become something of a celebrity, the sort of personage people pointed out on the boulevards. His social connections included *Le Grand Français.*

The second man was the financier Baron Jacques de Reinach, a short, stout, affable man about town, known for his political pull and his voracious interest in young women. Like Türr, he was foreign-born, but a German and a Jew, as would be made much of later. He had founded the Paris banking firm of Kohn, de Reinach et Compagnie and had become rich speculating in French railroads and selling military supplies to the French government. His dealings had been subject to some question, although as yet nothing serious had come of it.

Most important of the three was Lieutenant Wyse, Lieutenant Lucien Napoleon-Bonaparte Wyse, who was the illegitimate son of the first Napoleon's niece Princess Laetitia. Temporarily on leave from the French Navy, Wyse was twenty-nine years old. He did not look much like a Bonaparte. Tall and slender, he had an open, friendly face with a high forehead, blue eyes, and full beard. His mother, a sensational woman who had been known in every capital in Europe, was the daughter of Napoleon's wayward brother Lucien Bonaparte, Prince of Canino. Her early marriage to Sir Thomas Wyse, an Irish diplomat, had failed, but was never dissolved legally, and by the time Lieutenant Wyse was born, nineteen years had passed since she and her husband had even seen each other. Two illegitimate daughters had also resulted, magnificent-looking women, very much like their mother, one of whom married Istvan Türr (which made Türr and Wyse brothers-in-law). The other, known as Madame Rattazzi, was a literary figure of sorts and one of the most dazzling and publicized figures of the day. The father of the sisters was an English Army officer who had pulled Princess Laetitia from a pond in St. James's Park after she had attempted a public and rather ridiculous suicide at the time her marriage was breaking up. But the identity of the young lieutenant's father was never divulged, though naturally there was speculation on the subject and especially when the Panama venture commenced. The money he put into the syndicate had come from his wife, a wealthy Englishwoman.

It was Wyse who went to see de Lesseps and in the early stages de Lesseps appears to have found him a young man much after his own heart. Wyse would also be the sole member of the syndicate to subject himself to any physical danger or hardship.*

The initial plan announced by the Société de Géographie was for a series of definitive explorations and surveys, a binational, wholly nonpartisan effort, with the world's leading scientific societies participating. But all such talk ceased the moment the Türr Syndicate made itself available, offering to handle everything. Permission to conduct explorations within Colombian territory was secured by sending an

* In examining the relationship that developed between Wyse and de Lesseps, their kinship of purpose, the shared sense of adventure, the almost father-son spirit, the question inevitably arises: Might de Lesseps have been the unknown father? There is, however, nothing in the available record to suggest this was so. About all we can safely assume is that for a young man of such background, with his paternity in doubt and his aspirations so high, de Lesseps must have been an appealing figure and one to which he might very naturally wish to attach himself.

intermediary to Bogotá, and six months later, in early November 1876, an expedition of seventeen men sailed on the steamer *Lafayette*, flagship of the French West Indies line. Lieutenant Wyse was in command, assisted by another French naval officer, Lieutenant Armand Réclus, and their orders were to find and survey a canal route, but to confine their activities to Darien, east of the railroad, since the Colombian government had forbidden any intrusion along the railroad's right of way. In other words they were to look only in that area wherein the syndicate had a legal right to carry on its business, which was scarcely the broad-range perspective embodied in the Société's original proposal.

The party was gone six months, two of which were spent at sea. Everyone suffered from malaria, two men died in the jungle, a third died during the voyage home. Wyse returned thin and drawn and covered with tiny scars from insect bites. He was thoroughly discouraged and made little effort to hide it. Though they had managed to cross the divide, the terrain, the punishing heat, the rains, had defeated them. The best he could recommend—and purely by guesswork—was a Darien canal with a tunnel as much as nine miles in length.

De Lesseps was wholly dissatisfied. How instrumental he had been in planning the expedition, if at all, is not apparent. His role was supposedly that of an arbiter only. He was the head of the Société's Committee of Initiative.

At any rate, having heard the young officer's report, he declared it as good as worthless. He would agree only to a canal at sea level—no locks, no tunnels. Furthermore, he now knew where to build the canal. As he would remark later to a New York newspaperman, "I told Messrs. Wyse and Réclus when they made their report that there could be no other route than that of the railroad. 'If you come back with a favorable report on a sea-level canal on that route I shall favor it.' "

So Wyse and Réclus sailed again, accompanied by many of the same men who had been on the first expedition. And this time things went differently. Considering how much was to hang on their efforts, how much would be risked on the so-called Wyse Survey at Panama, it is interesting to see just how their time was occupied.

Landing at Colón and crossing to Panama City, Wyse assembled the necessary provisions and sailed for the Pacific shores of San Blas, where his efforts appear to have been half-hearted. Indeed, it is puzzling why he bothered at all, knowing de Lesseps' attitude. In three weeks, certain that no canal could be built at San Blas without a tunnel, Wyse ordered everybody back to Panama City.

Lieutenant Réclus was told to explore the Panama route, keeping to the line of the railroad, and since this was in violation of the agreement secured earlier, Wyse decided to go himself to Bogotá. Time suddenly was of the essence. On the first of April the president of Colombia, Aquileo Parra, would be retiring from office and President Parra was known to favor the Wyse-Türr enterprise.

Lieutenant Réclus, meantime, began an informal reconnaissance of the Pacific slope a few miles east of the railroad, assisted by a young Panamanian engineer, Pedro Sosa. It was, as Réclus himself noted in his diary, "not an exploration in the true sense of the word." It was more of a walk, a ride on the railroad even. Sosa became ill within a week. Then Réclus too was stricken with an excruciating earache, and so he called the whole thing off. On April 20, they were back in Panama City and ten days later, with no word from Wyse, Réclus sailed for France.

And that was the sum total of the Wyse Survey. The exploration of the Panama route that was "not an exploration" had occupied all of two weeks, four days. Wyse had played no part in it and in fact no survey had resulted.

By contrast, the American expedition of three years before had remained two and a half months in Panama and virtually all that time had been spent in the field. The Americans, more than a hundred in number, had run a line of levels from ocean to ocean, explored the Chagres watershed, and prepared maps, charts, and statistical tables. And the Government Printing Office in Washington had made most of these findings, except for the maps and plans, available in a document of several hundred pages, a document Wyse would be perfectly happy to rely upon. Such borrowing would pose no conflict presumably, since the material had been published in the spirit of open exchange of scientific information and the Americans had already rejected the Panama route.

It was well afterward, when he was safely back in Paris, that Wyse wrote of his mission to Bogotá, then one of the most inaccessible cities on the face of the earth. From Panama City to Bogotá was normally a journey of three, even four weeks, though the distance on a straight line was only about five hundred miles. Moreover, it was a vastly different world from Panama that one found on arrival—a gray stone city set on a tableland at 8,600 feet and hemmed in by two of the three tremendous ranges of the Andes that divide Colombia like giant fingers; a mild damp climate that seldom varies, skies often clouded; a solemn, impoverished populace clothed in black; a proud ruling class of bankers, scholars, poets, who spoke the most perfect Castilian to be heard in Latin America.

Because the Darien wilderness stood between Panama and the rest of Colombia, Panama was as removed as if it were an island, and Colombia could be reached only by sea, either by the Caribbean or the Pacific. One sailed first either to Barranquilla or to Buenaventura. The journey from Barranquilla to Bogotá involved a four-hundred-mile trip by river steamer up the Magdalena to a point called Honda, then another hundred miles over the mountains by horse or wagon. There were no railroads.

The other way, by Buenaventura, the route Wyse took, was shorter but considerably more arduous, covering nearly four hundred miles. Wyse went by horseback, traveling with one companion, a French lawyer, Louis Verbrugghe, the two of them in serapes and big Panama hats. The general direction, as Wyse wrote, was "*perpendiculaire*."

They reached Bogotá in just eleven days, during which they sometimes spent twenty-four hours in the saddle. They arrived unshaven, their clothes torn and filthy. Wyse was missing one spur and had broken the other so that it clanked disconcertingly as they walked along the streets. Hotels turned them away because of their appearance, as Wyse would tell the story. But the following morning, March 13, bathed, shaved, looking most presentable, Wyse met with Eustorgio Salgar, Secretary of Foreign Relations. On March 14 he saw President Parra, who was especially "well disposed" toward his proposition.

The newspapers in Bogotá, all closely tied to the party in power, the Liberals, took little notice of Wyse's presence in the capital. That the visit was one of the utmost importance to the future of Colombia, that Wyse was there in fact to settle the basic contract to build a Panama canal, a contract that could mean a world of difference to Colombia for centuries to come, or more immediately help solve the country's dire financial troubles, was in no way suggested. Possibly someone somewhere along the line had decided that a better bargain might be driven with the young man by playing down his importance.

On March 15, or just three days after his arrival, Wyse presented a draft of a contract. Everything was going as smoothly as could be hoped for. Five days later, having made only minor modifications, Salgar and Wyse fixed their signatures to the document, and three days after that, on March 23, 1878, President Parra, who had exactly one week left in office, did the same. Confirmation by the Colombian Senate took longer, but by mid-May, the concession at last in his pocket, Wyse was on his way back to Panama, going this time by steamer down the Magdalena.

At Panama City he learned from Pedro Sosa of the little that Sosa and Réclus had accomplished, yet took no time to do anything more. Rather, he wound up his affairs in the least time possible, sold off the supplies left over from the expeditions, made Sosa a gift of the surveying instruments, and departed. He seems to have felt obliged only to see Nicaragua—to travel the route the Americans had settled on—and it was another journey in record time. He crossed from San Juan del Norte, going by steamer up the San Juan, then over the lake. The Americans had "much simplified" his task, he was to report. In fact, their Nicaragua Expedition had been their largest and most extensive. To plot their canal line they had had to chop a path nearly the length of the entire valley, or more than twice the distance across Panama, and much of the time the men had worked in swamps in water up to their shoulders. Their survey was an accomplishment Wyse especially could appreciate. He himself paused only long enough to pick up a few rock samples.

From Nicaragua he went to Washington, but by way of San Francisco, another odd side of the story, since he could so easily have returned to Colón, taken a steamer to New York, and saved himself several thousand miles. The impression is that he wanted to appraise financial interest in San Francisco, the American city that stood to gain the most from the canal. But possibly he wanted only to take the transcontinental railroad, to ride like Phileas Fogg the "uninterrupted metal ribbon." Whatever his reasons, he can be pictured flying along in a Union Pacific parlor car, observing "the varied landscape" as Fogg had, checking his watch at the Great Salt Lake, or taking some air during the stop at Green River Station.

At the Navy Department in Washington he was received by Commander Edward P. Lull and A. G. (Aniceto Garcia) Menocal, authors of both the Nicaragua and Panama surveys. Lull had had overall command; Menocal, a Cuban by birth, had been foremost of the civilian engineers assigned by Admiral Ammen "to place the results of the work beyond the reach of criticism."

The conversation was cordial and for Wyse perfectly fruitless. The Americans showed great interest in his travels, and Wyse, who spoke excellent English, made much of their pioneering efforts in the jungle.

But it was their maps and plans that Wyse had come for and he was politely told that these were not available, that the department "did not feel disposed" to grant his request. He asked if he might pay his respects to Admiral Ammen, but Admiral Ammen, he was told, was not available.

So it was with the Bogotá contract only—the famous Wyse Concession—that Wyse sailed from New York; no survey of his own, not even a map of Panama other than one made by the railroad twenty-five years before. For the moment, however, the concession was enough. That its cash value could be phenomenal went without saying.

The agreement was this:

The United States of Colombia granted the Société Civile—the Türr Syndicate—the exclusive privilege, good for ninety-nine years, to construct a canal across the Isthmus of Panama. As a guarantee of their good faith, the grantees were obligated to deposit 750,000 francs in a London bank no later than 1882. It was required that surveys be made by an international commission of competent engineers, for which three years were allowed, and the grantees were permitted two additional years in which to organize a canal company, and then twelve years to build the canal.

Colombia in turn was to get 5 percent of the gross revenue from the canal for twenty-five years, 6 percent for the next twenty-five years, 7 percent for the next twenty-five years, and 8 percent for the final years of the concession. The minimum payment, however, was never to be less than $250,000, which was the same as Colombia's share in the earnings of the Panama Railroad.

Colombia conceded to the company, without charge, 500,000 hectares (1,235,500 acres) of public lands, in addition to a belt of land 200 meters (219 yards) wide on each side of the canal. The terminal ports and the canal itself were declared neutral for all time. At the end of ninety-nine years the canal would revert to Colombia.

Further conditions were stipulated, but the crucial ones were these:

The concession could be transferred (i.e., sold) to other individuals or financial syndicates, but under no circumstances could it be sold to a foreign government. It was left to the grantees to negotiate "some amicable agreement" with the Panama Railroad concerning its rights and privileges.

Once reunited in Paris, Wyse and Réclus quickly put together a plan to present to de Lesseps. It was for a sea-level canal following the line of the Panama Railroad and again they resorted to a tunnel as the essential feature. De Lesseps voiced no objections to any of it. Nor did he register any serious dissatisfaction with Wyse's so-called survey. The one dissenting voice at this stage was that of a young Hungarian engineer named Bela Gerster, who had served with Wyse on both expeditions and who pointedly refused to sign Wyse's final report. Gerster prepared his own minority report, but when he took it to a number of French newspapers none were interested in printing it.

Some loose ends had to be attended to before de Lesseps could convene his canal congress. He had to have a guarantee that the Americans would attend–their presence was essential to the prestige of the affair–and he needed a commitment from the Panama Railroad Company that there would be no problem over the "amicable agreement" required by the Wyse Concession. Actually, he wanted to buy the railroad. So back Wyse sailed once more, early in 1879, arriving at New York, where he saw the president of the Panama Railroad Company, a clever Wall Street speculator named Trenor W. Park. Standing up to greet Wyse, Park looked no larger than a twelve-year-old boy, but he had come as far as he had in the business world by making the most of every advantageous position, and at the moment he was in an extremely advantageous position, as he and Wyse both appreciated. The details of the Bogotá contract had become public knowledge by now, and if an amicable understanding could *not* be reached with Trenor Park, the major stockholder in the railroad, then obviously the contract was worthless.

It was within Park's power to decide whether Wyse or de Lesseps need go a step further with their plans.

Park was "not altogether reluctant" to sell the railroad. His price, he told Wyse, was $200 a share, or twice its market value at the moment. Park, it was understood, owned fifteen thousand shares.

In Washington next, Wyse not only succeeded in seeing Admiral Ammen, but was presented to the Secretary of State, William Evarts, and later to President Hayes, who expressed great interest in the forthcoming Paris congress. Evarts, however, seemed as suspicious as Palmerston had been about Suez. The ill-fated attempt by Napoleon III to make Maximilian emperor of Mexico had left Evarts, like many Americans, extremely uneasy about France and her aspirations in the Western Hemisphere and anything but trustful of anyone with the name Bonaparte, even so amiable a Bonaparte as Lieutenant Wyse. So it was a difficult interview.

At length Evarts agreed that the United States should participate in the congress but only Ammen and A. G. Menocal would be permitted to go as authorized delegates. They could join in the technical discussions—to "communicate such scientific, geographical, mathematical, or other information . . . as is desired or deemed important"—but they were to have no official powers or diplomatic function, no say concerning the canal policy of the United States.

Shortly afterward in Paris, sometime in the early spring of 1879, just before the opening of the congress, Charles de Lesseps met with his father in the office of Dr. Henri Bionne, one of the most respected figures in the Türr Syndicate.

At age thirty-eight, Charles was nearly bald, and with his dark brows and thick dark beard, he looked a good deal older than he was. Like his father, he was a man of great pride and natural courtesy. He was also a capable administrator and this, plus a good deal of common sense and a capacity for hard work, had won him wide admiration at Suez, where he had served as his father's principal aide. He was intelligent, rather than brilliant, careful, considerate, but with none of his father's glamour or his need for public acclaim. Charles was a chess player.

The demands on him at Suez had been heavy. His only child, "Little Ferdinand," had died in infancy of cholera at Ismailia in 1865. Still, he idolized his father no less than ever and remained his good right arm in numerous ways. Charles, as would be said later, was above all a devoted son. More, he was a son who knew his devotion was returned in full.

Charles was strongly opposed to the Panama venture and had been from the day Lieutenant Wyse first came to La Chesnaye to present his plan. To Charles the whole scheme was a kind of madness.

The account we have of the scene in Bionne's office is Charles's own, provided years later in a private memoir.

"What do you wish to find at Panama?" he asked his father. "Money? You will not bother about money at Panama any more than you did at Suez. Glory? You've had enough glory. Why not leave that to someone else? All of us who have worked at your side are entitled to a rest. Certainly the Panama project is grandiose . . . but consider the risks those who direct it will run! You succeeded at Suez by a miracle. Should not one be satisfied with accomplishing one miracle in a lifetime?"

Then, not waiting for a reply, he added: "If you decide to proceed with this, if nothing will stop you . . . if you want me to assist you, then gladly I will take whatever comes. I shall not complain no matter what happens. All that I am I owe to you; what you have given me, you have thc right to take away."

Ferdinand de Lesseps replied that he had already made up his mind. What he did not say, what perhaps he was unable to admit to himself just yet, was the extent to which his trust in Charles had influenced that decision.

3

Consensus of One

Great blunders are often made, like large ropes, of a multitude of fibers.

—Victor Hugo

I

The Congrès International d'Études du Canal Interocéanique, as it was formally titled, convened in Paris at two in the afternoon, Thursday, May 15, 1879. After centuries of dreams and talk, of hit-or-miss explorations and hollow promises, of little scientific knowledge, little or no cooperation among nations, leading authorities from every part of the world—engineers, naval officers, economists, explorers—were gathering under one roof "in the impartial serenity of science" to inaugurate *La grande entreprise*, greatest of the age. Or so it was being said.

The setting was the handsome new headquarters of the Société de Géographie, in the Latin Quarter, at 184 Boulevard Saint Germain, where rows of neatly spaced young chestnut trees, each fenced in ornamental iron, were in full leaf and crowds of bystanders gathered in the sunshine to watch the delegates alight from their carriages. De Lesseps had picked mid-May because it was the perfect time to be in Paris. He personally had issued every invitation. He had had final say on agenda, rules, the make-up of committees, even the entertainment. He would have nothing left to chance.

In all, 136 delegates entered through the great oak doors that opened onto the street. In addition to France and her colonies (Algeria and Martinique), a total of twenty-two countries were to be represented: Austria-Hungary, Belgium, China, Colombia, Costa Rica, Germany, Great Britain, Guatemala, the still-independent nation of Hawaii, Holland, Italy, Mexico, Nicaragua, Norway, Peru, Portugal, Russia, El Salvador, Spain, Sweden, Switzerland, and the United States. Among the Dutch delegates was the renowned Jacob Dirks, builder of the Amsterdam Canal. Sir John Stokes and Sir John Hawkshaw, equally distinguished engineers, had come over from London. The Germans had sent a general inspector of mines; the Russians, an admiral. (The Russians had actually been so little interested in the historic convocation that they had neglected to appoint a delegate, and none would have appeared had de Lesseps not cabled a last-minute reminder.) Colombia had sent a four-man delegation, one of whom was young Pedro Sosa; and the Mexican delegate, Francisco de Garay, was in such a rush not to miss de Lesseps' opening remarks that he left his baggage in customs at Saint-Nazaire and arrived on the Boulevard Saint Germain unshaven and still in his traveling clothes. At a nearby shop he picked out the appropriate attire (top hat, morning coat, gray gloves), had a barber sent in, then made his entrance with time to spare—a story that greatly pleased Ferdinand de Lesseps.

The American delegation, largest of the foregoing groups, numbered eleven, including Ammen, Menocal, and Commander Selfridge, plus delegates from the American Geographical Society, the National Academy of Science, the United States Board of Trade, and the City of San Francisco. And among the French were such recognizable personages as Jules Flachat, the explorer; Levasseur, the economist and geographer; Daubrée, president of the Académie des Sciences; Alexandre Gustave Eiffel, of the Société des Ingénieurs; and the very elegant Admiral de La Roncière-Le Noury. Finally there was de Lesseps himself, the star attraction, his young wife on his arm.

They gathered in the *grande salle* on the first floor of the Société, a beautifully detailed, cream-colored auditorium with a lofty arched ceiling, a small stage, and a seating capacity for nearly four hundred people. De Lesseps, his officers, and Admiral de La Roncière-Le Noury occupied the stage; the delegates filled the first five rows, while every remaining seat was taken by spectators, including, as no newspaper reporter failed to note, a surprising number of fashionable women in the feathered bonnets of the day. When de Lesseps stood up to bid all welcome, there was a storm of applause.

This first session, however, was purely ceremonial and amounted to little. De Lesseps offered a few pleasantries ("The presence of ladies at a scientific gathering is always a good omen . . ."), and Henri Bionne, who was to be secretary of the congress, read a rather tedious paper on the Société's prior interest in the canal idea. Then de Lesseps introduced those who were to head the various committees, hastily described the work of the committees, and read off the full list of delegates, asking each to rise in turn and be recognized. (The most prolonged applause was for the Chinese delegate, Mr. Li-Shu-Chang, first secretary of the Chinese legation in London, since China, as the newspapers explained, was expected to provide the labor to dig the canal.)

Several of the Americans were highly annoyed by all this. De Lesseps' remarks were obviously unprepared. The whole session had not lasted an hour and nobody but de Lesseps and Bionne had been heard from. Everything seemed too neatly and arbitrarily prearranged. Despite the emphasis on the numbers of nations represented, there was an obvious predominance of French delegates, most of whom seemed committed already, out of past loyalties or for reasons of personal ambition, to take whatever course the old man dictated. The more prominent French delegates, for example, included the former director general at Suez, Voisin Bey; Abel Couvreux, of the giant Couvreux, Hersent et Compagnie, a major Suez contractor; and Alexandre Lavalley, who had built the great steam dredges used at Suez.

Of the several committees, only one really mattered, the fourth, or so-called Technical Committee, which was charged with deciding where the canal should be built, what kind of canal it should be, and what it was all going to cost. It was the largest committee, the one de Lesseps himself would sit on, and of the 52 other delegates he had assigned to it, more than half were French. Indeed, of all the 136 delegates in attendance, a total of 73—well over a majority—were French and not a quarter were engineers.

Further, de Lesseps seemed bound to hurry things through in record speed. The congress, he had said in conclusion, should get on with its work "in the American fashion—that is to say, with speed, and in a practical fashion . . ." One week, he thought, should suffice.

Probably the least inhibited appraisal of the congress was that of the representative from the American Geographical Society, Dr. William

E. Johnston, a New York physician who described de Lesseps as "kind-hearted and obliging, but . . . ambitious also," and from the start was convinced that he and the other non-French delegates would count for little. Ammen and Menocal especially had no business even being there, he wrote. No plan other than that of the famous Frenchman and his compatriots stood the least chance of adoption.

Still, de Lesseps had welcomed the Americans with such warmth and courtesy that even Ammen could be seen to thaw. He had made Ammen the first of five vice-presidents of the congress; he had Ammen sit at his right hand; he insisted that Ammen and Menocal serve on the Technical Committee.

The labors of the smaller committees, which met elsewhere in the building, were minimal and of little significance. One group concluded that the canal would be opened ten years hence and accommodate an annual traffic roughly twice that at Suez. (In the committee's report it was stressed that such traffic should not be anticipated for the first year of operation, but de Lesseps would choose to disregard that.) Another group, estimating world commerce, a committee headed by Nathan Appleton, of Boston, the delegate from the United States Chamber of Commerce, met only three times and accomplished nothing. A committee on ship dimensions concluded that the canal need be no wider or deeper than Suez, and a committee on tolls, hamstrung to do much of anything without knowing what the Technical Committee would decide, made a gallant guess. With tolls set at $3 a ton, it was thought the canal could bring in a gross revenue of $18,000,000 a year, which was worked out to mean an annual net profit of no less than $8,000,000, or a return of 8 percent on a canal costing $100,000,000.

The deliberations of the Technical Committee, held in the auditorium, remained the focus of attention, the "impartial serenity of science" being pretty well shattered at the very first of these sessions on May 16.

The first speaker was to have been Admiral Ammen—another of de Lesseps' courtesies—but the trunk containing the reports and maps from the American surveys had been delayed somewhere between Liverpool and Paris. So it was Commander Selfridge who spoke instead and his subject was the Atrato River route, which should have been no problem, and would not have been, had Ammen been willing to let Selfridge simply have his say. Ammen, however, thought very little of any and all Atrato River schemes; nor did he wish anyone to get the impression that Selfridge spoke as an official representative of the American government; nor apparently did he like the idea that Selfridge was even in attendance; nor does he seem to have much cared for Selfridge personally. (What the issue was between them remains obscure.) Ammen openly challenged certain of Selfridge's claims and in no time a sharp and rather undignified argument resulted between the two: Ammen insisting that *he*, as the rightful head of the American delegation, should have the final say; Selfridge refusing to defer to the august admiral (now retired) and insisting with equal conviction that he had a perfect right to be heard and, further, that he could speak with authority since he at least had *been* there.

Selfridge would be asked to address the committee again in another few days. His explorations in Darien, his advocacy of a canal *à niveau*, his passable French, all made him a popular figure. (Later, the congress at an end, Selfridge would receive the Legion of Honor for his pioneering efforts in Darien.) But the Atrato scheme, though a "*projet sans écluses*," never really had a chance of attracting serious attention, as Ammen should have recognized, and it was put aside just as soon as Selfridge had had the opportunity to speak his piece. What *was* interesting to the delegates was the tone of the exchange between the two eminent Americans—to see how intensely, how passionately, such an issue could matter to fellow countrymen, fellow officers and gentlemen.

Ammen had his turn the following day, the crucial trunk having been located meantime, and was followed immediately by Menocal. The effect was stunning. "When it came to the turn of Messrs. Ammen and Menocal to give their figures and estimates of the different routes, a complete revolution took place," wrote Dr. Johnston. "The great body of able engineers who had come to seriously study the question without prejudice, were astounded to find that nobody in Europe knew anything about the question. The *exposé* of the American delegates was a revelation. . . ."

Ammen's part was a brief overall description of the American surveys, but the maps and plans he used to elaborate his remarks had an instant effect, since nothing of the kind had been seen before in Europe. Then followed the "technical exposition" on Nicaragua by Menocal.

A Nicaragua canal would involve fewer engineering problems than a canal at any other possible location, the audience was informed. The cost, based on an actual survey of the line, was so much less by comparison that for economy reasons alone a Nicaragua canal had to take precedence. The plan was for a lock canal, a sea-level canal at Nicaragua being out of the question.

The route of the canal was similar to that laid out by Vanderbilt's engineer, Orville Childs, in the early 1850's. The San Juan River would be made navigable by building several small dams and these would be bypassed with relatively short canal sections. Going west, up to the lake, there would be about forty miles of canal in total and ten locks. From the lake to the Pacific, a distance of only sixteen miles, there would be another section of canal with ten more locks descending back to sea level. The entire route, from Greytown to Brito, on the Pacific, would measure 181 miles, or more than three times the length of the Panama route. But as with the Childs' scheme, 56 of those 181 miles were already provided for by Lake Nicaragua and almost 70 miles of the San Juan could be made navigable for seagoing ships. So that left only 50-odd miles—58.23 according to Menocal's figures—of actual canal construction, or not much more than the length of a Panama canal.

The cost of such a canal he put at $65,600,000, a third less than the price being quoted by Lieutenant Wyse for his project.

It was a polished, confident performance lasting five hours, and it was made to look even better by the speaker who followed—Lieutenant Wyse. Menocal had spoken as one who had appraised all sides of the problem, seen to every detail, who had covered every foot of the ground on his own two legs. He was the thoroughgoing professional, the voice of experience. Wyse, by contrast, was often vague on details, unsure of his facts. He talked, said one delegate, as though he had dreamed up his entire plan without ever having left Paris.

Wyse, General Türr, Armand Réclus, and several others associated with them had been present from the first session, and though they were not accredited delegates, and so in theory had no real power or say, they were, as members of the Société de Géographie, perfectly at home in such surroundings and known to all. Wyse, as it happened, was the recipient of the Société's gold medal for that year, for his Darien explorations. Yet virtually from the moment he began to speak of his Panama project it was clear to a large number of delegates that he had little substantive knowledge of the terrain and that there was really no such thing as a Wyse Survey. And whereas Menocal had encouraged questions at the conclusion of his remarks and answered them to the satisfaction of every engineer present, Wyse was at a loss to defend his plan on even the most fundamental level. Specifically, he did not know what could be done about the Chagres River, which stood in the path of any canal taking the route of the railroad, or how, when he went down to sea level, he could cope with the twenty-foot tides of the Pacific.

When on Monday, May 19, General Türr and Lieutenant Réclus appeared before the committee and talked for several more hours, they contributed scarcely any more than Wyse had.

Tuesday, May 20, Menocal took the platform again. It was his professional judgment, as a result of three months in the Chagres valley, that any attempt to build a Panama canal would be disastrous. The absolutely unavoidable problem was the river. Any canal at Panama—a lock canal, a sea-level canal—would have to cross the river at least once, possibly several times. If a sea-level canal were cut through the river, the result, as anyone could readily picture, would be a stupendous cataract. The fall of the river into the canal would be 42 feet and this measurement was based on the level of the river in the dry season, when the river was only a few feet deep. In the rainy season the river could be instantly transformed into a torrent. It could rise 10 feet in an hour. At flood stage it could run as much as 36 feet deep, he said, and measure 1,500 feet across. The cost of controlling so monstrous a force—if it could be done at all—was beyond reckoning.

When he first went to Panama, in 1875, his own intention, he said, had been to plan for a sea-level canal. He had abandoned the idea as soon as he grasped the true nature of the Chagres River. Any plan that did not take the river into account was altogether unrealistic.

Lieutenant Wyse asked to be heard. Where had the speaker obtained his data? From actual surveys and from local authorities, Menocal replied. His figures had been obtained through surveys in the field.

The official American plan for Panama was for a lock canal that would dispense with the problem of the Chagres by going over it. Menocal had designed a colossal stone viaduct 1,900 feet long to carry the canal over the river at a point known as Matachín. The elevation of the canal at the viaduct—the summit of the canal, that is—was to be 124 feet, and to carry the ships to this height there were to be a total of twenty-four locks (an equal dozen in either direction from the summit). To build such a canal would cost $94,600,000, a figure that startled a large number of his listeners, since it was approximately what Wyse was claiming for the cost of a canal at sea level.

Menocal believed it to be as ingenious a solution as possible, considering the circumstances, but he had no heart for it. In good conscience he was unable to recommend a Panama canal of any kind. Even a lock canal, he emphasized, would always be threatened by possible floods, and he further warned that the deep cuts that would have to be made through the Cordilleras at that section known as Culebra—even for a lock canal—would be subject to persistent mud slides.

"The surprise and painful emotion on the part of those who had plans *à niveau*, and of their many friends in attendance, can hardly be conceived," wrote Daniel Ammen. "The fact stared them in the face that the plans which they had presented so confidently for adoption were absolutely impracticable."

"From this moment," observed Dr. Johnston, "the Congress became a real Congress and not a sham."

All together the Technical Committee was to consider proposals for fourteen different points on the map of Central America. Frederick Kelley's old San Blas plan was presented, for example. The Mexican delegate, de Garay, spoke for Tehuantepec. But these other options were rejected one by one. In less than a week the issue had come down to the Wyse plan for Panama and the American plan for Nicaragua, and to a great many delegates, having heard Menocal, the choice had been narrowed to Nicaragua. Had a vote been taken at the conclusion of Menocal's remarks on Panama, it is probable that the congress would have picked Nicaragua, as de Lesseps himself conceded privately. But with de Lesseps in charge nothing of the kind was even to be considered.

Behind the scenes he was extremely busy, talking to the French delegates in a manner that "would do credit to a modern American political boss." There could be no turning back. They could agree to no decision other than Panama. "That was the French route," wrote Dr. Johnston; "they had been manufacturing enthusiasm for that route; the bankers and the public would not give a cent to any route that was not patronized by M. de Lesseps and Lieut. Wyse. So that to abandon that route was to abandon entirely for France the glory of cutting the interoceanic canal, and that was not to be thought of for a moment."

By now, moreover, it was commonly understood that large sums of money were at stake. A Panama canal company had been formed in secret, it was rumored. "We were to be brought face to face with the singular spectacle of a congress which had become serious and honest, and which saw its way clear to the truth," observed Dr. Johnston, "and yet which was obliged to remain dishonest, and carry out the original plan, no matter by what means. . . . It was the game of 'I see you, and go you one better,' played by men who had no cards, but plenty of money."

Nor, it should be emphasized, were the warnings voiced solely by the Americans. A noted French engineer named Ribourt, one of the builders of the Saint Gotthard Tunnel, urged the delegates not to misjudge the magnitude of the undertaking. To cut through Panama *à niveau*, to dig a tunnel such as Lieutenant Wyse spoke of, would require not less than nine years of continuous labor, even if the work went on twenty-four hours a day. The cost, said Ribourt, would be at least twice what Wyse was saying. In the view of the revered John Hawkshaw a sea-level canal was physically impossible, since it would have to provide for the entire drainage of the Isthmus at that point. The tunnel being advocated would not be big enough to handle such a volume of water, he said, let alone any ships.

Wyse and Réclus were livid. Réclus could respond only with a rapid list of extraneous claims and countercharges. When the chair requested that he confine his remarks to the subject under discussion, Wyse all but shouted that their plans were being constantly attacked yet they were never given a chance to defend them. His manner, noted Daniel Ammen, was "very excited."

Wyse and Réclus, meantime, were working all hours making drastic revisions. The idea of a tunnel was dropped. Their canal would be an open cut from ocean to ocean. The Pacific tides, they announced, would be handled by a tremendous tidal lock at that end of the canal. The Chagres would be "diverted" into a man-made channel, although Wyse was less than clear as to how this was to be managed.

The week that de Lesseps had thought sufficient to settle all issues and problems had by now passed and a consensus seemed farther away, less likely than ever. So on Friday, May 23, he "threw off the mantle of indifference," as one delegate wrote, and convened another general session in the auditorium. "He is tenacious as well as able," observed Dr. Johnston, "and did not propose to suffer a defeat."

For the first time now he spoke at length, alone on the dais, a large map displayed behind him. The audience hung on every word and he spoke as though they were all his dearest friends, as confident of their eventual support as he was in his own preeminence in such matters. Walking back and forth freely before the map, he talked effortlessly, without pause, without notes. He was more like an actor on stage, radiant, virile, his ideas phrased in the simplest, most direct, and sensible-sounding terms.

One had only to look at the map to see that Panama was the proper place for the canal. The route was already well established, there was a railroad, there were thriving cities at each end. Only at Panama could a sea-level canal be built. It was really no great issue at all. Naturally, there were problems. There were always problems. There had been large, formidable problems at Suez, and to many respected authorities they too had seemed insurmountable. But as time passed, as the work moved ahead at Suez, indeed as difficulties increased, men of genius had come forth to meet and conquer those difficulties. The same would happen again. For every challenge there would be a man of genius capable of meeting and conquering it. One must trust to inspiration. As for the money, there was money aplenty in France just waiting for the opening of the subscription books.

He knew his audience and he delivered every line with perfect confidence in its effect. His audience adored him.

II

It was later that same day that another of the French delegates, one who had had nothing to say thus far, came to the front of the auditorium to deliver the most extraordinary pronouncement of the entire congress. A man of genius stepped forward then and there, in fact, although no one, not even de Lesseps, perceived this.

He was Baron Godin de Lépinay—Nicholas-Joseph-Adolphe Godin de Lépinay, Baron de Brusly—a small, bearded aristocrat who was a chief engineer with the Corps des Ponts et Chaussées (the French Department of Bridges and Highways), and a man known both for his brilliance and his ill-concealed disdain for those who failed to agree with him. He had devised an original answer to the Panama problem, including Panama's deadly climate, which he regarded as the most serious aspect of the problem. He was, as he told the delegates, one of the very few present who had had any actual experience with engineering construction in "the warm lands of tropical America." This, as he did not say, had been the building of a railroad between Córdoba and Veracruz, in 1862, during which a third of the labor force and two-thirds of the engineering staff died of yellow fever.

His solution was what Philippe Bunau-Varilla would call the "Idea of the artificial Nicaragua." Incredibly and tragically, the delegates paid him no attention. The Americans dismissed the plan as ridiculous. Menocal could hardly bring himself to mention de Lépinay's name in his report on the congress. Ammen referred only to the "plan," in quotes, as an illustration of the extremes some of the French had gone to in an effort to rescue the Panama route. Had the delegates reacted differently, had they taken de Lépinay seriously, the story of the canal could have turned out quite differently.

He acknowledged the truth of all Menocal had said regarding the

Chagres River. He himself had been considering the problem of a Panama canal for some years. The idea of digging down to sea level was thoroughly unrealistic if one understood the terrain and ought to be discarded without further fuss. Those who talked of diverting the Chagres River in some fashion were sadly misinformed and deceiving themselves. They were allowing the triumph at Suez to distort their capacity to see things for what they were. Suez and Panama must not be regarded as comparable, he said. The environmental conditions were opposite in the extreme. "At Suez there is a lack of water, the terrain is easy, the land nearly the same level as the sea; in spite of the heat, it is a perfectly healthy climate. In tropical America, there is too much water, the terrain is mostly rock, the land has considerable relief, and finally the country is literally poisoned." To act in the same manner in places of such opposite character, he declared, would be to "outrage nature" instead of to benefit by it, "which is the primary goal of the engineer."

His own plan—"the most natural method"—was brilliantly simple, a genuine stroke of genius, and, as time would tell, it was absolutely sound.

Like Menocal, he had concluded that the Chagres must be bridged, but instead of a stone viaduct at Matachín, he envisioned a bridge of water across most of the Isthmus. There would be two artificial lakes, with flights of locks, like stairs, leading up to the lakes from the two oceans. As Lake Nicaragua was the essential element in the Nicaragua plan, providing both easy navigation and an abundant source of water for the canal, so his man-made lakes would serve at Panama. Through engineering, in other words, he would create at Panama what already existed at Nicaragua.

What he was proposing was not really a canal at all in the conventional sense. His lakes would be created by building two huge dams, one at the Chagres near the Atlantic, the other on the Rio Grande, which flows into the Pacific. The dams would be built as near to the two oceans as the configuration of the land permitted. The Chagres dam, the largest, should be built, he said, at the confluence of the Chagres and the Gatun rivers, at a point called Gatun, and it would hold the largest of the lakes. The surface of the lakes would be eighty feet above sea level and the lakes would be joined by a channel cut through the mountain spine at Culebra, this being the only heavy excavation required.

The virtues of the plan were enormous. To begin with, it greatly reduced the amount of digging to be done. Further, it eliminated all danger of Chagres floods, since the river would feed directly into the lakes. The dams would control the river; the river would serve as an unlimited water supply for the lakes—for the canal. Thus the river would become the life blood of the system, rather than its mortal enemy.

The resulting passage, furthermore, would be a broad lake, rather than a narrow channel. Ships would be able to move at greater speeds and they would be able to pass one another without shunting into sidings, or tying up, as was necessary at Suez. Passage through such a canal would take no more than twelve hours, even if one were to figure the time in each lock at half an hour, when, in fact, actual time in the lock would probably run closer to fifteen minutes. And twelve hours was only an hour and a half more than could be expected for passage through the best-engineered sea-level canal at Panama.

Such a project could be completed in six years, he said, and at a cost of 500,000,000 francs ($100,000,000), including interest and overhead, but not including the cost of buying the Panama Railroad, which, he stressed, would be an essential step and a very sizable expenditure.

Most important of all, he said, was the saving the plan would mean in terms of human lives. As was understood by everyone in the audience, nearly all varieties of tropical fever and miasma were caused by "noxious vapors" released from the putrid vegetation and rank soil of the jungle. Any excessive disturbance of such ground, therefore, naturally meant the spread of disease in epidemic proportions. But since his scheme called for a minimum of excavation, there would be a minimum of disturbance during construction and the incidence of disease would be correspondingly small. Furthermore, once the canal was built, much of the poisonous terrain would be sealed off by the lakes, producing a long-lasting beneficial effect. To dig a canal *à niveau* in Panama, he said, would cost the lives of no less than fifty thousand men.

His ideas were eloquently expressed and uncannily prophetic, but the delegates did not think enough of them to grant him even a token discussion. The congress turned to other matters.

Two subcommittees—one on tunnels and another on lock canals—were in closed session through the weekend and on Monday, May 26, presented their conclusions. Menocal's Nicaragua canal was found to be perfectly practicable. Its cost was figured at $140,000,000 and the opinion was that it could be built in six years. The Wyse plan too was declared practicable. But the cost was put at $209,000,000. It was further stated that at least twelve years would be needed to build such a canal and all claims were qualified by a final explicit warning that construction of a Panama canal at sea level, as well as any measures designed to restrain the Chagres, presented many problems past reckoning.

Those delegates friendly to Lieutenant Wyse responded by asserting privately—in the halls outside the auditorium, over lunch in nearby cafés and restaurants on the Boulevard Saint Germain—that de Lesseps "*would positively refuse*" to lead the building of any canal other than one at Panama, a statement that had the desired effect of bringing a number of wavering delegates quickly back into line.

Commander Selfridge joined those attacking the Nicaragua plan and for the first time raised the issue of Nicaragua's history of earthquakes and volcanic disturbances. But two of the most prominent French members of the Technical Committee, Cotard and Lavalley, both former engineers with de Lesseps at Suez, sided with the Nicaragua forces, as did Gustave Eiffel.

Wednesday, May 28, following a late night at the banquet tables (at the Hotel Continental, overlooking the Tuileries Gardens), the language of several speakers became considerably less diplomatic than heretofore. Menocal in particular made no effort to conceal his mounting disgust. He had come to Paris, he said, to present serious proposals based on volumes of information gathered through great effort and at great cost. He and his colleagues had expected other delegates to present material of comparable character, and that from a proper consideration of all such data, serious people, professionals of proven competence, would make their decisions in a spirit of reason and impartiality. Instead, the American plans were being weighed on the same scale as were imaginary schemes traced on imperfect maps, some of them the result of a night's inspiration.

With that the oratory became highly charged. One French delegate, speaking for several hours, declared that it *must* be a sea-level canal, no matter what the cost.

The final recommendation of the Technical Committee, the decision the whole congress had been waiting for, was arrived at late that night amid tremendous confusion and excitement. Even in the stilted official account of the proceedings it is apparent that the session very nearly became a brawl. Twenty delegates, nearly half the committee, walked out before the vote was taken. In the end, only Ferdinand de Lesseps and eighteen others were willing to vote on a resolution, and of these, just three refused to vote as he wished.

Panama was pronounced the proper place for the canal and a sea-level canal was especially recommended.

It had been raining off and on for the past few days and it was raining again the following day, Thursday, May 29, when at 1:30 in the afternoon the full congress convened to hear the committee's report and to cast the final, historic vote. "The hall was densely crowded, many ladies being present," recalled Admiral Ammen; "about one hundred members or delegates and three to four hundred other persons. . . ." Dr. Johnston, who had no illusions as to how the vote would go, wrote bitterly:

> We had arrived at the moment of "sublime resolutions," of those "sublime resolutions" which have been the glory and ridicule of France; they were going to carry hundreds of millions of money abroad for the good of mankind in general. It would cost much money, but the money they had; it would require men of genius, but these also they had; the absurd barrier which nature had thrown up between the two seas was going to fall before the force of French genius and the power of French money.

French observers would recall the solemnity of the moment as de Lesseps read aloud the crucial resolution:

> The congress believes that the excavation of an interoceanic canal at sea level, so desirable in the interests of commerce and navigation, is feasible; and that, in order to take advantage of the indispensable facilities for access and operation which a channel of this kind must offer above all, this canal should extend from the Gulf of Limon to the Bay of Panama.

The necessary time for construction was fixed at twelve years—a finished canal by 1892—and the cost of construction was estimated at 1,070,000,000 francs, or $214,000,000. Supposing the interest payable in the meantime would amount to 130,000,000 francs, the total expenditure worked out to 1,200,000,000 francs, or $240,000,000—almost triple the cost of Suez.

It was a voice vote in alphabetical order. Henri Bionne called the names.

Daniel Ammen rose and abstained on the grounds that only professional engineers should be allowed to vote on such a proposition. Two other Americans, Nathan Appleton and Christian Christiansen, the latter from San Francisco, voted yes. The first no was sounded by the Guatemalan delegate. Daubrée, who had been chairman of the Technical Committee, voted yes, as did a former Suez engineer named Dauzats. Gustave Eiffel voted no.

Flachat, Hawkshaw, and Dr. Johnston decided to absent themselves from the proceedings. Alexandre Lavalley was also absent. When Godin de Lépinay was called, he got to his feet and looked about the large crowd. "Though unable to make my advice triumph, I will not abandon it. And in order not to burden my conscience with unnecessary deaths and useless expenditure I say 'no!' " When he sat down it was to a noisy chorus of jeers and booing.

Seventy-seven of the delegates had voted at this point and forty-three had voted yes. The next name in alphabetical order, everybody knew, was Ferdinand de Lesseps.

"I vote 'yes!' " he cried out, his voice filling the room. "And I have accepted command of the enterprise!" It was his first such public declaration and it electrified the house. The applause and cheering went on and on, interrupting the roll call for several minutes.

The rest went swiftly. Eli Lazard, of San Francisco, the Russian admiral, the Chinese delegate, a man from the Italian Geographical Society, declared themselves in the affirmative. A. G. Menocal abstained. As the final tabulation was being made, de Lesseps, looking immensely pleased, told the audience, "Two weeks ago I had no idea of placing myself at the head of a new enterprise. My dearest friends have tried to dissuade me, telling me that after Suez I should take a rest. Well! If you ask a general who has just won a first victory whether he wishes to win a second, would he refuse?"

To other audiences later he was to say that it was the overwhelming approval of the congress, the faces he saw before him, and especially the look his wife gave him that propelled him to make the decision, adding that to have backed down then would have been an act of cowardice. His wife, he would say, had been the foremost of those dearest friends who had tried to dissuade him. She had wished only that their life could continue as it was. So apparently she too had been swept up by the spirit of the moment.

There was absolute silence as the vote was declared: in favor of the resolution, 74; opposed, 8; abstaining, 16; absent, 38.

Delegates were on their feet cheering; women were waving handkerchiefs. It was as if an astonishing victory had been won. De Lesseps stepped forward and promised success; Admiral de La Roncière-Le Noury declared that the day marked the beginning of one of the greatest undertakings of modern times. "It seems to me that nothing could have been more glorious . . ." Henri Barboux, attorney for Ferdinand and Charles de Lesseps, would recall years later before a packed courtroom. "Might not one think of it as a council ordaining, after the lapse of seven hundred years, a new crusade?"

Ferdinand de Lesseps was a man accustomed to having his own way and he had not been disappointed. The congress, as its severest critics claimed, had been put on primarily to give the Wyse Concession a legitimacy, an authority, that it otherwise lacked and that it greatly needed to attract the necessary financial backing, as de Lesseps knew better than anyone. The grand international gathering had been conceived not to arrive at a consensus, but to provide an inaugural ceremony for a decision already made by the one delegate who mattered, Ferdinand de Lesseps. The objective from the start had been to ordain, to consecrate, the Wyse Concession, the Wyse plan, in full public view, with all possible ceremony, to give the appearance of an impartial, scientific, international sanction. The Americans with their maps and plans and convictions had come alarmingly close to spoiling the effect, but even they had been no match for "the first promoter of the age."

Virtually all of de Lesseps' blind spots, all the tragic errors of his way, had shown themselves in the course of the two weeks–the jaunty disregard of technical problems, the inability to heed, to *trust*, the views of recognized authorities if those views conflicted with his own, the faith that the future would take care of itself, that necessity would give rise to invention in required proportions and at the proper moment, the unshakable faith in his own infallibility.

But then all these same qualities had been fundamental to his success in Egypt, and combined with his love of people, his charm, these were what made him Ferdinand de Lesseps. And who then was to say that he knew more about building a canal, more about success in such grandiose undertakings?

The Americans went home furious and extremely skeptical that anything would ever come of the affair.

Why Ammen and Menocal had failed to vote no, to record their negative views in public when it counted, remained a puzzle to many and a disappointment to the handful who had. By way of explanation Dr. Johnston offered that "these delegates were met and surrounded during their whole stay with such a large hospitality, they were so dined and feted, that they will be excused for lacking the heart to look their entertainers in the face and pronounce so harsh a word as 'no.' "

A. G. Menocal, afterward, did an interesting analysis of the vote.

Though the yea votes were predominantly French, not one of the five delegates from the French Society of Engineers had voted for the proposal. Of those seventy-four delegates who did declare themselves for a sea-level canal at Panama, only nineteen were engineers and of those nineteen only one had ever set foot in Central America and he was young Pedro Sosa of Panama.

Ferdinand de Lesseps

CULVER PICTURES, INC.

COLLECTION OF GEORGES SIROT

BIBLIOTHÈQUE NATIONALE

ABOVE LEFT, Jules Verne

ABOVE CENTER, the second Madame de Lesseps, Louise Hélène Autard de Bragard

ABOVE RIGHT, Lieutenant Lucien Napoleon-Bonaparte Wyse

LEFT, Charles de Lesseps

CULVER PICTURES, INC.

LIBRARY OF CONGRESS

HARPER'S WEEKLY

ABOVE LEFT, Secretary of State William Evarts

ABOVE RIGHT, Admiral Daniel Ammen

RIGHT, American skepticism over the vast undertaking as expressed by Thomas Nast in *Harper's Weekly:* "Is M. de Lesseps a Canal Digger or a Grave Digger?"

BOTH PHOTOS: CULVER PICTURES, INC.

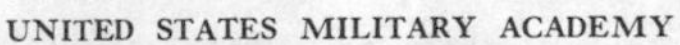
UNITED STATES MILITARY ACADEMY

ABOVE, headquarters of the Compagnie Universelle du Canal Interocéanique, Cathedral Plaza, Panama City

LEFT ABOVE, view of the Chagres River at the time the French arrived. The Panama Railroad is in the foreground; the village of Gatun is across the river.

LEFT BELOW, Front Street, Colón, as it looked during the French era

PANAMA CANAL COMPANY

LEFT ABOVE, Ferdinand de Lesseps with his entourage in Panama in 1880

LEFT BELOW, four unidentified French engineers and an unidentified companion. Probably three of the five died of disease.

BELOW, one of the giant French excavators upon which de Lesseps based his high expectations

UNITED STATES MILITARY ACADEMY

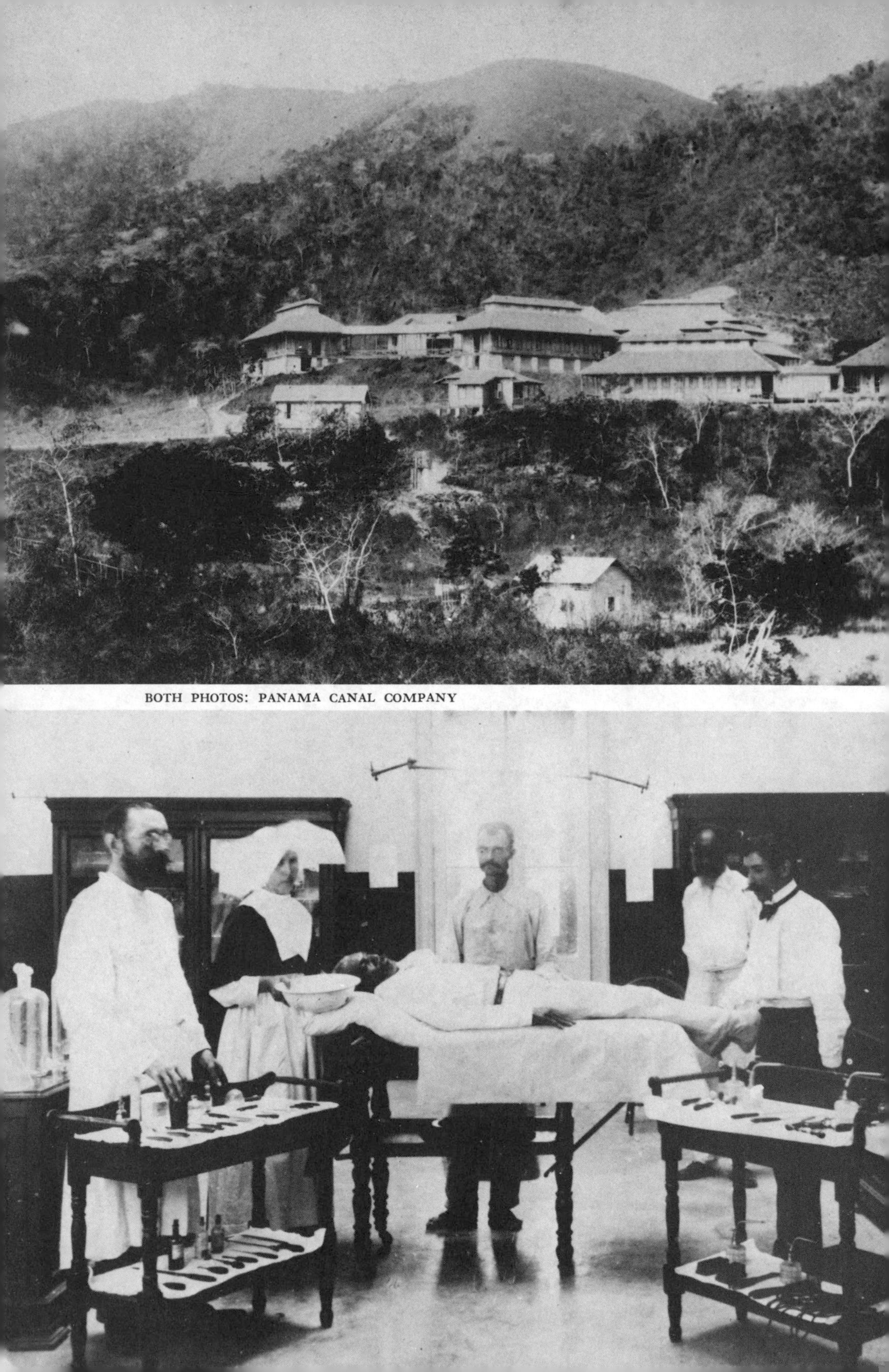

BOTH PHOTOS: PANAMA CANAL COMPANY

FROM MOSQUITO CONTROL, JOSEPH LE PRINCE, G. P. PUTNAM, 1916

ABOVE, crockery rings filled with water were used by the French to protect plants on the hospital grounds from the ravages of umbrella ants, but served also as perfect breeding grounds for *Stegomyia fasciata*, the yellow-fever mosquito

LEFT ABOVE, l'Hôpital Nôtre Dame du Canal, the French hospital on Ancon Hill outside Panama City

LEFT BELOW, operating room at Ancon sometime in the late 1880's

BELOW, one of hundreds of surviving death certificates from the Ancon hospital, this of a twenty-nine-year-old Frenchman who died of yellow fever in 1886

NATIONAL ARCHIVES

N° 631

CERTIFICAT DE DÉCÈS.

COMPAGNIE UNIVERSELLE DU Canal Interocéanique.
Service de Santé.
CIRCONSCRIPTION DE (a) Panama

Nous soussigné, Docteur en médecine, Médecin de la Compagnie Universelle du Canal Interocéanique à la circonscription de (b) Panama *Certifions que*

Noms Fouquet
Prénoms Jacques
Nationalité Français
Agent classé, temporaire ou journalier
Profession ou emploi maçon
Désignation de la Section ou de l'Entreprise Culebra
Domicile et résidence Culebra
Date et lieu de naissance 29 ans à
Etat-Civil Célibataire
Adresse de la famille à Cinq Mars (Indre et Loire)
Date de l'entrée à l'hôpital 2 7bre 1886
Date du décès 4 Septembre 6h matin
Cause du décès Fièvre jaune

A Panama le 4 Septembre 1886

LE MÉDECIN DE LA COMPAGNIE.
Signé: Dr de Meyrignac

COURTESY OF THE ÉCOLE POLYTECHNIQUE

Philippe Bunau-Varilla at the time of his graduation from the École Polytechnique

RIGHT ABOVE, West Indian labor gangs ride a raft specially devised by Bunau-Varilla for the underwater placement of dynamite charges prior to dredging. Long steel drills were driven by hand according to his mathematically calculated pattern.

RIGHT BELOW, French ladder dredge at work. It was upon machines of this kind that Bunau-Varilla rested his novel scheme for rescuing the French effort during its desperate finale.

BELOW, the hanging of Pedro Prestan at Colón, August 18, 1885, following the disastrous "Prestan Uprising"

PANAMA CANAL COMPANY

BOTH PHOTOS: PANAMA CANAL COMPANY

LEFT ABOVE, Baron Jacques de Reinach

LEFT BELOW, Georges Clemenceau

RIGHT ABOVE, Charles de Lesseps pleads his case in the Paris Court of Appeal in January 1893, at the start of the first of two sensational trials. The three other defendants seated behind Charles are, from left to right, Gustave Eiffel, Henri Cottu, and Marius Fontane. Henri Barboux, attorney for Ferdinand and Charles de Lesseps, is the small white-haired figure standing behind Charles's empty chair. This courtroom looks today exactly as it did then.

ALL ILLUSTRATIONS: BIBLIOTHÈQUE NATIONALE

RIGHT, a contemporary artist's conception of the bedridden Cornelius Herz, the "mystery man of Panama," sequestered in his hotel room at Bournemouth, England

UNITED STATES MILITARY ACADEMY

The abandoned château of the first Directeur Général, the much-publicized "*folie Dingler*"

An abandoned French excavator is overtaken by the jungle near Tabernilla.

PANAMA CANAL COMPANY

4
Distant Shores

> . . . and I maintain that Panama will be easier to make, easier to complete, and easier to keep up than Suez.
>
> —Ferdinand de Lesseps

I

With no further delay Ferdinand de Lesseps swung into action. In a matter of days he had organized a private syndicate of some 270 rich and influential friends who, for providing 2,000,000 francs, were to receive founders' shares at a bargain price, once a company was legally established. It was the same as he had done for Suez.

Next he bought out the Türr Syndicate for 10,000,000 francs ($2,000,000). Payment was to be half in cash, half in stock in the new company. The Wyse Concession was now his alone, as much as the Suez concession had been, and Wyse, Türr, and the rest realized a profit on their initial investment of more than 3,000 percent. Istvan Türr took his money and went off to negotiate a concession from the king of Greece to build a canal across the Isthmus of Corinth. Wyse and the others, however, were in to stay, convinced that the best was yet to come. Wyse was happily telling friends that de Lesseps had promised to put him in command of the work, as reward for his efforts. But as others had learned at Suez, de Lesseps was not one to share power or glory. He denied having guaranteed Wyse a role of any kind. The young officer had served his purpose and so now he was dropped—"betrayed" Wyse felt.

A prospectus was prepared for the Compagnie Universelle du Canal Interocéanique de Panama and de Lesseps was off on a whirlwind tour of the provinces to drum up 400,000,000 francs, twice what he had raised to launch the Suez venture. His support, he said, would come as it had before, from small investors, people from every walk of life. His self-confidence had never been greater. He would go to Panama, he said, to make a personal evaluation. He would tour the United States to explain his mission to the people of that great land. "M. de Lesseps is convinced that it [the canal] is the right thing," wrote the Paris correspondent of *The New York Times*, "and . . . the simple fact of his connection with it will secure that Archimedean lever of the nineteenth century, money . . ."

But it was not to be that simple. Times had changed since the Suez company had been organized. The powers of the financial world and of the press had advanced considerably and the new venture had already become the target of a calculated attack. Influential French bankers wished to demonstrate the indispensable nature of their services at such times, then to step in and take control of subsequent stock subscriptions, once de Lesseps' initial subscription failed, as they assured everyone it would. Influential publishers expected to be paid for their editorial backing.

Rumors were started in the Bourse that de Lesseps was in his dotage and ought not to be trusted with other people's money. It was claimed that at the first blow of a French pickax at Panama, the American fleet would arrive and massacre the workers. *Crédit Maritime* said the canal would cost so much that it would never pay a dividend. Another financial paper called the scheme a swindle and warned readers not to risk their savings in it.

The result was the failure of the stock issue, a failure so resounding that almost any man other than de Lesseps would have abandoned the whole plan there and then and spared himself any further humiliation.

Of 800,000 shares offered, a trifling 60,000—less than 10 percent—were purchased. Yet "the wonderful old man" appeared undaunted. It was obvious what had to be done, he said privately. The bankers would be invited to participate. "The financial organs were hostile," he explained, "because they had not been paid."

Those who had subscribed to the stock got their money back along with a new circular promising another issue the moment things were straightened out. Meantime, the firm of Couvreux, Hersent et Compagnie, one of his Suez contractors, was sending its best man to appraise the Panama route from end to end. "I have never feared obstacles and delays. Experience has proved to me that those who are too quick to believe have no deep roots."

On September 1 appeared the first issue of the *Bulletin du Canal Interocéanique*, an eight-page newspaper published for propaganda purposes that was to appear regularly thereafter twice monthly, its contents comprised largely of selected articles reprinted from French newspapers and magazines. (A similar journal, underwritten by Mohammed Said, had been published all through the Suez years.) The pledge to go to Panama was renewed. He would be accompanied by a party of internationally famous engineers—his own new International Technical Commission—who would be responsible for the final survey required by Colombia according to the Wyse Concession. To demonstrate that there was nothing to fear from the climate, he would also take several of his family.

When the representative of Couvreux, Hersent, an engineer named Gaston Blanchet, returned from Panama with a favorable report, de Lesseps announced that a crew of sappers would go next to make test borings along the projected canal line. That was in November. In December, with a large entourage, he boarded the *Lafayette* at Saint-Nazaire, leaving his affairs in the hands of Charles. Included were his wife, two young sons and a daughter—Mathieu, ten; Ismael, nine; Ferdinande, seven—a governess, Henri Bionne, Lieutenant Réclus, Gaston Blanchet, Abel Couvreux *fils* (the son of de Lesseps' old friend), the Dutch engineer Jacob Dirks, and about a dozen others. Expenses for the trip were to be met by Couvreux, Hersent et Compagnie.

The crossing appears to have been pleasant and without incident, other than an angry scene in front of everyone in the ship's salon between de Lesseps and Lieutenant Wyse, who had not been asked to make the voyage but had booked his own passage at his own expense. For the rest of the trip Wyse remained "apart and taciturn," while de Lesseps seems to have been quite literally the life of the party.

It was a voyage of approximately 4,595 miles, a great circle route, plotted, when weather permitted, by the brightest stars in the heavens, by Capella, the double star; by Altair and Vega; and afterward by Rigil Kent and Acrux, new stars in a new southern sky. The ship, a handsome, black two-masted steam sailer, made about eight to ten knots, and her progress into lower latitudes was marked by the steady decline toward the horizon of Polaris, the North Star, the great constant at the end of the handle of the Little Dipper. And low on the opposite horizon, meantime, off the port bow, rose the Southern Cross.

II

The *Lafayette* steamed into Limon Bay under a scorching sun, and with all passengers crowding her rail, on the afternoon of December 30, 1879, at the start of Panama's dry season. On the Pacific Mail wharf a little brass band was playing mightily.

The welcoming ceremonies were held in the ship's salon, moments after she tied up. It was proclaimed an occasion second only to the arrival of Columbus in Limon Bay. Señor J. A. Céspedes, chairman of an extremely sober-looking reception committee, made the first and longest of several speeches. De Lesseps, when it was his turn, responded in perfect Spanish, "very pleasantly, wearing the diplomatic smile for which he was noted," in the words of the one American present, a Panama Railroad official named Tracy Robinson. "When he spoke, the hearer would not fail to be convinced that whatever he said was true, or at all events that he believed it to be true."

After dark the town blazed with Japanese lanterns, and when the final burst of a fireworks display fell into the bay, de Lesseps came down the gangplank. Accompanied by a few friends and a small, noisy crowd—mostly ragged black children—he walked a while along Front Street, Colón's sole thoroughfare.

The following morning he was up in time to see the tropical dawn that comes all at once. With Madame de Lesseps he set off on an "inspection tour," their children, thrilled to be on solid ground again, racing ahead, climbing posts and stanchions, and astonishing the local citizenry, who, because of the long hair and velvet clothes, thought all three were girls. De Lesseps, fresh in a white linen suit, talked incessantly, concluding one remark after another with the assertion, "The canal will be made." The upper Chagres would be turned into the Pacific, thus ending floods in the lower valley. "The canal will be made." At the great cut at the summit, the work of many thousands of men would be handled by modern explosives. "The canal will be made." He was overjoyed by the morning air. Colón was a delightful place. "The canal will be made."

Yet it is hard to conceive of his being anything but terribly disappointed by Colón. Seen from a distance, from an inbound ship, the town appeared to float on the bay as if by magic. White walls and red roofs stood out against blue water and flaming green foothills. But close up, it was a squalid shantytown set on stilts, paint peeling. There was a stone church that the railroad's guidebook made much of but that would have been of little interest anywhere else. A variety of saloons and stores lined the east side of Front Street, facing the harbor. There were an icehouse, a railroad office, a large stone freight depot, two or three seedy hotels, and the "tolerable" Washington House, a galleried white-frame affair, which, like virtually everything else in sight, belonged to the railroad company. The railroad itself ran down the middle of Front Street, and in a park, or what passed for a park, in front of the Washington House, stood an ugly red-granite monument to the railroad's founders, Aspinwall, Chauncey, and Stephens. In a nearby railroad yard there was also a bronze statue of Columbus, an Indian maiden at his side, which had been a gift from the Empress Eugénie years before. But that was the sum total of Colón's landmarks.

The town had been built by the railroad on Manzanillo Island, a coral flat, no more than a mile by three-quarters of a mile in area, at the entrance to Limon Bay; and so there was open salt water on all but its southern side, where a narrow channel, the Folks River, divided it from the mainland. John Lloyd Stephens had christened the town Aspinwall, but the Colombians had insisted on calling it Colón, for Columbus, and so a silly dispute had been dragging on ever since. To most of the older Americans it was still Aspinwall.

Streets, barely above tide level, were unpaved and strewn from end to end with garbage, bits of broken furniture, dead animals. (One French visitor would write of walking ankle-deep in "*les immondices imaginables.*") Enormous dark buzzards circled interminably overhead, and the human populace, most of which was black—Jamaicans, by and large, who had been brought in to build the railroad—lived in appalling squalor. Disease and poverty, hopeless, bedrock poverty as bad as any to be seen in the Caribbean, seemed to hang in the air of back streets, heavy as the atmosphere.

The entire town reeked of putrefaction. There was nothing to do. It was as if a western mining camp had been slapped together willy-nilly in the middle of an equatorial swamp, then left to molder and die. Once, at the height of the gold rush, there had been a kind of redeeming zest to the place, and old-timers talked of such celebrated establishments of the day as the Maison du Vieux Carré, which specialized in

French girls. Now travelers disembarking to take the train dreaded spending an hour more than necessary.

There was, however, one quite pleasant section of the town, at the northern end of the island, near a tremendous iron lighthouse that could be seen from ten miles at sea. There the houses faced onto a white coral beach. Neat and freshly painted, with green lawns and surrounding palms, they were the quarters reserved for the white Americans who ran the railroad, and it was to one of them, the home of Tracy Robinson, that de Lesseps and his family were conducted, to judge for themselves the supposed privation of life in the American tropics. Robinson, a personable and intelligent man, had spent twenty years in Panama. He was fascinated by the country, liked the people and the life, and he was certain, as he told de Lesseps, that the great future of mankind was in the tropics.

About ten o'clock the Pacific Mail steamer *Colon* docked beside the *Lafayette*, bringing Trenor W. Park and a party of gentlemen from New York who were to join the tour. By 11:30, the introductions completed, baggage ashore, everyone climbed aboard a train standing on Front Street, its bright-yellow coaches bedecked with French and Colombian flags.

The new arrivals included several stockholders in the railroad, who by their own admission had come more for pleasure than business; a reporter from the New York *World;* a rotund and overbearing former Union Army engineer named W. W. Wright; and Colonel George M. Totten, one of the original builders of the Panama Railroad. Wright and Totten had agreed to serve on de Lesseps' Technical Commission. Wright, however, was a man of no particular reputation. It was Totten who got de Lesseps' attention. Totten had been in charge of the railroad all through the years of construction. He had weathered the heat, the bugs, the mud, political uprisings, stockholders' inspection tours, floods, fever, even a siege of yellow fever that brought him so near to death that his companions had his coffin ready and waiting. Now he was a leathery, white-bearded old figure with steel-rimmed spectacles. It is said also that he had a nice sense of humor, but a search of available sources reveals no evidence of it.

The brass band was pumping away again as the train rolled down Front Street, bell clanging, a cheering crowd chasing alongside in the brilliant sunshine. Then, after affording a brief open view of the glittering bay, the tracks turned into the jungle.

Of several surviving accounts of the tour, the most detailed is that by the *World* reporter, J. C. Rodrigues, who was as fascinated by the "lively Frenchmen" and their leader as the group was by the passing scenery.

> M. de Lesseps himself rode most of the way on the platform of the car on the rear of the small train. For half an hour that I was at his side I could witness the deep interest which he took in the luxuriant nature, which was to him so extremely novel. . . . Inside the car, however, there reigned more than tropical—simply torrid—enthusiasm. A yellow butterfly would cause a commotion in these excitable people. But you do not imagine what an event the first approach to the Chagres River was. The car was pandemonium. The train had to be stopped and the Chagres—the enemy—had to be inspected.

The point where they first saw the river was Gatun, a native village seven miles from Colón, at the confluence of the Rio Gatun, the place Godin de Lépinay had picked for his great Chagres dam. Before that, just out of Colón, they had passed Mount Hope, or Monkey Hill, as it was better known, a low rise on the left where during construction days the railroad had buried its dead. Then for the next several miles they had crossed a broad mangrove swamp on tracks only inches above the water. Between the swamp and Gatun, the growth of vegetation was as exuberant as any on earth. Giant *cedro* trees towered a hundred feet or more in the air, their smooth gray trunks like pillars of concrete. Trailing vines, blossoming creepers, scarlet hibiscus, orchids, crimson passionflowers, parasitic plants of every imaginable variety, hung wherever one looked. Bamboo crowded the tracks in clumps the size of a house. It was as if the train were running along the bottom of a narrow green canyon that went winding on and on with only a thin trace of bright sky to be seen straight up, in the gap between the crowns of the trees. Every so often there would be a sudden break—a patch of banana trees, a canebrake—but as quickly it would be gone again. So relentlessly did the jungle try to recover what it had lost to the railroad, the passengers were informed, that parts of the line had to be cleared several times a year.

At Gatun the entire population had gathered for the occasion, several hundred brown, square-faced, friendly-looking people, men in white linen and straw hats, women in loose-fitting muslin in a variety of sun-faded colors, children mostly naked, everyone smiling and waving. On the left side of the train was Gatun station, a two-story white-frame building with green shutters and picket fence that might have been transplanted directly from Massachusetts. On the other side of the train was the river, "now very low, running sluggishly," as

Rodrigues noted. The actual village was across the river. About fifty grass huts were scattered within a great bend in the river and in the forefront of a sun-flooded green savanna that reached to a range of darker-green hills, two, three miles in the distance.

Most of the passengers got out for a look, and the overwhelming green of the landscape, the intensity and infinite variety of green under a cobalt-blue sky, caught them unaware. Like so many before, they had come to Panama with little thought of being stirred by landscapes. That the place could be so breathtakingly beautiful struck them as a singular revelation. "*La plus belle région du monde*," de Lesseps exclaimed in a letter to Charles.

At Gatun the flags that hung over some of the train windows were taken away to give a better outlook. They were running along the valley of the Chagres now, where the river came down in big, wide loops, brown and unhurried—now that the dry season had returned—through intermittent patches of deep shadow and sharp, white sunlight. The river was on their right; their general direction was south and slightly east. They were still barely above sea level, climbing only very gradually, going "up" the valley (that is, against the current of the river) but "down" the map, as several of them needed to have explained.

They crossed miles of swamp, including the infamous Black Swamp, which supposedly was bottomless. It was not—Totten and his engineers had found bottom at 185 feet—yet the roadbed kept sinking there and had to be built up year after year. "Everything kept going down and down," an old-time employee would tell a Senate committee in Washington years later, "and they kept filling in and filling in."

There were more white station houses, all quite similar, neat, almost prissy, but often with names very much in keeping with the surroundings—Tiger Hill, Lion Hill, Ahorca Lagarto (which means "hanging lizard"). There was a glimpse of the Chagres again at Bohío Soldado ("place where a soldier lives") and at Frijoles ("beans"). Then, twenty-three miles from Colón, or just about halfway to Panama City, the train stopped and everyone was asked to get out. They had arrived at Barbacoas, an Indian word meaning "bridge," the point where the railroad crossed the Chagres. Only, at the moment, the bridge at Barbacoas was out of service.

The river here was swift and rocky, about three hundred feet wide and contained between high embankments. The bridge, a heavy wrought-iron structure set on stone piers, was more than six hundred feet in length and built forty feet above the river, or what had been presumed to be safely above the flood line. But in November, just weeks earlier, a violent "norther" had struck, bringing three days of torrential rain and the worst flood on record. In three days the Chagres had risen forty-six feet. Thirty miles of track had been under water and the bridge had been wrenched apart or out of line in several places. As future hydrographic studies would show, the discharge of the Chagres in the vicinity was normally less than 1,000 cubic feet per second in the dry season. In the rainy season, under normal conditions, the discharge would be ten times that—or more—with fluviograph readings of 10,000 to 13,000 cubic feet per second. But in the November flood, according to later studies based on the railroad's records, the flow of the river must have been nearly 80,000 cubic feet per second.

The river's drainage basin, from its headwaters to the Caribbean, was comparatively small—about 1,300 square miles, an area about the size of Rhode Island. Yet except for the dry season, virtually this entire basin was running water. The river originated in the steep jungle uplands miles off to the east, a "quick and bold" wilderness with mountains of two thousand to four thousand feet, where at the time of the Spanish conquest a legendary Indian chief, Chagre, had ruled. Even under average conditions, the runoff from such country was phenomenal. With abnormally heavy rains in the mountains, it was as if a dam had burst. And while the recent flood had been the worst since the railroad began bothering with records, the floods of 1857, 1862, 1865, 1868, 1872, 1873, and 1876 had been nearly as awesome.

The situation at Barbacoas should have been the clearest possible warning to de Lesseps and the others. The condition of the massive iron bridge was such that no through trains had crossed the Isthmus, no freight had moved between Colón and Panama City, in five weeks. Only by a crude arrangement of planks put across the breaks was it possible for passengers to walk over and transfer to another train. The river's violence, quite obviously, had been greater even than what A. G. Menocal had described in his speech before the Paris congress.

Readers of the *Bulletin du Canal Interocéanique* were to be told nothing of the broken bridge, however. The official account of the tour would contain only passing mention of an unexplained delay at Barbarcoas.

Most of them crossed single file, slowly, cautiously, amid much good-natured banter, the river sliding by forty feet below. A fuss was made over the safety of the de Lesseps children, who greatly enjoyed every moment of the experience, and then two or three of the Americans, after appraising the problem, decided to risk the crossing another day. Quantities of champagne had been available on the train since leaving Colón and this seems to have had a bearing on their decision.

On the other side stood a second train and beside it another official delegation, a dozen or so citizens from Panama City, all as formal as pallbearers. Among them were the president-elect of the province, Demaso Cervera, and a former president, Rafael Aizpuru, "a disreputable revolutionist," de Lesseps was told. Because of the heat, there was just one very short speech; then with several blasts of the whistle the journey resumed.

The river was on the left now as the train rolled smoothly along through open meadows. "In Suez we had to build everything," de Lesseps remarked; "here you already have a railroad like this. . . ." The first mountains came into view, small and bright-green and heaped up like the mountains in a child's drawing. Lunch was served—"with wines, etc., etc., and everything gave entire satisfaction."

Within the space of a few miles the railroad crossed the Rio Caimilo Mulato, the Rio Baila Monos, the Rio Culo Seco, the Rio Caribali, all tributaries of the Chagres. (Forty-seven and a half miles of railroad had required 170 bridges and culverts of 15 feet or more, 134 bridges and culverts of less than 15 feet, a statistic that gives some idea of the difficulties there had been in making headway in such half-drowned country.) Past Gorgona Station the train left the river again, taking a shortcut through steep red-clay embankments. Then it swung around a hill to meet the river at Matachín, another cluster of grass huts and the point where Menocal had proposed to build his giant stone viaduct. Again the train stopped, to be instantly surrounded by beaming brown faces.

Matachín was best known as the place where Chinese workers, hopelessly lost to "melancholia," had committed suicide en masse. *Matar* is Spanish for "to kill," it was explained; *chino*, the word for "Chinese." The fact that *matachín* is also a perfectly good Spanish word meaning "butcher" or "hired assassin," and that the place had been called that long before the railroad came through, did not seem to matter. Everyone who passed through Matachín heard the story.

To what extent de Lesseps and Totten discussed such topics, whether Totten was closely questioned on the death toll during construction of the road, or how much, if anything, may have been said about disease or bodies pickled in barrels, how much Totten may have been willing to admit, even to himself at this late date, is not known. The point he does seem to have stressed—the great lesson to be learned from his experience—was that everything, *everything*, had to be brought to Panama, including the men to do the work. The Panamanians themselves would be of no use. The poor were unused to heavy manual labor and were without ambition; the upper classes regarded physical work as beneath their dignity. There would be no home-grown labor force to count on, no armies of Egyptian fellahin this time. Labor had to be figured like freight, very expensive freight. Then every pick and shovel, every tent, blanket, mattress, every cookstove and locomotive, had to be carried by ship across thousands of miles of ocean. De Lesseps could count on Panama to provide nothing but the place to dig the canal.

Beyond Matachín the train left the Chagres bottom lands and entered the narrow valley of the Rio Obispo, largest tributary of the Chagres. After Emperador, or Empire, as the Americans called it, came the summit at Culebra ("snake"), or Summit Station. On January 27, 1855, at midnight, in the pitch dark and in pelting rain, the last rail had been laid. Totten himself had driven the last spike with a nine-pound maul.

Summit was ten and a half miles from Panama City, and on the rest of the ride, descending to the Pacific, the party looked out on scenery reminiscent of Chinese landscape painting, with feathery green conical mountains rising on every side. At one dramatic turn, a cliff of basalt seemed to hang precariously close overhead, the great crystals of the dark rock lying every which way. The route now followed the Rio Grande, "a narrow noisy torrent winding along through dense forests below the track." Its drainage was south, to the Pacific.

Paraíso, another native village, was tucked between high hills shaped like inverted teacups. Pedro Miguel and Miraflores followed, then a stretch of spongy lowlands, a brackish swamp with soil the color of coal, then, ahead, the bald top of Ancon Hill, overlooking Panama City. The train covered the last few miles with its whistle screaming, bell clanging. Cathedral towers and red tile roofs were in view ahead on the right, and dead ahead was the Pacific. At once everyone was cheering.

With all stops en route, the delay at Barbacoas, the trip had lasted six hours. In his letter to the *Bulletin*, de Lesseps said they did it in three.

III

The original city of Panama had been founded in 1519, or just six years after Balboa's discovery of the Pacific, and by an extraordinarily treacherous individual, Pedro Arias de Ávila, usually referred to as Pedrarias, who had been governor of Castilla del Oro, as the Central American isthmus was known, and who, to solidify his power, had Balboa beheaded on a trumped-up charge of treason.

A Cueva Indian word, Panama means "a place where many fishes are taken." For the Spanish, Panama became a marshaling point and clearinghouse for the most important crossroad in the New World, the *camino real,* or royal road, which was nothing more than a narrow, mean mule track cut from Panama to Nombre de Dios, then the one Spanish fort on the Caribbean side. The gold of the Incas, pearls, Bolivian silver—no one knows how many thousands of tons of treasure—went across to Nombre de Dios, to be picked up by Spanish galleons. And though Panama never became especially large—because of disease primarily—or achieved the fabled wealth pictured in some old accounts, its importance was considerable. The stone ruins of the original city, Old Panama, or Panama Viejo, still stood several miles down the bay. The site had been abandoned after the city was sacked and burned by the pirate Morgan in 1671, and the present Panama, a walled city, was begun three years later, at the head of the bay, on a narrow tongue of volcanic rock with water on three sides.

"Panama is a very miserable old town . . . fast crumbling to pieces," an American sea captain noted in his journal, at the start of the gold rush, having brought the *California*, the first of the San Francisco steamers, around the Horn. "The houses are miserable and going to decay and the churches are crumbling. . . ." The harbor was also too shallow for a ship of any size. He did, however, find the climate "delightful at evening and in the morning." Thereafter, for the next several years, the city had been a wide-open booming seaport wherein, as one disapproving traveler commented, "most of the people are deficient in the higher moral attributes." Now the pace was more what it had been centuries before.

Fire had ravaged the city again and again. As recently as 1878, nearly a third of it had burned to the ground. Streets were narrow, with hardly room for two carriages to pass, and shadowed by overhanging balconies. There was not one proper sewer, little sanitation of any kind. As at Colón, fresh water had to be collected in huge rain barrels that could be seen everywhere, or carried in from the country in jars on mules; and whatever its source, the water never looked particularly clean. Tuberculosis, smallpox, cholera, yellow fever, and malaria were all common. A Canadian physician named Wolfred Nelson, who took up residence soon after de Lesseps' visit, described the city as "simply awful."

Still, it was a considerable step up from Colón, and, unlike Colón, almost entirely Spanish in feeling—Spanish architecture, Spanish faces, Spanish traditions. The government of the province was in Panama City; the bishop of Panama resided there. The *Star & Herald*, in English and Spanish, appeared daily. There was an established society among patrician landowners and professional people whose family names could be found on the rolls of Balboa's and Cortez' companies. Panama City *was* Panama. The humidity was not quite so oppressive as on the Atlantic side; there was less rain. The climate was indeed "delightful at evening and in the morning"—just about ideal in the dry season with the trade winds blowing—and on moonlight nights the view of the bay from the Bóvedas, the old Spanish seawall and the city's "choice promenade," was one of the loveliest sights anywhere in the American tropics.

For de Lesseps' arrival, furthermore, an almost miraculous transformation had been worked. The local populace had been told to clean up the streets, to paint, scrub, or whitewash everything within sight of his path, or face a stiff fine. Such an air of neatness, according to one report, had not prevailed within the memory of the oldest inhabitant. To give the celebrated visitor the right impression, to see that he was properly honored and entertained, the local government had allocated its entire budget for the forthcoming year.

"The reception of M. de Lesseps at this town was something never to be forgotten," wrote J. C. Rodrigues in his first dispatch from Panama City.

> It seems that every one of the 14,000 inhabitants was at the railway station, shouting, struggling to get a glimpse of the distinguished guest. I doubt very much whether more than one twentieth of them knew the true importance and meaning of the occasion. But . . . [their] enthusiasm did not know any bounds. We may laugh all we want at their ways of expressing it, but it was a most genuine triumphal entry, this one of M. de Lesseps.

After the predictable speeches at the depot, a procession of carriages rolled off to Cathedral Plaza, along the Avenida Central, which was lined the whole way by an honor guard of little Colombian soldiers in white trousers, white tunics, and blue caps trimmed with red. The plaza was in the exact center of the city and was dominated by the old brown cathedral with its twin bell towers, the most imposing structure on the Isthmus. De Lesseps and his party were to be quartered in the handsome new Grand Hotel, also on the plaza, and so the entire square, lampposts, every window and doorway, had been hung with French flags—the perfect sign, de Lesseps would tell readers of the *Bulletin*, of "*nos bonnes revanches*."

There was a state banquet at the hotel that evening, followed by dancing and singing that went on through the rest of the night, spilling out into the plaza, which was lit by hundreds of Japanese lanterns. Almost nobody was able to sleep. But bright and early the following morning, New Year's Day, de Lesseps was up and dressed in full formal attire, all his medals pinned on, and parading across to the cathedral for the inaugural ceremonies of President Demaso Cervera.

Then he was off to the harbor, where a steam tug stood by to take perhaps a hundred people and a large supply of champagne and cognac three miles along the bay to the mouth of the Rio Grande, the projected Pacific terminus of the canal. Before leaving Paris, he had promised to strike the first blow for the canal on the first day of the new year.

Some six hundred people turned up, in addition to de Lesseps, his family, his Technical Commission, the bishop of Panama, and the boat was so late getting under way that they missed the tide and were unable to get anywhere near the appointed spot. "The whole fun seemed to be spoiled," Rodrigues remembered. De Lesseps, however, was "not a man to change plans." He climbed onto a wooden seat, as two men held him by each arm, and he called the passengers to attention, which was no easy task since the boat was pitching badly and the champagne and cognac had been distributed freely in the hot sun for nearly two hours. Wherever it was made, the first stroke—"*le premier coup de pioche*"—would be symbolic only, he said. There was no reason why it could not be done where they were, on the boat. His little daughter, Ferdinande, would deliver the historic blow.

The child then swung a shiny pickax, brought especially from France for the occasion, into a champagne box filled with sand, after which each of the Technical Commission took a swing ("*en signe de l'alliance des tous peuples qui contribuent à l'union des deux océans, pour le bien de l'humanité*"). The bishop, José Telesforo Paúl, blessed the work and the boat turned back to the city.

There were more banquets in the days following, more speeches, toasts, fireworks, a horse race, a bullfight. Between times the French visitors went fishing in the bay or strolled the Bóvedas or picked out Panama hats at the "emporium" of Vallarino & Zabieta across from the cathedral.

The hotel was the center of all activity. "Everybody meets everybody at the Grand Hotel," wrote Rodrigues. The food was "*à la Française*," as advertised, and the best to be had in Panama. The salon featured a "FIRST-CLASS PIANO PLAYER." Everything was new and clean. Large, airy rooms opened on to interior galleries that looked down on a cool interior court that served as bar and billiard room and was the place where most of Panama's business was transacted. The billiard tables, the largest the guests had ever seen, were busy at all hours; the bar, in their words, was "one of those vast bars that have such a place in American life." At a crowded roulette table adjacent to the bar, a croupier called out the winning numbers in Spanish, French, and English.

Once, for posterity, de Lesseps gathered everybody for a formal portrait. The photograph, though badly faded, has survived. They sit or stand in three rows, some holding their new hats, some with umbrellas, every man in coat and tie in a country where a light shirt can feel heavy. De Lesseps sits in the center of the middle row, in his white suit, looking handsome and a bit distracted. Totten is at his right, the seat of honor; Dirks, at his left, wears thick round spectacles that give him a strange popeyed look. It is not hard to imagine the occasion de Lesseps made of the sitting, the bit of ceremony that must have gone with the placement of each man.

Wyse, who is on the far right, holding a large umbrella, looks as if he is about to break into a smile or a sneer—it could be either. Gaston Blanchet, the Couvreux, Hersent engineer who stands in the center of the back row, is a tall, good-looking man with a big shock of dark hair. Rodrigues, also in the back row, end man on the right, is full-bearded and dapper and especially uncomfortable-looking in a heavy three-piece suit, complete with wing collar, stickpin, and watch chain.

Trenor Park is not in the picture, which may or may not say something about de Lesseps' feelings toward him. Nor is it clear what Park was doing all this time. We know only that he and most of the others who had come down from New York sailed for home shortly after the picture was taken. We have only his parting comments. He still saw no reason, Park said, why a sale of the road could not be arranged once the French company was organized, which was the polite way of saying once de Lesseps had the cash.

The Technical Commission got down to business officially at the Grand Hotel on the morning of January 6. The first meeting was brief. The canal was to be an open cut, de Lesseps reminded them. "And now, gentlemen," he said, "you see what you have got to do, go ahead and do it."

"From that time," General Wright would recall, "he left us entirely to ourselves—went out of the room and left us to consult. We were of different nationalities and different ideas as to how the work should be done . . ." Later in the day, de Lesseps, Wright, and Jacob Dirks—all three of them past seventy—took a train back to Matachín, and for an hour or more, under what de Lesseps benignly categorized as a "rather bright" sun, they drifted down the Chagres in a dugout canoe. They saw several drowsy alligators, and Wright, who spoke neither French nor Spanish and so was unable to converse with de Lesseps, decided after looking around that, indeed, "the best type of canal is obviously one at sea level."

A ball at the hotel the night of January 15 went on until one in the morning, when there was an enormous banquet, after which the music and dancing resumed until dawn. There was a day's outing to the island of Taboga, ten miles out in the bay. A French man-of-war, *Grandeur*, arrived. There was even a wedding one evening at eight at the cathedral, which was described as "gay with the presence of a multitude of the best of our Panama society." Gaston Blanchet was the groom; the bride was the daughter of the proprietor of the Grand Hotel, a stunning Panamanian girl whom Blanchet had found time to fall in love with during his previous inspection tour. Madame de Lesseps, "with great sweetness and expression," sang a selection from Gounod.

If Ferdinand de Lesseps was not having the time of his life all this while, he certainly left everyone with that impression. The Panamanians adored him, for his energy, for his "vivid interest in our rather dull Isthmus life" (as the *Star & Herald* said), for his beautiful children, his beautiful wife. ("Her form was voluptuous and her raven hair, without luster, contrasted well with the rich pallor of her . . . features," Tracy Robinson would still be able to recall nearly thirty years later.) "They really believe he is their man," wrote Rodrigues.

Rodrigues, who had the room next door to de Lesseps, described how at five in the morning de Lesseps' hearty laugh could be heard resounding through the halls of the hotel. Then, in the comparative cool of the early morning, the old man would be off on a fishing expedition or a hike with his sons. One morning, in the plaza, as Madame de Lesseps watched from the balcony of their room and a large and approving crowd gathered, he put a lively horse through its paces. Had there been a stone wall, as once there had been in Egypt, no doubt he would have attempted that as well.

"Then, M. de Lesseps is one of those men who know how to please," observed Rodrigues. "He begins by enjoying hugely those popular attentions, and because he wishes to retain them he tries to deserve them." Which was perhaps as good an explanation as anyone ever offered of why the old hero did what he did. His exuberance was irrepressible. At one point an elderly resident Frenchman told him that if he persisted with his plan there would not be trees enough on the Isthmus to make the crosses to put over the graves of his laborers. De Lesseps appeared unmoved. As Robinson wrote, "Nothing ever seemed for an instant to dampen the ardor of his enthusiasm, or to cloud the vista of that glorious future which he had pictured in his imagination."

Yet there is at least one clear sign in the record that the old hero saw more than he let on. In an amazing interview published a few years later, he told Emily Crawford, Paris correspondent for the London *Daily News* and the New York *Tribune*, that in fact he knew as soon as he traveled across Panama that the task would be far more difficult than he had been led to suppose, that the Wyse plan for a canal was a fraud—"daringly mendacious" were his words. "But," wrote Mrs. Crawford, "he was in for the enterprise, and as he thought it feasible, he meant to go on with it."

Less than twenty-four hours before de Lesseps was scheduled to sail for New York, his nine-man Technical Commission presented its final report. Three hundred pages in length, it was little more than a rubber stamp for what he had been planning. A canal *à niveau* was approved. A dam to hold the Chagres in check was recommended. There was to be a breakwater at Colón, as he wanted, a tidal lock on the Pacific end. Construction time was figured to require eight years rather than twelve, as declared at the Paris congress.

De Lesseps liked everything but the estimated cost—843,000,000 francs, a figure 357,000,000 francs below what had been estimated at the Paris congress. The new figure was still too high, he thought, and this in spite of the fact that now no interest on capital for eight years was included, or administrative expenses, or the sums due the Türr Syndicate and the Colombian government, or the very large expense of buying the Panama Railroad. Many such details had not been bothered with in the Paris estimate, it is true. However, the figure produced in Paris had included an extra 25 percent for contingent expenses, whereas the amount now added for contingencies was only 10 percent.

But what made the new estimate look even more remarkable was the further fact that the anticipated volume of excavation—the amount of digging to be done—had been increased by more than 50 percent (from 46,000,000 cubic meters, as reckoned at the congress, to 75,000,-000 cubic meters). Having seen Panama, having been over the ground, having decided that the job was to be half again greater than previously declared, the commission had *reduced* the total cost of the canal, then allotted less for unexpected difficulties. And de Lesseps' sole complaint was that their reductions were too timid.

"Our work will be easier at Panama than at Suez," he announced. To Charles he wrote: "Now that I have gone over the various localities in the Isthmus with our engineers, I cannot understand why they hesitated so long in declaring that it would be practicable to build a maritime canal between the two oceans at sea level, for the distance is as short as between Paris and Fontainebleau." Talk of the deadly climate, he said, was nothing more than the "invention of our adversaries."

None of the party had experienced any sign of ill health. The nearest thing to a crisis had been a case of sunburn suffered by Madame de Lesseps on the outing to Taboga.

On board the *Colon*, somewhere between Limon Bay and New York, seated quietly in his stateroom, de Lesseps took a pencil and went to work on the commission's report. By the time the ship reached the East River he had cut the estimated price by another 184,400,000 francs, or by nearly $37,000,000. The canal, he told the reporters who came aboard at New York, was going to cost no more than $131,720,-000. It was, everyone agreed, an impressive reduction from the $240,-000,000 predicted at Paris.

IV

To nobody's surprise he was front-page news the whole time he was in New York. Not for twenty years, the papers declared, had a foreign visitor been greeted by the city with such warmth and wholehearted appreciation. Newspaper articles made much of his striking physical appearance, the snow-white hair, the tropical tan, the youthfulness and intellectual vigor—all in notable contrast, it should be added here, to the claims made later that he was a dim, muddled old man by the time he first saw Panama. One reporter who had interviewed him years before at Suez found him "not a whit changed. The same marvelous bright eyes, the same earnest voice, the same sympathetic chuckle, personally magnetic as ever, erect, impulsive, and, if anything, younger." A writer for the *World* thought he looked about fifty-five.

It was his first time in the United States and, as at Panama, he enjoyed himself grandly. He strolled Fifth Avenue, took a ride on the El, went by elevator to the top of the Equitable Life Assurance Building on Broadway (a full six stories high), "inspected" the half-finished Brooklyn Bridge. The great Culebra Cut at Panama, he declared dramatically, would be as deep as the bridge towers were tall—274 feet!

In his suite at the Windsor Hotel on Fifth Avenue, talking rapidly but softly, and with numerous gestures, he assured reporters that his Panama plan posed no conflict with the Monroe Doctrine. The venture was a private enterprise and the American people especially should understand the virtues of private enterprise. The point was that he *welcomed* American investment. "I am but an executor of the American idea," he insisted, and in fact he would be happy to see a majority of the stock sold in the United States and to have the company's headquarters located in New York or Washington.

The tricolor flew over the Windsor as though a head of state were in residence. He was received by the American Geographical Society, and a reception given by the city's French community was attended by an estimated eight thousand people. Another night, wearing white tie and tails, he walked onto a stage banked with potted palms and told the American Society of Civil Engineers that he was a diplomat, not an engineer, but that he was honored to be welcomed as a colleague. As at all such occasions he spoke through an interpreter and extemporaneously. He never used notes, he explained, because he always spoke the truth, the truth required no preparation.

For a banquet at Delmonico's, the main dining room was decorated with French and American flags, a central chandelier had been transformed into an enormous floral bell from which ropes of evergreen intertwined with flowers ran to all corners, and every table had its ingenious confectionery centerpiece—Ferdinand de Lesseps in evening dress, shovel in hand, bestriding Africa; a sphinx; a Suez dredging machine; the steamer *Colon*. The menu was embossed with the de Lesseps coat of arms, and the all-male guest list, some 250 "notables," included Mayor Cooper, Andrew Carnegie, Jesse Seligman, Russell Sage, Albert Bierstadt, Clarence King, Octave Chanute, Abram Hewitt, Chauncey Depew, Peter Cooper, David Dudley Field. At about nine o'clock, just before the speeches began, Madame de Lesseps made a dramatic entrance onto the musicians' balcony, accompanied by Emily Roebling, wife of the crippled chief engineer of the Brooklyn Bridge.

The speakers were Alexander Lyman Holley, pioneer of the Bessemer process in America; John Bigelow, the diplomat and publisher; Dr. Henry W. Bellows, the famous Unitarian; and Frederick M. Kelley, who gave the main address. In terms of who was there, the things said, the setting and all, it was one of those marvelous moments around which a whole study of an age could be developed. (One of Bierstadt's vast western landscapes had recently been purchased by the Corcoran Gallery in Washington, for example; Carnegie's steel empire was by now the mightiest in the world due primarily to the innovative skills of Alexander Holley; a spiritual leader such as Henry W. Bellows could still speak, as he did, with perfect faith in "the white winged dove of commerce.") Nonetheless, as several editorials stressed the morning after, the occasion and the sentiments expressed were plainly in honor of the man, not his present scheme. New York was enormously interested in Ferdinand de Lesseps, interested—but guardedly—in his Panama canal.

"For what he *has* done we give him our most cordial welcome and the homage of our heartiest applause," declared Whitelaw Reid of the *Tribune*. "What he proposes to do is another matter . . ."

And so it was to be everywhere else he traveled in the United States. In Washington, where there were no banquets or receptions, the message was especially blunt. Everybody was perfectly cordial, to be sure, including President Hayes. On Capitol Hill, de Lesseps was received by the House Interoceanic Canal Committee and gave an eloquent plea for his project, then was invited to hear Captain James B. Eads, builder of the famous St. Louis bridge, present his proposal for a colossal ship railway across Tehuantepec. The plan was to hoist ships out of the water bodily, in huge wheeled cradles, and haul them overland with enormous locomotives pulling in tandem. An ingenious system of hydraulic rams, Eads explained, would push supports, or blocks, against the ship's hull. Each block would be equipped with a universal joint so that it would automatically conform to the shape of the hull at the point of contact, thereby distributing equally the weight of the ship. To put a six-thousand-ton ship into the cradle would take half an hour. The cradle itself would ride on twelve rails placed five feet apart and on 1,200 wheels (100 on each rail). The locomotives, five times as powerful as the largest then in existence, would have an average speed in crossing of ten to twelve miles per hour. The complete transfer of a ship from one ocean to the other, over a distance of 134 miles, would take sixteen hours. And to build such a system, Eads said, would cost only $50,000,000, about a third that of Ferdinand de Lesseps' canal.

De Lesseps' listened politely to all this and perhaps some of it touched a vital nerve. Such visions of ships being picked from the water like toys and towed over the mountains of Mexico might have been hatched by Jules Verne in his tower study at Amiens. In any event, he offered Eads his compliments and made a gracious exit, only to be confronted by reporters carrying the full text of a new Presidential message to Congress. The United States, Hayes avowed, would not surrender its control over any isthmian canal to any European power or combination of powers. Nor should corporations or private citizens investing in such an enterprise look to any European power for protection. "An interoceanic canal . . . will be the great ocean thoroughfare between our Atlantic and our Pacific shores and virtually a part of the coastline of the United States." The "policy of this country is a canal under American control."

The message was a clear and deadly serious repulse to de Lesseps' intentions. He could have protested, or he could have been diplomatic and evasive, or he could have said nothing for the moment. Instead he had Henri Bionne send off a cable addressed to Charles in Paris and, ultimately, to the *Bulletin:* "The message of President Hayes guarantees the political security of the canal."

He, *Le Grand Français,* had won another noble victory, that was the implicit claim. If the United States declared that no European power could function as protector over the canal, then this meant that the United States would fulfill that vital function. And with the United States watching over the enterprise during construction and afterward, investors could be certain of success. That was not at all what had been meant. "What the President said," wrote J. C. Rodrigues in his subsequent analysis, "[was that] he did not wish to see European corporations building canals in Panama; but M. de Lesseps was equal to the occasion, and consistent once more with all the plot of a play of which he was protagonist."

Then, like the veteran performer who knows how well physical movement alone can command attention, he was on his way, barnstorming across the continent by train, from Philadelphia to St. Louis to San Francisco, then back by way of Chicago, Niagara Falls, and Boston. Madame de Lesseps and the children meantime remained in Philadelphia, where she had relatives.

He made the trip in three weeks and he was news at every stop. His crowds were large, often very large, and they were always friendly. Frequently they were even as enthusiastic as the *Bulletin* claimed. He called Americans the most hospitable people on earth. In Chicago a heckler shouted a gibe about the Monroe Doctrine. "Here are twenty thousand of you Americans," de Lesseps responded. "Now explain to me how the Monroe Doctrine prevents my making the canal." He waited; no one answered. Then he patiently explained their Monroe Doctrine to them and why it had no bearing on his canal, adding, "I cannot agree with a town only a third my own age . . . which says that the thing is impossible."

"Hurrah! That's the boy we want!" somebody shouted and there was a long approving cheer.

In Washington the French minister presented a note to Secretary Evarts saying the French government was in no way involved in the de Lesseps enterprise at Panama "and in no way proposed to interfere therein or to give it any support, either direct or indirect."

De Lesseps and party sailed on April 1, 1880. He had been praised and feted right up to the final day, when there was a farewell luncheon at the home of Cyrus Field, who had laid the Atlantic Cable. Moreover, he had found the time in New York to put together what he was to call his Comité Américain to handle the sale of Panama stock in the United States. Three New York firms had agreed to participate—J. & W. Seligman & Company, Drexel, Morgan & Company, and Winslow, Lanier & Company. And yet for all this, his popularity, his vigor, there were no takers. He had sold no stock. Not a single American capitalist of consequence had expressed the least serious interest in his Compagnie Universelle.

The real money, he responded to his followers, would come from France. "It is in France alone, where one is in the habit of working for the civilization of the world, that I shall . . . raise the capital necessary . . ." And within no less than two hours after his arrival in Paris he was sitting beneath a large Venetian chandelier in the upstairs salon of Madame Juliette Adam expounding on his travels to a gathering of old friends and admirers.

The next weeks were packed with lectures, dinners, interviews, and he thrived on the schedule no less than he had on all the thousands of miles at sea, the countless new faces, the rich food and strange hotels and endless talk. Friends told him he never looked better. A doctor in one after-dinner speech accused him of jeopardizing the medical profession, since obviously the visit to Panama had resulted in the discovery of the fountain of youth.

So the odyssey of four, nearly five, months had ended in a resounding display of popular approval of exactly the kind he knew to be essential. By simply going to Panama, returning physically whole and hearty, he had worked a stunning transformation at home. His grave mistake was to underestimate his own success. His popular support now was far greater than he had any idea, and his misreading of that fact, ironically, was to prove nearly as fateful as his more obvious misreading of Panama itself.

5

The Incredible Task

For country, science, and glory.

—Motto of the École Polytechnique

I

The early years of the Compagnie Universelle du Canal Interocéanique were a time of soaring hopes, a considerable amount of misleading propaganda, and some real, if costly, progress. In Panama amazing things were accomplished by men whose devotion to the task was exemplary, often heroic. Their achievements were never quite what Ferdinand de Lesseps promised, and for all their noble intentions, they made serious mistakes almost from the beginning. Still, in view of the difficulties they had to face, the sheer magnitude of the task, the things they simply did not know, the repeated instances of personal suffering and tragedy, they did extremely well.

In Paris the sale of stock in the company—Ferdinand de Lesseps' second attempt to go public—turned out to be one of the most astonishing events in financial history.

La grande entreprise was to be the biggest financial undertaking ever attempted until then. Panama stock was to be more widely held than any ever issued before. And never had any strictly financial proposition inspired such ardent devotion among its investors. The explanation, of course, in good part, was that to most of them it was never a strictly financial proposition. Nothing so vital to French pride just then, nothing led by Ferdinand de Lesseps, could have been seen as that only. Of enormous importance also were the almost limitless expectations associated with venture capitalism—*pionnier capitalisme.* The talk was of "the poetry of capitalism" and of "the shareholders' democracy." The unprecedented response of the French people to de Lesseps had many sides, many psychological levels, and only later, in hindsight, would their allegiance seem blind or his leadership purposefully deceitful.

He put the management of the sale this time in the hands of Marc Lévy-Crémieux, vice-president of the powerful Franco-Egyptian Bank, who had been among his vigorous opponents during the first go-around. Lévy-Crémieux also sent his own man—an engineer—out to appraise the situation at Panama, and the man returned with the confidential report that the canal would never pay. Half a dozen companies would go down in ruin before any ship passed through Panama, the man insisted. But not a word was said of this publicly.

A consortium of commercial and investment banks was put together. The initial capital was set at 300,000,000 francs ($60,000,000). Clearly, this was nowhere near enough, but it was de Lesseps' decision. And it was a bad one.

Six hundred thousand shares were offered at 500 francs ($100) each. So for the average person, the investor de Lesseps was banking on, it was very expensive stock: 500 francs was nearly a year's wages for about half the working population of France. The terms, however, were tremendously appealing—only 25 percent down, with six years to pay off the rest. And added to this was the knowledge that Suez stock, which also had cost 500 francs a share initially, was presently listed on the Bourse at more than 2,000 francs and was paying dividends of 17 percent.

During the time of construction, shareholders were to receive 5 percent on their paid installments. Once the canal was completed, they were to get 80 percent of the net profits. The remaining 20 percent was to be divided among the "founders"—de Lesseps, Charles, those friends who had put up thc initial 2,000,000 francs—and the Lévy-Crémieux syndicate. Founders' shares in the Suez company, priced the same as these originally— 5,000 francs each—were currently valued at 380,000 francs!

"Subsidization" of the newspapers and magazines commenced the moment the syndicate was formed. Payments were made discreetly for the most part and the effect was pronounced. Almost without exception those papers that had so vociferously denounced the canal and its progenitor now praised it and him to the skies. Panama became "the magic word," *le synonyme de dividendes fantastiques.* "Capital and science have never had such an opportunity to make a happy marriage," said the *Journal des Débats. Le Figaro* said the canal would be built in seven years and de Lesseps could then proceed with some further titanic work. *La Liberté* declared that the canal had no more opponents. "Oh, ye of little faith! Hear the words of M. de Lesseps, and believe!"

Émile de Girardin, proprietor of *Le Petit Journal*, among the most powerful of French press lords and one of those who had been particularly unpleasant in his attacks on de Lesseps the year before, was now delighted to become a member of the company's board of administrators.

The price paid for such enthusiasm was high—1,595,573 francs, as near as could be determined by subsequent investigations—and the brokers' commissions for the flotation were higher still, in excess of 4,000,000 francs. For the bankers it was a perfect bonanza. Their commission on the stock sales was 4 percent, or 20 francs a share. Still, as some of the foreign correspondents in Paris were to note, such houses as the Crédit Foncier and the Rothschilds had steadfastly "refused to allow their counters to be used for such a flytrap."

The results were far beyond anyone's wildest hopes. Indeed, the success of the sale was such that it dramatically underscored how very much de Lesseps had misjudged the accumulative effect of all he now had working for him. He had asked for far too little money—less than half what even he was saying the canal would cost—and the irony is that he could have had all he thought he needed and more, right then at the start. Furthermore, he could have had it on relatively easy terms, and conceivably he might have succeeded without payoffs to the press. But he had failed to comprehend the difference that support from his former adversaries in the financial world could make, or, most importantly, the psychological impact of his trip to Panama.

Sale of the stock began December 7, 1880. Within three days, more than 100,000 people subscribed for 1,206,609 shares, or more than *twice* the number available. As a result many people had to be satisfied with considerably less than what they wanted.

As de Lesseps had forecast, it was the small investor who made the sale such a runaway success. Some eighty thousand people had bought one to five shares each. Only fourteen people owned a thousand shares or more. And about sixteen thousand of the shareholders were women.

Nothing like this had ever occurred before. The Paris-Lyon-Méditerranée railway, a huge corporation, with capital valued at 1,200,000,000 francs, and with 800,000 shares—or 200,000 more shares than were offered by the Compagnie Universelle—had but thirty-three thousand shareholders, or less than a third the number that the Compagnie Universelle was starting with. The first stockholders' meeting, in January 1881, had to be held in the Cirque d'Hiver, the circus building on the Boulevard des Filles-du-Calvaire, but even it held only five thousand people. All problems had been solved, de Lesseps said, all difficulties had been smoothed over.

The official incorporation of the company took place on March 3, 1881. De Lesseps was formally designated president; Henri Bionne was to be secretary general. Charles de Lesseps was named a director.

Salaries were remarkably low, considering the scale of the enterprise. De Lesseps would receive only 75,000 francs a year ($15,000); Bionne, a mere 18,000 francs ($3,600). But by contrast, interestingly, a salary of $25,000, or more than de Lesseps', would be paid to an American, Richard Wigginton Thompson, who had agreed to head the Comité Américain, which was to be among the most costly and fruitless of the company's operations.

Thompson, who had been Secretary of the Navy, was actually de Lesseps' second choice for the job. His first choice, General Grant, had been approached through Jesse Seligman, but had flatly declined the offer, a decision Grant explained in a letter to Daniel Ammen: ". . . while I would like to have my name associated with the successful completion of a ship canal between the two oceans, I was not willing to connect it with a failure and one I believe subscribers would lose all they put in."

The committee, as de Lesseps described it, was to mount opposition to the rival Nicaragua project—through the press, by lobbying in Washington—and to impress upon the American people the surpassing virtues of a canal at sea level. It was to supervise the purchase of American-made equipment and material for Panama. But its primary mission was to induce American investors to join in the Panama venture, and in that, as in most everything else, the committee proved just about useless. Thompson was the only member to do anything to justify his pay, chiefly as a lobbyist. The New York office served little purpose. Few investors were found. About all de Lesseps got for his bargain was the use of some impressive stationery, plus the freedom to mention, when need be, the name of somebody such as Jesse Seligman as among his backers.

Yet a fee of $300,000 was paid outright to the Seligman firm, and to each of the other two firms, and all three organizations would receive another $100,000 as time went on. The total cost of the Comité Américain would be $2,400,000.

So with brokers' commissions running to more than 4,000,000 francs, with "remuneration to banks" totaling about 1,800,000 francs, plus a commission of 11,800,000 francs for the Lévy-Crémieux syndicate, plus 12,000,000 francs to the American committee, then the 1,600,000 francs for "publicity," and another 750,000 francs for miscellaneous organization expenses, the grand sum for getting out the stock and getting the company under way came to something over 32,200,000 francs, or about $6,400,000.

Early in 1881, between his first and second stockholders' meetings, de Lesseps also purchased for 1,000,000 francs an office building at 46 Rue Caumartin, which stood back to back with the Suez Canal office on the Rue Charras. Ten thousand shares of stock that had been withheld from sale were turned over to the Türr Syndicate as previously agreed, and de Lesseps relinquished to the new corporation, at no charge, all rights and privileges obtained earlier from the Türr Syndicate. Thus the Wyse Concession now belonged to the Compagnie Universelle. The project was under way.

II

French civil engineers of the nineteenth century were an exceptional breed and justly proud of their heritage. It was the French who pioneered in the use of pneumatic caissons for bridge foundations and who perfected the use of wrought-iron I-beams for building construction. Les Halles, the famous Paris market building, the Menier Chocolate Works, the stunning Galérie des Machines at the recent Paris exhibition, were recognized as bold and innovative structures suggesting infinite possibilities for the future of architecture. It was in France that reinforced concrete had first been tried and French engineers remained preeminent in its use. Over all, the French were preeminent in civil engineering in general, and French technical schools, like French schools of medicine, were the finest in the world.

There were two varieties of engineers in France, other than the military engineers: the *ingénieurs de l'état*, who had been trained first at the École Polytechnique, then at the École des Ponts et Chaussées;

and the *ingénieurs civils*, whose school was the newer École Centrale des Arts et Manufactures. Of these, the graduates of the École Polytechnique were the cream of the crop, an elite technical class of a kind, who had vast influence in the bureaucracy, and, consequently, over the whole economic life of France.

Just to be accepted at the state-run Polytechnique was a supreme honor. A rigorous entrance examination excluded all but the most brilliant applicants. And from the moment a young man walked through the gray stone portals on the Rue Descartes, he knew that nothing less than intellectual superiority and full devotion to *La Patrie, les Sciences, et la Gloire* were expected of him. Founded in 1794, the school was the chief scientific creation of the Revolution. Napoleon hailed it as his hen with the golden eggs, and the symbol of hen and eggs had been carved into keystones and made the centerpiece of the stained-glass skylight over the largest lecture hall, lest any student forget. It was not an engineering school, but a military school—rigid rules, tight-fitting blue uniforms, swords, the traditional *bicorne* for parade-ground ceremonies—and devoted to the study of pure science. The curriculum provided what was essentially a classical secondary education plus what was then the most advanced mathematical education in the world.

Upon finishing at the Polytechnique, the highest-ranking graduates generally went on to the Ponts et Chaussées—an *école d'application*—from which they emerged as engineers in the service of the state, as builders of bridges, highways, harbors, or as officials with the state-run railroads. Early in the 1860's, after a thorough study of the system, an American authority on education, Henry Barnard, declared it gave France the best-trained corps of civil and military engineers of any nation. Lavalley, builder of the Suez dredges, Sadi Carnot, Godin de Lépinay, had come out of the Polytechnique, as had Ferdinand Foch, and as would, in another era, Valéry Giscard d'Estaing.

The *ingénieurs civils*, the graduates of the École Centrale, were the engineers of private enterprise and closer to American engineers in spirit. Gustave Eiffel was an outstanding example; another was William Le Baron Jenney, a Chicago architect and engineer, who was to build the world's first skyscraper.

But all French engineers, and those from the Polytechnique especially, regarded themselves as men of *science*. Their creations were the result of abstract computation. The Americans, in the French view, were merely adroit at improvisation, which, however inspired or ingenious, was nonetheless of a lower intellectual order. The bias concerning American engineers was not wholly justified. Still the essential American spirit *was* improvisation. It was the attitude expressed in a remark attributed to the engineer John Fritz, who, upon building a new machine, is supposed to have said, "Now, boys, we have got her done, let's start her up and see why she doesn't work."

But at Panama the French had to improvise—or rather they had to learn to improvise under pressure. And they had no past experience to go by. Virtually everything had to be learned by trial and error, and their chief difficulty as time went on was the fearful cost of their errors. The experience at Suez was little help. Probably they would have been better off in the long run had there been no Suez Canal in their past. For despite all de Lesseps told the press and his public, Panama had only one advantage over Suez: the distance to be covered. Everything else at Panama was infinitely more difficult. Panama was an immeasurably larger and more baffling task than Suez, just as Godin de Lépinay had warned.

There was, to begin with, the fundamental geology of the Isthmus, a subject that had been given scarcely a fraction of the study it deserved. At Suez the digging had been mostly through sand. The climate at Suez had been hot, but dry; the climate at Panama, eight months of the year, was not only hot, but heavy, smothering, with a humidity of about 98 percent. At Suez there had been the problem of bringing enough water to the canal site to sustain the labor force; the annual rainfall at Suez had been about nine inches. At Panama the annual rainfall could be measured in feet, not inches; ten feet or more on the Caribbean slope, five to six feet at Panama City.

Suez was as flat as a tabletop, with a maximum elevation along the canal line all of 50 feet above sea level. Panama was covered with steep little mountains, and the maximum elevation on the canal line would prove to be 330 feet. There was the Panama jungle. And there was the Chagres River, which still stood directly in the path of the canal.

Questions of housing, labor supply, and health had to be faced. John Bigelow, who would visit the Isthmus later to appraise the French effort, wrote, "There probably was never a more complicated problem—a problem embarrassed by a larger proportion of uncertain factors—presented to an engineer. . . . Every step . . . is more or less experimental."

Ferdinand de Lesseps would never see it that way, however. "It is," he informed his stockholders, "an operation the exact mathematics of which is perfectly well known. . . ." Couvreux, Hersent had built Suez; Couvreux, Hersent and exact mathematics would build Panama.

He could never quite put Suez out of his mind and his engineers in the field had had no experience in the tropics and had not been trained to improvise. So they not only had to learn as they went along, which would have been difficult enough, but they had to learn to learn as they went along and to unlearn nearly everything that had been supposedly "taught" at Suez.

The first group arrived at Colón on the faithful *Lafayette* at the end of January 1881. There were some forty engineers, headed by Gaston Blanchet and by Armand Réclus, who was the new general agent of the canal company, and his assistant, Louis Verbrugghe, the lawyer who had gone to Bogotá with Lieutenant Wyse. Several of the engineers had brought their wives with them and there were various festivities at Panama City to honor the occasion. There is, however, nothing to the story that Sarah Bernhardt had also come out on the *Lafayette* or that she "presented a drama in the wretched little box of a playhouse that was then the only theater in the city." Bernhardt's one and only visit to the Isthmus came later, in 1886.

On February 1, Réclus sent a two-word telegram to Ferdinand de Lesseps that was to thrill newspaper readers all over France: "*Travail commencé*"—"Work begun."

Réclus had overall control, as things were organized, but Blanchet, as the ranking official for Couvreux, Hersent, was the one in charge of the actual work. De Lesseps' contract with the construction firm was for two years and the work was to be done on a cost-plus basis; that is, any costs above the estimates were to be met by the canal company.

As someone to set things in motion, Blanchet was a good choice. He was decisive and forceful and his marriage had smoothed the way in Panama society. His strategy was to spend a year preparing to dig. He wanted to cut a path four hundred feet wide the whole way across the Isthmus, nearly fifty miles, to provide enough open space for the most accurate siting of the canal line as possible. But right away the idea was rejected by the Paris office. He had to content himself with clearing a strip of only fifty feet, which proved too narrow to do what he wanted, so, eventually, the entire line would have to be redone as he had intended in the first place.

With a force of black and Indian laborers hired locally, he proceeded to chop his way from Colón to Panama City on a line that crossed the looping brown Chagres no less than fourteen times. This in itself was an enormously difficult and dangerous task. Immense trees and all their tangled undergrowth had to be taken down by hand.

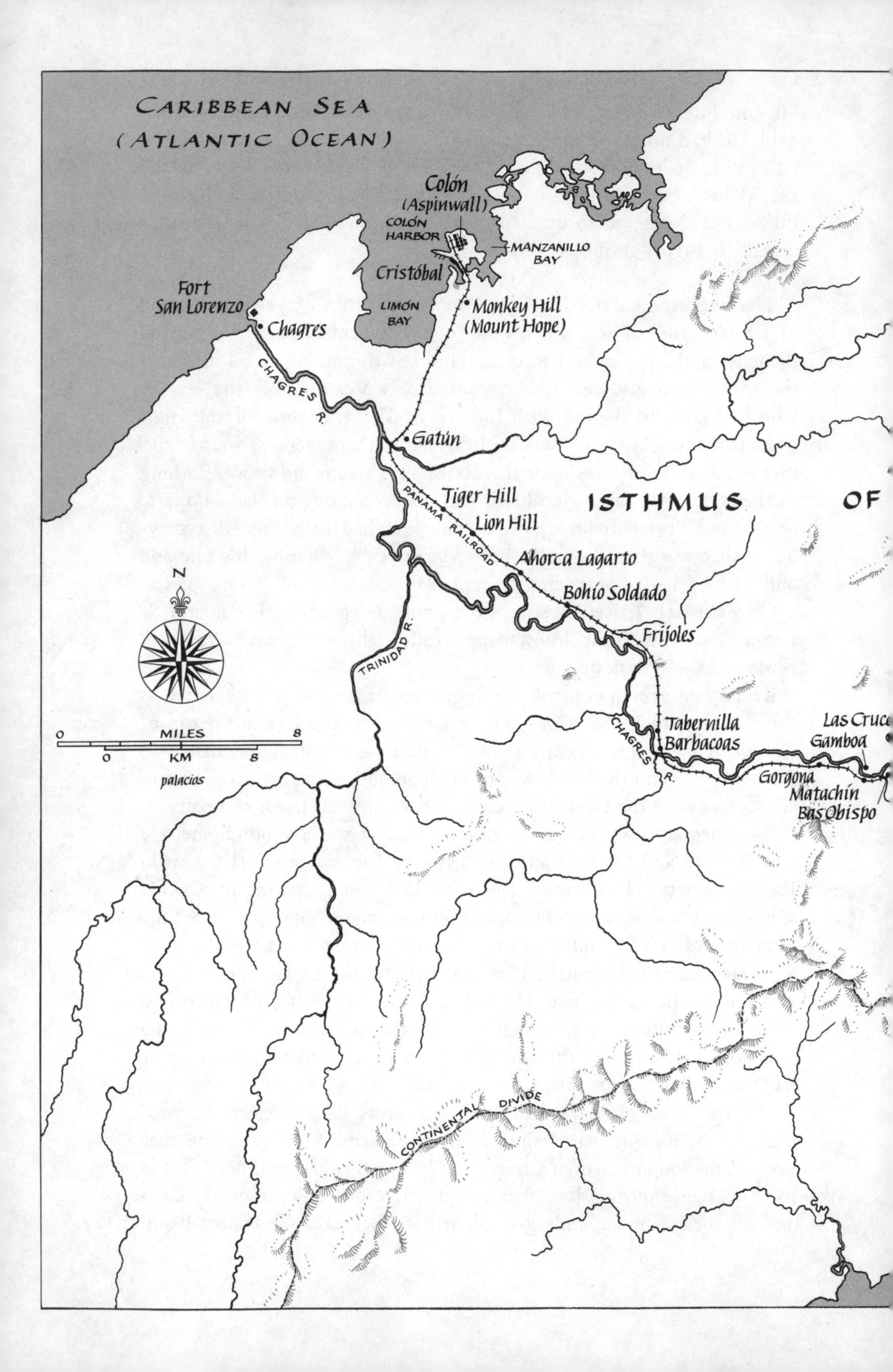

CARIBBEAN SEA
(ATLANTIC OCEAN)
Colón
(Aspinwall)
COLÓN HARBOR
MANZANILLO BAY
Cristóbal
Fort San Lorenzo
Chagres
LIMÓN BAY
Monkey Hill
(Mount Hope)
CHAGRES R.
Gatún
PANAMA RAILROAD
Tiger Hill
Lion Hill
ISTHMUS OF
Ahorca Lagarto
Bohío Soldado
Frijoles
TRINIDAD R.
Tabernilla
Barbacoas
CHAGRES R.
Las Cruce
Gamboa
Gorgona
Matachín
Bas Obispo
N
0 MILES 8
0 KM 8
palacios
CONTINENTAL DIVIDE

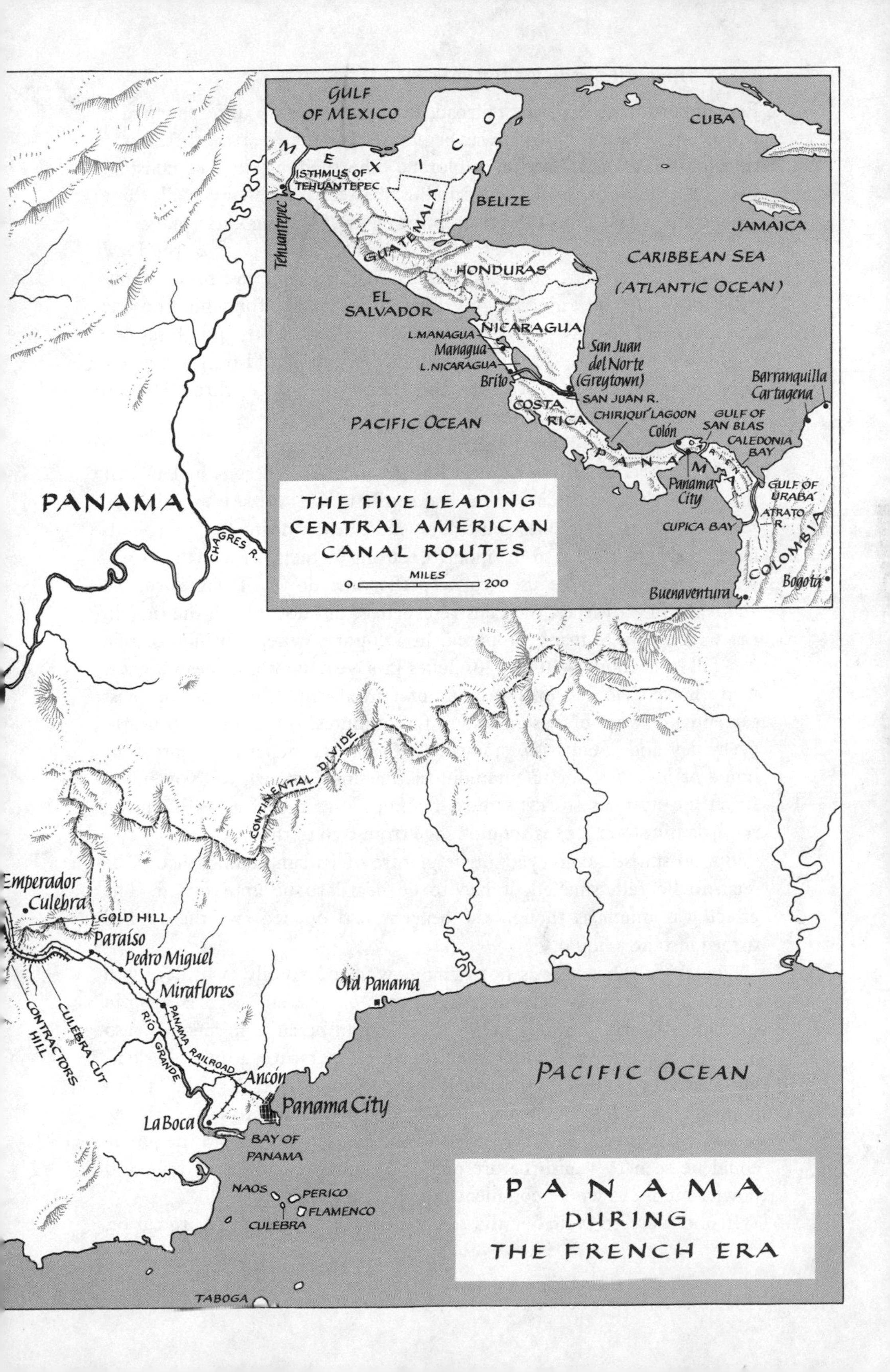
THE FIVE LEADING
CENTRAL AMERICAN
CANAL ROUTES
MILES
0
200
GULF OF MEXICO
M E X I C O
ISTHMUS OF TEHUANTEPEC
Tehuantepec
BELIZE
GUATEMALA
HONDURAS
EL SALVADOR
NICARAGUA
L. MANAGUA
Managua
L. NICARAGUA
Brito
San Juan del Norte (Greytown)
SAN JUAN R.
COSTA RICA
CHIRIQUI LAGOON
Colón
GULF OF SAN BLAS
CALEDONIA BAY
DARIEN
P A N A M A
Panama City
GULF OF URABA
ATRATO R.
CUPICA BAY
COLOMBIA
Bogotá
Buenaventura
Barranquilla
Cartagena
CUBA
JAMAICA
CARIBBEAN SEA
(ATLANTIC OCEAN)
PACIFIC OCEAN
PANAMA
CHAGRES R.
CONTINENTAL DIVIDE
Emperador
Culebra
GOLD HILL
Paraiso
Pedro Miguel
Miraflores
CONTRACTORS HILL
CULEBRA CUT
RÍO GRANDE
PANAMA RAILROAD
Old Panama
Ancón
Panama City
La Boca
BAY OF PANAMA
NAOS
PERICO
FLAMENCO
CULEBRA
TABOGA
PACIFIC OCEAN
PANAMA
DURING
THE FRENCH ERA

Except for the path of the railroad, the jungle was no different than it had been when the railroad was begun, or, for that matter, than in the time of the earliest Spanish explorers. The men worked in constant fear of poisonous snakes (coral, bushmaster, fer-de-lance, all three among the world's most deadly reptiles) and of the big cats (puma and jaguar). Days and nights were made a living hell by *bichos,* the local designation for ticks, chiggers, spiders, ants, mosquitoes, flies, or any other crawling, buzzing, stinging form of insect life for which no one had a name. The only tools were machete and ax, and the jungle, as one engineer wrote, was "so thickly matted that one could only see a few yards in any direction. . . ." Beyond Gatun the line cut through seven miles of marshy flats and swamp.

And before the job was finished, the rains had begun.

There were, as all newcomers learned, but two seasons in Panama: the season Ferdinand de Lesseps had seen and the wet season. The dry season, with its clear skies and trade winds, began normally about mid-December and lasted four months, during which, in Panama City, water carts had to be used to keep the dust down. Then, abruptly, about the first of May, the rains returned. It did not rain all the time in the wet season, as many supposed. In a country where an inch of rain can fall in an hour, 120 or 130 inches in a year may not mean a great many more than 120 or 130 hours of rain all told. Some of the most torrential downpours lasted only a few minutes. But it did rain nearly every day and it never *just* rained. At Colón six inches in twenty-four hours or less was not uncommon. In the single month of November, when the heaviest storms struck, rainfall along the Chagres basin—on the Atlantic slope, that is—could range from two to three feet.

But no statistic conveyed a true picture of Panama rain. It had to be seen, to be felt, smelled; it had to be heard to be appreciated. The effect was much as though the heavens had opened and the air had turned instantly liquid.

The skies, when it was not raining, were nearly always filled with tremendous, towering clouds—magnificent clouds, and especially so in the light of early morning. Then there would be an unmistakable rush of wind in the trees, a noticeable drop in temperature, a quick darkening overhead followed by a sound that someone likened to the "trampling of myriad feet" through leaves. In villages and towns everyone would instinctively dash for cover. From the hills at Culebra the jungle could be seen to vanish before onrushing silver cataracts of rain, and howler monkeys would commence their eerie ruckus.

If one were to wait out the storm beneath a corrugated iron roof, the sound was like that of a locomotive. Often these storms became violent thunderstorms, with lightning "so stunning," wrote one American, "it just makes a person feel as though he were drunk." And then, while the trees still tossed and roared, the rain would be over—in an instant. The sun would be out again, fierce as ever. Everything would glisten with rainwater and the air would be filled with the fecund, greenhouse smell of jungle and mud.

By May the canal line had been cleared from Colón to Panama City and Colón had become a beehive of building and of ships unloading. A sawmill went up, along with fifty prefabricated houses that had been shipped from New Orleans. Crates of equipment and vast quantities of material were piling up the length of Front Street.

Great gaps in the jungle were cleared for intermediate towns at Gatun and Emperador. (The one at Gatun was to be called Lesseps City.) There were to be machine shops at Bohío Soldado, labor camps at San Pablo (just beyond Barbacoas) and at Matachín. Barracks for black workers were set on high concrete footings, a precaution against floods and rats. The buildings were large enough for fifty bunks each. Well-seasoned lumber was used, and the design was sensible for the climate, with long verandas and plenty of windows.

The "cottages" for the white technicians were also as comfortable and as well constructed as conditions would allow. They were built near the water, along the eastern shoreline in what was to be a new community called Christophe-Colomb (later renamed Cristobal). They were one-story buildings, all very much alike, white with green shutters, each enclosed by verandas, and generally there was a Yucatán hammock slung at one corner of the front veranda. Everything considered, the location was ideal. At night, with a full moon flooding the white beach and a breeze coming in off the water, a young newly arrived French engineer might well find Panama all that he had dreamed.

At Panama City, the company bought the Grand Hotel and set up headquarters. From a second-floor office overlooking the plaza, Armand Réclus wrote regularly to Paris, as instructed, to report on local politics, employee morale, his own daily problems. To maintain an adequate labor force seemed nearly impossible. In this first year only about ten out of every one hundred newly arrived laborers remained on the job after six months. But contrary to later accusations, the well-being of the men was regarded as a priority responsibility. "We must make certain that the personnel suffer no privations and that their welfare is looked into," Charles wrote to Réclus. "You will always find us disposed to approve any measures that you may have to propose in this matter." The great hope of the de Lesseps', father and son, was to establish a "Panama family." "Everything you can do to ensure the well-being of the personnel, including their pleasures, will be immediately approved," Charles would advise. "Do all you can so that off-duty there are no bosses and employees, but only members of the same family united by sentiment." And at the bottom, his father added in his own hand, "This is an excellent letter and I am one with it."

A hospital, to be known as L'Hôpital Notre Dame du Canal, was being planned for a spacious site on Ancon Hill, overlooking the city and bay. The physician in charge, Dr. Louis Companyo, was the former head of the sanitary division of the Suez Canal. There were to be several handsome buildings, with good ventilation and comfortable verandas, set among magnificent gardens. There would be a full-time staff of doctors and nurses, the Filles de la Charité. A smaller hospital would be built on the northern shore at Colón, to take advantage of the sea breezes; and a hotel on the island of Taboga, a rambling, filigreed white ark, was to be converted into a sanitarium for convalescents.

It would be told later how the French had plunged into Panama blithely disregarding the threat of disease, and how hopelessly primitive their medical facilities were. But the intentions expressed repeatedly in personal correspondence between Panama and Paris, the efforts taken in Panama, the money spent by the canal company, all belie this. De Lesseps had once faced a cholera epidemic in Egypt; he had lost a wife and a son to disease; he was no fool, however frequently his public declarations concerning health conditions in Panama would appear to prove otherwise. The facility at Ancon, which was to include some seventy buildings by the time it was finished, would cost $5,600,000, a staggering sum in that day. Another $1,000,000 was spent on the hospital at Colón, nearly $500,000 on the Taboga sanitarium. Dr. Wolfred Nelson, the Canadian physician who had opened an office in Panama the previous year, a man who was to be severely critical of almost everything the French did, wrote, "The canal hospitals on the Panama side are without doubt the finest and most perfect system of hospitals ever made within the tropics." William Crawford Gorgas, writing some thirty-five years afterward, was to appraise the Ancon complex as "a very much better institution than any in the United States . . . at the same period carried on by a firm or corporation."

The effect of the climate on tools, clothing, everyday personal items, was devastating. Anything made of iron or steel turned bright orange with rust. Books, shoes, belts, knapsacks, instrument cases, machete scabbards, grew mold overnight. Glued furniture fell apart. Clothes seldom ever dried. Men in the field finished a day drenched to the skin from rain and sweat and had to start again the next morning wearing the same clothes, still as wet as the night before. Without laundry facilities, a clean shirt or fresh pair of trousers were luxuries beyond compare.

Panama was "a hell upon earth," an English traveler on the Panama Railroad once observed; besides, he said, it was "overrun with Yankees." And for those French officials struggling to establish system and order to their efforts, the Yankees who ran the Panama Railroad were proving to be as large an aggravation as anything they had to face. To judge by the correspondence of Armand Réclus, the railroad people seldom if ever did anything as he wished, and since the railroad was the sole means of transportation and communication, the results were maddening. When nothing moved on the railroad, nothing moved on the Isthmus. If there were delays, if shipments were held up, lost, damaged, the effect was felt all down the line.

Réclus saw more than poor or indifferent management or simple bad luck as the root causes of his troubles. It was all, he believed, part of a diabolic scheme to force de Lesseps to buy the railroad at an inflated price. "I am persuaded that this, in effect, is their plan," he informed Charles. The Americans were merely following "orders from New York to do everything to create the greatest possible difficulty for us." The only conceivable solution was the one the railroad company wanted. "It is necessary that we become the absolute masters of the railroad," and that, he emphasized, could only be done by buying the road outright as quickly as possible.

Trenor Park by now owned even more stock in the line than he had before—and he had raised his price, should the French still be interested. The $200-a-share figure quoted initially had been advanced to $250 a share. Park insisted on full payment in cash.

It was a holdup, a great many people felt, but there was little that could be done about it. He had de Lesseps in a corner. "It is necessary at any price to settle the question of the railroad," Réclus pleaded again in desperation, "because on its possession or not depends the accomplishment of the canal."

So in June 1881, after drawn-out negotiations between Paris and New York, the sale was agreed to. The canal company bought some

68,500 of the existing 70,000 shares, which at $250 a share came to more than $17,000,000. In addition the company took over a sinking fund amassed by the railroad toward the eventual amortization of its bonded indebtedness of some $6,000,000. So all told the little stretch of track cost over $20,000,000, which was more than equal to a full third of the company's resources. On a per-share basis the stock actually wound up costing $292 at a time when the true par value was less than $100.

For all that the legal status of the road remained the same. It was still an American company, incorporated under the laws of the state of New York; its franchise from the Colombian government remained unchanged. Trenor Park, who personally cleared approximately $7,000,000 on the transaction, did step down as president of the line. However, the man who replaced him was John G. McCullough, his son-in-law.*

Furthermore, the old Bidlack Treaty, the 1846 treaty between Colombia and the United States, was as much in effect as ever. The railroad's fundamental right of transit still rested on Article XXXV of the treaty, and to guarantee uninterrupted traffic on the line, as well as Colombian sovereignty on the Isthmus, remained the obligation of the United States. So an American military presence would continue, in the form of gunboats standing off Colón and Panama City.

At a stockholders' meeting in June, de Lesseps explained the purchase in straightforward, businesslike terms. He asked for approval to borrow the money to pay the bill, plus another 300,000,000 francs, which, with the company's present capital, would give him, he said, an ample amount to build the canal. The stockholders approved.

Gaston Blanchet, meantime, had led a surveying party far up the Chagres, to begin work on the first serious maps and surveys. They were the advance guard and they made a striking picture—intent, tanned faces under white sun helmets, pistols at the belt. They chewed on Havana cigars as they squinted into the brass eyepieces of surveying instruments. They slapped at the interminable mosquitoes; they picked scorpions the size of a hand from their boots in the morning. They shot alligators, some twenty feet in length, and brought back the stripped pelts of jaguars. And they were extremely good at their work.

* Trenor Park would have little chance to enjoy his new fortune. A year and a half later, in December 1882, en route from New York to Colón, he died on board the ship. The cause of death was reportedly an overdose of sedatives.

Copies of their surveys, compilations of the data accumulating, were sent off to Paris, and as a detailed picture began to materialize in the office on the Rue Caumartin, de Lesseps called in a new superior advisory board, still another technical commission, to give an opinion on all plans. None of these men was to take part in making the plans, or in the preparation or control of contracts; they were merely to give an opinion. And of course the mere fact that they were gathered, that they were known to be sitting as a jury over all technical decisions, had considerable public-relations value.

The important point is that de Lesseps, once again, would get exactly what he wanted; he would follow his own lead and they would nod in agreement and go along with him as willingly as his stockholders had, raising no serious objections about anything, which must be viewed as another testament to his powers of persuasion, rather than any lack of perception on their part. Unmistakably these were men of eminence and ability. At the head of the group was Lefebure de Fourcy, inspector general of Ponts et Chaussées. Jacob Dirks, Daubrée, Voisin Bey, participated again. There were six chief engineers of Ponts et Chaussées. One man was the chief-of-port at Marseilles. Another was an admiral. Yet none was willing, or bold enough perhaps, to challenge de Lesseps' judgment or to take seriously the inevitable cost of a sea-level canal. Later it would be charged that de Lesseps never listened to his engineers. But in fact it was the other way around; it was they who were listening to him.

"Perhaps no other man ever possessed to such a marvelous extent the power of communicating to other minds the faith and the fervor which animated his own," a writer for the *Illustrated London News* once observed.

III

By summer of 1881 there were two hundred French or European technicians and clerical help on the Isthmus and some eight hundred laborers at work—making test borings with great, cumbersome steam drills, building barracks and hospitals, assembling and testing newly arrived equipment.

But by summer it was also apparent that yellow fever had returned to the Isthmus. The wet season was traditionally the time of sickness and this year had been no exception. Several cases were reported in May. Then in the second week of June the first canal employee died of yellow fever, another of those incidental details not featured in the *Bulletin.*

On July 25 an engineer named Étienne, a graduate of the Polytechnique and one of the ablest of the young technicians, died at Colón—of "brain fever," supposedly—and was hastily buried at Monkey Hill that same afternoon. On July 28 Henri Bionne died.

Bionne's death would be attributed in Paris to "complications in the region of the kidneys." But on the Isthmus the story would be told for as long as the French remained. He had arrived from France to make a personal inspection for de Lesseps, and several of the engineers had arranged a dinner in his honor at the employees' dining hall at the camp at Gamboa. It was a festive evening apparently. Bionne, the last to arrive, had come into the hall just as everyone was being seated. One of the guests, a Norwegian woman, was exclaiming with great agitation that there were only thirteen at the table. "Be assured, madame, in such a case it is the last to arrive who pays for all," Bionne said gaily. "He drank to our success on the Isthmus," one engineer recalled; "we drank to his good luck. . . ." Two weeks later, on his way home to France, Bionne died of what the ship's doctor designated only as fever, not yellow fever. The body was buried at sea.

"The truth is that the climate . . . like all hot climates, is dangerous for those who underestimate its effects . . . and who fail to observe the principles of hygiene," explained the *Bulletin.* Yellow fever was not prevalent in Panama, the paper assured its readers, though "unhappily" a few laborers had been victims of the disease.

The British vice-consul at Panama, young Claude Coventry Mallet, decided out of curiosity to join one of the surveying parties in the upper reaches of the Chagres. The expedition consisted of twenty-two men. Within a few weeks everyone but Mallet and a Russian engineer named Dziembowski was sick, whether of so-called Chagres fever or yellow fever is not clear. The expedition returned, in any event, and of the twenty men who went into the hospital ten died. Mallet and Dziembowski returned to Cathedral Plaza feeling no ill effects, however. Mallet, who told the story later, said they agreed to meet for lunch the following day and that Dziembowski asked for a loan to buy a new suit. When Dziembowski failed to show up for lunch, Mallet went around to the canal offices to ask his whereabouts. The Russian, he was told, had died of yellow fever at three that morning and had been buried at dawn in a new suit of clothes.

There were more deaths as the summer wore on, but in October, speaking before a geographical congress at Vienna, de Lesseps said there were no epidemics at Panama and that the few cases of yellow fever had been "imported from abroad."

Then, in November, a few days after he had returned from a particularly strenuous exploration of the upper Chagres, Gaston Blanchet died, apparently of malaria. The importance of Henri Bionne to the operation in Paris had been considerable and his death had been a heavy personal blow for Ferdinand de Lesseps, but Blanchet was the driving spirit of the enterprise in the field, and his loss would be felt for a long time.

How many died that first year is uncertain. The official company estimate on record is about sixty. Malaria, which is an entirely different disease from yellow fever, probably accounted for a great many of the fatalities then as later. The fact is that more people would die of malaria at Panama than of yellow fever, notwithstanding the popular impression to the contrary.

Malaria, the most common of tropical diseases and the one endemic disease at Panama, takes many forms and went by many different names on the Isthmus: *calentura*, miasma, the shakes, the chills, *paludisme*, ague, pernicious fever, putrid fever, intermittent fever, and, in its most virulent form, Chagres fever. Historically, malaria was the world's greatest killer and it was confined to no one geographical area. Only the year before, there had been a serious epidemic in New England. But in places such as Panama, malaria never went away. The prevailing attitude was that everyone got a dose of it sooner or later. Among the native population, infection usually began in childhood.

The typical malarial attack began with terrible chills, uncontrollable shivering, and chattering teeth, the spell lasting perhaps fifteen minutes, sometimes more. Often the shivering of patients in a malaria ward would be so violent that the room could actually be felt to tremble; a single bed would move on the floor.

The chills would be followed by high fever and a burning thirst. As the fever fell off, the patient would break out in a drenching sweat. For those who survived, the experience was unforgettable. With the passing of the fever, the patient was left feeling totally debilitated, mentally as well as physically. Acute depression usually set in, the "melancholia" that was so well known in Panama.

And the patient could be stricken again. Indeed, it was considered impossible ever to recover fully from malaria so long as one stayed on in such country. But by the same token, a patient could move to some distant, seemingly safe climate and still experience a return siege of malaria, which was perhaps the most insidious characteristic of the disease. John Lloyd Stephens, as noted, was struck down by malaria in the spring of 1852, recovered sufficiently to return to New York, only to die of a recurrence of the disease in October.

There was no such thing as an immunity to malaria. With yellow fever it was different. A person had yellow fever only once. Either he lived or he died. If he lived he would never get it again. Malaria could be a lifelong infirmity, and if the first dose did not kill, the second, third, or fourth could.

Yet in the tropics, malaria was taken as an inevitable fact of life, part of the landscape. Yellow fever, by contrast, came and went in vicious waves, suddenly, mysteriously. In those places where it was most common—Panama, Havana, Veracruz—it was the stranger, the newcomer, who suffered worst, while the native often was untouched. Wherever or whenever it struck, it spread panic of a kind that could all but paralyze a community. It was a far more violent and hideous thing to see; a more gruesome way to die.

The mortality rate among those who contracted the disease could vary enormously, from 12 or 15 percent to as much as 70 percent. Generally speaking, however, a yellow-fever patient in Panama in the 1880's had a less than fifty-fifty chance of survival. As with malaria, the patient was seized first by fits of shivering, high fever, and insatiable thirst. But there were savage headaches as well, and severe pains in the back and the legs. The patient would become desperately restless. Then, in another day or so, the trouble would appear to subside and the patient would begin to turn yellow, noticeably in the face and in the eyes.

In the terminal stages the patient would spit up mouthfuls of dark blood—the infamous, terrifying *vómito negro*, black vomit. The end usually came swiftly after that. The body temperature would drop, the pulse fade. The flesh would become cold to the touch—"almost as cold as stone and [the patient] continues in that state with a composed sedate mind." Then, as a rule, in about eight to ten hours, the patient would die. And so great was the terror the disease generated that its victims were buried with all possible speed.

Blacks and nonwhites were somewhat less susceptible to malaria than were whites. But while it was commonly believed among whites, and repeatedly published on supposed scientific authority, that all blacks were naturally immune to yellow fever, they were not. Panama, famous as "the white man's graveyard," was in fact deadly territory for any nonimmune of any race or color. Many blacks, lifelong residents of Caribbean islands or coastlands, had an immunity resulting from previous mild cases of the disease, usually during childhood. Modern medical research also indicates that the common tropical disease known as dengue, or "breakbone fever," can also have the resulting effects of an immunity to yellow fever. But no human being ever achieved an immunity to malaria; there was no such thing as a *natural* immunity to yellow fever, and if many blacks had been made immune to yellow fever before reaching Panama, there were vastly more blacks at work than whites, so the number of nonimmune blacks on the Isthmus was always quite large. Black laborers died of both malaria and yellow fever and no less miserably than the whites.

Yellow fever—yellow jack, *fièvre jaune*, *fiebre amarilla*, the "American plague"—had been a terror of seamen for centuries. A single case on board ship could mean death for the entire crew. The legendary *Flying Dutchman* was founded on the story of a ship condemned to haunt the seas after yellow fever broke out on board and no country would permit the ship in its harbors. The Philadelphia yellow-fever epidemic of 1793 had been as savage as an attack of bubonic plague and doomed the supremacy of Philadelphia among the cities of North America. Recently, in 1878, in Memphis, Tennessee, more than five thousand people had died of yellow fever and the estimated financial loss, due to the entire cessation of commerce, was upward of $100,000,000.

Historically, the disease had played a critical role in Central America and the Caribbean since the first known outbreak in Barbados in 1647, and ironically, the French had already seen one New World dream fail disastrously, in good part because of the disease. Napoleon, with plans for an American empire of his own, had sent a military expedition of twenty-five thousand men under his brother-in-law General Leclerc to Haiti in 1801 to put down the black insurrection led by Toussaint L'Ouverture. With that accomplished, the French army was to have occupied New Orleans and Louisiana. But yellow fever cut through the veteran troops like no enemy imaginable; thousands died, including Leclerc, and this was a major contributing factor in the ultimate triumph of the black patriots. Haiti achieved independence, and Napoleon, thoroughly disenchanted with his American venture, decided to sell all of the Louisiana Territory to the United States.

There was still no known remedy or palliative for yellow fever. The medical profession stood helpless. For malaria, however, there was quinine, the bitter, colorless powder made from the bark of the cinchona tree, a palliative the Indians of Peru had known for centuries. Quinine was distributed freely among the French in Panama and was taken regularly in preventive doses, usually at meals and mixed with wine to kill the dreadful taste. Nobody knew why quinine worked, but it did. The one big problem with it was that a heavy dose could cause vomiting and headaches, or, worse, a horrendous ringing in the ears that rendered the patient deaf.

The word "malaria" was from the Italian *mal'aria* ("bad air"), and it had been widely agreed long since that bad air, "noxious effluvium"—poisonous marsh gas in particular—was the cause. The French for malaria, *paludisme*, was even more specific, being derived from a word meaning "marsh fever." This miasma theory, as it was called, had been undisputed for centuries and seemed perfectly logical since the disease prevailed in hot, low-lying country where the humidity was high, the growth and decomposition of vegetation extremely rapid.

Yellow fever also was believed to be airborne, but filth was supposedly its source—sewage, the putrefying carcasses of dead animals, all the distasteful human and animal waste to be found in the streets of Colón or Panama City. The greatest source of contamination supposedly was the patient himself, and to touch his clothing, his soiled bedding, anything he had come in contact with, meant almost certain death—hence the mortal fear of the body after death and the quickest possible burial.

The night air was thought to be especially dangerous in any area infected by fever of any kind. It had been observed also that the wind had an effect. People spoke of yellow-fever winds, for example. At Panama City, south winds, those blowing in from the marshy lowlands near Panama Viejo, were regarded as especially deadly.

There was, however, another theory—even as early as 1881.

Dr. Josiah Clark Nott was a general practitioner in Mobile, Alabama, and it is one of those extraordinary coincidences of history that he happened to be the doctor who, in 1854, attended Amelia Gayle Gorgas at the birth of her son, William Crawford. In 1848 Dr. Nott published a paper in the *New Orleans Medical and Surgical Journal* in which he made the fantastic claim that malaria and yellow fever were undoubtedly conveyed by insects and possibly by the mosquito. His mention of the mosquito was only in passing. His main point was that the spread of the disease could not be explained by any laws governing vapors or gases. But in 1854 a "traveling naturalist" for the Paris Museum, Lewis Beauperthuy, then in Venezuela, concluded that malaria and yellow fever "are produced by a venomous fluid injected under the skin by mosquitoes like poison injected by snakes." Swamps and marshes spread sickness, he said, not by the vapors they exuded, but because mosquitoes bred there. In Washington, a Dr. Albert Freeman Africanus King, a professor of obstetrics, had arrived at the same conclusion.

King was a well-known figure. On the night of Lincoln's assassination, he had been in Ford's Theater and was the first physician to reach the dying President. His mosquito theory was not to be formally presented until 1882, however, and when he suggested how malaria might be eradicated from the capital, many people, understandably, took the whole thing as a jest. The way to do it, he wrote, was to encircle the city with a wire screen as high as the Washington Monument. Still, well before the French engineers arrived at Panama, King had worked out the means for reducing the spread of the fever—by window screens, the drainage of swamps and pools, and the destruction of the insects by special traps.

Others as well had all but hit on the solution. Amazingly, buried in the reports of the Nicaragua Expedition of 1872–1873 and the Panama Expedition of 1875 are two small notations by John Bransford, a Navy doctor who accompanied both expeditions. He had observed that the mosquito netting provided by the Navy afforded notable protection against fever or miasma of all kinds—"by straining the air of germs and moisture."

In 1880, the very year de Lesseps launched his Compagnie Universelle, a French doctor on the staff of a military hospital in Algeria, Dr. Alphonse Laveran, discovered the presence of tiny crescent-shaped bodies wriggling in a blood sample taken from a malaria patient. Incalculably minute, they were detectable only under the strongest microscope, but he had little doubt that they were living organisms and it dawned on him that here was the cause of malaria. He described his discoveries in a letter to the Académie de Médecine in Paris and published a small monograph.

Laveran's claims were not accepted, however, any more than were the theories of Nott or King. The miasma theory had been fixed in people's minds for generations. In addition, there now appeared a rival claim that seemed to support the miasma theory with scientific fact. Two doctors working in Rome, a German named Klebs and an Italian, Tommasi-Crudeli, had isolated a bacteria from the soil of a malarious region that when injected into a rabbit produced a malaria-like fever. The phenomenal discoveries of Pasteur and Koch had educated everyone to the role of bacteria, that whole other world beneath the microscope, and so it was widely accepted that a bacterium—*bacillus malariae*—was the long-sought cause of malaria.

(Ronald Ross was at this time in his early twenties and newly enlisted in the Indian Medical Service, having barely passed the entrance examination. An indifferent student, a physician because it had been his father's ambition for him, he was as yet mainly interested in music and poetry.)

But in 1881, the year the ill-fated Gaston Blanchet began chopping his path across Panama, the mosquito theory was voiced once more, and with greater conviction than ever, by a Havana physician.

Dr. Carlos Juan Finlay was the son of a Scottish father and a French mother. He had been educated in France and at the Jefferson Medical College in Philadelphia, and having practiced medicine in Havana for twenty-odd years, he had concluded that yellow fever was not only transmitted by the mosquito but by a specific mosquito—a silvery, comparatively noiseless household variety, *Stegomyia fasciata* (later to be called *Aëdes aegypti*). Out of some eight hundred known varieties, he had picked this one as the carrier of the disease.

Finlay was an ingratiating hawk-nosed individual who looked out on the world through gold-rimmed spectacles and spoke with a lisp, the result of a childhood case of chorea. He was a linguist, an amateur historian, and he was a tireless worker. But he had failed to produce any proof of his theory—he was not very good at research—and his professional peers in Havana gave no encouragement. He applied innumerable mosquitoes that had bitten yellow-fever patients to healthy persons, yet no case of yellow fever ever resulted. Thus it was his own work that appeared to bring the most discredit to his theory.

Like Nott, Beauperthuy, and King, Finlay had the right idea about mosquitoes, and with astonishing precision he had singled out the right variety. Gorgas was to call it a splendid example of medical clairvoyance, "a beautiful manifestation of scientific imagination." However, it made little difference. Finlay was utterly ignored. At the Ancon hospital, Dr. Girerd, chief surgeon of the canal company and a "profound microscopist," set up a system of experiments whereby he examined the blood of new workers upon their arrival, then again in another month or so, when, invariably, he found the supposed malarial bacillus.

And all the while, in the lovely gardens surrounding the hospital, thousands of ring-shaped pottery dishes filled with water to protect plants and flowers from ants provided perfect breeding places for mosquitoes. Even in the sick wards themselves the legs of the beds were placed in shallow basins of water, again to keep the ants away, and there were no screens in any of the windows or doors. Patients, furthermore, were placed in the wards according to nationality, rather than by disease, with the result that every ward had its malaria and yellow-fever cases. As Dr. Gorgas was to write, had the French been consciously trying to propagate malaria and yellow fever, they could not have provided conditions better suited for the purpose.

But if malaria and yellow fever were airborne, if plague could come or go with the wind, if the slimy pools and swamps along the railroad and the suffocating back streets of Colón and Panama City were the sources of deadly night airs and miserable death, it was also "known" that not everyone was in equal jeopardy. Fever struck according to a discernible pattern. Some people stood a better chance of surviving than did others, as countless examples attested. Simply stated, the odds on one's survival were in direct proportion to one's moral fortitude. The clean, blameless life was the long life in the tropics. Confidence, courage, belief in one's destiny, a "disdain of peril," as Philippe Bunau-Varilla would say, also mattered enormously. Debauchery, sins of the flesh, moral or physical cowardice, were sure paths to ruin.

There were some, to be sure, who held that a dose of whiskey or rum was as dependable a palliative as quinine. (Bourbon and mustard seed was a popular "infallible specific" for yellow fever.) But few old-timers on the Isthmus subscribed to such theories. "Many foreigners have fallen victims to fear rather than fever," Tracy Robinson wrote, "while many others have wrought their own destruction by drink, which . . . has killed, directly and indirectly, more than the entire list of diseases put together . . ." (Robinson had arrived at this conclusion, he said, after trying both abstinence and "moderate indulgence.") Dr. Nelson was "firmly of the opinion that the people who best resist such climates and make the best fight against disease, are the total abstainers."

Like numbers of North Americans on the Isthmus, Robinson and Nelson had a low regard for the manner in which the French were conducting themselves off-hours. Nelson, from the suffering and death he was to witness professionally, would develop an abiding hatred of Ferdinand de Lesseps—"The Great Undertaker," he would call him. But many of the French engineers were the most puritanical of all in their views, and nearly everyone was profoundly shaken whenever the death of some notably upright person seemed to make a mockery of such views. "Certainly his moral character was above reproach," wrote one bewildered, grieving French engineer of another who had died of yellow fever the first year.

In the United States especially, the death toll among the French would be attributed largely to moral decadence. One of the American railroad contractors, for example, would tell a congressional committee of seeing with his own eyes piles of discarded wine bottles in Colón that were as high or higher than a two-story house. Joseph Bucklin Bishop, a prim New York newspaperman who was to spend a decade in Panama, wrote that the French years had been a "genuine bacchanalian orgy." Colón was a "veritable sink of iniquity. . . . Champagne, especially, was comparatively so low in price that it 'flowed like water,' and . . . the consequences were as deplorable as they were inevitable." Bishop was to be Theodore Roosevelt's official biographer, and his views on the French in Panama, expressed in one of the popular early books about the canal, would have a broad and lasting effect.

The most frequently quoted summation was by James Anthony Froude, the reigning English historian and biographer of the day, who declared that "in all the world there is not, perhaps, now concentrated in any single spot so much swindling and villainy, so much foul disease, such a hideous dung-heap of moral and physical abomination as in the scene of this far-famed undertaking of nineteenth-century engineering." According to Froude the place was overrun with cardsharpers and "doubtful ladies." "Everything which imagination can conceive that is ghastly and loathsome seems to be gathered into that locality. . . ." Froude, however, was speaking only from what he had been told during a visit to Jamaica. He had been urged to go on to Panama, he wrote, to see for himself, "but my curiosity was less strong than my disgust."

For all the underlying self-righteousness of such (for then) lurid descriptions, they were probably justified. We have no eyewitness account of what went on; no private diaries of professional gamblers or confessions of "doubtful ladies" have come to light. But the general tone can be imagined. Tracy Robinson, who must have seen a good deal of life during his years on the Isthmus, was appalled by the spectacle. "Vice flourished," he wrote. "Gambling of every kind, and every other form of wickedness were common day and night." Issues of the *Star & Herald* are filled with reports of barroom brawls and riots. In the first year alone there were half a dozen murders among the canal workers. At Gatun, for example, the night of May 25, 1881, a Dutch employee went wild and stabbed two men to death in their sleep, then vanished into the jungle.

As to the consumption of wine there is little doubt. It was phenomenal—and for understandable reasons. The French were accustomed to wine with meals and wine happened also to be a great deal safer to drink than the local water, as even Joseph Bucklin Bishop conceded. The bottle dumps at Colón were every bit as high as a house. The foul alley behind Front Street was actually paved with wine bottles turned bottom-side up and became famous as "Bottle Alley." Nearly a hundred years later construction workers and amateur archaeologists would be turning up French wine bottles.

Gambling was widespread, and prostitution appears to have flourished from the start. The three most thriving industries were gambling houses, brothels, and coffin manufacturing. To signal the arrival of new "ladies of leisure" on the Isthmus, a code announcement was flashed along the railroad's telegraph line: "*langoustes arrivées*" ("lobsters arrived"). And the women, like the labor force and the technicians, came from every part of the world. If there was one obvious characteristic of the so-called French years that would be misunderstood in time to come, it was this cosmopolitan quality of society at every level.

By the end of 1881 there were two thousand men at work, including the technical staff and office help. Any thought of reliance on local labor had been put aside. Some of the laborers were from Colombia, some from Venezuela and Cuba. The vast majority, however, were English-speaking blacks from the West Indies—from Jamaica mostly. Subsequently some five hundred black Americans would come down from New Orleans and other Gulf ports of the United States. So among the actual laborers the language was English, not French.

Also, more white Americans were involved than was ever fully realized in the United States. White American technicians arrived along with equipment purchased in the United States. Nearly all the mechanics were Americans. American contractors arrived, bringing their own people, and the Panama Railroad was run by Americans—engineers, conductors, stationmasters, telegraph operators.

There were German, Swiss, Russian, Italian engineers, Dutch and English contractors. The Gamboa camp had a Belgian cook. Looking back, Tracy Robinson could recall no country that was not represented.

IV

The actual digging of The Great Trench—*La Grande Tranchée*—began at Emperador on Friday, January 20, 1882, with much champagne and dynamite. Thereafter the work at Emperador proceeded by steam shovel and by pick and shovel—mostly by pick and shovel—and it moved faster than expected, the ground in the vicinity being unusually soft. It was again the dry season.

In February, Couvreux, Hersent et Compagnie agreed to subcontract the dredging of the Atlantic end of the canal to an American firm, Huerne, Slaven & Company, and later, in November, a second American firm, the Franco-American Trading Company, was signed to start at the Pacific end. Presently other small subcontractors, several of them American, appeared and went to work. Excavation was under way at Culebra, Monkey Hill, Gorgona, and Paraíso. At Colón a gigantic earth platform was built out into the harbor from spoil brought from Monkey Hill. The hospitals were completed; the first fire engine arrived.

But it was a year marked by repeated and entirely unexpected setbacks, beginning with the sudden resignation of Armand Réclus and ending with the complete withdrawal of Couvreux, Hersent from the Isthmus and from all further responsibility for the project. In between, the death toll mounted alarmingly, and the Isthmus was struck by an earthquake.

Réclus' decision to quit was never explained. The best guess is that he had about reached the breaking point, trying to cope with what he described as "the disorder of details." Whatever the facts, he returned to Paris to serve as a "consultant" to de Lesseps, his real usefulness ended. Until a suitable replacement could be found, Louis Verbrugghe, a lawyer, not an engineer, became the ranking official in the field. When the new man, Commodore Richier, finally arrived, he proved no more capable of mastering "the disorder of details" than Réclus and soon he too quit.

The most horrendous and immediate problem for anyone in command was the volume and diversity of equipment in use. The display was terribly impressive and terribly confusing—thirty-odd steam shovels, three thousand flatcars and dirt trucks, fifty locomotives, steam launches, tugs, coal lighters, dredges, hundreds of rock drills, pumps, some eighty miles of railroad track—and this was only the beginning. Most of the machinery arrived in parts and had to be assembled at Colón. Most of it was also the best available at the time. The fashion later among American politicians and writers would be to ridicule the European-built machinery and various items ordered by the French engineers for their tropical empire—including "ten thousand snow shovels" to a land "where snow never ever has fallen."

The problem with the equipment was not its quality, but the bewildering variety of it. The track put down by Couvreux, Hersent had a different gauge than that on the Panama Railroad. French-made railroad cars came in differing sizes and gauges. French and Belgian locomotives, though built like a watch in workmanship, some with all-copper fireboxes, had such a rigid wheelbase that they required track built to the most exacting standards; otherwise, as one American noted, "they just went off and started for somewhere else."

The French "plant" was in effect something of a mechanical Noah's Ark, with every imaginable species represented. To get it all working efficiently, according to some kind of harmonious system, seemed nearly impossible. And while in a few instances certain tools and machines proved of little or no use in the tropics, there had been no certain way of knowing that in advance. No ten thousand snow shovels were ever sent out to Panama, as later charged—only a thousand shovels that looked like snow shovels but were in fact specially designed for scooping the ash out of steam-shovel boilers, a use for which they were ideally suited.

The first shock of the earthquake occurred at 3:30 the morning of September 7, and though it lasted but a fraction of a minute, it was the longest and worst ever experienced on the Isthmus. At Panama one of the two towers on the great cathedral crashed through the roof near the main entrance, while a big part of the front wall toppled into the plaza. The Cabildo, or town hall, was wrecked, and the walls of the Grand Hotel were so badly cracked that it was feared another tremor would bring the building down.

Wolfred Nelson, who lived in an annex of the hotel, said it was difficult to see anything at first. He had jumped from his bed and rushed out into the plaza. "It was black with people who had . . . got in the open and away from buildings that were expected to fall. There was still a little light, and the moon was in its last quarter. The hum of voices and excitement was something astonishing. There they were, people of all classes—black and white—some dressed, and some very hastily dressed, and some had brought chairs with them." One elderly lady, duenna of an old, distinguished family, was found dead sitting in her chair, the victim of a heart attack.

Damage along the railroad was extensive. In some places the roadbed had sunk as much as ten feet, leaving rails torn and twisted. At Colón, starting at the freight depot, a fissure in the earth, inches wide, ran some four hundred feet down Front Street.

There was another violent tremor the next morning, again before daylight, and the sense of panic this time was worse. All told, five people were killed, including the old lady in the chair. A week was lost getting the railroad back in running order. Cable communication with Jamaica (and the United States) was not resumed for another month.

Of greatest concern among the French officials, however, was the psychological damage the news might have among investors in France, since Panama was supposed to be safe from such natural convulsions. But when the news reached Paris, de Lesseps simply promised that there would be no more earthquakes and one cannot help but wonder if his deceptive propaganda was becoming self-deceptive.

More progress had been made at Panama in the first two years, he told his stockholders, than there had been at Suez in the first six years. And who was to refute such a claim? Those closest to the financial side of the company had already seen their founders' shares soar from 5,000 francs to 75,000 in the over-the-counter market. The press remained enthusiastic. The public had every confidence that all was well. A first bond issue, to meet the cost of buying the Panama Railroad, had been heavily oversubscribed. Nor was belief in de Lesseps by any means limited to the French. "With $30,000,000 already invested in the enterprise," reported the New York *Tribune*, "and with applications for shares showering him from all quarters of France . . . he can now reckon with confidence upon the resources required for so vast a scheme. He can get the money, and unquestionably he has the genius requisite for surmounting the engineering difficulties. Englishmen and Americans may as well reconcile themselves to the situation."

At the end of the year, when it was suddenly announced that the great firm of Couvreux, Hersent was retiring from the field, leaving the work entirely in de Lesseps' hands, he again stood unfazed, his leadership unchallenged. Such news could well have been a mortal blow to almost any other venture. His public composure and poise were total.

By its contract Couvreux, Hersent had every legal right to back out. The contract had been drawn up in an atmosphere of monumental mutual trust—for instance, it named de Lesseps as among those who could arbitrate any misunderstandings that might arise—and now the parting was carried off with comparable equanimity. The partners Couvreux and Hersent declared themselves honor-bound to say that the excavation could be carried forward more effectively without them. The canal could proceed faster and at less cost, it was said, by parceling the work out to a number of smaller contractors, each specializing in a particular task, an arrangement partly in effect and showing excellent results. The canal company henceforth should merely supervise the work on its own.

To this de Lesseps obligingly agreed; the contract was not renewed, and Couvreux and Hersent were out in the clear.

The real reason for the break, however, appears to have been rather different, as revealed by subsequent investigations conducted by the Chamber of Deputies. The death of the resourceful Gaston Blanchet had been a disheartening blow to Couvreux, Hersent et Compagnie. Nor was there anyone else in the firm of comparable ability who was willing to go to Panama and take Blanchet's place. But much more important was the realization, after two years, that the canal could never be built in anything like the time or for anywhere near the cost foreseen by the exuberant *Grand Français*, whose glowing declarations often as not were derived from figures and forecasts supplied by Couvreux and Hersent.

They could, of course, have made public their disheartening view of the situation. But they chose not to, as they later explained, out of respect for Ferdinand de Lesseps and so as not to add to his burdens. "The truth is," reads the report issued later by the Deputies committee, "that during the trial period Couvreux and Hersent had been able to form a shrewd idea of the difficulties of the enterprise but were unwilling to undermine the [canal] company's credit by a frank admission of the motive behind their retirement."

For the Compagnie Universelle the situation was really quite serious, and it is hard to imagine anyone in de Lesseps' position failing to go to Panama as soon as possible to determine to his own satisfaction what should be done. However, he saw himself as the company's major asset—its sole asset—and he believed, as did the financial interests involved, that his visible presence in Paris, at the helm, was essential. Appearances, as always, mattered enormously. There must be no sign of alarm, nothing to suggest that the contractors' defection had been either harmful or indicative of some deeper, fundamental flaw or anything other than a perfectly natural administrative reorganization.

Also, he had other demands on his time just then. British troops had seized Alexandria, ostensibly to protect the khedive's government, and de Lesseps had rushed to the scene in a futile, single-handed effort to keep the British from occupying the Suez Canal zone. Later he was in London to negotiate with more success an agreement covering Suez operations, only to rush back to the Avenue Montaigne to be present for the birth of his tenth child.

Twenty years before, at Suez, "with jealous personal authority," he had taken a direct interest in everything that went on. Now, more often, it was Charles who took the initiative, who handled the numerous small decisions that had to be dealt with daily—Charles, who had never set foot in Panama. "With your good judgment you will arrange things as they should be," reads a line from one of his notes to Charles; "everything you do will have my approbation."

Thus it was Charles, rather than his father, who departed for Panama in the wake of the Couvreux, Hersent defection—Charles and a new chief engineer, the first Directeur Général.

6
Soldiers Under Fire

"We are, gentlemen, soldiers under fire; let us salute the comrade who falls in the battle, but let us think only of the fight of tomorrow and of victory."

—PHILIPPE BUNAU-VARILLA

I

Jules Isidore Dingler—pronounced Danglay in French—was not impressive-looking. In his mid-forties, he was short and bald. He had small, round shoulders, a soft, round face, soft blue eyes, and a drooping mahogany-colored mustache. He might have been a bank clerk or a provincial wine merchant. The appearance suggested neither initiative nor resolution and the appearance was deceiving.

In his student days at the Polytechnique he had been a shining star, finishing near the top of his class and going on to the Ponts et Chaussées. As an engineer for the state he had risen rapidly to become a *chef des Ponts et Chaussées*, a very thorough professional accustomed to the multitudinous demands of large-scale public works. To Charles de Lesseps and his father, he seemed as qualified as anyone could be for the task at Panama; and unlike the three or four others whom they had approached with an offer, he alone had been willing to go. Concerning his own final decision in this regard, it would be said that he had an ambitious wife, that hers had been the deciding voice.

For the next two years, from early 1883 until the summer of 1885, he was to direct the largest, most ambitious engineering effort the world had as yet seen. His decisions were not to be the best always. Before sailing from France he also made the unfortunate declaration that once on the Isthmus he would prove that "only drunkards and the dissipated take the yellow fever and die there." Still it would be a long time before a more effective chief engineer would be dispatched to Panama—not until Theodore Roosevelt sent John Stevens in 1905—and Jules Dingler was to pay a dreadful price for his devotion to the work.

Yet subsequent histories of the canal would have little to say for Dingler. Quite unjustly, his memorial in most accounts would be the big frame villa built for him on Ancon Hill—"*la folie Dingler.*"

Charles and Director General Dingler reached the Isthmus on March 1, 1883, and were occasion for the predictable round of banquets and spirited oratory. The work was entering its "Second Great Stage"—"The Period of Construction" had begun. Champagne corks popped and Charles, sounding remarkably like his father, promised progress on all fronts. Ferdinand de Lesseps himself, declared the son, would return to oversee the work and a hundred chairs were shoved back in the dining room of the Grand Hotel as everyone rose to drink to the health of *Le Grand Français* some four thousand miles away in Paris.

Charles stayed on for another month. Dingler got directly to business in the office upstairs in the hotel, where by now the effects of the earthquake had been largely mended. He would begin by restoring order and confidence, both sadly lacking since the departure of Couvreux, Hersent. Paper work was in disarray. To date, one French writer observed, it had been an enterprise of passionate pioneers and mediocre accountants. Dingler was an organizer. Responsibilities needed clarifying; the work load had to be distributed. So at the outset a number of individuals accustomed to the comparative ease and convenience of the head office found themselves arbitrarily reassigned to one of the camps in the jungle. The word spread that the new man had no aversion to stepping on toes and at first chance several of those individuals most offended would take their revenge by spreading stories of the royal comforts Dingler had arranged for himself at company expense.

Dingler was not merely contemptuous of laggards and incompetents, but regarded them as cowardly, disloyal, less than true Frenchmen. "The purge continues," he would inform Charles nearly a year later. "I can well imagine that in Paris you are getting echoes of the complaints of the victims . . . [but] I never [act] until I am sure of facts." Later, again to Charles: "It was put into [their heads] that I had come to the Isthmus to martyrize them. Today they must realize that I have hatred towards no one except the idlers and the traitors."

Having inspected the entire line, having examined all completed surveys, reports on soundings, he prepared a master plan for the canal, the first that had been made in all this time and in fact the only one ever made by the French.

Like all their surveys and maps, the plan was in the metric system. The line from Colón to Panama was 74 kilometers, including a deep-water channel into the Bay of Panama ending near the island of Naos. The bottom width of the canal was to be 22 meters (72 feet); the depth, 9 meters (29½ feet).

To confine the Chagres, a tremendous earth dam, one of the largest ever built, possibly 48 meters (157½ feet) high, was to be stretched across the river valley at Gamboa, several miles above the Barbacoas bridge. The river was the heart of the matter, Dingler wrote; it was "the great unknown." He also had no doubt of success—"it only requires that we quadruple our efforts, which is absolutely possible." The incoming tides from the Pacific could be handled by a tidal lock that would maintain a constant water level in the canal from Colón to Panama.

His most important change was to reduce the slope of the cutting—that is, he declared that the sides of the canal would have to be sloped back far more than previously foreseen, a change of tremendous consequence since it increased the so-called "cube" of the total excavation by 60 percent.

His estimate was that the final amount of earth and rock to be removed would amount to 120,000,000 cubic meters. This was 45,000,000 cubic meters more than the Technical Commission had estimated, 74,000,000 cubic meters more than what had been prognosticated at Paris in 1879. Indeed, the difference between Dingler's estimate and that made at the beginning in Paris was equal to the total amount of excavation required for the entire Suez Canal. Yet when, in the early fall of 1883, he returned to Paris to review the plan with de Lesseps and the advisory board, it was calmly approved in total. Notwithstanding so radical a reassessment of the task, de Lesseps declared no change either for the completion date or the projected cost. Everything was proceeding quite smoothly as planned, he said.

In May alone Dingler signed seventeen new contracts for excavation. Orders for equipment went out to Belgium, France, the United States.

The numbers of steam shovels, locomotives, and flatcars in use were to be more than doubled, even tripled, in less than two years' time. Warehouses were built, machine shops, locomotive roundhouses, coal depots, a half mile of new docks. He was spending big money now. By September of 1883 the work force was increased to ten thousand men. By the end of the year there were thirteen thousand on the payroll. The harbor at Colón had become so crowded that inbound freighters sometimes had to wait weeks for a turn to unload.

Before he was finished, Dingler would sign up nearly thirty contractors, but the most impressive show was put on by Huerne, Slaven & Company, later known as the American Contracting and Dredging Company. The firm's association with the canal predated Dingler, but it was in April 1883, about the time he was getting things in rein, that the monstrous Slaven dredges arrived.

Prosper Huerne, one of the partners, was a San Francisco architect who had contracted to build some of the French work camps and supervised repairs on the Grand Hotel. Slaven was the galvanizing force in the organization and a fascinating sample of the sort of individual such an undertaking could attract. Actually there were two Slavens, Moses and his brother, H. B. (Henry Bartholomew). They were Canadians who had settled in San Francisco, where H. B. established a drugstore. Moses, we are told, was a "mechanical engineer," which could have meant any number of things. But H. B., the druggist, was the one in charge, and it was he who carried on at Panama, gathering in a fortune, after Moses died.

Neither of the Slavens nor Prosper Huerne knew the first thing about building a canal, and they never let that bother them. Hearing of the enormous contracts being let out by the French "and determined to have a finger in the canal pie," H. B. had sent off bids for several miles of excavation and the bids were accepted. He found a financial backer in New York, a banker named Eugene Kelly, who had never met H. B. and knew nothing about him until H. B. walked into his office. Kelly put up $200,000.

The famous Slaven dredges were built and launched at Philadelphia, and getting them to Colón was a harrowing experience. Each machine resembled an immense wooden tank, square at both ends, about 120 feet long and 30 feet wide. At sea they had all the sailing qualities of a medium-sized barn. So long as the weather was calm, there was relatively little trouble, but towing one in a gale became a nightmare, especially for the men stationed on the dredge itself.

The first of the machines to arrive had been tied up at Colón only a week when it burned to the water's edge, leaving nothing but a blackened hulk. After it were to come the *Comte de Lesseps*, the *Prosper Huerne*, the *Nathan Appleton*, the *Jules Dingler*, and miraculously none was lost en route. Once fitted out, with their booms and chains and iron buckets, they might have been fantastic war machines. Each dredge was powered by several steam engines, the largest to turn the great wheels by which an endless chain of iron buckets was kept in motion. The buckets, with a capacity of one cubic meter, ran to the top of a wooden tower, like a moving flight of stairs. At the top a blast of water washed the earth out through pipes, or "chutes," four feet in diameter, that extended, like great dangling arms, 180 feet on both sides, or far enough to be clear of the working site.

The smaller engines were used to run the powerful force pump that sent the blast of water to the top of the tower, or to move the huge dredge forward, or to swing it from side to side, or to hoist or lower huge legs, or spuds, "by means of which she walked step by step into the material to be excavated."

"The towers were from fifty to seventy feet high," Tracy Robinson would recall, "and I often climbed one and another, and stood fascinated and thrilled upon the summit, watching what seemed more like some intelligent antediluvian monster revived."

The Americans who ran the dredges—Crawford Douglas, Nathan Crowell, Captains Ward, Morton, Bardwell—were tough, independent men who lived on board, where they hung out their wash. A few had brought their wives with them, even children; a few had brought women who were merely listed as wives. Once everything was in order, smoke poured from stacks like those at a factory. The noisy bucket chains ran day and night.

For the new arrivals at Colón, the Slaven operation was the first and most impressive visible sign of actual canal construction, something Ferdinand de Lesseps had "very dextrously" considered (in the view of a correspondent for *The Times* of London). The gigantic American contrivances churning away at the front door, so to speak, were bound to have a favorable effect. De Lesseps had an abiding faith in machines and spoke often of how Alexandre Lavalley's dredges had revolutionized the work at Suez. He had every confidence that at Panama still more extraordinary machines would work an even more astonishing success.

The wonderful thing was that the American dredges *did* make progress, and rapidly, starting inland from the mud flats of Limon Bay. Later, farther inland, difficulties would increase, the pace would slow as the ground became less easy to work, and of necessity the price would rise. All told the Slaven firm would be paid more than $14,000,000 for its efforts. What its profits were remained a secret, since the company's books were kept in New York. Long afterward, a French investigating committee would conclude that the firm cleared $7,000,000. In any event, H. B. was never to return to his drugstore.

The Atlantic end was the easiest part of the work, and the progress there in the mud flats would have been the most conspicuous whichever firm had been fortunate enough to get that assignment. Still the Slaven firm alone, out of the two hundred-odd contractors that were ultimately involved, completed its allotted task on schedule and would account for as much of the total excavation as the five other largest contractors combined. Among the more curious facts about the French canal at Panama is that about a third of it was dug by Americans.

Excavation continued along the entire line, the work organized in three divisions: Limon Bay and the lower reaches of the Chagres represented the first division; the second took in the upper Chagres and the hills between Matachín and Culebra; the third ran from Culebra to the approaches to the Bay of Panama. At the head of each division was a French engineer and there were a dozen or more contractors at work under any one of these men. In the lower Chagres and in the channel on the Pacific side the work was done almost entirely by dredges. In the first division, for example, a Dutch firm, Artigue et Sonderegger, had twenty dredges at work. These too were ladder dredges, Belgian-made and not so large or powerful as the Slaven machines, but more efficient and extremely well built. The dredges used off Panama City were a self-propelling marine type, constructed like a ship. They had been built in Scotland and came out to the Isthmus under their own steam.

In the uplands the work was done by steam shovel, pick and shovel, and wheelbarrow. It was there the army of black workers were concentrated, where in these first years, progress was made largely by hand, as at Suez. Wages were regarded as extremely good, about $1 to $1.50 a day, more money than most of the men had ever dreamed of making. Each worker was required to do a specific amount in a day—so many buckets of earth—but he could work at his own speed and do more if he wished, his pay being computed by the bucket.

Lieutenant Raymond Rodgers, an officer from an American gunboat stationed offshore, made a tour of the work in 1883 and described the canal as "fairly begun." He had watched the dredges in action; he had been to the top of the fluviograph at Gamboa, a picturesque, brightly painted tower where watch was kept of the temperamental Chagres and where, on a small platform enclosed by a fancy gingerbread railing, he had been able to look out over the treetops as his French hosts expounded on their plans. At Matachín he watched a force of men drill and blast through solid rock. He was astonished by the "immense amount of machinery and material now on hand" and by the courtesy he was shown. A special train was put at his disposal. He was given maps, statistics. At Culebra a barefoot gang of workers stopped long enough to pose beside the most conspicuous piece of American equipment in view, an Osgood & McNaughton steam shovel made at Albany, New York. But seen from a nearby hilltop where Lieutenant Rodgers climbed for a panoramic view, the same machine was a mere toy, the line of excavation nearly lost in the tossing green hills.

Visitors were told that the rate of progress would soon exceed 1,000,000 cubic meters a month, then 2,000,000 cubic meters. In fact, the present rate was considerably less than 200,000 cubic meters—146,000 in May, 156,000 in June. There was, moreover, one irrefutable cloud in the sky.

According to the company's records 125 employees died in 1882, more than twice the number given for the first year. In 1883 there were 420 recorded deaths, or almost eight times the number given the first year. Yet such figures can be taken as only suggestive. Patients in the company hospitals were charged $1 a day, nearly a day's wages. While the company covered this expense for its own employees, all but a fraction of the labor force worked for the contractors, not the company. Aware of what hospital expenses could amount to, familiar with the mortality rate inside the wards, the contractors were reluctant to finance such care and would even discharge a man at the first sign of illness to avoid the responsibility. Among the workers themselves the hospitals were regarded with abject horror, the common belief being that if a patient did not have malaria or yellow fever when he entered, he would very shortly. A hospital permit was considered little better than a ticket to the graveyard.*

So for all these reasons the majority of the sick never went near a hospital, and consequently, the majority of deaths never appeared in the record books. Dr. Gorgas would calculate that for every recorded

* Such fears were well founded. Physicians who were at Colón and Panama City during these years would later state that more than 75 percent of all hospital patients had malaria.

death in the French hospitals there were at least two more outside that were not counted. In other words, the given casualty figures have to be multiplied by three: the toll in 1883 was closer to 1,300 than to 420.

It was the suffering and death of individuals, rather than aggregate numbers, that most affected those around them.

In the fall of 1883, when Dingler returned to the Isthmus after reviewing his plans with de Lesseps, he brought his family—Madame Dingler, a son, a daughter, and the daughter's fiancé. This, stressed the *Bulletin*, was the best possible proof of the director general's perfect confidence in Panama. They moved into a large, comfortable house on the Avenida Central, just off Cathedral Plaza, the Casa Dingler, as it would be known henceforth, which was supposed to serve temporarily until the more elaborate quarters were built. They were a family of avid equestrians and Dingler, who enjoyed a little show, had arranged that each be provided with magnificent mounts brought over from France. (The diminutive ponies used locally would never have satisfied any self-respecting European horseman.) It was the dry season and there were family excursions into the hills, accompanied by servants with enormous picnic hampers. One old photograph shows his daughter, a pretty, dark-haired girl who appears to be about eighteen, sitting sidesaddle in full skirt and a little Panama hat.

But in January the daughter contracted yellow fever and died within a few days. Dingler was overcome with grief. "My poor husband is in a despair which is painful to see," his wife wrote to Charles de Lesseps. "My first desire was to flee as fast as possible and carry far from this murderous country those who are left to me. But my husband is a man of duty and tries to make me understand that his honor is to the trust you have placed in him and that he cannot fail in his task without failing himself. Our dear daughter was our pride and joy."

The death of the young woman had a profound effect on everyone, canal officials, workers, the local citizenry. Bishop Paúl presided over the funeral in the crowded cathedral, and with the cathedral's great discordant bells tolling, Dingler and the fiancé rode at the head of a long procession to the cemetery.

A month later Dingler's son, age twenty-one, showed signs of the dreaded disease. In three days he too was dead. Some weeks after, it was Dingler who wrote to Charles:

> I cannot thank you enough for your kind and affectionate letter. Mme. Dingler who [knows] that she is for me the only source of affection in this world, controls herself with courage, but she is deeply shaken. . . .

We attach ourselves to life in making the canal our only occupation; I say "we" because Mme. Dingler accompanies me in all my excursions and follows with interest the progress of the work.

Presently the fiancé died, also of yellow fever. By summer, forty-eight officers of the canal company had died of yellow fever alone, and according to one American naval officer, laborers were dying at a rate of about two hundred a month.

Still the work went ahead. Travelers crossing by train were amazed by the spectacle. It was true, they wrote; a canal really was being built at Panama. Buildings were going up almost everywhere one looked. Hundreds of acres of jungle were being chopped back to make room for more. Millions of dollars' worth of equipment was being unloaded at Colón. More and more young French recruits were arriving, more engineers, more doctors, nurses, more boats from Jamaica, their decks solid with black men. By May upwards of nineteen thousand people were at work and the payroll was running to 200,000 francs ($40,000) a day.

In Paris the fearful death toll was no longer secret, despite the uniform silence of the press on the matter. Too many parents had been informed of the death of their sons. Among professional engineers the tragedies of the Dingler family were taken especially to heart. Older faculty members at the École des Ponts et Chaussées were now privately advising graduates not to go to Panama, saying it would be suicidal. Still there was never a shortage of able volunteers. Indeed, the young men who came over the sea to Colón were the pick of the best-trained technicians. For them the canal was a stirring opportunity, a "Cause"—grand in scale, glorious in concept, *French*—and they sailed as if to battle, as they themselves said repeatedly. They were warriors bearing the banner of France. Discomforts, dangers, the likelihood of a miserable death on the wrong side of the world, these, wrote Philippe Bunau-Varilla, only "exalted the energy of those who were filled with a sincere love for the great task undertaken. To its irradiating influence was joined the heroic joy of self-sacrifice for the greatness of France."

Philippe Bunau-Varilla merits a great deal of attention. Everything considered, he is one of the most fascinating figures in the entire Panama story, as important and controversial as Ferdinand de Lesseps, as time would tell. And it is fair to say, as his admirers have, that without him there would have been no canal at Panama. Because he survived the so-called French years and wrote extensively, drawing on his experiences, he also provided the fullest account we have of the French effort seen from the point of view of the elite young French technician, the man upon whom, presumably, the fate of the enterprise, not to mention the national honor, rested. And though his declarations of faith in the task, his ardor for the historic civilizing mission of France, would seem foolishly high-blown in another later age, more like lines from a melodrama of his time, they appear to have been both sincere and representative.

In his last year at the Polytechnique the young man had sat spellbound in the front row when de Lesseps, newly returned from his triumphant Panama tour, came to lecture on the great task ahead. In 1884, having finished at the École des Ponts et Chausées, and having served briefly in North Africa, he sailed for Colón on the *Washington* (sister ship of the *Lafayette*) at the same time Jules and Madame Dingler were returning from a home leave.

He was a small man. He stood only five feet four, which was shorter even than Dingler, and probably weighed no more than 130 pounds. However, he had a square, high brow, a good chin, extremely pale blue eyes, a luxuriant dark-red mustache, and his posture was always perfect. He was, as well, proud, ambitious, phenomenally energetic, blatantly self-confident, and, for all that, quite likable, in an eager and direct way. His age was twenty-six, a fact even the mustache failed to camouflage.

When he became a celebrity in the United States years later, it would be said that he was from a prominent, wealthy Paris family. But according to his registration records at the École Polytechnique he was the son of Pamela Caroline Bunau and of a *père inconnu*, unknown father, which can only mean that he was an illegitimate child. He was registered, moreover, as plain Philippe Jean Bunau. His mother, the records show, was the widow of someone named Varilla, but apparently Philippe was born well after Varilla's death, or at least long enough so that she was obliged to give her son her maiden name. The widow Varilla is also recorded as *rentière*, meaning she had some kind of income or pension of her own. However, it could not have amounted to very much, since Philippe is listed as a scholarship student. In addition, he is registered as a Protestant and by the time he finished the Polytechnique he had added Varilla to his own name.

No sooner had he reached the Isthmus than he was made a division engineer in charge of operations at Culebra and the Pacific end, which immediately set him off as somebody to watch. (Dingler appears to have been as impressed by him on the voyage out as was he by Dingler.) The advance thereafter was to be remarkable.

Why in the name of God would he want to go to Panama? the old librarian at the École des Ponts et Chaussées had asked in Paris. "As an officer runs to it when he hastens to the battlefield," Bunau-Varilla answered, "and not as the coward who flees from the sorrows of life." Once there, seeing one compatriot die after another, he would exhort the living that they were soldiers under fire who think only of the victory to be won. Disdain of peril was the surest safeguard. There was nothing good men and true could not accomplish when committed to a Noble Task. He saw them all as figures in a romance, embarked on what he was to call The Great Adventure of Panama. They were more themselves, better men, in this wild field of combat.

For Bunau-Varilla, for all the younger engineers, Dingler remained an inspirational figure. Dingler was "bold, loyal, scientific, and stimulating." Dingler bore his suffering with grave dignity and courage. Dingler was determined to succeed. The exorbitant salary he was supposedly receiving, the various trappings of position he fancied, seem never to have offended the sensibilities of any of them.

Liveried servants were in attendance at the offices on Cathedral Plaza. Elaborate stables had been built. Horses and expensive carriages had been imported from New York for staff use. The stablemaster was a full-fledged baron. For inspection tours back and forth on the railroad there was a special Pullman car that supposedly cost $42,000. And on Ancon Hill stood the famous private villa, nearly completed, an imposing structure with mansard roofs and spacious verandas.

Wolfred Nelson reported Dingler's salary to be $50,000, that one French engineer had a pigeon house put up at a cost to the stockholders of $1,500, that another official had a private bathhouse built for $40,000, again at company expense.

The most colorful source for the supposed extravagances perpetrated under Dingler was a little book published by an embittered stockholder named Henri Maréchal, who visited the Isthmus the winter of 1884 and who enjoyed spreading nonsense.

In one part of the jungle he had seen men at work building beautiful avenues and ornamental clearings, "a kind of miniature Bois de Boulogne, where the officials entertain at charming picnic parties and make daily pleasure excursions on the company's horses. . . . Ladies, possibly somewhat too swarthy but not too strictly virtuous, render these jaunts more agreeable and are repaid for their services by being carried on the company's payrolls as laborers." He declared that since the dump cars sent from Europe proved to be too high for the average workman to reach, the company, at great expense, had sent a delegation to Mexico to investigate a tribe of giants whose existence had been reported by some practical joker.

The truth was considerably less sensational. The Pullman car, though hardly a necessity, was not the sumptuous affair pictured in later accounts and seems to have been little used. The horses too were seldom used and were soon sold off. The house, though very grand by Panama standards, probably cost about $100,000–not $1,000,000, as later claimed–and because of the tragedies in the Dingler family it was never lived in. Dingler's salary was $20,000, not $50,000.

That some ladies "not too strictly virtuous" may have been carried on the payroll is certainly possible. Simple mismanagement was conspicuous enough, according to dozens of reports; the company was swindled repeatedly in small ways. One common deception concerned the delivery of coal at Colón. When a coal ship arrived, only part of the cargo would be landed, but vouchers were made for the full amount. The ship then departed, to return again with what supposedly was another load, for which another voucher would be given, the result being that the company paid for the same shipment twice, even three times.

And certainly money was wasted on needless, even foolish material comforts. But for those then struggling against the jungle and the heat, life was never easy and often extremely grim. In retrospect there is even something pathetic about Dingler's gestures toward a semblance of civilization as he knew it. How grand could one Pullman car have been on a railroad forty-seven and a half miles long?

The intensity of the boredom these men faced after hours, the longing for home, can be imagined. There were no restaurants or cafés of quality, no theaters in the city, no galleries, no libraries, never a concert. A walk on the old seawall, as one of them recalled, was as pleasant as could be expected, "but after one has strolled up and down it every day . . . for several months . . . it ceases to provide more than mild diversion." Even to sit and read at night could be a misery, since the smallest lamp or candle drew swarms of insects.

What praise and respect they did get would be a long time coming–from the American engineers who were to follow years later, none of whom ever deprecated the French work, despite all that appeared in the papers, all that had been long since fixed in the public mind. Among critics of the work, as an American naval officer observed, not one in a hundred would have the courage to go out and stay in such a place. The French engineers, he said, were "young, zealous, and energetic . . . and no one can appreciate more than these men the difficulties that lie in their path. Instead of censure and detraction, they deserve the highest praise and respect."

But the image of vain, spendthrift, immoral officials squandering company funds, heedless of the misery of others, blind to the handwriting on the wall, was to be too useful an image later, in France no less than in the United States.

II

The task to be faced daily in the field grew ever more horrendous. Presumably the work would go more smoothly with time and experience, but not so. Hard as the engineers pushed, as seasoned as many of them became to the sweltering climate, the incessant rains, the going was always more difficult than before, the technical problems ever larger and more perplexing. Maddeningly, some problems seemed quite insoluble.

The river remained the worst of these. Success, as the best of the French engineers understood perfectly, depended on somehow containing and controlling the Chagres, yet it remained, in Dingler's phrase, "the great unknown." The dam he proposed to throw in its path at Gamboa was a reasonable solution, but only in the abstract. No adequate rock formation had been located upon which to found such an enormous structure. Nobody had devised a realistic means for handling the tremendous overflow there would be when the river was in flood. After one storm in May of 1884, the fluviograph at Gamboa had recorded a rise of ten feet in twenty-four hours. On July 18 and 19 of the same year, the river came up fourteen feet.

Visitors were told that plans for the dam were not available as of the moment. One French engineer privately declared that the whole idea was hopeless. As the work went bravely on, as the river responded to the turn of the seasons, as the elder de Lesseps kept insisting in Paris, Micawber-fashion, that something would turn up—his man of genius with the perfect answer—nobody on the Isthmus honestly knew what in the world might be done. So this most vexatious of problems was simply put aside.

More immediate and much more discouraging were the slides in the cut through Culebra, which grew steadily worse the further the excavation progressed. For those in charge, they were the most infuriating part of the entire undertaking. Nearly twenty years later an American named S. W. Plume, an old man by then, would shake his head in dismay as he tried to describe for a Senate committee the troubles the

French had encountered at Culebra. He had spent a lifetime building railroads and canals throughout Central and South America, but never had he seen anything like Panama "in the French time." He had been employed by the Panama Railroad and kept an eye, he said, on just about everything that went on. Once Dingler had asked him to make a personal inspection of the operation at Culebra.

"The whole top of the hill, sir, is covered with boiling springs," he would recall. "It is composed of a clay that is utterly impossible for a man to throw off his shovel once he gets it on. He had to have a little scraper to shove it off." Nothing they had tried had kept the hill from sliding. "It won't stay there . . ."

Why? he was asked.

"The rainy season will saturate the earth and it will slough off."

"Did it do so while you were there?"

"Yes, we had a cut right alongside of where the canal was going to be built and it sloughed off, not only over the top of our track, but we found it was going to be so expensive to move it that I cut the track away there and laid another one. And a year or so afterwards the same thing took place and I laid another track, and where the present track is there are two underneath.

". . . when I was there at Culebra that week, my house was up on the hill about four hundred to five hundred feet from the canal and I got up one morning and come out and the land had gone off and left a crack there two to three feet wide, and I did not say anything, but I knew what it was. . . . The whole side of that mountain is going down into that canal. . . . Every rainy season, whenever it rains a little, the earth becomes saturated and it slides right off on this strata of blue clay."

"It slides on the blue clay?"

"It slides on the blue clay."

In somewhat more precise terms the Culebra uplands can be described as a disorderly combination of several geological formations, some sedimentary, others volcanic in origin—a generally unstable combination that was bound to mean a great deal of trouble. The oldest of these formations dated from the geological time period known as the Oligocene, making them roughly thirty million years old. Probably the Isthmus had its beginnings in the Oligocene as a string of islands in a shallow sea. A land bridge formed, and in the long geological periods that followed, this land bridge sank back below the sea at least four different times. In the late Pleistocene, the epoch of the glaciers—yesterday in geological time—the land was elevated to several hundred feet above its present level, then subsided again to perhaps thirty feet below the present level. The uplifting that followed began within a thousand years of the arrival of the canal builders (as evidenced by freshly raised old sea beaches) and the uplifting was still going on.

The whole history of the ground underfoot, wherever one went on the Isthmus, was of change and instability. Within the forty-plus miles between Colón and Panama City was a total of seventeen different rock formations, six major geologic faults, five major cores of volcanic rock.

A formation at Culebra, one taking its name from the hill itself, was found to consist of beds of soft, dark shales, marls, and carbonaceous clays—the blue clays S. W. Plume remembered—of beds of limestone and sandstone, sandwiched among thin layers of lignite. Such material drilled and blasted readily enough, as the engineers discovered, but then other formations were composed largely of volcanic or igneous rock—of dark, fine-grained basalt, of andesite or diorite or the glassy rhyolite, all rocks much like granite. The great volcanic core of Culebra Hill, for example, was solid basalt.

It was endlessly fascinating terrain to a geologist, but for the engineer it was an unrelieved nightmare. The worst troubles were in what was called the Cucaracha formation, composed chiefly of dark-green and reddish clays, lava mud flows, gravel, some shales. The first of the Cucaracha slides occurred on the eastern side of the Culebra Cut, where the uppermost layers of porous clay, layers overlying relatively impervious rock stratum, were from ten to forty feet thick. In the rainy season these clays became thoroughly saturated, slick and heavy, with a consistency of soap left overnight in water. But the saturation stopped at the underlying rock, and the build-up of water created a slippery zone along the whole plane of contact. If that plane happened to be tilted toward the Cut, then it was merely a matter of time until the clay began to move, by simple force of gravity, down into the Cut. Tremendous masses of the upper stratum would let go with all the effect of an avalanche, carrying with them whole sections of track, steam shovels, anything caught in the way.

In the dry months the sides of the Cut remained reasonably stable. But always with the return of the rains the slides resumed. (And to grasp the magnitude of the problem, one must always keep in mind what those rains meant—thirteen inches in the month of June 1884, for instance, sixteen inches in August, ten inches in September, twenty-two inches in October.)

In an attempt to alleviate the problem, the French dug an extended system of drainage ditches parallel to the Cut to channel the rainwater away from the exposed slopes. But such efforts had little lasting effect. Year after year hundreds of thousands of cubic meters of mud and rock came thundering down into the open Cut, blotting out months of work. Everything that came down had to be laboriously cleared away. Progress slowed, or stopped; contractors' estimates had to be drastically altered.

Removing the mud and debris was only part of it, for the one sure way to prevent further slides was to keep slicing the sides of the Cut back and back; that is, to stabilize the slopes by making them less steep, flattening them out until they had reached an angle of repose, the point at which the material would remain at rest of its own accord. Yet as far back as they cut the slopes, it was never enough. The amount of digging involved, furthermore, was always greater than one might imagine, for the reason that the canal was being dug through a saddle between steep hills. So as the Cut was made steadily broader at the top, its sides, against the bordering hills, rose steadily higher. Or to express it another way, every foot added to the width of the Cut at the top increased its depth as measured from the brow of the Cut.

This meant that the volume of excavation, the total cube, was being compounded steadily and enormously. The deeper the Cut was dug, the worse the slides were, and so the more the slopes had to be carved back. The more digging done, the more digging there was to do. It was a work of Sisyphus on a scale such as engineers had never before faced.

Simple mathematics made the prospects appear overwhelming. Prior to Dingler's initial reappraisal of the situation, all estimates on the quantity of excavation to be done were figured on an angle for the sides of the canal of one on one—one meter back for every one meter deep—a slope of 45 degrees in other words. In actual practice it appeared as though the sides would have to be one on four—four back for every one deep. So if Culebra Hill was 339½ feet above sea level and the canal was to be 29½ feet deep, this would give a total depth to the Cut of 369 feet. The breadth of the bottom was to be 72 feet; the breadth at the water line, 90 feet. If from that point upwards, with sides sloped back at one on four, then the final Cut would have to be three-quarters of a mile across!

To further complicate matters there remained the very basic problem of what to do with the mountains of rock and earth being excavated, and it was a problem the French failed to solve.

Their method of excavation at Culebra was first to carve off any intervening hilltops along the projected line, then start their giant steam excavators, their steam shovels, and their labor gangs working along in the direction of the line, digging down in a series of stepped terraces, each about sixteen feet wide and sixteen feet deep—the depth to which the excavators could reach. (These machines looked much like a big railroad car with tall smokestacks and long iron ladders that hung down on one side, each ladder supporting an endless chain of dirt buckets. They worked on the same principle as a dredge, but were borne by tracks rather than water.) The spoil was then hauled away by trains of little dump cars (Decauville dump cars with a carrying capacity of 4½ cubic meters) to some convenient adjacent valley, where, from improvised tracks run along the brow of the valley, the spoil was dumped over the side until it built up sufficiently below to create a terrace. The track was then taken up and relaid below. So in time the dumping grounds, like the Culebra Cut, became a vast series of long, horizontal terraces of raw-looking mud.

As solutions went, this system was quick and economical, which was what the contractors wanted. And the contractors, as Bunau-Varilla noted, were "absolute masters when it came to choosing their method of work."

The trouble was that the plan was fundamentally flawed. The terraced dumps were less stable even than the slopes of the Cut. When the torrential rains struck, whole terraces slipped out of line, track was dislodged, buried. The entire system broke down. Excavation would have to stop until things were back in order and the dirt trains could start running again, as hundreds of men with crowbars and shovels struggled knee-deep in the gummy morass.

Natural watercourses were blocked, water gathered in great pools, acres of new swamplands were formed—all perfect breeding grounds for mosquitoes.

The time lost, the effort wasted, grew to alarming proportions. But for some strange reason the French never figured a better way. It never dawned on them that digging the Cut was more a problem of transportation—of moving the spoil out of the way—than of actual excavation. They never saw that the Panama Railroad was the key, which is especially ironic considering the heavy price that had been paid to get control of the railroad. That de Lesseps had neglected to send to Panama a single specialist in railroads was among his gravest errors.

Predictably, as progress grew more difficult, contractors, and especially the smaller ones, grew obstinate, peevish, or quit outright. Invariably the next contractors to take on the same tasks wanted more advantageous terms. Since few of them could afford the kind of equipment needed, the canal company bought the machines, which were then rented out. And as much machinery as there was, it was never enough or, often, it was not exactly suited for the particular job at hand.

The canal company, in addition to the machines, was also obliged to furnish—that is, to deliver to the Isthmus—the necessary labor and to provide adequate housing. The contractors then had only to pay the men's wages. Any failure on the part of the canal company to provide either men or machines could give a contractor excuse enough to back out of his contract if it was proving unprofitable, or to hold on to a profitable arrangement irrespective of his failure to perform as agreed. The large Anglo-Dutch Company—its formal name was Cutbill, de Longo, Watson, and Van Hattum—had the Culebra contract and was bound to remove 700,000 cubic meters a month. As yet it had managed to remove 100,000 cubic meters in a month. Still it hung on, subcontracting the most troublesome tasks to more and more small operators. One high hill on the western slope of the saddle soon had so many different contractors laboring away that it became known as Contractors Hill.

Across from this same hill, on the eastern side of the Cut, stood Gold Hill, so named and widely known because supposedly it was one of the canal company's greatest assets—enough to offset all mounting costs. According to one prospectus issued in Paris, company officials had been informed that "this mountain is full of gold and it is believed that the ore from this place alone will be worth more than will be the total cost of the canal construction."

By October 1884 there were 19,243 employees at work, of whom 16,249 were blacks. To order and distribute supplies, to keep watch on contractors, to keep the books and see to the needs of this labor force, naturally required a small army of clerks, paymasters, stenographers—six to seven hundred in office help—most of whom were French. And French bureaucracy, it was found, could flourish no less in the jungle than at home. File clerks were given the title of Keeper of the Archives. Among the supplies being landed at Colón were crates weighing hundreds of pounds filled with nothing but pen points. "There is," wrote Wolfred Nelson, "enough bureaucratic work, and there are enough officers on the Isthmus to furnish at least one dozen first-class republics with officials for all their departments."

Now, too, more seriously, observers with technical backgrounds had begun reporting that the actual work was not going so well as supposed. Completion of the canal according to present plans was highly doubtful, Navy Lieutenant Robert M. G. Brown informed his superiors in Washington. A writer for the *American Engineer*, after several months on the Isthmus, figured that at the present rate of progress twenty-four years would be needed to finish the canal and charged that the French press was being bribed to withhold the truth. In October 1884 a Captain Bedford Clapperton Pim, of the British Navy, reported in a confidential memorandum to the American Secretary of the Navy that de Lesseps' dream of a canal at sea level was plainly impossible. Pim's only praise, after an extended tour of the work, was for "the gallant employees who have struggled manfully to carry out the wishes of their chief . . ."

On New Year's Eve, 1884, the last of Jules Dingler's family, Madame Dingler, died of yellow fever. It is not known how long her agony lasted, but the morning following her death, though so grief-stricken he could barely speak, Dingler was at his desk at the customary hour. Later, following the funeral, he took all the family's horses, including his own, up into one of the mountain ravines and shot them.

The toll in human lives was growing ever more ghastly, unlike anything anyone had foreseen, except possibly Godin de Lépinay. Eighteen eighty-five was to be the worst year. Probably more people died then than at any other time during the French regime. In the years to follow, the ravages of yellow fever, malaria, typhoid fever, smallpox, pneumonia, dysentery, beriberi, food poisoning, snakebite, sunstroke, were only a shade less appalling. Ordinarily on the Isthmus, yellow fever came and went in cycles of two to three years. Now, unaccountably, it never went away and there was not a thing anyone could do. Malaria, ever present as always, remained the deadliest killer.

New arrivals, unaccustomed to the climate, suffered worst. The files of the Panama *Star & Herald* carry the obituaries of individual French officials who had been on the Isthmus so brief a time as to be scarcely known. A new French consul, Paul Savalli, died on July 25, 1885, having only arrived at his post. Louis Frachen, a young engineer of the École Centrale who had come to do a special inventory of equipment in use, died miserably of yellow fever on August 10. Two division chiefs named Petit and Sordoiliet, who had sailed from France on the same ship, died of yellow fever on the same day. They had been on the Isthmus two weeks. There were scores of others. Of one dead engineer, Henri Berthaut, the *Star & Herald* could say only that at age twenty-six he "gave great promise of attaining distinction in his profession."

Bunau-Varilla estimated that of every one hundred new arrivals at least twenty died, and of those who survived, only about twenty were physically strong enough to do any real work; "and many of that number had lost the best of their intellectual value." (Whether he was speaking of all new arrivals, black as well as white, is not clear.) Others calculated that of every four people who came out from France at least two, often three, died of fever.

But these were neat mathematical averages, whereas numerous individual experiences recalled later were far more tragic. One French engineer told William Gorgas of sailing to Colón with a party of seventeen young Frenchmen. In a month he was the sole survivor. Of thirty-three Italian workers who arrived in 1885, twenty-seven were dead within three weeks. In October 1886, thirty French engineers arrived at Colón and in less than a month thirteen had died of what de Lesseps once called the "supposed deadliness" of Panama. There were times when the death toll from all causes ran to forty a day.

Bunau-Varilla would write of ships riding at anchor in Colón harbor without a soul on board. Their crews were all dead of yellow fever. Of the initial group of French nuns who had come to serve in the hospital at Ancon—twenty-four Daughters of Charity, wearing the big white, winged coifs that had earned their order the affectionate name "God's geese"—only two had survived, one of whom, by great fortune, was the head nurse, or Mother Superior, Sister Marie Rouleau. A woman of extraordinary courage and stamina, Marie Joseph Louise Rouleau had entered the order in 1868, when still in her early twenties, at the hospital at Versailles; and in 1877, or several years in advance of the first French engineers, she had been sent to Panama. Throughout the worst of the yellow-fever years, indeed throughout all the French era in Panama, she would remain a leading figure, known to everyone.

A correspondent for the New York *Tribune* reported that the mortality rate in the hospital wards—roughly 75 percent—was a subject never discussed in the presence of a patient. There were "no long faces" at the Ancon hospital and Sister Marie was "one of those rare women whose personal zeal is contagious." At the foot of each bed hung a card giving the patient's name, his job, the nature of his illness. Rarely was the real disease listed if it was known to be fatal. *Fièvre jaune* was usually put down as gastritis.

So swiftly were patients dying, so desperate was the need for bed space, that in his final minutes of life, a dying man sometimes saw his own coffin brought in. It was even claimed that the bodies placed in the coffins and carried off were not always without life. When one exhausted chief physician on the staff was sent home to France for a rest, he was in such an unsettled mental state that he had to be locked in his cabin.

For the sick who never made it to the hospital—for the vast majority, that is—the end was frequently even more gruesome. The accusation that black workers were sometimes disposed of in the dumping grounds—simply rolled down an embankment, then buried beneath several tons of spoil—appears in several accounts and is undoubtedly based on fact.

The following account by the *Tribune* man, a guest in a Panama City hotel, is also a reliable one most likely, the buzzard and all:

> . . . Sitting on your veranda late at night you see the door of the little adobe house across the way open. The woman of the house, who lodges two or three canal employees, peers cautiously out into the street, re-enters the house, and when she comes out again drags something over the threshold, across the narrow sidewalk, and leaves it lying in the dirty street. When she closes the door again there is no noise but the splash of tide. . . . Soon it grows lighter. A buzzard drops lazily down from the roof of the cathedral and perches on something in the street. The outlines become more distinct. You walk down, drive away the bird who flies sullenly back to his watchtower, and stand looking in the quick dawn of the tropics at what was yesterday a man—a month before a hopeful man, sailing out of Havre. He is dead of yellow fever.

From Colón the Panama Railroad ran a regular funeral train out to Monkey Hill each morning. "Over to Panama," S. W. Plume would recall in his memorable testimony, "it was the same way—bury, bury, bury, running two, three, and four trains a day with dead Jamaica niggers all the time. I never saw anything like it. It did not matter any difference whether they were black or white, to see the way they died there. They die[d] like animals."

The rate of sickness throughout the French operations (as opposed to the mortality rate) would be as impossible to determine accurately from surviving records as the mortality rate; but it was extremely large. A conservative estimate made later by American physicians was not less than a third of the total force at any given time. So in a year such as 1884, with more than nineteen thousand at work, probably six thousand were sick.

Company doctors advised staying out of the hot sun and to avoid getting wet, which was about like advising an arctic explorer to avoid the cold. New arrivals were warned against the night air and were told not to eat fruit. A few doctors were frank to say that it did not matter greatly what a person did or ate or drank and that nobody understood the cause of the fevers anyway. With sardonic unanimity it was agreed among physicians and employees alike that the only safe course was to get out, to leave the Isthmus as quickly as possible.

One of the many who did was the painter Paul Gauguin, whose entrance and exit date from a later time, but whose feelings about the experience were no doubt shared by hundreds of others. Gauguin came out from France with another young painter, Charles Laval, in 1887. It was Gauguin's first attempt to escape the atmosphere of Europe. The dream was to buy some land on Taboga and live "on fish and fruit for nothing . . . without anxiety for the day or for the morrow," as he wrote. But he was broke by the time he reached Colón, and so like countless other drifters who wound up on the Isthmus—tropical tramps, as they were called—he took a job on the canal as a common laborer. (Though the vast majority of the labor force was black, the company would hire anyone who looked the least fit and willing.) The ordeal of swinging a pick all day in such heat was like nothing he had ever experienced. "I have to dig . . . from five-thirty in the morning to six in the evening, under tropical sun and rain," he wrote to his wife. "At night I am devoured by mosquitoes." His partner, Laval, had been making money doing portraits of canal officials, but Gauguin would have none of it, since only portraits done "in a special and *very bad* way" would sell.

Land on Taboga, he discovered, like land anywhere near the canal, was priced far beyond reach. He felt himself weakened—"poisoned"—by the wet heat and he took an ardent dislike to the Panamanians. At one point he was arrested for urinating in public in Panama City. His defense, that the street was nothing but an open sewer anyway, failed to sway the arresting policeman who marched him across town at gunpoint to pay a fine of one piastre (four francs). His one desire thereafter was to earn enough money to leave and in another month he was happily sailing for Martinique.

III

Central to the health problem all along, the French recognized, was their lack of jurisdiction over the two cities through which everything and everyone had to pass and wherein a sizable number of their employees lived and worked. Without jurisdiction, there could be no control over sanitation in either Colón or Panama City, and as things stood, sanitation at even the most primitive level was still virtually nonexistent in both places. Colón—port of entry for all new recruits from France, for the thousands of workers from Jamaica, for all shipments of food from New York, and where everybody took the train to Panama City or to points along the way—grew more vile by the year. Compared to Colón, wrote one French journalist, the ghettos of White Russia, the slums of Toulon or Naples, would appear models of cleanliness. There were still no proper sewers in Colón, no bathrooms. Garbage and dead cats and horses were dumped into the streets and the entire place was overrun with rats of phenomenal size. And since yellow fever was understood to be a filth disease, Colón was looked upon as its prime breeding ground.

Where the French did have control, the contrast was striking. Their town of Christophe-Colomb, side by side with Colón, was neat and clean, as different as if separated by a hundred miles.

Then early in 1885 tragedy struck, taking everyone by surprise and eliminating the sanitation problem at Colón in about the most thorough fashion possible. On March 31, with a strong wind blowing out of the north, the town went up in flames.

The fire was the climax of what was to be called the "Prestan Uprising," a brief reign of terror that was set off by another bloody affair in Panama City, the work of the former Panamanian president, Rafael Aizpuru. Though the French had been uneasy from the beginning about the incendiary quality of Panamanian politics—it will be recalled that de Lesseps was warned his first day on the Isthmus about Aizpuru—the violence of what happened seems to have caught them completely off guard.

Pedro Prestan was a tiny Haitian mulatto with a deep-seated hatred for foreigners, white men and white North Americans most especially. It was a feeling shared by Aizpuru apparently. Still there appears to have been no direct connection between the two. As with most political upheavals on the Isthmus, the situation had its origins in the politics of distant Bogotá and was somewhat complicated. In essence, here is what happened.

Rafael Núñez, a major political figure for years, a former Liberal turned Conservative, had been elected to the presidency of Colombia and this had touched off insurrections in a number of places, including Cartagena and Buenaventura. Government troops stationed on the

Isthmus were rushed off to Cartagena and Buenaventura, leaving at Colón only a token force of a few hundred men. With Panama City thus unguarded, Aizpuru, a Liberal who had been waiting his chance, led his "army" of some 250 men into the city and, after much destruction and loss of life, seized control. The loyal troops at Colón, under an officer named Gónima, then crossed by train to drive Aizpuru out. Aizpuru took refuge in the hills for the time being, but at Colón, with Gónima gone, Prestan went into action.

At this stage Prestan's band probably did not number more than a dozen barefoot men armed with machetes and perhaps a pistol or two. Yet with the French all about and an American gunboat, *Galena*, sitting in the harbor, he commandeered the Colón prefecture and arranged substantial "loans" from several Front Street merchants. He moved swiftly, displaying, as most everyone had to concede later, considerable ability as a leader. Nobody did a thing to stop him.

On the morning of March 29 the Pacific Mail steamer *Colon* arrived from New York with a contraband shipment of arms consigned to "order." Prestan and his men marched to the Pacific Mail wharf and demanded that the weapons be turned over. When the order was refused by the Pacific Mail superintendent, an American named William Connor, Prestan seized him and five other Americans—the general agent of the steamship line, the American consul, the superintendent of the Panama Railroad, and two officers from the *Galena*, one of whom was sent out to the ship to tell his commanding officer that no hostages would be released until Prestan got the guns. Should the American commander try to land any men, Prestan said, he would kill the hostages and every other American in Colón.

The ship's commander, Theodore F. Kane, was in a difficult position. The presence of the *Galena* in the harbor was a routine matter (as part of the 1846 agreement with Colombia), but he was under instructions not to intercede in local matters without express orders from Washington or in the event that railroad property or services were plainly in jeopardy. Across the Isthmus, in the Bay of Panama, two more American ships stood by, *Shenandoah* and *Wachusett*, with similar instructions.

The American consul saw himself in a far more difficult position. Convinced that Prestan meant what he said, he ordered the Pacific Mail agent, a man named Dow, to surrender the arms. Hearing this, Prestan let the men go. Only now, before Prestan knew what was happening, Commander Kane brought the *Galena* up to the wharf, boarded the *Colon*, and towed her a safe distance out into the harbor. The incensed

Prestan made Connor and Dow prisoners again, which is the way things stood by nightfall.

At first light the next day Kane landed a force of a hundred men. Yet to the aggravation of the French officials and the American railroad people, all of whom were in a great frenzy over what might happen to their property, he refused to take Prestan or to intervene in any fashion. Prestan's force by now had increased to several hundred. Word of this reached Panama City, and by dark, Gónima's troops were on their way back by train.

To avoid a pitched battle in Colón, the railroad superintendent, George Burt, required that Gónima disembark with his troops at Monkey Hill. So before dawn the morning of March 31, taking Connor and Dow along as a shield, Prestan marched out to Monkey Hill. A savage battle back and forth across the tracks lasted about an hour, during which the two American hostages managed to fade into the jungle. Then Prestan was on the run. Falling back on Colón, he set fire to the city and in a few hours there was little left but heaps of smoldering ashes. Only the brick offices of the railroad, the steamship offices, the stone church, and a fringe of buildings along the beach were still standing. Eighteen people had been killed. Perhaps eight thousand had been made homeless.

The seesaw effect continued. At the moment Prestan was being routed at Monkey Hill, Aizpuru was again on the rampage in Panama City at a cost of another twenty-five lives. Again triumphant, he declared himself Panama's supreme authority.

A desperate George Burt called on the American naval officers to do something. Aizpuru's men had been tampering with the railroad's switches, he said, and they had boarded and robbed a train. Prestan had ripped up track, cut the telegraph lines, derailed an engine. So with this, men were landed from the *Shenandoah* and the *Wachusett* and order was quickly restored all along the railroad. On April 10, the *Tennessee* and the *Swatara*, under the command of Rear Admiral James Jouett, arrived at Colón with a battalion of Marines.

In Panama City, crowds gathered in Cathedral Plaza to watch the Americans parade about in their smart uniforms, wheeling a Gatling gun this way and that. Only once was there any real excitement. A fight broke out between some of the local citizens; the Gatling was fired across the plaza at an elevation to clear the tops of the buildings and the plaza was emptied in seconds.

Troops were sent across the Isthmus on improvised armored cars, flatcars fitted out with half-inch steel boiler plate and more Gatling guns. Marine guards in white sun helmets were posted at the Barbacoas bridge and at Matachín, but nothing more happened along the line. In Colón fifty-eight people were rounded up by government troops, tried for treason, and shot. Many, it would be said later, had been quite innocent.

In Panama City, Aizpuru claimed to have adequate force to maintain order on his own, but his invitations to the American officers to come to the Governor's Palace to confer were ignored. He then offered to declare Panama independent if the United States would recognize his government, but this offer too was ignored. "You have no part to perform in the political or social disorder of Colombia," Admiral Jouett had been explicitly instructed in a telegram from the Navy Department before landing, "and it will be your duty to see that no irritation or unfriendliness shall arise from your presence on the Isthmus." He was to protect American interests without offense to the sovereign, Colombia.

On April 24 Aizpuru met with the American officers at the Central Hotel and surrendered. A few days later more Colombian troops arrived by ship and in another week most of the American forces had been withdrawn from the Isthmus.

Prestan, who had fled into the jungle after firing Colón, was captured and brought back to await trial. The grim job of cleaning up, of tending to the injured and homeless, the whole effort of rebuilding Colón from scratch—a job that would be accomplished with amazing speed—fell largely to the American railroad officials.

The canal company had suffered no physical damage to speak of. Close as it was to Colón, Christophe-Colomb had not been touched. When, during Aizpuru's brief reign, a mob of looters swept up the road to the Ancon hospital, Sister Marie took a big umbrella, gathered up her skirts, and went out to meet them at the gate to the hospital grounds. "Listen to me, you," she said. "Someday you will be sick yourselves, and if you trouble us now, do you think we will have you in the hospital? No! Now go!" And back down the road they had gone.

Dingler reported to Paris, "Our works have continued to function in the usual manner." But the uprising had had its effects on the French enterprise, and some were quite serious. At Culebra, government troops had broken into a company barracks and slaughtered a number of Jamaicans, claiming afterward that the Jamaicans had fired on them first. A savage riot broke out among the black workers, more lives were lost, and hundreds of men walked off the job and took the next boat to Kingston, where the company's recruitment efforts henceforth would be extremely hampered.

As for the blame for what happened at Colón, the French and the Americans agreed. Commander Kane, it was charged, could have averted the entire tragedy by simply grabbing Prestan the morning he started on his rampage, something the prudent Commander Kane understood he had had no right to do.

Prestan and Aizpuru were dealt with in due course. Wearing a black suit and derby, Prestan was marched to the tracks on Front Street and was hanged before one of the largest crowds ever seen in Colón. Aizpuru was more fortunate. He was taken to Bogotá, tried, fined, and sentenced to ten years in exile.

Politically, things quieted down. The crisis passed, and seemingly without significant aftereffects. But, in fact, the rebellions in Panama and the other provinces marked a critical turning point for Colombia. The long-range repercussions would be considerable. To strengthen his position, President Núñez would proclaim a new constitution, with all real power centered in Bogotá. The nine provinces of Colombia, Panama included, were to be headed hereafter by governors appointed by the federal government—by Bogotá.

Also of importance as time would tell was the presence through all that had happened on the Isthmus of three observant parties whose personal roles had been relatively minor, but who would not forget what they had seen and the lessons to be drawn. Philippe Bunau-Varilla was one. Another was Dr. Manuel Amador Guerrero, a physician employed by the railroad. The third was Captain Alfred Thayer Mahan, of the *Wachusett*, who had landed with one of the Marine guards to protect the railroad.

What most impressed Bunau-Varilla and Dr. Amador Guerrero was the degree to which events had been shaped by the mere presence of American naval strength. In back of everyone's mind throughout had been those ships and the question of what their commanders might or might not do given the state of affairs ashore. And it was only a few months later that Captain Mahan, having been appointed to the faculty of the War College at Newport, Rhode Island, would commence to develop a series of lectures on the influence of sea power.

Things began coming apart now. In the office overlooking the plaza, his last reserves of strength nearly exhausted, Jules Dingler had become so short-tempered and abusive of his staff that several key people, including one division head, resigned. Late in August, close to physical and mental collapse, Dingler himself gave up and sailed for France, a lone and defeated man. He had left all his family buried in Panama. He was never to return again.

His place was filled by Maurice Hutin, the next in line. Hutin too was ill and worn out and in another month he quit. That left Philippe Bunau-Varilla. So, a year after his arrival, at age twenty-seven, Bunau-Varilla found himself acting head of the entire effort. For the next several months, until another director general was recruited, he would seldom have more than two hours' sleep a night.

He moved into Dingler's office, determined, as he later said, to be all de Lesseps had been at Suez. "Men's energies are spontaneously influenced by a chief who is inspired by a sincere faith in the ultimate triumph of a difficult undertaking," he would lecture later in one of his books. "They take their place in regular order, like particles of iron around the pole of a powerful magnet." When a new French consul general was ushered into the office to meet him and expressed surprise at finding so young a man in a position of such vast responsibility, Bunau-Varilla suggested that the new consul general learn to judge men according to their ability.

Elsewhere along the line there was widespread bitterness. Failure was in the wind and people were looking for somebody to blame. Lieutenant William Kimball, an officer from the *Tennessee* who accompanied Bunau-Varilla on a tour of the work, wrote of rampant suspicion of Americans. Bunau-Varilla was unfailing in his courtesies, Kimball noted, but other French officials made little effort to disguise their feelings. American contracts, American machinery, American technicians, were no longer wanted. French mechanics accused their American counterparts of trying to sabotage equipment in order to stop the work so that the United States could take over. Lieutenant Kimball attributed such talk to the depression caused by malaria.

His remarks were contained in a special intelligence report transmitted to the chief of the Bureau of Navigation in Washington. It was an extremely interesting document. He estimated that not more than a tenth of the work had been accomplished, a perfectly accurate estimate as it happens. "Unforeseen and vexatious, as well as stupendous and apparently insuperable, difficulties are constantly occurring." At Paraíso he saw a slide so large that it came down completely intact, the grass on top undisturbed, and with such force that it carried the entire distance across the Cut.

Hospital facilities, extensive as they were, were not enough. Food prices were high. Workers, black and white, were fearful of more political violence. Black workers were leaving faster than they were being replaced, going home to spend their money "before they are killed by the climate." But, Kimball emphasized, loss of human life would never be a deterrent in itself. Money was what counted. Human life was "always cheap."

7

Downfall

Faithful to my past, when they try to stop me, I go on.

—FERDINAND DE LESSEPS

I

On April 23, 1885, three weeks after news of the Colón fire reached Paris, Ferdinand de Lesseps donned the green robe of the Académie Française. With all traditional solemnity, in the small ceremonial hall beneath the great dome of the venerable Institute, he achieved the ultimate honor in French life. He belonged now to the forty "Immortals," the chosen of chosen.

We are told that "such a galaxy of celebrities had rarely gathered" and that de Lesseps spoke with the air of a conquering general. Great works were never easy, he affirmed. Nothing was easy in this world, especially the useful. Skeptics and doomsayers and character assassins were what one had to expect. The world was not without evil. "The Arab proverb says, 'The dogs bark, the caravan passes.' I passed on."

Yet at this crowning moment the dogs had scarcely been heard. While talk of the sums being spent by Jules Dingler was commonplace in financial circles and Henri Maréchal's diatribe had given rise to considerable gossip, no publication of influence within France, no individual of importance, had had a derogatory word to say as yet.

De Lesseps' own popularity seemed as invulnerable as always. Though Panama shares had begun to slip quite noticeably on the Bourse, though the company's credit was being questioned for the first time, virtually none of the small shareholders were selling out. At the annual stockholders' meeting in July, he was cheered again and again; and like some figure from Shakespeare, he gave his audience visions of heroic contest waged against an exotic, distant wilderness. They were part of the crusade; they were one with him in this great and good and terribly difficult work. The triumphs of the engineers—*their* engineers—were real; the task was better than half finished—"the efforts actually put forth may be considered as more than half the total efforts necessary" was his exact claim and a total fiction. (Only about a tenth of the canal had been dug, as the American officer Kimball rightly judged.)

The completion date, he also said, was being deferred somewhat and the original cost estimate of 1,200,000,000 francs, as set by the Paris congress, was being adopted. But such announcements went unchallenged. Nobody questioned or protested anything he said. When a man called for a formal investigation of the company's management, there was no one in the entire hall who would even second the motion.

Not until later that summer did the talk of failure begin to sound serious. Writing in the *Économiste Français* on August 8, the highly regarded financial editor Paul Leroy-Beaulieu, known to be an old friend of de Lesseps', warned that unless the canal company was reorganized, "we shall see the most terrible financial disaster of the nineteenth century." The New York *Tribune*, among other foreign papers, began calling the canal a "gigantic wild-cat speculation" and lamented that the poor innocent stockholders were unable to go out and see the truth for themselves.

The most blistering attack was a series of articles carried by the London *Financial News*, a series written by J. C. Rodrigues, the American reporter, who had had a tremendous change of heart since his tour of Panama five years before. Like de Lesseps, Rodrigues had not been back to the Isthmus in all that time. Nor had he bothered to talk to anyone connected with the canal, anyone whatsoever with firsthand knowledge of the situation there or inside the offices on the Rue Caumartin. The articles in fact contained no new information. But by combing through the published observations of other journalists, by looking through what de Lesseps and the French press had been saying from one year to another, by shrewdly noting all that had *not* been forthcoming from the canal company in the way of solid data on costs and the volume of excavation actually accomplished (rather than what was anticipated), he had come up with a picture of willful deception at every turn. The French newspapers were giving de Lesseps their unstinting support, he charged, because editors and reporters were being bought wholesale. This was perhaps the tenth time the same accusation had appeared in print in the past year, but like others who had made it, Rodrigues had no proof or testimony to back it.

His statistics were taken from the *Bulletin*, from de Lesseps' letters to stockholders, from public pronouncements, and the like. It was shown that the projected completion date had been a fantasy from the beginning, as had every projected cost figure. No plan for the Chagres dam existed on paper as yet, after five years. The canal was doomed because the company was going to fail. The enterprise would be defeated in Paris, that is, not on the Isthmus, and it was only a matter of time. "The whole thing is a humbug, and has been so from the start."

In October, as the articles were being published as a book in the United States, Panama stock hit a new low of 364 francs. Then in the first week of December a raging "norther" swept across the Caribbean and struck Colón, smashing eighteen ships onto the shore and destroying much of the waterfront. Fifty seamen were killed. Tremendous rains fell and the Chagres, rising thirty feet in a few hours, flooded miles of railroad and canal diggings. No trains could get through. To inspect the line after the storm, Philippe Bunau-Varilla had to go most of the way by canoe and wrote afterward of gliding past half-drowned trees the tops of which were black with millions of tarantulas.

The effect of such news on the skeptics at home may be imagined. Yet de Lesseps, sanguine as always, reminded everyone that there had never been a year at Suez without a crisis. The Suez venture had been called all the same names; as had he. The caravan *would* pass on.

In May he had spoken for the first time of lottery bonds to guarantee the canal. And at the July stockholders' meeting he had asked for and received a show of approval. Now he talked of little else. Canal bonds would be sold with numbered tickets attached, some of which, the winning tickets, would be worth large cash prizes. In the final year of the Suez Canal, when an issue of conventional bonds failed to provide funds sufficient to finish the work, just such a lottery issue had saved the canal. Moreover, those same bonds were now worth almost twice their original value. To finish "promptly" at Panama, he said, would take another 600,000,000 francs. The only thing needed was government authorization for a lottery issue. So as summer turned to autumn the Chamber of Deputies was swamped with petitions signed by thousands of Panama stockholders–a grand, spontaneous show of faith, said the Nyons banker, Ferdinand Martin, who started the campaign. And to ease any possible apprehensions over the situation on the Isthmus, de Lesseps declared he would go again to look things over.

Impressed by the flood of petitions, the Chamber appointed a committee of deputies to study the lottery proposal and a noted civil engineer, Armand Rousseau, was selected to go to Panama and report back to the committee. Rousseau departed before de Lesseps did. Whether he wished specifically to avoid the old man's irradiating influence is not clear, but he did sail on the *Lafayette* at the same time as Charles, who was taking a new director general, thirty-five-year-old Léon Boyer, to relieve Philippe Bunau-Varilla.

Bunau-Varilla, meantime, had decided to resign from the canal company, but would stay on in Panama, in the employ of a private contractor, as head of the major excavation at Culebra.

When he departed for this, his second, tour of the Isthmus–for his first actual look at the Panama canal–Ferdinand de Lesseps was eighty years old. And in the minds of his thousands of shareholders this was the critical figure in the equation, more important than any stock prices or excavation statistics. It was not a company they believed in, or even a canal through Panama, so much as one exceptional human being. For them *he* was *la grande entreprise*. So, baldly put, the question was, How much longer could the mortal hull last and perform?

His large and much publicized retinue included technical advisers, company officials, and special guests (delegates from half a dozen French chambers of commerce, a German engineer, an Italian diplomat, an admiral in the Royal Navy, the Duke of Sutherland), and everyone went at company expense. His wife and children remained in Paris. Interestingly, there was no one along this time from the first expedition. Bionne and Blanchet were dead; he and Wyse had not spoken for years; Jacob Dirks, Colonel Totten, Trenor Park, they too were dead by now.

In the end, when everything was in ruin, few would regard him very kindly for such efforts as this. The dazzling faith would seem almost maniacal and pathetic in hindsight. It would be forgotten how magnificent he was in so many respects. The man of younger days, the Hero of Suez, would be the one eulogized. Yet of all the performances in that long, glittering career, there is nothing quite comparable to the final drive to succeed at Panama. The falling star blazed very bright indeed.

They reached Colón on February 17, only days after Armand Rousseau had completed his studies and departed. The stay on the Isthmus this time lasted just two weeks, and the intentions throughout, in all de Lesseps said and did, were fundamentally the same as six years earlier—to be inspirational, to foster courage, confidence, and to draw attention to himself. In this, once again, he succeeded grandly. In the army of laborers strung out across the Isthmus, he had crowds unlike anything since Egypt. For the thousands at work the mere sight of him was electrifying. "You are for us," said one engineer in a welcoming speech, "the venerated chief around whom we rally, ready at all times to sacrifice even our very lives to assure your triumphant success in your present great and glorious work."

"With hearts and minds like yours," de Lesseps responded, "everything is possible."

Again he had chosen the dry season. He would appear, as before, in brilliant sunshine. For one inspection tour he went on horseback. "M. de Lesseps," reported one member of the party, "always indefatigable, rode at the head of the caravan. I saw him gallop up the hillside at Culebra, amid a roar of approval from blacks and whites, all astounded by so much ardor and youthfulness." Years afterward young American engineers relaxing in comfortable new clubrooms would listen to old Panama hands tell how he wore a flowing robe of glorious colors, "like an Oriental monarch."

There were parades and fireworks at Panama City. A triumphal arch in the plaza was emblazoned, "Glory to the Genius of the Nineteenth Century." Little girls presented bouquets. He made speeches; he drank toasts. He sat through banquets that would have finished off most men his age. He danced all night at a grand ball in his honor.

His accommodations on the Atlantic side were in an imposing frame house facing the water at Christophe-Colomb, built especially for his use and known ever after as de Lesseps' Palace. In Panama City he stayed at the palace of the bishop, where the guardian at his door was a dog called Bravo, a local phenomenon of unknown breed and origin which had first attached itself to Dingler some years before, then Hutin, then Bunau-Varilla, and most recently to Charles de Lesseps.

Through long, stifling mornings, as noisy crowds filled the plaza, he conferred with his engineers in the canal offices. He toured the hospital, machine shops, labor camps. Bunau-Varilla set off a tremendous charge of dynamite for his benefit, then made a little ceremonial presentation of one small bit of the rock—a "thousandth part of one-millionth of the little mountain which he had seen raised in the air . . ." De Lesseps, with somewhat less ceremony, would later give it to the Académie des Sciences.

Bunau-Varilla was kept close at hand to answer all technical questions. Charles likewise was never very far. Charles was "very clear headed and capable," noted John Bigelow, an emissary from the New York Chamber of Commerce who had joined the party, and the friendship Bigelow struck up with Bunau-Varilla was to last a lifetime.

All in all, it was a brave show. Morale had been restored on the Isthmus to a degree no one there would have thought possible. Boyer, the new head, had been installed. The guests enjoyed themselves throughout, with the exception of Bigelow, who came down with something briefly and thought he was dying of yellow fever. Their host enjoyed himself supremely. When it came time to go, all parties (Bigelow also by then) were in perfect health.

The one dissatisfied individual seems to have been Philippe Bunau-Varilla, whose devotion to the work was no less for having been relieved of command, but who felt he had been slighted by *Le Grand Français*. In their private talks de Lesseps had expressed nothing but praise and gratitude for all the young man had done, yet said nothing to that effect publicly. "Any homage paid to any other personality but himself seemed to steal a ray from his crown of glory," Bunau-Varilla would write years later, still resentful.

On reaching home the delegates from the French chambers of commerce expressed unqualified confidence. Unquestionably the canal could be built—provided de Lesseps had the funds needed (provided the government let him have his lottery). Bigelow, too, in a long, detailed, and entirely fair appraisal, concluded that the canal could and would be built, for the reason that "too large a proportion of its cost has already been incurred to make retreat as good a policy as advance." He had been bothered by the heat; he remained deeply troubled by the great number of lives being lost, a subject scarcely mentioned during the tour. But like Lieutenant Kimball, he concluded that the critical, unanswerable question was cost, not the cost of construction simply—and any figure offered, even by the most experienced engineers, was pure conjecture—but the cost of the money itself. "Till the money is secured, and the cost of getting it is ascertained, it would be about as safe to predict the quarter in which the winds will be setting next Christmas day at St. Petersburg . . ."

Bigelow was a prominent figure, in addition to being highly intelligent and observant, a former part-owner (with William Cullen Bryant) of the New York *Evening Post*, a former American minister to France, lawyer, and scholar. He had taken the tour very seriously, preparing in advance long lists of questions for the French engineers. What impressed him beyond everything else was the magnitude of the effort, and this, like a certain foreboding, can be read between the lines throughout his report. The task had no parallel in history he said. Americans had far too little appreciation of what the French were attempting. Once, to his astonishment, Charles had told him that the United States would eventually have to take financial control, as England had done at Suez. It was a remark made in confidence—and with sadness, one would imagine—and Bigelow said nothing of it, other than in his diary. But in his report, published well before any of Captain Mahan's theories, Bigelow did write that whenever or by whatever manner the canal was completed, the great beneficiary would be the United States, for the canal would "secure to the United States, forever, the incontestable advantage of position in the impending contest of nations for the supremacy of the seas."

De Lesseps, as soon as he reached Paris, made a predictable declaration of faith in the speedy completion of the work. Yet in virtually the same breath, and as smoothly and cheerfully as if he had been announcing some favorable turn of events, he also conceded that Panama was a more difficult undertaking than Suez—ten times more difficult.

On March 29, in an interview with Emily Crawford, correspondent for the London *Daily News*, he further claimed that the baffling issue of the Chagres was at last resolved: ". . . we have changed the whole course of the river and made it run on the other side of the mountains altogether." Undoubtedly he meant that the river had been rerouted on paper, which it had, and actual excavation for a vast diversion channel had begun. That, however, was not the impression given by Mrs. Crawford's article, which, as she noted in conclusion, he read and vouched for prior to publication. The distinct impression was that the river no longer intervened, that it had been neatly placed somewhere safely out of the way, and thus success was assured.

Meantime, as the government, indeed, as the whole of France, waited for the other prognosis—the one from Armand Rousseau—the canal company was having trouble paying its bills, despite another issue of conventional bonds.

The Rousseau Report was released in May, and to the French public, accustomed to unwavering support for the canal in the press, to the shareholders, long fortified by the *Bulletin* and de Lesseps, its impact was considerable. Rousseau, a former *chef des Ponts et Chaussées*, was a man of the highest repute and his views were taken as entirely forthright, which indeed they were.

To abandon the canal now would be unthinkable, he had concluded. It would spell disaster not merely for thousands of shareholders, but for French prestige. If the present company were to drop the work, a foreign one—unnamed—would surely take it up. The canal company therefore ought to be given some kind of moral assistance by the government. He favored the lottery. But then came the crucial passages. Completion of the canal was possible; however, completion with the resources anticipated or within the time announced appeared extremely doubtful *unless* the company would agree at once to radical modifications of its plans. He did not specify what ought to be done, because he believed his charter did not authorize him to do so. But there was little doubt as to what he had in mind. The one radical modification possible was to abandon the sea-level plan while there was still time.

In sum, he was saying that the hopes entertained by Ferdinand de Lesseps were without foundation—in terms of time, money, and, most important, the canal he had been selling the French people all these years. A canal *à niveau* had been the axiom of de Lesseps' conception since before the Paris congress. It had been his reason for picking Panama in the first place. So it was easy enough to draw from the report the dark and unsettling conclusion that the whole plan had been a stupendous mistake from the beginning.

In quick succession came two more opinions, both filed at the request of the canal company. The first was from another respected engineer, a man named Jacquet, whose candor was courageous in view of the situation and the mood among those paying for his services. Having toured the work, he declared that a sea-level passage was unattainable and urged the building of a canal with locks along the same path.

The second opinion, sent from Colón, was written by Director General Boyer, whom Bunau-Varilla would recall as one of France's most gifted sons. Boyer had brought sixty engineers out from France. Within a few months they had nearly all fallen by the wayside, sick, demoralized, or dead. Then, only a week or two after his report reached Paris, came the news that Boyer too was dead, another victim of yellow fever.

His was the most disquieting judgment of all, in that he spoke as the company's ranking technical authority. Like Rousseau and Jacquet, he regarded the sea-level canal as impossible within the limitations of available time and money. It had taken him but a short time on the scene to reach this view.

Charles knew that his father must give way, but if his father saw this, he never let on publicly. Rather, he now insisted that it *must* be a canal at sea level—a great new "Ocean Bosporus" was his favorite expression—saying that it could be achieved in just three more years. His will was iron. All critics were enemies. What secret, unspeakable dread he may have felt—if any—what premonitions of disaster haunted his private hours, will never be known. But he certainly had all the facts at his disposal and the truth of the situation was plain as day: with the excavation in its fifth year, the contractors had managed to extract little more than a quarter of the total excavation anticipated by Dingler's reckoning.

The Minister of Public Works, Charles Baïhaut, moved that a lottery bill be acted upon without delay, but with the cloud of the Rousseau Report hanging over them, the deputies on yet another special lottery committee now voted to postpone further discussion and called for an audit of the canal company's accounts. Outraged, de Lesseps refused to comply. Further, he wanted his application for a lottery withdrawn. "I am postponed, I do not accept the postponement," he exclaimed in a letter to his stockholders. "Faithful to my past, when they try to stop me, I go on. . . . I am confident that together we will overcome all obstacles and that you will march with me . . ."

But by no means had he dismissed the lottery from his plans, the defiant words notwithstanding. His immediate next move was another conventional bond issue, and although the results appeared perfectly respectable—some 90 percent of the offering was taken—the company was paying too much for too little. It was becoming a dangerous, costly trend. In total the company was presently paying out a staggering 750,000,000 francs, or $15,000,000, just for annual interest on its borrowed money.

Paris was filled with conflicting claims and rumors emanating from the Bourse or from the offices on the Rue Caumartin, or from around the corner at the Suez offices, where de Lesseps still spent much of his regular day. Philippe Bunau-Varilla, home since spring, was telling visiting Americans of the money he had behind him, claiming he had a new concept that would revolutionize the work and save the canal. He had returned from Panama after barely surviving an attack of yellow fever, resigned his position with the company, and was currently the guiding spirit of the reorganized Artigue, Sonderegger et Compagnie, now the Culebra contractor.

But others were talking of the resignation of the secretary general of the canal company, Étienne Martin, because he regarded the contract with Artigue, Sonderegger as so outrageously advantageous to Artigue, Sonderegger. His replacement was Marius Fontane, the company's publicity agent. For every rumor of collapse and bankruptcy, there was another story, always on good authority, that the government was prepared to rescue the canal come what might, that the company's great hidden strength was its political influence. It was noted, for example, that the Minister of Public Works, Baïhaut, who had set the lottery bill in motion, had declared his support for the lottery, irrespective of what Rousseau might conclude. Baïhaut, known as "the man with the beautiful wife," was an outspoken moralist (an officer of the Society for the Promotion of Good) and a popular topic, since until recently the beautiful wife had been married to his best friend.

Management of the company's financial affairs, moreover, had fallen to new hands. Marc Lévy-Crémieux, the banker who directed de Lesseps' initial stock successes, had died. Serving now as the canal's chief financial agent was Baron Jacques de Reinach, of the original Türr Syndicate. The ebullient little baron had friends in high places—Adrien Hebrard, publisher of *Le Temps;* Jules Grévy; General Boulanger, the glamorous new Minister of War; the Radical leader Georges Clemenceau; Premier de Freycinet.

Émile de Girardin's *Petit Journal,* the most popular paper in France, remained conspicuously loyal to the "cause of Panama." As Emily Crawford was to note, the paper's chief editorial writer "puffed Panama" to such a degree that he was "carried above the concert pitch of the paper by the heat of his enthusiasm." Mrs. Crawford, the widow of an English newspaperman, was among the ablest foreign correspondents of the day, a handsome middle-aged Irishwoman who knew just about everyone in power and "did not worry about being conventional."

De Lesseps remained as active as ever, a familiar figure on the boulevards, still fathering children, still to be seen riding in the Bois with a troupe of the older ones. He appeared to be without a problem. According to one story, a group of salesmen struck up a conversation with him on a train somewhere outside Paris and failing to recognize him asked what his line might be. "Isthmuses!" he said. He was introducing ship canals and after Panama he would build the Kra Canal across the Malay Peninsula.

In October, largely in response to a malicious rumor that he was terminally ill, he crossed the Atlantic still one more time—his last—to participate in the unveiling of the Statue of Liberty. At a rollicking celebration in New York Harbor, October 28, 1886, with cannon booming and tug whistles screaming on every side, he stood with President Grover Cleveland on a flag-draped platform set at the foot of the colossal statue.

"Soon, gentlemen, we shall meet again," he said, "to celebrate a peaceful conquest. Good-bye until we meet at Panama . . ."

II

The façade of an Ocean Bosporus was maintained for another year. In January of 1887 de Lesseps' Advisory Commission formally convened to consider the possibility of a lock canal, with the net result that a subcommission was appointed to look into the matter. The subcommission would not meet until the fall.

Charles de Lesseps returned to Panama in March, taking still another delegation of experts to give still another appraisal, and the delegation returned with the unanimous opinion that the sea-level plan must be dropped at once if disaster was to be averted. The company quietly put this report into safe storage.

Every day the decision was postponed was a day of enormously expensive wasted effort on the Isthmus. Cash reserves in Paris were shrinking rapidly. And not until the sea-level plan was scrapped could there be any hope of government action in support of the lottery.

Had de Lesseps decided on a lock canal in the fall of 1886, had he gone to his stockholders then with a new plan, instead of sailing off to New York, the outcome of *la grande entreprise* might have been quite different. Possibly the dream could have had a different ending had he but spent the first part of 1887 preparing his public for the change in plans. This he did not do, however. Apparently he kept thinking that somehow, some way, the crisis could be resolved, that some miraculous turn of fate would save the sea-level canal. It is hard to imagine what turn of fate he possibly had in mind, but then he remained, as before, one who saw his own existence and all that he did as part of a glorious cosmic pattern.

Meantime, consideration was being given to a temporary solution devised by Bunau-Varilla. He wanted to make one kind of canal in order to dig another kind. It had been the accepted engineering wisdom that dredges were pretty nearly useless against rock. At Suez, for instance, when the dredges struck underlying rock between the Bitter Lakes and the Red Sea, the engineers had put earth dams across the cut and drained that section so it could be excavated "in the dry." Consequently at Panama the assumption had been that only soft strata could be removed by dredging. But when the Slaven machines hit rock at a point called Mindi, near Colón, Bunau-Varilla conceived a technique whereby the rock could be blasted into pieces of an exact size for the dredges to cope with. The system rested on a particular mathematical placement of underwater charges, and once perfected, it was no more expensive than conventional dry excavation, or such was his claim. Later on, at Culebra he had taken the idea another step. Artificial lagoons were built at either side of the saddle. Earth dams were thrown up and water was piped in. The dredges were then brought up and floated on these lagoons and the excavation proceeded.

His proposal now was to carry the idea to its ultimate conclusion: Subdivide the whole line of the canal into a series of such artificial pools and unite these with locks; in other words, build a lock canal upon which to float the dredges and let the dredges eventually transform that canal into an uninterrupted passage at sea level. He would use water, rather than railroad track, to transport his excavation machinery, and to carry the spoil away. Such a system would be little affected by rains or landslides. The cut at Culebra need only be made half as deep for the time being. The locks could be removed two by two as the dredging progressed. And since the locks would be built to accommodate conventional ships, regular canal traffic could begin as soon as the locks were ready. So ships could be passing to and fro as the work proceeded, the tolls going far to meet the cost of the work.

Bunau-Varilla was an extremely high-powered, persuasive individual, as future events would bear out; and he appears to have convinced most of the engineers and technical advisers that the scheme could actually work. The genius of the proposal, however, its enormous value at the moment, was not in its technical ingenuity. It was the fundamental precept that a lock canal need be only a *transitional* step toward the old ultimate goal of a channel *à niveau.* It represented no betrayal of the dream. It offered de Lesseps an *honorable* alternative. There need be no promises broken, no semblance of retreat or failure.

Another bond issue was tried, the third in little more than a year. It produced a sum de Lesseps quickly pronounced sufficient to carry on for another two years. But again the company had paid dearly for the money and by autumn its financial position was desperate.

It was only then, with his back to the wall, that de Lesseps at last did what had to be done.

The subcommission of his Advisory Commission met and endorsed the temporary lock canal; a new set of plans was rushed into presentable form. At the end of October he used an invitation to speak at the Académie des Sciences as opportunity to prepare the public for the momentous change. Saying nothing of the plan itself, he announced that the canal would be far enough along by 1890 to permit the passage of twenty ships a day. Annual receipts, he said, would be 100,000,000 francs.

Panama stock, in steady decline since late summer, now fell to a new low of 282 francs.

On November 15, de Lesseps sent off two letters. One was to the Minister of Finance asking once again for authority to sell lottery bonds. The other, addressed to his shareholders, contained the dramatic announcement that "as of this morning" Alexandre Gustave Eiffel had been engaged to design and build the locks that would open Panama to the ships of the world.

The new canal was to follow the same line. There were to be two huge flights of locks, with five locks to each flight, at Bohío Soldado on the Atlantic side and between La Boca and Paraíso on the Pacific side. They would be single locks—that is, locks capable of handling ship traffic in one direction at a time—and the dimensions of each were to be 180 by 18 meters (590 by 59 feet). Passing through the full flight, a ship would be lifted to, or brought back down from, a summit level of 49 meters (roughly 161 feet), which would be well above the level of the Chagres at flood stage. So instead of a canal with a bottom 29½ feet below the level of the sea, it was to be one 161 feet above. The total excavation required was reckoned to be only 80,000,000 cubic meters (instead of 120,000,000, as Dingler had figured), which, it was said, left only about 34,000,000 cubic meters still to go (instead of 74,000,000, using Dingler's estimate as the base figure).

The official cost of the completed canal was also revised, making this the fourth time since the Paris congress. The estimate arrived at by the congress, it will be recalled, was 1,200,000,000 francs; at Panama in 1880 the Technical Commission had cut that to 843,000,000 francs; next, on reaching New York, de Lesseps had announced that it would be 658,600,000 francs. In July of 1885 it was back to what it had been to begin with—1,200,000,000 francs. Now, for a lock canal 161 feet above sea level, the price was declared to be 1,654,000,000 ($331,200,-000).

To build a Panama canal the French public to date had supplied something over 1,000,000,000 francs. Now de Lesseps was declaring in the same old positive, confident tones that another 600,000,000 francs were needed to open a canal of a kind he had long since educated them to regard as inferior and therefore unacceptable.

Recruiting Eiffel had involved some fast footwork on both sides and a contract that Eiffel and company officials alike thought best kept confidential. De Lesseps considered Eiffel's name a golden touch. It was as though, or it could *appear* as though, that "man of genius" was making his entrance in the final thrilling act, just as he, de Lesseps, had foretold so many years before: Eiffel, the most brilliant engineer in France, suddenly famous as progenitor of the gigantic iron tower being started on the Champ de Mars. Conceived as the centerpiece for another Paris exposition scheduled for 1889, the tower was to be the tallest structure on earth. The plan had caused an uproar—many Parisians foresaw their city disfigured by an iron monstrosity—but to the vast majority of the public, the tower, like the canal at Panama, was a bold affirmation of French genius, French supremacy in the art of civilization. So for Eiffel to step forth now and join forces with Ferdinand de Lesseps seemed the perfect, brilliant stroke, and the announcement had an especially energizing effect on de Lesseps, whose response to such dramatic turns, even when self-contrived, was invariably infectious.

Eiffel's locks, the gates and operating apparatus for which were to be manufactured in France, were in fact modified versions of the locks he had designed for his Nicaragua plan ten years earlier, prior to the Paris congress. Also, construction of the lock basins was to be delegated to another contractor, a new firm organized by Philippe Bunau-Varilla. Still, Eiffel believed in action and decisiveness. He did move quickly. By January his engineers were on the Isthmus. In another month, excavation for the lock basins was under way.

For all this, however, Panama stock kept falling.

Months slipped by as a new premier, Pierre Tirard, refused even to submit the lottery request to the Chamber of Deputies. "From all information received through other channels than the Company," reported *Économiste Français*, "it is clearly shown that the situation of the undertaking is getting more and more hopeless. . . . The year 1888 will certainly see the liquidation of the Company. The lottery bonds can do nothing towards meeting such necessities."

On March 4, 1888, several deputies introduced a bill permitting the company to go ahead with the lottery. How quickly the Chamber might act on the bill or what its chances were remained open questions. De Lesseps, however, was in the midst of offering still another issue of bonds—this made eight times in eight years that he had gone to the public for money—but with a proviso that they could be converted into lottery bonds should such a bill be passed into law. It appeared that he had been given a saving boost, and as always with every chance bit of good news, he made as much of it as possible.

Even so the sale was a fiasco. Not a third of the bonds were sold. The lottery was no longer the best hope for salvation, it was the last hope. There was no money to be had by conventional means.

The parliamentary process that followed was painfully slow. The Chamber had to vote first on whether to consider the bill, then the bill had to be appraised by a committee of eleven duly appointed deputies. Weeks passed. All signs were that the committee would defeat the bill by a vote of six to five, until one member, a hero of the Franco-Prussian War, Charles François Sans-Leroy, switched his vote at the last minute and the bill was approved by a margin of one. By now it was mid-April. The debate in the Chamber began, with loyal Panama supporters packing the public galleries. Every speaker for the bill was vigorously cheered; anyone who attacked it, or even questioned the idea, was roundly jeered or interrupted, which caused several of the bill's more vocal opponents to grow even more impassioned and bitter. "The ruin is getting on fine!" declared one irate member from the tribune. "Scarcely more than fifty per cent remains to be lost."

On April 28, the Chamber approved the lottery by a wide margin and instantly the company's stock soared. In the Senate, on June 5, the bill was again approved. On June 8, three years after de Lesseps first proposed the lottery, the new law went into effect.

It was none too soon. So desperate had the company's situation grown that Charles de Lesseps had been forced to negotiate a hurried loan of 30,000,000 francs from two of the largest commercial banks, the Crédit Lyonnais and the Société Générale. As security he had been required to put up the Panama Railroad stock. Moreover, the banks had insisted that they handle the sale of a large block of the lottery bonds, in the event the bill passed, for which they were to receive extremely handsome commissions.

It is conceivable that what happened next could have been avoided. Many of those involved would always think so.

By the new law the company was authorized to borrow a total of 720,000,000 francs. Six hundred million francs were to be used to finish the canal according to the lock plan, while the remainder was to go into government securities to guarantee payment on the bonds, as well as the cash prizes.

There was nothing in the law stipulating how the issue should be staged, and at the canal company and the two banks, opinions were sharply divided. Baron Jacques de Reinach argued for scheduling the sales in successive installments, rather than going for everything in one blow. Until then, his position was, the public had supplied the canal company with some 150,000,000 francs a year on the average. If the lure of cash prizes proved strong enough even to double that, then the most that could be expected was 300,000,000 francs, or less than half the amount that was wanted. The safest plan therefore was to try for smaller parts spread out over a year or more.

The banks thought otherwise. Primarily it was Henri Germain, the august founder and chairman of the board of the Crédit Lyonnais, who thought otherwise. The strength of the whole lottery concept, he insisted, was in the size of the cash prizes. Since subdividing the issue would proportionately reduce the prizes, he was against it. "The prudence, the maturity of judgment of the man who endowed France with one of its greatest banking concerns, seemed to exclude the possibility of any error of judgment," Bunau-Varilla would write. That Germain also had the railroad stock in hand, that he had the company by the throat, as a matter of fact, was doubtless of great importance. And it was Charles de Lesseps, not de Reinach, who had responsibility for the final decision.

So the bonds were to go on sale in one gigantic issue—2,000,000 bonds redeemable in 1897 at 400 francs ($80) and priced at 360 francs ($72). The sale would start June 20, run six days, and there were to be lottery drawings every two months thereafter, year after year. Prizes in one year would range from as high as 500,000 francs ($100,000) down to 1,000 francs ($200).

A stupendous publicity campaign was launched at a cost of more than 7,000,000 francs, a figure, like many others, that was not to be divulged until long afterward. The intensity of feeling for and against the company was enormous. Throughout France people talked of little else.

On the morning the bonds went on sale, somebody—his identity remains a mystery—put a telegram on the wire in Paris to every major provincial city, to London and to New York, announcing the death of Ferdinand de Lesseps. There was no truth to it and the company issued an immediate denial, but the damage was done. Two days later, bear raiders dumped Panama shares on the market, driving the stock into sharp decline. Company securities could now be bought on the open market for nearly 100 francs less than the bonds being offered.

Defeat appeared inevitable and it was. Of the 2,000,000 bonds offered, less than half—800,000—were sold. Receipts for the company amounted to 255,000,000 francs, or considerably less than half of what was wanted. The cost of the issue had been enormous—31,000,000 francs—and since the law required that 120,000,000 francs be set aside for interest and for the prize fund, that left the company with little more than 100,000,000 francs.

By all rights, by every established rule of the game, that should have been the end of the Compagnie Universelle. The bubble should have burst then and there and certainly it would have had de Lesseps conceded defeat. His faith, however, was in people. He saw the arithmetic differently. (Édouard Drumont, who was to be his most vicious critic, asserted that the whole problem all along was that de Lesseps never learned how to add and subtract.) While it was true that only 800,000 bonds had been sold, they had been subscribed for by 350,000 persons. That was 350,000 men and women who still believed, who were still behind him and the honor of France. That was also *three times* the number of people who had subscribed to the 1880 stock issue, which had been such an unprecedented demonstration of the "people's capitalism." The bond sale had shown that after eight years, after all the setbacks, the expenditures, the loss of life, his popular following was greater than ever!

In the words of one contemporary English writer, "M. Lesseps soon showed he was not dead, and he was speedily laboring with great energy to repair the effects of the blow. . . ." A less admiring editor of the New York *Tribune* wrote of his "unscrupulous audacity."

He would launch a campaign to sell the remainder of the lottery issue. Again he would take his case to the people of France. Striding onto the platform at a mass meeting of shareholders in August, he was cheered as never before in his career. His mere presence was a magnetic charge; he was inspiring, heartrending, commanding, all these at once; he was the voice of authority, the ageless living emblem of French verve and *grandeur*. The Suez Canal had brought 2,000,000,000 francs to France, he told them; the Panama Canal would bring 3,000,000,000. "All France, it may be said, is joined in the completion of the Panama canal. Actually more than six hundred thousand of our compatriots are directly interested in the rapid success of the enterprise. If each of them will take two lottery bonds or get them sold, the canal is made!"

His call was for one all-out, climactic assault. The proposition was like that of a chain letter: a bonanza only if everyone participates, only so long as no one breaks the spell. To line up new investors, security holders throughout France were organized into hundreds of committees, these supposedly an expression of a huge ground swell of popular support. He and Charles set off on a grueling cross-country tour to Lyon, Nîmes, Marseilles, Bordeaux, to Nantes (where the Eiffel locks were being built), to Lorient, to twenty-six cities in October and November, spending long nights on trains, listening to advice, shaking hands, sitting for photographers, answering questions, exhorting local committee heads, enduring endless formal banquets. It was billed as a lecture tour, but de Lesseps, who turned eighty-three on November 19, was unable any longer to sustain the old exuberance. It was Charles who gave most of the speeches this time, while his father sat on the stage like some national monument. On cue he would rise and add a few words.

The price of the unsold bonds was cut once, twice, three times, to a bargain price of 320 francs. A buyer need only put down 90 francs ($18) and could pay the balance in eight monthly installments.

Only once does the strain of the ordeal seem to have affected the old hero. He and Charles had taken their campaign across to England, for an appearance before the British Association at Bath. Afterward came reports of personal bitterness, reports ignored by the French press but picked up in Panama by the *Star & Herald*. In London he implied that there had been more to his labors than met the eye. He threatened "to publish an account of every step he had been forced to take in the Panama Crusade," steps that had involved certain high-placed figures in the French government.

If it was an attempt to frighten the government at home into saving the company at the last minute, it did not work.

The remaining lottery bonds went on sale November 29. The closing of the issue was set for December 12. The condition of the sale was that unless 400,000 of the bonds were taken, the subscription would be annulled. "I appeal to all Frenchmen," wrote de Lesseps in a final stirring call. "I appeal to all my colleagues whose fortunes are threatened. . . . Your fates are in your own hands. Decide!"

The offices on the Rue Caumartin became the focus of a tremendous emotional tension that reached to every part of France. For hundreds of thousands of people the fate of the company meant the difference between a chance of real security for once in their lives and absolute financial disaster. If the company were to fail, it would indeed be, as Paul Leroy-Beaulieu had said, the largest, most terrible financial collapse on record, a stupendous event historically; but for the vast majority of those who had stood behind Ferdinand de Lesseps all these years, investing their life's savings, often borrowing heavily, mortgaging land, selling off family treasures–jewelry, pictures–to invest more, it would very simply mean a personal disaster of almost unimaginable proportions.

Again bear raiders made an all-out attack and Panama stock plummeted more than 100 points in less than two weeks, from 270 francs to 165 francs. By December 8, lottery bonds on the Bourse were selling for 260 francs, 40 francs less than what de Lesseps was asking.

December 11, the next to the last day of the sale, the large public hall on the ground floor of the company's offices was filled with investors, "flushed and excited," as Emily Crawford wrote, "but willing to stake their last penny on the hope of retrieving their fortunes." They were, she said, like desperate gamblers, their hopes highest at the point when their losses had become greatest. "Strangers met and mutually strengthened their faith with words of comfort. A man who ventured to express doubts as to the possibility of the canal found the place too hot for him." That same day Panama shares on the Bourse closed at 145, down another 115 points in three days.

The next day, the final day of the sale, the scene at the canal offices was even more frenzied, as a still larger, more excited crowd packed into the public hall. At about four in the afternoon, de Lesseps appeared, causing a sensation. A way was cleared and he climbed onto a table at one end of the room. He motioned for silence.

"My friends," he cried out, "the subscription is safe! Our adversaries are confounded! We have no need for the help of financiers! You have saved yourselves by your own exertion! The canal is made!"

Tears were streaming down his face. Then he was being helped from the table and the crowd seemed delirious. People were crying, cheering, embracing one another. He began reaching out for hands. Several women tried to kiss his clothing.

No details were available on the success of the sale. But by nightfall reports swept through the city that more than the required 400,000 bonds had been sold in Paris alone, that Marseilles had taken another 86,000.

The throngs that pressed into the Rue Caumartin the next morning, December 13, had come for a victory celebration. Orders were still pouring in, it was announced when the company opened its doors. The subscription had reached an astounding 800,000. There were repeated shouts for de Lesseps to appear; everyone was pressing toward the table where he had stood the day before. Perhaps an hour went by. Then the figure that entered, moved through the crowd, and climbed onto the table was seen to be Charles de Lesseps.

"Do you wish to see Monsieur de Lesseps?" he called, and there was a roar of approval. ("Yes! We want to see that good Monsieur de Lesseps," said an elderly woman standing beside Mrs. Crawford.)

"My father will always be happy to see you, but I suppose you all wish for some information."

He waited for the cheering to stop.

"We are sitting at an important meeting of directors which I have left for a moment to come here. I do not know what decision may be taken by this meeting, but I am willing to tell you whatever I know."

The crowd was suddenly still. He had barely to raise his voice to be heard.

"I will be perfectly open with you, only do not hold me responsible if you learn anything else tomorrow. If you would like to wait for another hour, I will let you know the full result of our deliberations, but would you rather know at once what I can tell you?"

According to Mrs. Crawford, whose account of the scene is the most vivid available, everybody assented to this. He asked what sort of information they wanted, she wrote, "and being told in reply that they wished to know the result of the subscription, he went on in a deliberate tone."

The subscription, he said, had thus far reached a total of 180,000 bonds. "This being below the minimum fixed by Monsieur de Lesseps, we will commence returning the deposits tomorrow. You see, I am telling you exactly how things are."

People began muttering that . . . yes, this was the best thing; yes, there must be another subscription. It was as if they did not yet understand the meaning of what he had said. Many people were so dazed they stood frozen, saying nothing at all, little expression on their faces. When someone finally asked how the picture could possibly have changed so drastically overnight, Charles answered: "My father is younger in spirit than I. His remarks were made on the strength of a hopeful report that I made to him. The result is bankruptcy or the winding up of the company."

The morning after, December 14, the company suspended payments and petitioned the government for a three-month moratorium on bills and interest, so that a new company could be organized to continue the work. The news was immediately put on the wires and within hours newspapers around the world carried the story of "The Great Canal Crash." At the Palais Bourbon, on the other side of the Seine, the Chamber of Deputies convened at once and the following day, December 15, 1888, turned down the proposal by a vote of 256 to 181. Within hours the appropriate court, the Tribunal Civil of the Department of the Seine, appointed three temporary receivers to administer the company's affairs.

It was a reporter for *Le Figaro*, arriving at de Lesseps' home just ten minutes after the vote in the Chamber, who told him how the vote had gone. De Lesseps turned dreadfully pale, the man wrote afterward, and could only whisper, "It is impossible! It is shameful!"

The pallor and the loss of words were but momentary, however. Instinctively the old reflexes responded. He was in motion again, issuing statements, talking of new schemes. The company was in wreckage, the government had turned its back; the long battle was ended and he had been crushed. It was Sedan again for France, yet he refused to accept that—he was incapable of accepting that.

For his family and friends the next weeks were an agony, as he drove himself and others in a final, hopeless attempt to pick up the pieces and rally his forces. Demonstrations of popular support were staged throughout the country. At one in Paris, in the Palais d'Hiver, a huge ice-skating rink, five thousand of the faithful turned out. "Shall we pledge ourselves, each according to his means, to aid this great enterprise by purchasing new shares of Panama stock?" one speaker had cried and the response was thundering, "*Oui! Oui!*"

H. B. Slaven arrived in Paris and there were daily conferences on the Rue Caumartin. It was announced that the major contractors would continue all essential operations for the time being, working on credit, since any abrupt cessation of the effort would mean the certain ruin of machinery worth millions of francs (machinery all belonging to the company), as well as tremendous damage to the unfinished excavation.

A new company would be launched, de Lesseps said, and in late January new stock actually went on sale. The idea never had a chance, of course; of 60,000 shares, all of 9,000 were sold.

The official end came on February 4, 1889. In accordance with a desire formally expressed by shareholders in the original company, the Tribunal Civil appointed a liquidator. The Compagnie Universelle du Canal Interocéanique was no more.

8

The Secrets of Panama

"What have you done with the money?"
—ÉDOUARD DRUMONT

I

It was nearly three years later when the Panama scandal broke wide open, rocking France to its foundations. Between times, the great Universal Exposition of 1889 had been staged beneath Gustave Eiffel's gargantuan tower, and French political life went along little changed from year to year, one ministry succeeding another, despite the flaming oratory, despite the Boulanger crisis. General Boulanger, "the strong man," having sat out his chance to seize power, having escaped to Brussels with his adored mistress, Madame de Bonnemains, had also, soon after her death, shot himself at her graveside.

Panama, to be sure, had remained a major topic. Some 800,000 French men and women had been directly affected, the savings of entire families had utterly vanished. People who could ill afford to lose anything had lost everything. Still, no panic had been touched off when the company went under. There were no demonstrations in the streets May 15, 1889, the day the liquidator ordered that the work be halted on the Isthmus. Instead, shareholders submitted their grievances by formal petition, in polite, written pleas for redress through government action. Tempers cooled; rumors of fraud and political payoffs were denied or discounted or simply grew stale. When the liquidator established a special committee to go to Panama and estimate the cost of finishing the canal, many shareholders actually took heart, convinced that the government was about to rescue them.

Among those who had been distrustful of the Panama proposition all along and remained clear of it, there was the feeling that such was life for the unwary, that sheep were there to be shorn. For the rest, for nearly everyone as time passed, there was the feeling that Panama was best put behind. The prospect of the tragedy being compounded by a sensational and ruinous scandal was neither anticipated nor desired by the public at large. And very possibly there would never have been an *affaire de Panamá* had it not been for the country's leading anti-Semite, the strange, secretive Édouard Drumont.

Edouard Drumont, a devout Roman Catholic, a professed lover of history, had observed the world about him and concluded that the sickness of modern France—by which he meant France since "*La Débâcle*"—was finance capitalism and that the nation's most treacherous human foe was the Jew. The Jew by his very nature, said Drumont, had no sense of justice, none of the finer sensibilities that made civilization possible. Jews were carriers of disease, born criminals and traitors, who could be recognized by their "crooked nose, the eager fingers, the unpleasant odor." He had said this and much more in *La France Juive* (*Jewish France*), a book of more than a thousand pages that appeared in 1886, that ran to more than a hundred editions and made the author famous—feared, despised, secretly admired—throughout the country.

Drumont was black-haired and spectacled, with a thin, hooked nose, and a thick black beard and mustache that together masked the whole lower half of his face so effectively that his mouth was all but hidden. Many people thought he looked Jewish. His wife was dead; he had no children; he kept his money hidden in secret nooks and crannies about his house. By profession he was a journalist, and until the appearance of *La France Juive*, he had been tremendously frustrated by lack of recognition, having tried his luck as a traveling salesman and a novelist but without success.

That his elephantine tract received such phenomenal attention was in itself a point of fascination, since anti-Semitism had been rare in France. Beyond Paris it was hardly known. Drumont's assertion was that there were 500,000 Jews in France; in reality there were perhaps

80,000. But it was the time of the pogroms in Poland and the Ukraine, and those refugees who had found their way to France, though comparatively few in number, were highly conspicuous and had aroused anxiety among some elements of French society, including French Jews. More important, there was the growing belief that finance capitalism had become a conspiracy, that the country was in the grip of the financiers, and that in the face of such power, the small shopkeeper or the ordinary workingman counted for little.

It was for the Jewish monarchs of finance—the Rothschilds, the Ephrussi—and those Christians who courted the favor of such people that Drumont reserved his worst venom. Saint-Simonianism was declared to be nothing more than a device of the Jews to lift themselves out of the ghetto. The Franco-Prussian War had been engineered by Jews. If examined closely, all failures in modern French society could be traced to the Jewish capitalist system, Drumont asserted.

So with the fall of the Compagnie Universelle—the greatest of ventures thus far in finance capitalism—he naturally began looking into things.

The initial result was another book, *La Dernière Bataille* (*The Last Battle*), a self-styled "history" of *la grande entreprise* that appeared in 1890 and became a runaway best seller. How could a forthright government fail to audit the canal company's books? Drumont demanded. How could the likes of Ferdinand de Lesseps be permitted to walk about a free man? "This evil doer is treated like a hero. The poor devil who breaks a shop window to steal a loaf of bread is dragged . . . before the judge of a criminal court. But into this affair, which has swallowed up almost a billion and a half [francs], there has been no investigation whatever; not once has this man been asked: 'What have you done with the money?' " *Le Grand Français*, he cried, was in fact a great fraud, a cheat and liar, a fountainhead of corruption. He painted a vivid and greatly distorted picture of extravagant luxuries on the Isthmus and took special pleasure in a vicious, personal assault on Jules Dingler. Dingler's house had cost $1,000,000, he insisted. "This man, who seems to have endured heavy afflictions but who was a stranger to every sentiment of justice and humanity, was hated so bitterly that the death of his wife became the occasion of a merry festival. Champagne flowed in torrents . . ." Sixty percent of the workers had died, he claimed. The death count could not have been less than thirty thousand. "The Isthmus has become . . . an immense boneyard. . . ."

Presently, he founded his own newspaper, an illustrated anti-Semitic daily called *La Libre Parole* (*Free Speech*). His chief lieutenants included such individuals as Jacques de Biez, who enjoyed asking priests if it was true that Christ was a Jew ("Drumont doesn't mind," he would say, "but I can't swallow it!"), and the Marquis de Morès, an aristocratic psychopath whose wife was Medora von Hoffman, daughter of a Wall Street banker. On one occasion de Morès, who was a crack shot, had stood in for Drumont in a duel and killed a French Army officer, a Jew who had challenged Drumont as a result of certain insulting remarks about Jewish officers in *La Libre Parole*. Drumont and de Morès were accused of staging the whole episode, as a kind of execution, and de Morès was tried for manslaughter but was acquitted.*

La Libre Parole was no runaway success, however, not, that is, until September of 1892, when Drumont broke a series of sensational stories under the title "The Secrets of Panama," these signed "Micros." The "Jewish plot" at the heart of the Panama tragedy had at last been found and enough else had begun to happen meantime to make the public sit up and take notice.

The chief magistrate of the Paris Court of Appeal, Samuel Périvier, under orders from the Minister of Justice, had appointed a court counselor and an expert accountant to audit the canal company's books and to interrogate various former officers, including Ferdinand and Charles de Lesseps. The charge was fraud and breach of trust.

Ferdinand de Lesseps had made an unforgettable appearance before the court counselor, a man named Henri Prinet. For nearly two years de Lesseps had been living in seclusion. In the early months of 1889, after the company failed, he had tried manfully to assist in the liquidation process, while at home he went through the customary motions as head of the family. From the upper floors of the big house on the Avenue Montaigne, he—all of them—had been able to watch Eiffel's tower rise higher and higher as the spring of 1889 and the opening of

* This same Marquis de Morès had also challenged a young American rancher to a duel not very long before this, when de Morès was establishing himself as a cattle baron in the Bad Lands of Dakota Territory. The American was Theodore Roosevelt, who responded by informing de Morès that he harbored no ill will toward him but that he would be willing to face him if de Morès insisted, whereupon de Morès let the matter drop. In his days in the Bad Lands, de Morès had built a thirty room mansion on a high bluff overlooking the Little Missouri and was known behind his back as "the crazy Frenchman." He lost most of his wife's money in the venture, because of, he asserted, the Jewish-controlled beef trust.

the exposition approached. But by that summer he had become a distant, bewildered old man. In recent months he had been confined to his bed under the care of a physician.

Yet on the day he was to appear before the court counselor, he suddenly revived. It was as if something had clicked on inside the ancient head and suddenly he was himself again. Charles afterward described what happened:

> The doctor . . . expressed the opinion that it would be very imprudent for my father to go out. . . . Nevertheless the meeting had to take place sooner or later; I hoped that one conference would suffice and . . . so I thought it better for him to suffer the shock immediately. . . . My father rose from his bed and . . . said: "I shall go." He dressed, and by the time he reached M. Prinet's he had apparently recovered all his strength; he remained three quarters of an hour . . . and when he left his face radiated charm and energy as it always did under difficulties.

It is not known what de Lesseps said in the interview, other than that he defended his management of the company and the canal itself. On returning home, however, he went straight to his bed and for three weeks he hardly stirred, saying nothing to anyone other than to tell his wife a day or so later that he had had the most horrible dream. "I imagined," he said, "I was summoned before the examining magistrate. It was atrocious."

The homes and offices of company officials, and of Gustave Eiffel, were gone through by police and documents were seized. On January 5, 1892, in response to persistent prodding by a hotheaded young Boulangist named Jules Delahaye, the Chamber of Deputies voted unanimously for "resolute and speedy action" against all those involved in any foul play in the Panama business. Next, the court-appointed accountant who had been auditing the company's books reported that though he found no sign of company officers profiting personally, he nonetheless thought several of them were indictable for misuse of funds and for willfully deceiving the public.

But these were faint tremors compared to the impact of the so-called "Micros" articles in Drumont's paper, the first of which appeared September 10. "Micros," as was later divulged, was a pseudonym for Ferdinand Martin, the banker from Nyon who had organized the first petitions for a lottery in 1885. That great show of popular support, as Martin would testify, had in fact been wholly conceived and organized in the offices on the Rue Caumartin. He had been paid for playing his part, but not enough in his view, and as a result he had had a falling out with Charles de Lesseps the following year. The articles for Drumont were his way, six years later, of evening the score.

The articles charged that twenty members of the Chamber of Deputies and the Senate had been bribed by the canal company to vote for the lottery bond bill and that the prime fixer for the company was Baron Jacques de Reinach. A director of the company, Henri Cottu, had been involved, as had the publisher of *L'Économiste Pratique*, a man named Blanc. The actual cash deliveries had been made by one Léopold-Émile Arton.

Arton, or Aron, as he was also known, was a flashy, out-and-out swindler, a former sales agent for the Société de Dynamite, the explosives trust, who had once unloaded a bad shipment of dynamite on the unsuspecting canal officials and as a reward had been placed in a secretarial position in the Société's head office. Thus established, he had then managed to embezzle somewhere in the range of 4,600,000 francs, and at the moment, with a warrant out for his arrest, he was no longer to be found in France.

Almost daily *La Libre Parole* provided additional bits and pieces, bearing down heavily all the while on Jacques de Reinach. It became the most eagerly read paper in France. Other Paris papers quickly took up the scent. To a powerful figure such as Arthur Meyer, publisher of *Le Gaulois*, a royalist paper, the scandal was the long-awaited chance to topple the Republic, and so every new revelation was given full play, irrespective of the fact that Meyer was himself a Jew.

When *La Cocarde*, another right-wing paper, carried an interview with Charles de Lesseps that appeared to confirm the "Micros" revelations, a new Minister of Justice, Louis Ricard, felt compelled to act. De Reinach, Cottu, Blanc, and Charles de Lesseps were summoned for interrogation. All but de Reinach either denied the allegations or refused to reply; the publisher Blanc was so convincing that he would not be bothered again. De Reinach conceded to the interrogator, Henri Prinet, only that he had given Arton 1,000,000 francs to use for publicity and that he personally had distributed large sums to the press—in excess of 3,000,000 francs—but that he had certainly never bribed anyone in a position of public trust.

Convinced of the baron's guilt, Prinet ordered the commissioner of police to see de Reinach at his home at once and to confiscate all papers relating to his Panama dealings. For some unexplained reason, however, the commissioner allowed several days to pass before doing anything. When he appeared at de Reinach's door in the fashionable Parc Monceau quarter, he was told the baron had gone to the Riviera. So no papers were obtained.

Drumont's charges, meantime, had become far more specific and detailed. It was plain that he had found a new inside source, somebody who knew precisely how the Panama business had been run. With the public outcry mounting, the Ministry of Justice was under tremendous pressure to act. Delahaye and others in the Chamber were demanding a parliamentary investigation. But Premier Émile Loubet and his Cabinet held back, urging prudence, urging patience, hardly daring to make a move, living only day by day and fearing the worst. None could say how many old friends might have been involved, who might be destroyed were the avalanche to let go.

In October the public prosecutor, one Jules Quesnay de Beaurepaire, proposed a civil suit for damages, instead of a criminal trial. He had no wish to see Ferdinand de Lesseps in prison. His purpose, he stated, was to provide some restitution to the ruined shareholders rather than to impose "the sterile penalty of imprisonment upon an octogenarian in his dotage." So now the government had to decide, and at length, on November 15, a special meeting of the Cabinet was convened. Premier Loubet, small and spotlessly groomed, excused himself from any part in the decision, saying it should be the prerogative of the Minister of Justice. (Such deference was a quality hitherto unknown in Premier Loubet.) The decision had already been made, declared the Minister of Justice, Ricard, who was large and fat and had creamy-white side whiskers. He had ordered that criminal charges be brought against Ferdinand de Lesseps and the others thus far implicated.

But on Monday morning, November 21, the morning the summonses were to be served, Paris awoke to the stunning news that Baron Jacques de Reinach had been found dead.

The papers were vague on details. It was known only that when his valet went to wake him as usual at seven Sunday morning, he found on entering the bedroom "a member of the family" who turned and said the baron was dead. The "member of the family," as later disclosed, was the baron's nephew Joseph Reinach (without the *de*), an influential editor, and it was he who reported the death first to Adrien Hebrard, of *Le Temps*.

The rest of the day was one nobody would forget. The summonses were served that morning as scheduled, on Ferdinand and Charles de Lesseps, Henri Cottu, Marius Fontane (former secretary general of the

Compagnie Universelle), and Gustave Eiffel. At the Palais Bourbon the word spread that Delahaye would speak in the afternoon, and when another deputy, seeing Delahaye walk by, stopped him to urge restraint, Delahaye responded, "Do not leave the sitting, there will be a big explosion."

At five o'clock, when Delahaye rose from his place and started down to the speaker's stand, all seats in the semicircular red-plush tiers of the Chamber were taken. Premier Loubet and his ministers were seated down front; the two levels of the public galleries were solid with spectators. Newspaper reporters and the Chamber's veteran silver-chained, frock-coated ushers could recall no moment quite like this one.

Delahaye mounted the eight steps to the rostrum, just below and in front of the president of the Chamber, Charles Floquet. Delahaye, Deputy for Chinon and a member of the extreme right, was known as a "good hater." He was spare and athletic, with sleek black hair and an upswept handlebar mustache nearly the size of a sickle blade. He seldom ever smiled and there was a decided squint in his right eye. It was a face, wrote his fellow deputy and fellow Boulangist Maurice Barrès, that bespoke "inflexible cruelty." Whether Delahaye and Drumont were working together is not entirely clear, but Barrès, among others, would claim they were.

"I would stake here my honor against yours," Delahaye began. He would give no names, but behind the canal company there had been an "evil genius": the directors duped the public, the evil genius duped the directors.

"Name him, name him!" several voices shouted from the packed chamber.

"If you want names, you will vote an inquiry," Delahaye answered.

He charged that 3,000,000 francs had been distributed in the Chamber, that 150 deputies had been bought. He had seen the list. At once there was a violent uproar. "The names! The names!"

There were only two kinds of deputies, Delahaye exclaimed, those who took the money and those who did not. Floquet, who was also a former premier, was now on his feet directly behind Delahaye. "You cannot come into this house and accuse the entire body," he thundered down at the Deputy. Again there were angry cries for names.

"Vote the inquiry," Delahaye shouted.

When he started for his seat the great room was in wild disorder. He was hissed at; deputies banged their desks to add to the uproar. The

Premier next ascended the steps to the tribune but for several minutes was unable to speak against the noise. Such irresponsible charges, he cried, stemmed solely from uncontrolled political passions—that is, the old Boulangist faction was trying to destroy the Republic. Assuredly, light must be shed on so grave a matter; of course his government would hide nothing.

The Chamber voted the inquiry. A committee of thirty-three members was named and the debate raged on. At another sitting of the Chamber, two days later, one honored member became so excited he collapsed and had to be carried out. Men wept as accusations were hurled from the tribune. Fist fights broke out in the aisles.

Much of the country and the foreign press could scarcely believe what was happening. George Smalley, star reporter of the Battle of Antietam, who was now London correspondent for the New York *Tribune*, cabled his home office that as fraught with recklessness and venality as the Panama business must have been, "we can with difficulty be induced to believe that it has utterly debauched public life in France."

There were wild rumors concerning de Reinach, the evil genius of Delahaye's speech, whose body had been taken from Paris and buried immediately after the required twenty-four-hour delay. It was said that he had taken poison. (*Le Gaulois*, the smart society paper, described his final agony in such exquisite detail that it was as if someone from the paper had actually been in the bedroom.) It was said that he had been poisoned by someone, that he had been murdered in his sleep. It was said that he was alive and out of the country. The coffin was empty, exclaimed one deputy in a speech.

A family physician had attributed the death to "cerebral congestion"; a regular certificate of death had been signed by a city doctor. Still the Chamber rang with cries for an autopsy. The critical issue had been found; the line had been drawn. And the Ministry of Justice held fast. The baron's body, Louis Ricard announced, could be exhumed only if there was a clear suspicion of murder, and having no such suspicion, he refused to step beyond the law.

But it was too late; it was, as Philippe Bunau-Varilla would write, "the beginning of a convulsion." On November 28 the Chamber voted overwhelmingly to proceed with an inquiry into Jacques de Reinach's death. And with that the Loubet government fell.

The week following, the president of the Republic, Sadi Carnot, called upon Alexandre Ribot, Loubet's foreign minister, to form a new government.

II

No one ever got to the bottom of the Panama Affair and no one ever will. The Chamber's own committee of inquiry, the much-publicized Committee of Thirty-three, held 63 sessions; it received 158 depositions, compiled more than 1,000 individual dossiers. Its final report fills three ponderous volumes. But time and again the fact-finding stopped short of facts that might prove too embarrassing or destructive. Old colleagues were protected. Barrès, by no means an impartial observer, but a keen judge of human nature, wrote, "The committee fell into the mistakes of all inexperienced courts of inquiry. They were unduly affected by the skillful emotional show of a lot of sly old sinners who appeared before them."

De Reinach's lips were permanently sealed of course. His death, the pivotal event in the unfolding story, would also remain one of the most puzzling of several unsolved mysteries.

The coffin was exhumed four days after the new government took office. A large party of physicians, police, and newspaper people went out to the baron's country place at Nivillers and stood about in a snow-covered cemetery to witness the unearthing. The body was in the coffin, but the autopsy proved nothing since by then the vital organs had so decomposed that the cause of death was impossible to determine.

No sooner had the mutilated remains been returned to the grave than the fundamental design of the scandal—as then perceived—was drastically altered by the introduction of an entirely new character of even more sinister cast. To the astonishment of everyone, de Reinach was now revealed to have been but nominally the villain of the piece. At once the affair became more complex, more fascinating, and far more sensational politically, because now it implicated the most formidable figure in French public life, Georges Clemenceau.

The new character in the plot had been involved since the time when the Türr Syndicate had been organized. He was Cornelius Herz —known to his associates as "*Le Docteur*"—and to the decided satisfaction of Édouard Drumont and the rapidly growing numbers who saw things as Drumont did, he was both a foreigner and a Jew. He was unknown to the public and to most of the press when his name first turned up. The papers could report only that he was an American who somehow or other had been awarded the Grand Cross of the Legion of Honor and who had a standing in the electrical industry.

The rest of his story came later, pieced together by reporters in

Paris and San Francisco, and seldom had a more perfect charlatan been discovered. The man was shrewd, daring, utterly charming when it suited his purpose. Though he had attained his initial fortune and social acceptance as a physician and "man of science," the best evidence was that he had neither a medical degree nor little, if any, substantive scientific knowledge.

Herz was born at Besançon, in eastern France, in 1845, which made him forty-seven at the time his name leaped into print. His parents were German Jews who, when he was three years old, emigrated to New York, where the father became a packing-box manufacturer. At thirteen Cornelius entered the College of the City of New York; in 1864 he was graduated at the bottom of his class. Presently he returned to France to study medicine, and by the time of the Franco-Prussian War he had acquired enough background—or said he had—to qualify as an assistant surgeon in the French Army.

After the war he went to Chicago, where his parents had relocated. Later he turned up on the staff of Mount Sinai Hospital in New York, but in a few months he was on the move again. Officials of the hospital had begun investigating his background and found, in the words of their report, that he had "very ingeniously avoided taking the examination for house physician and surgeon," and that "his supposed graduation from the University of Paris was fictitious." Whenever asked for his diploma, Herz said it had been destroyed in the Chicago fire.

He arrived in San Francisco with one volume of medical terminology (a twenty-six-year-old edition of John Mayne's *Dispensatory and Therapeutical Remembrancer*), a few hundred dollars, and a dark-haired American wife, Bianca Saroni. Presenting himself as a specialist in diseases of the brain, he opened an office filled with electrical gadgetry. "He was a man of the world, apparently well equipped for his profession," the San Francisco *Evening Bulletin* would later report, "yet with a sanguine, sky-scraping temperament that made him soar above men and to seek wonderful and world-stripping achievements. A dozen valuable inventions were his—a hundred marvelous scientific processes were to be worked out by his genius."

His practice grew rapidly until 1877 when, with his wife and two infant daughters, he suddenly departed for France. As later disclosed, he had gone off with several checks from a retired brewer, a former patient, amounting to $80,000, notes the brewer had made out to Herz when Herz had him under hypnosis. In New York, as his ship was about to sail, Herz cashed a check from another patient, this for

$30,000, which supposedly he was to invest in a French electrical scheme. A San Francisco electrician had been taken for $13,000; a physician who had been in a partnership with Herz was out $20,000.

In Paris he became known as the successful young American with large plans. Edison and Alexander Graham Bell were personal friends, he said. Some of Edison's ideas were actually his own. He founded a respected scientific and industrial review, *La Lumière Électrique*. He established telephone service between Paris and Versailles, invested in a variety of speculative, quasi-scientific ventures, including the Türr-Wyse scheme. Within an amazingly short time he seemed to know everybody who counted—Charles de Freycinet (four times premier), Hebrard of *Le Temps* (who introduced him to Clemenceau), Emily Crawford, Ferdinand de Lesseps, Boulanger, President Jules Grévy.

De Freycinet had been the one who had arranged for him to receive the Grand Cross of the Legion of Honor. The decoration, as de Freycinet was to explain subsequently, and to the satisfaction of very few, had been conferred at the request of the scientific community.

In Emily Crawford's phrase, Herz understood "the inner lines" of French politics and by far his most valuable friendship was with Clemenceau, with whom he had much in common. Clemenceau too had begun his career as a physician; Clemenceau had lived in the United States and married an American girl. It was a friendship that Herz knew how to advertise and that Clemenceau would later find expedient to forget. Herz supplied the money to launch Clemenceau's newspaper, *La Justice*, and put still more into the paper as time went on, perhaps as much as 2,000,000 francs. At one point Clemenceau even appointed Herz the guardian of his children in the event of his death.

Clemenceau, the impassioned republican, the fiery voice of "*revanche*," editor, atheist, teetotaler, the most aggressive orator in the Chamber of Deputies, occupied a unique position: he was feared by everybody—on the left, extreme left (his own Radical party), center, right, extreme right. Harsh, even ruthless, brilliant, he made other politicians seem dull-witted, flabby in spirit. He was the "Tiger," and with his taut physique, his bristling eyebrows and yellowish complexion, he looked the part.

As for Herz, he was the most ordinary-looking of men. "Everything is ranged against me," he is said to have once remarked, "even my own appearance." Anyone trying to remember him at the scene of a crime would have had trouble thinking of a single distinguishing feature. There was a full mustache, a mustache of the kind then being worn by most every male Parisian past the age of twenty. There was a round face, a generally bland expression, a thinning hairline. He was short, a bit overweight; he could have been any of a dozen men on any busy street in the city. Only the eyes, it is said, suggested the energy and cunning within.

The first public mention of Herz in connection with the Panama scandal came as a result of an appearance before the Deputies committee by a banker who had once been an employee of the firm of Kohn, de Reinach. Anthony Thierrée said that in July 1888 de Reinach had deposited with him a single check issued by the canal company for 3,390,000 francs. In return de Reinach had asked for twenty-six separate checks of differing amounts equivalent to the same sum. The checks were made out to "Bearer." It had seemed a perfectly normal transaction. Though the banker first claimed to have no records of the checks, the stubs were soon confiscated from the bank's vault and on all but one were the plainly written initials or first several letters in the names of prominent recipients, as well as the amounts of the individual checks.

The stubs were the first solid evidence in the case and they caused a sensation. The checks were to be traced through banks all over Europe. The sums made out to various government officials and legislators ranged from 20,000 to 195,000 francs (from $4,000 to $39,000). The recipients included such personages as Senator Albert Grévy, brother of the former president; Senator Léon Renault, one of the most esteemed legal minds in France; and a deceased Minister of Agriculture. The single illegible stub was for 80,000 francs and the identity of the recipient became another of those tantalizing mysteries that would sell enormous numbers of newspapers and keep the rumors flying in the marble corridors of the Palais Bourbon.

But the largest amount—two checks for an even 1,000,000 francs each, or $400,000—had gone without any explanation to Cornelius Herz, who, it was now learned, had removed himself to London.

A few days after the committee obtained the check stubs, a lawyer for Herz, a dandified and rather notorious former Paris prefect of police, Louis Andrieux, produced for the committee a photograph of a note containing a list of names, a photograph obtained from Herz. The names were the same as those on the stubs, but the note referred to still another check for more than 1,000,000 francs that had been broken down by the unsavory Léopold-Émile Arton and delivered to 104 members of the Chamber of Deputies. The two largest payoffs, the photograph further disclosed, a check for 250,000 francs and another for 300,000 francs had gone to none other than Charles Floquet and to Deputy Charles Sans-Leroy, the man whose last-minute vote had cleared the lottery bill for action in the Chamber.

The note appeared to corroborate everything that Jules Delahaye had charged in his momentous attack. Overnight the public fastened on to one expression, one tangible, understandable image—the check-taker, the *chéquard*. A new word had been added to the French language. "That fatal word became the topic of every song, gibe, anecdote, and demonstration," wrote a contemporary chronicler. "All along the boulevards itinerant vendors sold songs and broadsheets, 'Who Hasn't Had His Little Check,' . . . Comedians, cabaret singers, and everyone else found the *chèque* a mine of inexhaustible satire."

As the committee pressed on with its investigation, and later, when the actual trials got under way with witnesses testifying under oath, the picture of what had been going on began to change dramatically. Bit by bit Herz emerged, like some crucial but long-concealed figure discovered beneath the surface of a familiar painting. For all those absorbed in the scandal—for just about all of France—it would be impossible thereafter to think of Panama and not think of Cornelius Herz.

There was testimony to the effect that Charles de Freycinet had extracted campaign contributions from the Panama company. A deputy who had served with Charles Sans-Leroy on the lottery-bill committee said he too had been offered money—100,000 francs—to vote for the bill, by a professed emissary of Charles de Lesseps, and that when he refused, the offer had been tripled. Charles de Lesseps denied any knowledge of such an offer, but when it was charged that former Minister of Public Works Baïhaut ("the man with the beautiful wife") had been paid 1,000,000 francs to give the lottery bill his support, Charles de Lesseps made no denial.

Adrien Hebrard admitted that he had entered into a secret partnership with Gustave Eiffel, in Eiffel's bid for the Panama contract. Eiffel, moreover, had paid Hebrard 2,000,000 francs and had paid de Reinach nearly the same amount for their influence.

Then Andrieux, Herz's lawyer, dropped the astounding news that Drumont's inside source in the weeks following the "Micros" articles had been none other than de Reinach. Indeed, had it not been for de

Reinach, Drumont's despised Jew, the paper would have run short of material. The bargain had been that Drumont would keep de Reinach out of the columns of the paper so long as de Reinach kept Drumont supplied with incriminating Panama details. Andrieux knew because he had been the go-between.

Drumont, called to appear as a witness, happened to be in jail—he had at last been convicted on a libel charge—and he refused to appear unless his sentence was suspended, a proposition the courts refused.

But the thing that was most perplexing to the investigators was the gathering evidence that de Reinach, supposedly the arch crook in the plot, had been keeping nothing for himself. He who was bleeding the canal company of millions—to "subsidize" the press ostensibly, but in fact to pay off a great many others as well—he who was getting regular kickbacks from several other Panama contractors besides Eiffel, was in turn being bled by someone else. And the someone was Herz.

Herz and the baron had known each other since Herz first arrived in Paris from San Francisco. De Reinach had been instrumental in obtaining government contracts for various Herz enterprises. It was de Reinach who had brought Herz into the arrangement with Istvan Türr and Lieutenant Wyse. But at no time thereafter, not once in all the years the canal was being attempted, had Herz lifted a finger to help the Compagnie Universelle. He had never performed a single identifiable service for the company—legal or illegal—to warrant compensation; yet, as near as could be figured, he had received millions of francs through de Reinach, possibly as much as 10,000,000 francs, or $2,000,000. In addition, the company had made at least one direct payment to Herz of 600,000 francs.

The arrangement had been this. At the time of Ferdinand de Lesseps' first campaign for a lottery issue, Herz had persuaded the old man and Charles to advance him the 600,000 francs for the good he could do them politically—the expenditure was entered in the company's books under "publicity"—and to agree to a 10,000,000-franc payoff if and when a lottery bill was passed. The elder de Lesseps had even blithely agreed to give Herz a written copy of the secret agreement, which put the company pretty much at Herz's mercy until 1887 when Charles recovered the incriminating paper and burned it.

Apparently the final 10,000,000-franc payment was to have been made through de Reinach, who was to take a cut for himself in repayment for money Herz owed *him*. But Herz had suddenly turned the tables on de Reinach. When the lottery bill was passed in June of 1888, the company refused to give de Reinach the 10,000,000 francs to give to Herz, the company's position being that Herz had done nothing and so deserved nothing. Herz, who was then in Frankfurt, immediately informed the directors by telegram that de Reinach must "pay or be destroyed . . . I shall wreck everything rather than be robbed of a single centime; take warning, the time is short."

So it was then that the directors turned over to de Reinach the check for 3,390,000 francs, which de Reinach took to the bank and converted into the various checks to "Bearer," the two largest of which went to Herz. From that point on Herz had been blackmailing de Reinach, and thereby the canal company, for everything he could get.

What hold Herz had on de Reinach remained obscure, although there were innumerable theories concerning various dark secrets in de Reinach's past, things so dreadful that he had been willing to go to any lengths to keep Herz silent. A favorite theory was that de Reinach had committed treason in order to advance himself socially or financially—the sale of state secrets to Italy possibly, or to the British Foreign Office—and that Herz had made it his business to know the details.

At one point de Reinach had threatened to expose Herz if he did not leave him alone, but it was a hollow threat. At another time, according to Louis Andrieux, de Reinach became so desperate that he hired an assassin to do away with Herz, a charge that was never substantiated or denied.

The story of de Reinach's last night alive was revealed by degrees, beginning with an article in *Le Figaro*. On hearing of Ricard's decision to proceed with criminal charges, de Reinach had rushed back from the Riviera, and on the fateful evening of November 19 he had gone to see Herz at Herz's home near the Bois. But he had not gone to see Herz alone, the paper reported. There had been two others with him, Finance Minister Maurice Rouvier and Georges Clemenceau, whose mission was to plead for mercy in de Reinach's behalf. When this side of the story broke, the response in the Chamber, and in the cafés, was a host of new theories about de Reinach's death: he had been poisoned by Rouvier, by Clemenceau, by Herz.

Rouvier resigned at once. It was true that he had gone with de Reinach, he told his stunned colleagues in an emotional speech. It had seemed the humane thing to do; de Reinach was being driven insane by the newspapers and he thought Herz had the power to silence the attacks. The explanation satisfied almost no one, however, and the fall of a figure of Rouvier's stature—he was the one brilliant financial expert in the government, as well as a former premier—sent shock waves through Paris and set off a panic on the Bourse.

Clemenceau told a similar story and without so much as a trace of remorse. It was at *Rouvier*'s urging, he said, that he had agreed to go along that night; *Rouvier* had thought it important. The conference in Herz's study had been brief. De Reinach had been extremely agitated, "his face crimson and his eyes popping out of his head." Herz had said it was too late to silence the papers; he considered the subject closed. Herz was perfectly cool and controlled, Clemenceau recalled, and wholly unsympathetic.

Later in the evening de Reinach went alone to see his nephew Joseph, then, still later, to see two young women whom he kept in respectable style in the Rue Marbeuf. Apparently he reached his own home about one in the morning and was found dead six hours later. The consensus in Paris was that he had taken poison and probably that is what happened.

It was Adrien Hebrard who informed Herz of the death, immediately after Hebrard had been informed by Joseph Reinach. So Herz knew before almost anyone and Herz departed for London that afternoon—that is, the afternoon of the Sunday before the arrests were made. On his way to the station he called on Andrieux to report that Joseph Reinach was "at work" on his uncle's papers and could be trusted to leave nothing embarrassing lying about.

Herz was cause for torrents of invective in the Chamber, naturally enough. But Herz was out of the country and would stay there presumably, so in a large sense Herz was a straw man. The real target was Clemenceau.

How had such an individual as Herz, a foreigner, a quack doctor, risen so far so fast? In back of de Reinach was Herz, but who was in back of Herz? Who was the powerful, unseen patron of "this little German Jew"?

"Now, this indefatigable and devoted intermediary, so active and so dangerous, you all know him, his name is on all lips," exclaimed Deputy Paul Déroulède, self-appointed superpatriot and fanatic Boulangist. "But all the same not one of you would name him, for there are three things that you fear, his sword, his pistol, his tongue. Well, I brave all three and I name him: it is . . . Monsieur Clemenceau."

Instantly the Chamber became a shouting mob. One deputy jumped up and cried that Herz was an agent of the Foreign Office, a British spy. Yes, yes, Déroulède proclaimed, and Clemenceau had been his colleague.

Clemenceau responded with cold fury and looking directly at his accuser proclaimed him a liar: "*Monsieur Paul Déroulède, vous en avez menti.*"

The inevitable duel followed, in a paddock at the Saint Ouen race track outside Paris on a chill, gray afternoon a few days before Christmas. Déroulède, tall and somber, stood bareheaded; Clemenceau, who was known as an expert marksman, kept his hat on. It was agreed in advance that photographers could be present and that dying words would be faithfully recorded. They each fired three times; they each missed three times. After the final volley, Déroulède was seen to examine a corner of his coat with "an appearance of apprehension."

So by the year's end, in the less than four months since Édouard Drumont commenced his disclosures in *La Libre Parole*, a government had fallen; three former premiers had been named in the plot, along with two former ministers and two prominent senators; more than a hundred deputies or former deputies stood accused of taking payoffs; there had been one probable suicide, a panic on the Bourse, a much-publicized duel. The sinister Herz had become a subject of worldwide fascination and there was the growing conviction that France had been the victim of a diabolic conspiracy.

About the only missing ingredient was a *femme fatale* and the suggestion of prominent government officials involved in illicit pleasures of the flesh. But then that also was added shortly by the Paris correspondent for *The New York Times*. In a long dispatch on the front page of the *Times*, January 15, 1893, appears a reference to the dazzling courtesan Léonide LeBlanc. It is not very much, only enough to imply a great deal, and there is no way either to verify it or to enlarge upon it. According to the *Times* report, "Her house in Paris was the center of the whole Panama intrigue and at her dinners these incriminated ministers, deputies, and editors met the cashiers of the rotten enterprise. She herself feathered her nest luxuriously out of the haul. . . ." She was also, according to the *Times*, "so braided up on every detail" of all that had gone on that nobody would dare say a word about her, and especially no one in the Chamber, "since more than half its members were guests under her roof [in] those lavish, hospitable days." And as it turned out, no more was said of her.

III

Ferdinand de Lesseps remained unaware of events unfolding in Paris. Though somewhat improved physically, he passed his days seemingly oblivious of the world at large, wanting only peace. He had been removed from the city to the seclusion of his country place, where Madame de Lesseps made certain that he saw no one who by some slip might give things away. Bundled into a double-breasted seaman's jacket, a smoking cap on his head and a fur-lined robe over his knees, he spent hours staring into a log fire. Only on New Year's Day when Charles failed to appear for the traditional family gathering did he become suspicious, demanding to know what had happened. He was told and the effect was devastating; but then he lapsed into the slow, silent decline from which he was not to recover.

The arrest at Charles's apartment on the Avenue Montaigne had been handled with such discretion by the police that his wife was unaware of their presence. But like Marius Fontane, Cottu, and former Deputy Sans-Leroy, Charles was to be treated thereafter as if he were a dangerous public enemy. They were taken to the Mazas, the old metropolitan jail opposite the Gare de Lyon, and were put in separate cells. No visitors were permitted, other than their attorneys; the few times they were taken from the jail to testify in advance of the trials, they went handcuffed and under heavy guard, riding in the sort of van used to transport common criminals.

For Charles the ordeal was to be a long one. He would be tried twice, in two separate courts on different charges. The first trial was for fraud and maladministration, for "fraudulent maneuvers to induce belief in unreal schemes, and to raise imaginary hopes of the realization of a chimerical event." The second trial was for corruption of public officials—political bribery. And since the elder de Lesseps was to be excused from appearing, because of his health and advanced age, Charles would bear the entire weight of the defense.

The first trial commenced in the Paris Court of Appeal on January 10, 1893, and lasted four weeks, two days.* On the eve of the first session the mood of the city was strange and unsettling. People by now, people of every political hue, were openly questioning the entire structure of French society. An atmosphere of general distrust per-

* Since Ferdinand de Lesseps was a Grand Officer of the Legion of Honor, he could be tried only in an appeal court—that is, without a jury—and this meant that the others accused had to be tried in the same court.

vaded the whole of France. "The situation in Paris grows more ominous day by day," reported Smalley of the *Tribune*. There was talk of a royalist coup; military units near the city were kept on alert. The Italian ambassador informed his government that France was on the brink of revolution.

"Panama" had become a universal term of abuse, and, for many, a battle cry. On the night of January 6, *La Libre Parole* had staged a large anti-Semitic rally at the Tivoli Vauxhall. Jews had created Panama, exclaimed the main speaker, the Marquis de Morès, and it was the Jews who were rejoicing now at the ruin of French honor. When several hundred spectators rose in angry protest (as the rest of the audience cheered wildly), a riot broke out. Chairs were smashed; people were beaten to the floor and trampled.

This was not the "real France," the still-incredulous Smalley cabled New York. "The real France is the France of M. Pasteur," he urged his readers to bear in mind. ". . . It is the France of Baron Alphonse de Rothschild, who makes the new year welcome to the poor by his gift of a million in charity. It is the France not of Panama but of the French who rejoice to think themselves . . . the one great republic of Europe."

Large crowds gathered outside the beautifully wrought iron gates of the huge, gray Palais de Justice and an unmistakable air of apprehension filled the courtroom from the moment the prisoners were brought in. Périvier, the chief magistrate, and the four other judges and the prosecuting attorney entered in black robes and round gold-encrusted black caps. They were seated, the caps were placed on the benches before them, pencils were picked up, Périvier nodded to the clerk who then read the indictments.

Charles, the first witness, rose from the prisoners' benches, a line of four folding chairs placed directly in front of the judges. Behind him, also standing now, was his attorney, Henri Barboux, who, like all others of his profession present, wore the traditional black gown and white, starched bib and collar of the *avocat*. Charles appeared in good health, even "full of energy," despite a month in solitary confinement, but he would remain noticeably circumspect, even a shade pompous. He showed, as one reporter observed, little of the "élan that would have made him a more sympathetic figure."

The others took their turn in due course: the gray, bookish Fontane, who appeared frightened and maintained that he had done only what he had been told; the impeccable Cottu, who nervously twisted a black mustache; Gustave Eiffel, vigorous in speech, handsome, near-sighted, whose lawyer was Pierre Waldeck-Rousseau, a future premier of France. Eiffel, though not imprisoned like the others, had been charged with making ill-gotten profits from a contract by which he had been guaranteed, among other things, an enormous cash advance and an enormous cash indemnity should the company fail to provide the necessary machinery for his locks within a ridiculously brief time. In the aggregate (as estimated by the Deputies committee), the company had paid out more than 74,000,000 francs to Eiffel, from which his profit had been tremendous. As an example, the indemnity paid on account of undelivered lock machinery was 18,000,000 francs, whereas his own outlay for such equipment amounted to less than 2,000,000 francs.

It would be said in Eiffel's defense, however, even in the committee's report, that the famous engineer had done nothing dishonest himself; he had merely had the advantage of a "curious" contract. Moreover, had the entire project been completed, the locks built, his profit would have diminished greatly.

In reply to questions from the bench, Eiffel said he had been deeply touched by Ferdinand de Lesseps' call for his help, but that it was with de Reinach that he had entered into negotiations. "Oh, I hesitated," Eiffel recalled. "I hesitated a long time."

A decrepit old man from Nîmes, a witness for the prosecution, who was deaf and nearly blind, gave a moving description of how he had invested the last of his savings in Panama upon seeing Ferdinand de Lesseps at one of the lottery bond rallies. Then four other similar "victims" were brought in by the prosecution, and the last of them, after telling the court he had been ruined, recalled that Ferdinand de Lesseps had personally counseled him to hang on to his Panama shares. "It is terrible," remarked the chief magistrate. "Everybody here pities you."

No witness questioned the good faith of Charles. Armand Rousseau pointedly dismissed all thought that Charles could have harbored fraudulent intentions. And since the political side of the scandal was not at issue in this trial and no political celebrities were paraded in to testify, Charles remained the central figure. His intention from the beginning, he said, had been to serve his father to his best ability; he was proud to have played that role and to have stayed at his father's side to the end. He affirmed that Baïhaut, who also had been arrested by this time to await criminal prosecution, was paid to introduce the lottery bill and that de Reinach had been too.

"You wasted the millions of your stockholders intentionally," charged the prosecutor, a bald, fierce-looking man named Ráu.

"With as much intention as one hands over one's watch at pistol point," Charles answered.

"And how did you understand that Baron de Reinach used these enormous sums?" Judge Périvier broke in.

"In remunerating financiers and, without doubt, senators, deputies, and ministers. Others also assisted de Reinach," Charles said. He had taken care not to ask what was done with the money.

"That is, you gave them the dirty job which you preferred not to do yourself, but provided them with the means of doing!"

Charles made no reply and as the trial wore on he grew increasingly solemn and diffident. At times his voice was so low that only those in the first rows could hear him.

The prosecutor talked of "the greatest fraud of modern times." Addressing the bench he exclaimed, "You will not hesitate to punish these criminals, who in order to attract millions have had recourse to every maneuver, every fraud . . . I demand the most stringent application of the law."

The plea for Ferdinand and Charles de Lesseps, delivered by the small, white-haired Henri Barboux, was a classic example of nineteenth-century courtroom oratory lasting the better part of two days. With rolling phrase and mounting emotion, he called upon the court to recall just who Ferdinand de Lesseps was, what he had done in his life, what he had meant to France, what he still meant in the hearts of millions, including those very people who had suffered most from the Panama failure. Hours passed, marked by a large, gilt-edged clock high on one wall. The longer he spoke, the greater the tension became. Imagine the response, he said, were the old hero to enter the room at this very moment to speak in his own defense, and much of the audience appeared to catch its breath, half expecting the rear doors to swing open at any instant.

How could his clients be charged with fraud when they had taken nothing to enrich themselves?* If they had miscalculated the expense

* Neither de Lesseps was a man of wealth, as attorney Barboux would substantiate for the court and as time would bear out. Charles's total assets amounted to less than 400,000 francs, less than $80,000. And while it was true that Ferdinand de Lesseps had at one point sold his Panama founders' shares for 1,400,000 francs, he had invested 1,778,000 francs, including part of his wife's savings, in the canal, all of which was lost. So he too had suffered financially from the collapse. His avowed disinterest in ever making money from the venture was genuine; he could have, several times along the way, just as at Suez, but he had not.

of the undertaking, what great projects had ever cost what was originally estimated? His clients had never been alone in their faith that the great enterprise would succeed. The one sin *Le Grand Français* might be found guilty of was excessive optimism. But it was only the optimist who succeeded in this world. Pessimists were never anything but spectators.

The Court of Appeal passed judgment on the morning of February 9, the room "packed to suffocation." The five defendants were found guilty as charged and the sentences were unexpectedly severe. Ferdinand and Charles de Lesseps were each sentenced to five years in prison and fined 3,000 francs. Fontane and Cottu were sentenced to two years and also fined 3,000 francs. Eiffel, though acquitted of complicity in swindling, was found guilty of misusing funds entrusted to him. He was sentenced to two years and fined 20,000 francs.

Several days later, in the evening, and escorted by two police officers, Charles left the Mazas prison on a special pass. Ferdinand de Lesseps had been demanding to see his son, whom he accused of abandoning him. His anxiety was such that his doctor had become extremely concerned.

Charles, his wife, and the two policemen arrived at the little railroad station nearest La Chesnaye well after midnight. A carriage was waiting and they drove another fifteen miles through the dark countryside, arriving at the château around three in the morning. At first light Charles went in to see his father.

He found the old man awake but still in bed. On the bedside table were a number of Paris papers, all of them a year or more out of date.

"Good morning, Father. I have been able to leave my work and here I am."

"Ah, Charles," the old man responded. "Is there nothing new in Paris?" Then he kissed his son, repeating simply, "Ah, Charles! Ah, Charles!"

When the family gathered for lunch in the huge dining room with its great, carved seventeenth-century buffets, Ferdinand de Lesseps entered slowly with a cane and took his customary place at the head of the table. No one explained to him who the policeman was sitting beside Charles, and he never asked. Once he smiled at his son and seemed about to say something.

In the afternoon Charles and the family and the two policemen took a long walk; then, after dark, the prisoner was returned to Paris.

• •

Charles was a different man after that. It would be said that his manner during the first trial had been carefully contrived, the strategy being that discretion, even silence on certain matters, would be rewarded with a light sentence. There may be some truth to the interpretation. However, it does not seem quite in character. More likely the months in prison, the personal humiliation of the trial, and the visit to La Chesnaye had been a momentous inward journey from which he had returned with a profoundly different view of himself and his responsibilities.

His deportment during the second trial was still that of the perfect gentleman, only now there was an unmistakable edge of outrage; that and the utter composure—even ease—of a man with little more to lose, with no one left to protect, who has decided quite literally to have his day in court.

Again the setting was the Palais de Justice, but now in the larger Court of Assizes and before a jury. Charles and Fontane were charged with distributing bribes; Baïhaut, Sans-Leroy, three relatively unknown deputies, one inconsequential senator, and a go-between named Blondin, a former employee of the Crédit Lyonnais, were charged with accepting them. Charges against those more celebrated figures implicated thus far—Floquet, de Freycinet, Senators Grévy and Renault, various deputies of note—had all been dropped for lack of evidence. So the important politicians were not to be tried. Nor was Ferdinand de Lesseps.

The payoff artist Arton was another absent party and a dreadful embarrassment to the police, for in recent weeks Arton had led two Paris detectives, one an aide to the head of the Sûreté, on a wild chase across half of Europe, from Budapest to Bucharest to Jassy to Nuremberg to Prague to Magdeburg to Hannover, then back to Budapest, where the detectives lost the trail. Where he was now nobody could say.

Cornelius Herz had also been added to the list of accused this time. But he was sequestered in a small seaside hotel at Bournemouth, England, refusing to return to Paris on the grounds that he, like the elder de Lesseps, was too ill to travel. Nor did there appear to be much that could be done. To secure total privacy, Herz had rented the entire hotel (no great problem at that season of the year). Two men from Scotland Yard sent to arrest him on an extradition warrant affirmed that he was indeed an extremely sick man. He was placed under house arrest until several eminent London physicians, including the queen's own Dr.

Russell Reynolds, could be dispatched to Bournemouth to give an opinion. To a man, they certified the gravity of Herz's condition.

But what if Herz was putting on an act? What if, in fact, he was an agent for the Foreign Office, as had been charged, and so the whole thing was an act?

Paris was filled with such conjecture, and the same questions come readily enough to mind even now. But if it was an act, then Herz must have been something of a theatrical genius or the new government of Premier Alexandre Ribot must have been in on it as well; for a succession of French physicians came next to Bournemouth at the invitation of the Foreign Office and they too reported the patient to be in a ghastly state, and mentally as well as physically.

So Herz, still under arrest, remained in his suite overlooking the English sea. No one could see him, no one could get him to divulge a word—not until Emily Crawford appeared and talked her way through the guarded front door and up to Herz's bedroom. The interview was the only one Herz was ever to permit and it raised questions concerning the Panama Affair that never were to be answered.

Herz, she reported, was suffering from Bright's disease (an acute inflammation of the kidneys), complicated by an unnamed "malady of the nervous system." He would never come out of the hotel alive she prophesied.

She found him in bed, propped up with pillows and covered with furs and blankets. She was shocked by the change in him, but "the light hazel eye had lost none of its electrical brilliance . . . The clearness and vigor with which he expressed himself was amazing."

He told her that de Reinach had been involved in a vast European intrigue, the object of which was a "readjustment" of the alliances that then bound the central powers and to "fill the pockets" of a syndicate of politicians who were working under the direction of de Reinach.

> They were to have divided among themselves a tremendously big sum which was to have been obtained as a commission on a state loan issued in Paris under the auspices of M. Rouvier and by means of "virements," or transfers of credits voted by Parliament [the Chamber of Deputies] from the War and Public Works Departments to the Foreign Office, which was to pay the members of the syndicate.

The story was like something from the *Arabian Nights*, she wrote. "But . . . I could not regard it as fanciful. Dr. Herz was the key to worse scandals than the Panama one. . . ."

The incriminating documents were in a safe place in London, Herz said. The reason he refused to return to France was that he had been charged with treason and espionage in the Chamber, and this meant he could be tried behind closed doors. His sentence, almost certainly, would be one of long penal servitude—Devil's Island.

"It smells bad in here," one spectator is reported to have said the morning of March 8 as the second trial began. "Yes," answered another, "it stinks of scapegoats."

The judges now were in magnificent red robes, and the parade of witnesses included Floquet, de Freycinet, Clemenceau. Remarks made by some witnesses also struck the audience as uproariously funny for the first time. Sans-Leroy, for example, declared on the second day that the 200,000 francs he deposited in his bank account just after the committee vote on the lottery had been part of his wife's dowry. His attitude, he insisted after the judge called for order, was always "that of a member of a committee who wished to be enlightened," which sent the audience into another convulsion of laughter.

Then, that same day, Charles Baïhaut burst into a long, agonized confession, head down, voice cracking. The once exceedingly self-righteous minister told how he had obtained 375,000 francs as a down payment for his support of the lottery measure. "For fifteen years I served France faithfully as deputy and as minister and led an irreproachable life. Even now I cannot understand how I could have sinned."

Yet for all this, Charles de Lesseps was again the one around whom everything turned in a drama that held the nation spellbound. His account now was at once open and lively and immensely interesting. Spectators were immediately conscious of the change. "His intelligence, his ability, his dignified bearing, all made a marked impression . . ." wrote the French historian André Siegfried. "He appeared chiefly as someone who had been struggling against a gang. He had undertaken an impossible task, and had done so against his own better judgment, and yet he had tried to fight on to the bitter end. But the sharpers had got the better of him . . ."

He could have put the blame on his father at any time, but this he never did, not even by inference. His position, simply stated, was that neither he nor any official of the Compagnie Universelle had set out to bribe anyone; rather, they had been the repeated victims of extortion. Everybody had wanted a cut. The company had been told to pay for political support, for influence on the Bourse, for the willingness *not* to discredit its claims—or face the consequences. Newspaper reporters, financial advisers, people who merely knew people who supposedly could help or do harm—"They seemed to rise up from the pavement. We had to deal with their threats, their libels, and their broken promises."

At one point, when Judge Pilet Desjardins told him to "cut it short," Charles calmly replied, "No, I have time enough. All this is necessary to my defense."

Powerful financiers, he continued, could not force anyone to buy stocks or bonds, but they certainly could prevent them from doing so. He described the initial overtures made by Herz, who had talked of the "improvements" he could obtain in the company's standing with the government. "We should have preferred that he had not come to us, but . . . it was better to do that which would make him our ally instead of that which would make him our enemy." Herz had taken him to visit President Grévy at Grévy's country estate, where Herz, Charles said, had been received as a friend of the family. "I was then convinced that he was a man we must reckon with."

"Your duty as a man of honor was to show such a fellow to the door," interrupted the judge.

"But we could not make an enemy of the sleeping partner in *La Justice*," he answered.

He recalled how the first sale of lottery bonds had been wrecked by anonymous telegrams announcing his father's death. "Subscriptions stopped and we appealed to the courts for the punishment of those who had sent the telegrams, but there was no prosecution of the offenders. We were obliged to protect ourselves. . . . The financiers showed us how to resort to those methods which are now matters of general knowledge. They said: 'Unless you pay the money to all the banks under the influence of Girardin [Émile de Girardin, owner of *Le Petit Journal*] you will have all the newspapers in Paris against you.' We still held out against such methods, the newspapers attacked us, and finally we were driven to paying out enormous sums right and left . . . and this mode of procedure was encouraged by the government."

"Leave the government alone," the judge responded sharply, which brought a great outburst from the audience—shouts of "Why not the government?" "Give us the truth"—and the judge ordered that the room be cleared.

Charles testified subsequently that he had decided not to pay de Reinach and that de Reinach had threatened to take the company to court, warning that a public scene would bring the company to its knees. But Charles had made up his mind to run that risk and would have, he said, had not Floquet, de Freycinet, and Clemenceau sent for him, one by one, to tell him—"for the good of the Republic"—to pay de Reinach off and keep the waters smooth. Boulanger had been on the rise, Charles explained, and they were fearful of the consequences should the Panama company suddenly collapse. "They were very polite about the matter. They did not take me by the nape of the neck. . . ."

When their turn came de Freycinet and Clemenceau denied any part in the affair. Charles Sans-Leroy said he had no idea how his initials happened to be on the incriminating check stubs. Accused by the prosecution of selling his vote, Sans-Leroy, a large and extremely homely man in pince-nez glasses, replied with perfect equanimity, "Prove it."

Charles de Lesseps was probably telling the simple truth as he knew it through the length of this, "The Great Bribery Trial." Yet neither Floquet, nor de Freycinet, nor Clemenceau, nor Rouvier, nor anyone of importance was ever prosecuted. No newspaper publishers or reporters were brought to judgment. Those deputies and the one senator on trial were acquitted, Sans-Leroy as well. The single political figure to be convicted was Baïhaut and that was only because he had confessed.

The jury delivered its verdict March 21, 1893. Charles and Blondin, the intermediary in the arrangement with Baïhaut, were found guilty with extenuating circumstances. Charles was sentenced to a year in prison, Blondin to two years. Baïhaut's sentence was for five years, the forfeiture of all civil rights, a fine of 750,000 francs, and full repayment of the 375,000-franc bribe. If Baïhaut were to find himself unable to meet these payments, Charles de Lesseps would be held accountable.

Charles de Lesseps was alone in maintaining his composure through the long reading of the sentence.

From everything that was said during the course of the two trials and from the mammoth report issued by the Deputies committee when its investigations ended in June, a few generalizations and one or two further facts of interest can be drawn.

The total amount paid out to de Reinach, Herz, and different political people, either directly or through de Reinach or Arton, can only be approximated. De Reinach, for example, received some 4,500,000 francs for his handling of the flotation syndicates, plus another 3,000,000 francs for "publicity," plus nearly 5,000,000 more at the time Herz was threatening to "wreck everything." That makes 12,500,000 francs—$2,500,000—that the company paid to Jacques de Reinach alone. Some of that was perfectly legitimate theoretically (for his part in the various security flotations); a good portion of it (according to the check stubs) went to fix various politicians; much of it, perhaps even all the rest of it, went to Cornelius Herz. But Herz is known to have received 600,000 francs directly from the company. Baïhaut got 375,000, Floquet obtained another 250,000, Sans-Leroy almost certainly got 300,000. And undoubtedly there were others. But how many? Perhaps there were more than a hundred, as the brave, unpleasant Delahaye had charged. No one will ever know. But it seems reasonable to conclude that the total sum paid out for political influence and for "friendship" on the Bourse could not have been less than 20,000,000 francs, or roughly $4,000,000. Conceivably it could have been a great deal more than that.

Payments to the press, beginning with the first stock flotation in 1880—a subject about which little was said during the trials—were reckoned by the committee to have been between 12,000,000 and 13,000,000 francs. No less than 2,575 different French newspapers and periodicals had shared in the company's beneficence. Some little fly-by-night publications had even been founded for the sole purpose of getting in on the take. In addition to such giants as *Le Temps* and *Le Petit Journal* (which received the largest sums), the full list included such publications as *Wines and Alcohols Bulletin*, *Bee-keeper's Journal* and the *Choral Societies Echo*.

Frequently payments were made to a particular editor or writer (for example, checks written to Arthur Meyer, of *Le Gaulois*, amounted to 100,000 francs); and often as not, and especially in the early years, the confidence these men expressed in the Panama enterprise, their faith in Ferdinand de Lesseps, were perfectly genuine. One man who did several Panama articles for a fee of 1,000 francs per article became so thoroughly sold by what he wrote that he invested all his savings in Panama stock and as a result lost everything.

Nor, it should be noted, was there anything strictly illegal or even unorthodox about such practices. What impressed the committee most, in fact, was the extent of services rendered for money invested. As large a sum as 12,000,000 to 13,000,000 francs might seem, it represented only about 1 percent of the company's total expenditures. Of course, it was regrettable, the committee declared, that the press had need to resort to such practices, but such were the realities of survival.

Of those convicted, Baïhaut suffered the most. He was put in solitary confinement in a prison where inmates were made to wear a hood whenever they were taken from their cells. Only after three years of this did the courts and the public decide he had been punished enough.

Gustave Eiffel, the only engineer to have been stained by the scandal, would be cleared later of having done any "dishonorable" act by a special committee of inquiry convened by the Grand Chancellor of the Legion of Honor. But his career as a builder was finished; he would thereafter apply himself to wholly different work in meteorology and aerodynamics.

De Freycinet, Floquet, and Rouvier would recover from their disgrace in time and be recalled to office. Loubet eventually became president of the Republic.

For Georges Clemenceau, the future savior of France–*Le Père de la Victoire* during the First World War–the next elections (those of 1893) were a disaster. The voters had cast their own verdict on his part in the affair and it would be nine more years before he made a successful return to public life. His own standard interpretation of the scandal was that it had been engineered by the Boulangists, and that the only reason they descended on Herz was that Herz had refused to give them money. When he re-emerged to save the country in 1917, Clemenceau would be seventy-six, as old as de Lesseps had been when he set out to redeem French honor after Sedan.

As for Cornelius Herz, he spent the rest of his life inside the hotel at Bournemouth. How much or how little truth there was to the things he told Emily Crawford cannot be determined. The secret cache of incriminating documents was never found. He died in 1898, taking his side of the story with him.

Léopold-Émile Arton was eventually discovered living peacefully in London. He was returned to France, tried, convicted, and sent to prison. Some years after his release, he committed suicide. Yet he had achieved an immortality of sorts. Spoofs of his flight from the detectives became the delight of the Paris music halls and among the most fascinated observers at his trial was Henri de Toulouse-Lautrec, who did a series of rapid pencil sketches of the proceedings.

Gustave Eiffel never went to prison because in June of 1893 the Cour de Cassation, the supreme court of France, overruled the verdict of the Court of Appeal. Eiffel, Fontane, Cottu, Ferdinand and Charles de Lesseps were all acquitted on a technical ground: the summonses for their arrest, issued November 21, 1892, had come more than three years after the most recent of their alleged crimes and so, the court ruled, they were entitled to immunity under the statute of limitations.

Since the decision did not apply to the recent sentence by the Court of Assizes, Charles still had the one-year sentence to serve. But the months he had spent in the Mazas were deducted, and after becoming seriously ill, he was moved to a hospital, where he remained for the duration of his sentence. He was released in September 1893.

Charles's troubles did not end there, however. Because Baïhaut was unable to make good on his fine and indemnity, Charles was ordered in 1896 to pay nearly 900,000 francs in Baïhaut's behalf. Unable to produce any but a small part of that amount and faced with another prison sentence if he did not make full payment, he fled the country and remained in London in self-imposed exile. Not until January of 1899, when the government at last agreed to accept a partial payment, did he return to Paris. By then it had been ten years since the fall of the canal company and Ferdinand de Lesseps had been dead for four years.

With family and friends and in all the remaining years of his life, Charles refused to speak of Panama. "He would not talk about it," recalled an adoring nephew, "never, never, never, never." And in the view of those who knew him best, he was regarded no less than ever as the most honest and admirable of men. The Suez company had kept him on its board of directors even during his time in prison. "He was a very honorable man, you know, the old-fashioned sort of thing," the nephew would say. "And I am absolutely certain—I don't know about the whole story, it's very complicated—I'm certain he would never have done anything he thought dishonorable. That's positive."

Charles had been with his father at the end. It happened the year following Charles's release from the hospital. Madame de Lesseps and the rest of the family were also present and death came very quietly for the old adventurer. He died at La Chesnaye, in his second-floor bedroom facing south, late in the afternoon on December 7, 1894, three weeks after his eighty-ninth birthday.

The body was taken up to Paris by train for burial in Père Lachaise Cemetery. There was no grand funeral procession; there were no crowds at the graveside services, only the family, a representative of the Société de Géographie, one very old boyhood friend, and the directors of the Suez Canal Company. The Suez company paid all the funeral expenses. In the eulogies the word "Panama" was never mentioned.

IV

The extraordinary venture had lasted more than a decade. It had cost, according to the best estimates, 1,435,000,000 francs—about $287,000,000—which was 1,000,000,000 francs more than the cost of the Suez Canal, far more in fact than had ever before been spent on any one peaceful undertaking of any kind.

The number of lives lost, a subject that had been strangely avoided throughout the Affair, had not been determined, nor was it ever to be with certainty. Dr. Gorgas, from his analysis of the French records, would conclude that at least twenty thousand, perhaps as many as twenty-two thousand, died. Possibly that is high, but it remains the accepted estimate.

For France to have suffered such a massive financial and psychological defeat so soon after Sedan seemed a cruel, undeserved turn of fate. Even Bismarck lamented that so heavy a tragedy had overtaken so gallant a people. And the surge of anti-Semitism that Édouard Drumont unleashed was soon to spill over into the appalling Dreyfus Affair.

It had indeed been a blunder on such an inordinate scale, a failure of such overwhelming magnitude, its shock waves extending to so very many levels, that nobody knew quite what to make of it; and as time passed, the inclination was to dismiss it as the folly of one man, Ferdinand de Lesseps, about whom markedly different views evolved.

A popular conception was of the flamboyant enthusiast who began with limitless faith in his own omniscience, but reverted to his worst instincts the moment the scheme began to founder. That he had fallen in with the likes of de Reinach and Herz was, by this view, natural enough, since he was as accomplished a swindler as any of them. France, the world, had been taken in, according to a great many attorneys and business people who claimed to regard the Panama effort "by the ordinary rules of financial probity," no more, no less.

To many American writers he had been the leading performer in a comedy of the absurd—"dancing and pirouetting in the front of the stage blissfully unconscious, apparently, of everything except his own capers." Later, in Panama, it would be commonly understood among American canal workers that he had died in an insane asylum.

For a surprisingly large part of France, he still remained the beloved grandfatherly hero of old; "ancient and honorable," but sadly lacking the power of sober analysis or even common sense—like all creative geniuses. His submission to the demands of financiers and crooked politicians had been, by this interpretation, as innocent as his disregard for what the engineers called practicalities. His gaze had been on his star, and his star, this time, had failed him. To debate his tragedy was to debate the stars. It was a view that bequeathed innocence by making him something of a simpleton. Monumental naïveté had been both his making and his unmaking. And destruction at the end for such a spirit thus became no less inevitable or blameworthy than it had been for, say, Joan of Arc, such being the real world's reward for sainted madness.

But as events receded farther into the distance, he became something rather different. He was seen more and more as the tragic victim of earthly forces beyond his control: of the satanic jungle; of ambitious technical advisers willing to say anything, conceal anything, to satisfy their own selfish ends; of unscrupulous financiers (who to many people would be forever regarded as unscrupulous *Jewish* financiers). The fatal mortal flaw according to this interpretation had been to grow old. Once during the *affaire de Panamá*, a newspaper had suggested quite sympathetically that it might have been better had Ferdinand de Lesseps died earlier, at the peak of his career, and Madame de Lesseps had written a moving reply that was quoted widely, then and for years to come.

"I will not protest against this unchristian sentiment," she wrote, "except to say that its author can have given no thought to the wife and children who deeply love and revere this old man and to whom his life, however frail it may be, is more precious than anything in the world. It is no crime to grow old."

So the corollary assumption was that he would have succeeded had he not grown old, that he would have repeated Suez at Panama had he still been the de Lesseps of Suez, at the height of his manhood and in possession of his famed "powers."

There was a degree of truth, of course, in all such interpretations. In the main, however, they were delusions. The real man had been infinitely more complex, his motives far more ambivalent, the personality filled with many more contradictions, than implied by any simplistic answers. He was both the most daring of dreamers and the cleverest of back-room manipulators. He was the indestructible optimist, believing to the depths of his soul that goodness and right invariably triumphed in the long run; and he was perfectly capable of deceit and of playing to the vanity and greed in other men. He was a trusting, decent, endearing man who could confide to a reporter several years after the canal was under way that he had known from the start that there would be trouble, who could blithely inform the press that his engineers had redirected the entire course of the impossible Chagres, who could tell his adoring stockholders on the eve of the final, inevitable collapse that success was theirs.

Arteries were hardening in the old system, no doubt, but to argue that age was his undoing is to disregard too many other factors of importance. His age, furthermore, became an apparent problem only toward the end when the cause was already lost. Until then it was the display of youthfulness that so captivated his following, that impressed so shrewd and impartial a close-hand observer as John Bigelow. Indeed, it could be as readily argued that his curse was the *failure* to decline, his inability to look and act his age. It is no crime *not* to grow old, Madame de Lesseps might have said. Again and again things could have gone differently, more prudent or realistic views might have prevailed, had he been incapable any longer of playing on his powers—to charm, to flatter, to inspire, to sweep good men onward, contrary to their better instincts, using nothing but the phenomenal force of personality. Men who *did* know how to compute realistic excavation schedules, men who *had* experienced Culebra "in the wet," serious expert engineers at the top of their form, had listened and agreed and gone ahead as he wished time after time.

The root sources of his downfall had been apparent since the Paris congress of 1879: the insistence on a sea-level passage through country he knew nothing about, the total disinterest in conceptions other than his own, the refusal to heed voices of experience, the disregard for all data that either conflicted with or that appeared to vitiate his own cherished vision; but none of these would have mattered greatly had it not been for that extraordinary ability to inspire the loyalty and affection of individual human beings at every social and intellectual level.

From the technical standpoint the tragedy hung on the decision to cut through at sea level, to make another Suez Canal. Such a task at Panama was simply too overwhelming, if not impossible. The strategy did not suit the battleground.

The handwriting had been on the wall a good three to four years before the money was gone. With the equipment then available, even a lock canal of modest dimensions would have been an enormously difficult and costly task. But had he and his technical advisers decided to make it a lock canal even as late as 1886, at the time of his second tour of the Isthmus, there probably would have been a French canal at Panama, death, disease, jungle, geology, costs, and de Lesseps' advanced age all notwithstanding. The size of the locks being contemplated would have made the canal obsolete in relatively little time, but the canal would have been built.

As for any possible complicity on his part in the less-than-noble practices that went on behind the scenes, there is no real mystery. He was neither innocent nor a simpleton. He was involved in bribing the press, in the Herz compact, indeed he was the one who crossed that line at the very beginning at the time of the first successful stock issue. His public pronouncements, his *Bulletin*, were replete with misinformation, misleading statistics, promises that he knew to be beyond realization. In his "dashing, off-hand way [he] lied any amount to interviewers," as Emily Crawford said. He was determined to build the canal, to succeed again, to be all that his adoring multitude believed him to be. As the situation worsened, he had agreed to desperate measures to gain time, to postpone disaster. When Charles said in court that he himself had done what he thought he had to do, he was undoubtedly speaking for his father as well. "What would you have decided in our place?" Charles had asked.

The fundamental mystery one comes down to in the end is the endlessly trumpeted faith of Ferdinand de Lesseps in success. Was all this the skilled and quite conscious deception of a grand imposter? Or was it the self-deception of a vain old fool who had been captured by his past success? These are the implicit questions in nearly all that has been written about the man.

The evidence is that it was something else again.

At heart, by nature, by every instinct in his body, Vicomte Ferdinand de Lesseps was a rainmaker. He was, as Masefield said of Shakespeare, "the rare unreasonable who comes once in ten generations." And it had been on that fundamental ground that Henri Barboux had rested his defense. "Beautiful illusions!" the attorney had exclaimed at the high point of his sonorous two-day oration. "That is what the Attorney General would call all great adventures which do not succeed. But humanity has need of such illusions. And when a great people is no longer kindled by them, then it must resign itself to be but a stolid ox, head bowed to earth."

But the crucial point is that de Lesseps was a rainmaker to the nineteenth century: he himself was no less bedazzled than anyone by that era's own new magical powers. An enormous part of his appeal, perhaps the very essence of his appeal, was the fact that he was a nontechnical, nonscientific spirit, the most human of humanists. It made it possible for people to take him to their hearts. And yet it was he who had, at Suez, succeeded in bringing science and technology to bear for one noble, humanitarian purpose; and after that it had been very difficult to doubt his word or distrust his vision. From Suez on, as he himself once said, he enjoyed "the privilege of being believed without having to prove what one affirms." It was this that made him such a popular force and such a dangerous man.

His was not "the faith that could move mountains," as was written or said by so many who never troubled to look at what he had been saying repeatedly since the Paris congress. Not at all. His was the faith that the mountains could be moved by technology. He was as much bedazzled by the momentum of progress as by his own past triumph. "Science has declared that the canal is possible, and I am the servant of science," he had remarked at the Delmonico's dinner in 1880. Wondrous new machines would save the day, he told his stockholders again and again. Men of genius would come forth, by which he meant technicians and scientists—workers in physics, mathematics, soil engineering, chemistry, tropical medicine, hydraulics—things about which he knew little or nothing, but which he counted on. He had the nonscientific, nontechnical man's faith that science and technology would "find a way." That was his faith; that had been his experience. Of the 75,000,000 cubic meters excavated at Suez, 60,000,000 had been removed by machines in the final four and a half years of the work. In the years since, he had seen the use of dynamite and nitroglycerin become widespread. He had witnessed the miracles achieved by Pasteur. So in the largest sense, his tragic folly had been to misjudge the momentum of progress: he had felt certain the machines, the medicines, whatever it took, would be ready in time and he was wrong. And one cannot help but feel that in the end he drifted into that last dim stage of his life haunted by an awful sense of betrayal.

It can also be said, and with certainty, that nothing whatever would have been attempted or accomplished at Panama had it not been for Ferdinand de Lesseps, a point missing from the postmortems of the 1890's, largely since the actual work itself had been either forgotten or was assumed to be utterly without value. In France, as André Siegfried observed, no one seemed to recall that Panama had had anything to do with the building of a canal. "In the end one almost believed that The

Company had hardly done anything at all in the isthmus . . ." The money, declared *The Times* of London, was "as clean gone" as if it had been sunk in the North Atlantic.

Nobody talked of the hospitals that had been built, the offices, storehouses, and dock facilities, the living quarters and machine shops; the maps, plans, surveys, and hydrographic data that had been assembled; the land that had been acquired or the Panama Railroad. And the fact that more than 50,000,000 cubic meters of earth and rock had been removed from the path of the canal, an amount equal to two-thirds of the total excavation at Suez, was virtually forgotten. All had been in vain was the prevailing, unchallenged attitude; the defeat of the old pioneer had been total.

As it happens, the commission appointed by the liquidator to appraise the work had returned with an encouraging report: the amount accomplished was "very considerable"; the plant was "in a good state of preservation"; the lock canal could be completed in about eight years. With an eye to the future, the liquidator had also arranged an extension of the old Wyse Concession, by sending Wyse back to Bogotá. The concession was declared valid until 1903 on the condition that a new French company should be organized to carry on the work, and on October 20, 1894, just seven weeks before the death of Ferdinand de Lesseps, a Compagnie Nouvelle du Canal de Panama had been formally incorporated.

Yet few people took any of this very seriously. The jungle was said to have already reclaimed most of what it had lost to the French engineers, and further, an American canal at Nicaragua was regarded as a certain thing, irrespective of the fact that one American attempt in Nicaragua—by the Maritime Canal Company, which had been chartered in 1889—had already gone down in defeat. It had been an underfinanced affair that collapsed with the Wall Street Panic of 1893.

A canal was beyond the capacity of any purely private enterprise; that much now was plain. It must be a national undertaking. The United States appeared to be the one nation ready to mount such an effort, and if the American people had drawn one overriding conclusion from the French disaster, it was that the place not to build a canal was Panama. The failure of the French—"the greatest failure in modern times"—was above all a lesson in *geography*. They had gone down to defeat not merely because they were French (and therefore incompetent, impractical, and decadent) and led by Ferdinand de Lesseps, but primarily because they—he—had chosen the wrong path.

American correspondent Richard Harding Davis wrote in *Harper's*

Weekly that in all probability Panama had a curse on it. He had gone to see for himself in 1896, and he judged it "unholy ground." It was, he wrote, as if some evil spirit haunted the Chagres bottom lands. He was astounded to see the care with which French equipment and machinery was still being maintained. The armies of black laborers had since departed—returned home at the expense of the Jamaican government—but locomotives stood safely on blocks, oiled and cared for, he reported, as if on display at the Baldwin Works. In machine shops "each bit and screw in each numbered pigeon-hole was as sharp and covered as thick with oil as though it had been in use that morning."

Other writers traveling through Panama had found melancholy themes in the hulks of abandoned French machinery lying belly up in wayside swamps. But to Davis such devoted care and attention were more pathetic. "For it was like a general pipe-claying his cross-belt and polishing his buttons after his army has been routed and killed, and he has lost everything, including honor."

In time to come, he wrote, when the Americans built the Nicaragua canal, Panama would remain one of the greatest ruins on earth, a relic of swindle and death and of the tragic old man who had been so misguided as to believe in a Panama passage.

BOOK TWO

Stars and Stripes Forever 1890-1904

9

Theodore the Spinner

. . . the universe seemed to be spinning round and
Theodore was the spinner.

—Rudyard Kipling

I

On a summer day in the year 1901 there was, as the guidebooks said, no pleasanter place in Washington to sit and pass the hours than Lafayette Square. In the shade of a southern magnolia or a flowering Chinese paulownia (or perhaps an elm or a beech planted by Jefferson) one could watch the flow of traffic along Pennsylvania Avenue or contemplate the north façade of the White House; or try to fathom—as nearly everyone did—what marvelous bit of ingenuity kept the equestrian bronze of Andrew Jackson in such uncanny equilibrium.

Flower beds were carefully tended, paths swept clean. Tourists came and went, and pretty girls on their noon hour passed by in twos and threes, wearing the wide-brimmed straw hats and crisp white shirtwaists that had become the fashion.

Especially satisfying was the sense one had of being at the very center of things. It was the nearness not just of the White House but of the elegant private residences fronting on the other three sides of the square, of the Arlington Hotel, the Cosmos Club, the easy proximity of the Metropolitan Club, the Treasury Building, and that great baroque pile, the State, War, and Navy Building, that made it such a rarefied and endlessly fascinating world within the world of Washington.

On the east side of the square, next door to the Cosmos Club, lived Senator Hanna–Number 21 Madison Place, the "Little White House." At the buff-colored Cosmos itself, once the home of Dolly Madison, could be found such luminaries as Alexander Graham Bell or Professor Samuel Langley of the Smithsonian. The Arlington, diagonally across from the Cosmos, on H Street, was the city's largest "distinguished hostelry." Virtually every President since Grant had been accommodated there the night before his inauguration.

Secretary of State John Hay, who had first come to Washington as Abraham Lincoln's private secretary, and Henry Adams, that cultivated lineal descendant of two Presidents, lived in adjoining houses at the corner of H and 16th streets, just across from beautiful little St. John's Episcopal Church. A comparatively new addition, built in the eighties, this Hay-Adams edifice was the one "unconventional" note on the square. It appeared to be one massive red-brick bastion with trimmings of light-colored stone, innumerable windows, imposing stone steps, and dark carved doors set within deeply shadowed archways–all trademarks of Henry H. Richardson, the most brilliant architect of the day. It seemed the safest possible refuge for the two fragile gentlemen who resided within, both of whom were looked upon as national treasures of a sort. Adams wryly referred to his address as the only position of importance he had attained in life and he reigned there over the nearest thing thus far to an American salon. To be asked to breakfast at 1603 H Street was to have "arrived."

For John Hay, author of the "Open Door" policy in China, his house was little more than a block from his office at the State Department or from the Metropolitan on 17th Street, the city's most fashionable club, or from the French embassy, the large yellow house beside the Metropolitan. "Life," wrote Adams, "is a narrow valley, and the roads run close together." It was a view one might well have conceived from so privileged a vantage point.

But at summer's end, on September 6, 1901, the comparative tranquillity of Lafayette Square, like the whole order that had evolved in Washington, ended when two shots from a .32-caliber revolver were pumped point-blank into the unsuspecting William McKinley at Buffalo, New York. He had gone to attend the Pan-American Exposition and was standing in the Temple of Music beside a potted palm shaking hands with a long line of people, one of whom, a deranged young anarchist named Leon Czolgosz, stepped forward, his right hand wrapped in what appeared to be a bandage. Eight nights later McKinley was dead.

"Now look!" Mark Hanna is said to have exploded on hearing the news. "That damned cowboy is President of the United States."

The sudden advent of Theodore Roosevelt in the White House was to mark the most dramatic shift in Presidential style and attitude since the inauguration of Andrew Jackson, the first avowed "man of the people," when tubs of liquor had been put out in Lafayette Square to divert an overjoyed mob from the White House grounds. Roosevelt's own inaugural was a rushed, solemn little ceremony held in an overstuffed Victorian parlor in Buffalo. But it can be said that the twentieth century truly began when he took the oath of office.

At age forty-two he was not only the youngest President in history, he was an entirely novel figure in American politics—an eastern Republican with national appeal (phenomenal national appeal, as the campaign had shown). Where McKinley had been Midwestern, "of the plain people," "TR" was big-city gentry, raised among nursemaids and gilded mirrors. He was a Harvard-trained, Harvard-sounding reader of books (two a day on the average); he was the Rough Rider, author, historian . . . a bird watcher! . . . and the most tireless political warrior the country had ever encountered. As the Vice-Presidential candidate he had been seen in twenty-four states, traveled twenty-one thousand miles, made nearly seven hundred speeches, all in one tour, while William McKinley, as was his custom, kept to the shade of his front porch in Canton, Ohio.

Violent fate in the form of Leon Czolgosz had put Roosevelt in power at a time when the country was prospering, just as Mark Hanna had promised; when his party was in control of Congress; when the national spirit was expansive, confidence boundless; when the average American felt "400 percent bigger" than he had before the turn of the century, as Senator Chauncey Depew observed.

And he had every intention of exercising power as it had not been in a very long time. "I did not care a rap for the mere form and show of power," he would write, "I cared immensely for the use that could be made of the substance."

The first weeks in office would remain a vivid memory for all who were on hand. "He strode triumphant among us," recalled Lincoln Steffens, "talking and shaking hands, dictating and signing letters, and laughing. Washington, the whole country was in mourning, and no doubt the president felt that he should hold himself down; he didn't; he tried to, but his joy showed in every word and movement." To Harry Thurston Peck, the literary critic, he was "a stream of fresh, pure, bracing air from the mountains, to clear the fetid atmosphere of the national capital." He himself, at the end of his first week, confided to Henry Cabot Lodge, "It is a dreadful thing to come into the Presidency this way; but it would be a far worse thing to be morbid about it. . . ."

He saw more people, he handled more paper work, he cut more red tape in the next several months than anyone who had ever held the office. And he adored the role. No man ever had a better time being President.

There were some, to be sure, and particularly within his own party, who were considerably less than ecstatic over the prospect of such a person in power. Hanna was the outstanding example. "We need not tell our readers that up to this time we have discovered in Mr. Roosevelt very little cause for serious rejoicing," declared the conservative Washington *Post*. "He has at all times been far too theatrical for our taste." Even the venerable Henry Adams, who had found Roosevelt the Vice President "breezy and a tonic," returned home gravely unsettled by his first social evening with Roosevelt the Chief of State. Everything at the White House had been too informal for Adams, the meal indifferent and badly served. Worse, Roosevelt had lectured *him*, the former Harvard professor. "As usual Theodore absorbed the conversation," wrote a disgruntled Adams to a friend. "If it tired me ten years ago, it crushes me now . . . really, Theodore is exasperating. . . ."

But for reporters and the reading public he was a dream come true. He would give a Presidential view on any subject any time. The monologues were likened to Niagara Falls. To get him to *listen*, the story went, it was best to see him about 12:40, just before lunch, when he was being shaved.

He was the first President to call his official residence the White House (rather than the Executive Mansion), the first to be known by his initials, the first to take up tennis, which he played badly but with explosive verve, the first to be photographed jumping on horseback. (When the photographer missed his shot, the President gladly obliged by jumping several times again.) He also brought to Washington the large, young, and exuberant family that was to dominate the popular imagination in ways that had never been known or that would never quite be equaled again. Edith Carow Roosevelt looked so youthful driving about the city in her carriage that she was sometimes mistaken for her stepdaughter, seventeen-year-old Alice; and Alice, as the country quickly discovered, was a "handful." The five other children, the eldest just turned fourteen, seemed wholly unaffected by the aura of their new surroundings. Visitors were to encounter Roosevelt offspring racing the White House halls on stilts. A Cabinet meeting would have to be halted temporarily due to the noise overhead. The President himself, it became known, was in the habit of "looking in" on the children before state dinners, by which he meant a terrific pillow fight.

It all seemed to agree with him, as did everything in life. He had acquired some poundage in recent years, but physical bulk was in style for men of position, and he was by no means fat. He stood only five feet eight inches tall, yet most people, when they saw him for the first time, were struck by how big he seemed. His frame was big, his neck and shoulders were big, and he stood with his shoulders thrown back, which gave him an even more imposing look. His weight during the time he was President was something over two hundred pounds. "His walk," recalled William Allen White, "was a shoulder-shaking, assertive, heel-clicking, straight-away gait, rather consciously rapid as one who is habitually about his master's business."

Mainly Theodore Roosevelt was interesting, interesting as no President had ever been. He was someone who would make things happen.

II

The obvious differences in age and nationality aside, there were striking similarities between Theodore Roosevelt and Ferdinand de Lesseps. Both were the products of cultivated, worldly families. Both were raised on the ideal of patriotic service and the heroic exploits of adventurous kinsmen. There is the common love of the out of doors, of shooting, and of horses; the common joy in children, books, theatrics, popular acclaim. In his boundless love of life, his immensely attractive animal vitality, Theodore Roosevelt might have been a direct descendant of Ferdinand de Lesseps. There is even a kind of continuity to such traits as they were sometimes despised for—craftiness, self-glorification, self-deception.

Nor was Roosevelt ever anything but positive about the need for a Central American canal to rival Suez. "No single great material work which remains to be undertaken on this continent is of such consequence to the American people," he declared in his first message to

Congress. Whether he or any of those around him suspected then that the canal would become the great material set piece of his Administration, as well as the work in which he would take the most personal pride, or that it would be the subject of more controversy than anything else he did while in office, is impossible to say. But his eagerness to get on with the job was unmistakable.

Roosevelt, however, looked upon the canal quite differently than de Lesseps had, differently, in fact, than nearly everyone. It was very well for others to talk of it as the dream of Columbus, to call it a giant step in the march of civilization, or to picture as de Lesseps so often had its immeasurable value to world commerce. Roosevelt was promoting neither a commercial venture nor a universal utility. To him, first, last, and always, the canal was the vital—the *indispensable*—path to a global destiny for the United States of America. He had a vision of his country as the commanding power on two oceans, and these joined by a canal built, owned, operated, policed, and fortified by his country. The canal was to be the first step to American supremacy at sea.

All other benefits resulting, important or admirable as they might be, were to him secondary.

His guiding light in this regard, the beloved prophet and teacher, was a tall, spare, beaked, painfully shy, deadly serious naval officer and scholar, who looked like a predatory bird. As bald nearly as an egg, with pale hooded eyes, Alfred Thayer Mahan had been a member of the faculty at the Naval War College at Newport, Rhode Island, when Roosevelt, years before, had been invited to lecture there on one of his specialties as a historian—the War of 1812. The two had liked each other instantly and remained fast friends and earnest correspondents. And for some fifteen years, first in the War College lectures developed following his Panama experience, then in his famous book, *The Influence of Sea Power upon History*, as well as in magazine articles and private correspondence, Mahan had been preaching a strident, uncompromising canal doctrine. His role as teacher and prophet had been a factor of the greatest importance, giving the old dream of a Pacific passage a meaning it had not had before.

Like Mrs. Stowe, earlier in the nineteenth century, Mahan had happened out of the blue. Born at West Point, New York, in 1840, he was the son of Dennis Hart Mahan, a noted professor at the Military Academy who had taught Grant, Sherman, Lee, Jackson, and who was the author of a mathematics text familiar to a whole generation of cadets, including several who were eventually to build Theodore Roosevelt's canal. The younger Mahan's naval career had been undistinguished, however. He and his father agreed that he might have done better in some other profession. By the time he was appointed to the staff of the War College, after thirty years in the service, he was still, in his own words, "drifting on the lines of simple respectability as aimlessly as one ever could." An Annapolis classmate would subsequently remember him as the most intellectual man he had ever known, yet nothing out of the ordinary had ever happened to him; he was not an especially able line officer—he was never able to do knots, the square knot was the "top of his ability"—and he had written nothing to indicate any literary gifts or penetrating grasp of world history.

His world-shaking *Influence of Sea Power upon History*, the result of four strenuous years "in the closet," as he said—reading, writing, rewriting—was published in May of 1890 by Little, Brown & Company. The essence of his views was contained in the first ninety pages. By tracing the rise and decline of past maritime powers, he had arrived at the extremely simple theory that national greatness and commercial supremacy were directly related to supremacy at sea. This, he declared, was the towering truth of history. Like many earthshaking concepts, it was not exactly original; numbers of his own contemporaries in the Navy had been thinking along similar lines for some time. He, however, had developed the thesis historically, and *that*, he also asserted, no one had done before. Also, like many such iron-willed theorists, he had a knack for making his case so that it seemed indisputable.

In England, predictably enough, the book was taken as gospel and had its earliest success. Clad in dress uniform, wearing a sword beneath red-silk academic robes, the author received honorary degrees from both Oxford and Cambridge and later dined with the queen at Buckingham Palace.

Kaiser Wilhelm II had telegraphed his friend Poultney Bigelow, son of old John Bigelow, to praise the book: "It is on board all my ships and constantly quoted by my captains and officers." On the other side of the world Mahan was adopted as a text for the Japanese military colleges.

Not to be outdone by Oxford and Cambridge, Harvard and Yale conferred honorary degrees, and in the United States Senate such powerful "expansionists" as Lodge and John Tyler Morgan were immediately won over. "It is sea power which is essential to every splendid people," Lodge lectured the nation from the Senate floor.

Most important, however, was the overwhelming effect on the ambitious young man with the eyeglasses and the flashing teeth who was then serving on the Civil Service Commission. Roosevelt, it is a matter of record, was the first person of influence to read the book and to grasp its import. Probably not another ranking political official in the country had ever heard of Mahan at the time the book appeared. But for Roosevelt, who received one of the first copies and who wrote the first major review for the *Atlantic Monthly*, the prophet had arrived. The book, he immediately informed Mahan, was "very much the clearest and most instructive general work of the kind with which I am acquainted. . . . A *very* good book . . ."

If ever there was a disciple ideally suited, by interest and temperament, Roosevelt was it. In the long introduction to his opus, Mahan had lamented that conventional historians seldom knew anything about the sea. It was because of this that the "profound determining influence of maritime strength" had been so long overlooked. Roosevelt had no such blind spot. He had been fascinated by ships and the sea since childhood. Two uncles on his mother's side had been in the Confederate Navy. His uncle Irvine Bulloch was a midshipman on the fabled *Alabama*, and the accepted family story was that he fired the last gun in the battle with the *Kearsarge*. James Dunwoodie Bulloch was a Confederate admiral and an exceedingly resourceful Confederate operative in England during the war who arranged the building of the *Alabama*. In his own travels with his parents, Roosevelt had crossed the Atlantic several times, and on one trip had sailed through the Suez Canal. His first published work, *The Naval War of 1812*, had been started when he was still an undergraduate. Furthermore, he had acquired a fundamental conviction that life is a struggle and life among nations no less than life among man and beast. He believed in military strength, the military virtues; he deplored pacifists, he said, as he deplored men with "shoulders like champagne bottles." He was, as every American youngster would come to appreciate, the champion of the strenuous life, the once near-sighted, asthmatic little boy who had willed himself to be the world's leading proof of "the rugged fighting qualities."

Roosevelt's determination to have a canal can be dated from the appearance of *The Influence of Sea Power* in 1890, which, very interestingly, was the same year the Census Bureau declared there was no longer any land frontier. The Caribbean Sea was the American Mediterranean, wrote Mahan, and like the Mediterranean, it demanded a canal. The canal was the thing to bestir "the aggressive impulse," and turn the American people from their "peaceful gainsaying" ways. With the isthmian barrier broken, the Caribbean would become not simply a prime commercial crossroads, but a vital military highway. The United States would require Caribbean bases, "which by their natural advantages, susceptibility of defense, and nearness to the central strategic issue [the canal] will enable her fleets to remain as near the scene as any opponents."

The problem, as Mahan explained it, was that thus far the nation had been too well supplied with its own resources, too complacent in its self-sufficiency.

So the canal, "the central strategic issue," was to be the great redeeming task. It would shake the country out of its naïveté, release it from myopic concerns. It would breed an international, expansionist spirit. It would breed ships, coaling stations, naval bases, colonies afar. It would create an American navy. "Whether they will or no," Mahan wrote in the December *Atlantic Monthly*, "Americans must now begin to look outward." His head was filled with American armadas steaming to distant and glorious horizons.

Roosevelt was thirty-one years old at the time Mahan's book appeared and had already made a place for himself among the leading figures in Washington. He would expound on his views at length during evenings at the Cosmos Club, for example, and to the rapt delight (appropriately) of the young English writer Rudyard Kipling, who used to drop in about half-past ten with the express purpose of hearing the expansive young American go on. "I curled up on the seat opposite," said Kipling, "and listened and wondered, until the universe seemed to be spinning round and Theodore was the spinner."

In an "entirely confidential" letter written from Washington in 1897, Roosevelt told Mahan that the Nicaragua canal should be built "at once" and, in the same breath, that "we should build a dozen new battleships." By then, through the influence of Lodge, who had been primed by Mahan, Roosevelt had been made Assistant Secretary of the Navy and had entered upon his duties characteristically, as if accompanied always by a band playing Sousa's "Stars and Stripes Forever." He visited shipyards, poked his nose into technical matters, from ordnance to dry docks, went out on maneuvers. From a richly carved desk in the State, War, and Navy Building, with John Paul Jones looking down from a gold frame and a big, glass-cased model of Dewey's flagship, *Olympia*, standing within arm's reach, he mapped global strategy and fired off letter after letter to congressmen and newspaper editors, urging more ships, improved weapons. "Gradually," he would recall, "a slight change for the better occurred, the writings of Captain Mahan playing no small part therein."

Lodge, Roosevelt's closest friend and greatest admirer in the Senate, was saying that the canal would make Hawaii a necessity. Senator Morgan declared that Cuba was needed as well, because of its position in relation to the canal. To Roosevelt, Lodge and Morgan were uncommonly "far-sighted," a favorite accolade of Captain Mahan's.

Home from the Cuban war a few years later, Roosevelt told a Chicago business club in his rasping falsetto, "We cannot sit huddled within our own borders and avow ourselves merely an assemblage of well-to-do hucksters who care nothing for what happens beyond." Such a policy would be self-deluding and disastrous. (It might have been Mahan himself speaking.) ". . . if we are to hold our own in the struggle for naval and commercial supremacy, we must build up our power without our borders. We must build the Isthmian canal, and we must grasp the points of vantage which will enable us to have our say in deciding the destiny of the oceans of the east and west."

A naval base had been established at Cuba. Hawaii had been annexed. Puerto Rico, Guam, and the Philippines had been acquired, and the canal had become an enormously popular cause largely as a result of an incident early in the war, the celebrated "Voyage of the *Oregon*."

The *Oregon*, one of the first true battleships, had made Mahan's and Roosevelt's case for them about as effectively as anything could have. The ship had been in San Francisco when the *Maine* blew up in Havana harbor and victory in the Caribbean was said to depend on her. Her orders from Washington were to proceed at once around the Horn. So on the morning of March 19, she had steamed off on a perilous race of 12,000 miles—instead of some 4,000, had there been a Central American canal. For the next two months the whole country waited in mounting suspense. There were long, ominous periods of silence, weeks when the ship was "lost from communication." Then came rousing dispatches from some point in Peru or Chile. The excitement kept building, every American was caught up in it.*

From Rio north the gleaming white ship was cleared for action and repainted a dull battle gray. Then just over the equator, approximately

* One of many popular renditions of the story, "The Race of the Oregon," by John James Meehan, went as follows:

> Lights out! And a prow turned toward the South,
> And a canvas hiding each cannon's mouth,
> And a ship like a silent ghost released
> Is seeking her sister ships in the East. . . .
>
> When your boys shall ask what the guns are for,
> Then tell them the tale of the Spanish War,
> And the breathless millions that looked upon
> The matchless race of the *Oregon*.

on a line with the mouth of the Amazon, there occurred an amazing crossing of paths. The *Oregon* steamed by the tiny sloop *Spray*, a random speck in the empty sea, upon which, sailing all alone, was Captain Joshua Slocum, of Massachusetts, then on the last leg of the first solitary cruise around the world. ". . . I saw first a mast," he wrote, "with the Stars and Stripes floating from it, rising astern as if poked up out of the sea, and then rapidly appearing on the horizon, like a citadel, the *Oregon!*" Signals were exchanged and Slocum learned for the first time that his country was at war.

On May 24, sixty-seven days after leaving San Francisco, the *Oregon* was spotted off Palm Beach, Florida, and the news was flashed across the country. She had arrived in time to play a part in the Battle of Santiago Bay.

Though the voyage was hailed as "unprecedented in battleship history," a triumph of American technology and seamanship, it was the implicit lesson of the experience that would matter in the long run. "By that experience," wrote Mark Sullivan, social historian of the era, "America's vague ambition for an Isthmian canal became an imperative decision." As a demonstration of the military importance of the canal, it had been made to order.

Still, of those impelling new reasons for the canal produced by the Spanish war, none counted for so much in Washington as the acquisition of the Philippines. The Philippines, Roosevelt foresaw, would affect America's future more than any other result of the Spanish war. He was not an imperialist, he insisted. It was inconceivable to him that Americans could ever be viewed as imperialistic. In all the United States he had never met an imperialist, he once said before an audience in Utah. He was personally offended by the charge. Expansion was different; it was growth, it was progress, it was in the American grain. He was striving to lead his generation toward some larger, more noble objective than mere moneymaking. ("For after all," the revered Mahan wrote, "if the love of mere glory is selfish, it is not quite so low as the love of mere comfort.")

To each generation was allotted a task, Roosevelt knew. "I wish to see the United States the dominant power on the shores of the Pacific Ocean."

Roosevelt was governor of New York when he first thrust himself into the actual shaping of policy concerning the canal. The contribution was uninvited and was an extreme aggravation to Secretary of State John Hay.

In 1898, the war in Cuba over, McKinley had directed Hay to begin negotiating a new canal treaty with Great Britain, to supplant the old Clayton-Bulwer Treaty, which, after nearly fifty years, still remained a diplomatic stumbling block to any substantive support of a Central American canal by the United States government. Hay and the British ambassador, Sir Julian Pauncefote, had made rapid progress. Tied down with its own unpopular Boer War in South Africa, by now disenchanted with Central America as a "sphere of influence," the Foreign Office was ready to bow out of a partnership in building the canal, quite willing to sign the task over to the Americans.

According to Hay's proposal, the United States was to have the right to construct and operate the canal, which, like Suez, was to be "free and open in time of war as in time of peace, to vessels of commerce and of war of all nations, on terms of entire equality. . . ." The United States could keep order along the route with its own police, but there were to be no fortifications. The agreement was signed in Hay's office on February 5, 1900.

That was the *first* Hay-Pauncefote Treaty and for a few days John Hay felt he had achieved a milestone. McKinley too spoke elatedly of "the great achievement." But Hay had chosen to ignore the Senate. No one on the Hill had been shown a draft of the treaty, nor had he bothered to describe its provisions to anyone on the Foreign Relations Committee. "When I sent in the Canal Convention," he later explained to McKinley, "I felt sure that no one out of a mad house could fail to see that the advantages were all on our side." The rumblings commenced quickly enough, principally over the concept of a neutralized canal, a subject seldom questioned before. Suez had long since established the precedent of neutrality. The concept was in keeping with the old American policy of freedom of the seas. In addition, there was substantial naval opinion that if the need ever arose, the canal could be quite properly defended from bases at San Juan and Pearl Harbor.

Senator Lodge was the "first to flop," in Hay's words. The British had given up nothing, Hay was told; they had simply agreed to let the United States spend the money and do the work. John Tyler Morgan, another "force" on the Foreign Relations Committee and now head of his own Senate canal committee, concurred.

Then from Albany came the most shrill denunciation of all, which, to add to Hay's exasperation, was played across page one of the New York papers no less than if it had been a major policy pronouncement.

George Smalley, former London correspondent for the New York *Tribune*, now Washington correspondent for *The Times* of London, was the one who rushed across Lafayette Square to give Hay first word of Roosevelt's attack. "You can imagine to what extent the fat is in the fire!" wrote a bemused Henry Adams. "If Hay is beaten on his treaty he will resign; if he doesn't resign, he will certainly hamstring Teddy. Won't it be fun?"

For his own part, Hay sent an icy response to Albany, declaring that such matters ought not concern a mere governor.

The mere governor would be heard all the same. "I do not see why we should dig the canal if we are not to fortify it so as to insure its being used for ourselves and against our foes in time of war," he wrote to Captain Mahan. To Hay he insisted that the treaty was in fact a step backward and "fraught with very great mischief." He asked the Secretary to consider the case of the *Oregon.* Had a canal of the kind the treaty guaranteed been in existence in 1898, the *Oregon* could certainly have reached the Atlantic more quickly; but the advantage would have been far outweighed by the fact that the Spanish fleet would also have been at liberty to use the canal to prey on the Pacific Coast or to go after Dewey in the Philippines.

"If that canal is open to the war ships of an enemy it is a menace to us in time of war; it is an added burden, an additional strategic point to be guarded by our fleet. If fortified by us, it becomes one of the most potent sources of our possible sea strength."

Roosevelt's view was the popular one and opposition to the treaty gathered rapidly. In the Senate, Morgan noted that England had once done everything short of war to prevent the canal at Suez, but then took it over after the work was completed. Allegedly this could again be the intent.

To add to Hay's burdens, meantime, his friend Adams, who had since departed for Paris, lectured by mail that the whole balance of world power rested on the two isthmuses. Suez was settled, but who was to say what the consequences might be if the Kaiser were to make a move for Panama? Five minutes of negotiation in Paris would be enough, said Adams, to guarantee the completion of the French canal.

The Senate refused to ratify the treaty without amendments. Hay was beside himself. Overly sensitive by nature, he was stunned by the attacks on the treaty, taking everything said about it quite personally. It was his first experience with "filthy newspaper abuse." He was certain he was in the right, and he had assured Pauncefote that the treaty would be acceptable. A career dedicated to the resolution of Anglo-American difficulties appeared to be going up in smoke.

He handed McKinley his resignation, which McKinley calmly refused. "We must bear the atmosphere of the hour," the President said. "It will pass away." And like many of McKinley's instinctive responses, it was the right one.

When the British refused to accept an amended version of the treaty, Hay, to his enormous credit, picked up the pieces and began over again. Negotiations with Pauncefote resumed; this time the Senate was kept apprised. By late summer of 1901, shortly before McKinley left for Buffalo, Hay was able to report that much progress had been made. He had worked on harder than ever, and despite personal tragedy and recurring premonitions of doom. In June his elder son, Del Hay, McKinley's private secretary, had been killed when he fell from an open window at New Haven, while attending a Yale commencement. "I have hideous forebodings," Hay wrote Adams. "Good luck has pursued me like my shadow. Now it is gone . . ."

And then had come the shattering news from Buffalo. His world, his career, his usefulness, all had ended, he wrote to Roosevelt. But he also saw Roosevelt as a "young fellow of infinite dash and originality," as he confided to a friend on the day of McKinley's death, and when Roosevelt arrived in Washington with the funeral train the night of September 16, Hay was among the first to come forward on the crowded station platform to pay his respects.

Hay was a man who generated lasting affection. The French ambassador, Jules Jusserand, would remember him as "modest withal, never trying to push himself to the front, speaking in subdued tones and scarcely opening his lips when uttering a memorable saying or shrewd humorous remark." Nearly three-quarters of a century later, over tea, Mrs. Alice Roosevelt Longworth, at the mere mention of his name, would say simply, "Oh, *dear* little Mr. Hay . . ." He was, as well, many things Theodore Roosevelt was not—fastidious, subtle, self-effacing, a public official who lost sleep over speeches that had been written perfectly in advance. To Roosevelt he was "the most delightful man to talk to I ever met." His only problem, to Roosevelt's way of thinking, was a "very ease-loving nature . . . which made him shrink from all that was rough in life."

But on the station platform that September night, Roosevelt implored Hay to remain as Secretary of State. They stood together only a moment, Roosevelt with his hand on Hay's arm, both men in black, wearing high silk hats, the noise of the station drowning out their words to everyone but themselves. He told Hay that he must stick by him—it was a command, Roosevelt said—and Hay, deeply touched, said he would.

So it was Hay after all who put his signature to what was to go down in history as the *second* Hay-Pauncefote Treaty, the first important treaty of Roosevelt's Presidency.

This time the clause forbidding fortification had merely been omitted. The United States was to be free to do whatever was necessary to protect the canal "against lawlessness and disorder" and the unwritten understanding was that this in fact authorized fortification. Roosevelt, Lodge, and Morgan were quite satisfied and there was never any serious doubt about the fate of the document after that.

On the morning of November 18, 1901, the portly, white-haired Pauncefote was ushered into Hay's large office at the south end of the State, War, and Navy Building. No special fuss was made. It was not even generally known that the British ambassador was in the building until he had been with the Secretary for about an hour. Then two elderly Negroes—William Gwin and Edward Savoy, State Department messengers who had attended countless such occasions—were asked in. Gwin held a silver candlestick which contained the taper used to burn the red wax for the seals. Savoy would apply the wax. Hay and Pauncefote signed their names. The seals were fixed. "If there was anything approaching ceremony it was putting out the candle," observed a reporter. "It is never blown out . . . but must be snuffed out with a silver extinguisher."

At the White House Theodore Roosevelt declared himself "*Delighted!*"

III

Like John Hay, the British Foreign Office, Lodge, Captain Mahan, like the editors of virtually every major newspaper, like all but a tiny minority of his countrymen, Theodore Roosevelt had been operating on the assumption that the canal was to be built in Nicaragua. In none of his numerous speeches on the subject, for example, had he ever even used the word "Panama." (He had either referred to the Nicaragua canal or the isthmian canal, never to a Panama canal.) And like everybody else in Washington, or everyone who understood how things worked there, he looked to Senator John Tyler Morgan as not merely the ultimate authority on the subject but someone with whom cooperation would be mandatory. Only a few weeks after becoming President, Roosevelt had written to Morgan, "You know the high regard I have for you. . . . I particularly wish to see you and consult with you about various matters; and I hope, my dear Senator, you will understand that I desire earnestly to hear from you about every appointment as well as every question of public policy, and that wherever possible I shall pay the utmost heed to your advice."

Morgan was chairman of the Senate Committee on Interoceanic Canals, the Morgan Committee, as it was more commonly known, which included several extremely interesting and influential figures: Spooner, of Wisconsin, who was as fine a speaker as anyone then in Congress; William Harris, a burly, imposing man, who had an engineering background and had actually seen something of Central America; and Senator Hanna, who was regarded, with reason, as the most important man in American politics, Roosevelt not necessarily excluded. But it was Morgan who ran the show; Old Morgan, of Alabama, who at age seventy-seven qualified as one of the most powerful and interesting figures in American politics.

Morgan did not look like much. He was small and frail, a dry little stick beside a man like Hanna. His hair and mustache were as white as paper, his scrawny neck several sizes too small for the inevitable wing collar. He was known as one of the old-time characters on the Hill. A lawyer from Selma, Alabama, he had led a cavalry charge at Chickamauga and survived to become a brigadier general. He had been elected to the Senate first in 1876 and had been serving without interruption ever since. Friend and foe considered him the most intellectual of Democrats (as Hoar was the most intellectual of Republicans), and to judge by performance, rather than appearance, his career was anything but in the decline. No member of the Senate, irrespective of age, worked harder.

Morgan's efforts over the years had been largely constructive. He was watchful, uncompromising, fiercely independent, nearly always irritable. He was also scrupulously honest. Never had he been known to vote on anything for reasons other than his famous "principles," some of which, such as those concerning relations between the black and white races, were viewed as shamefully out of date. His handwriting, a savage, consistently illegible scrawl, was known all over town, as was his sense of humor, which was a bit like that of Mark Twain, whom he resembled to a degree. "A lie," he was once heard to declare on the floor of the Senate, "is an abomination unto the Lord and an ever-present help in time of need."

To cross him in any fashion was considered extremely dangerous. "Senator Morgan was an extraordinary man in many respects," wrote Shelby Cullom, chairman of the Foreign Relations Committee. "He had a wonderful fund of information on every subject . . . He was one of the most delightful and agreeable of men if you agreed with him . . . but he was so intense on any subject in which he took an interest, particularly anything pertaining to the interoceanic canal, that he became almost vicious toward anyone who opposed him."

The two greatest pleasures in Morgan's life, it was commonly said, were work and a good fight.

The interest in the canal dated from his first years in the Senate. He knew the reports of every surveying expedition to Central America, the findings of the several successive canal commissions since the Grant Administration. It was John Tyler Morgan, everyone knew, who had worked longest and hardest for congressional support for the ill-fated Maritime Canal Company, who had been the author of several canal bills, who had done more to inform the public, heard more testimony, read more, asked more questions, and had more information on the entire subject of an interoceanic passage than any figure of either party. The canal was the dream of his life and he was as certain as he could possibly be that it must be a Nicaragua canal. Nicaragua, in the popular phrase, remained "the *American* route" and his long, frequently lonely fight to have the canal built there had made him a national figure.

The canal would be his monument, Morgan was often told by admiring colleagues. He, however, was not interested in prestige. He wanted no monuments, he wanted the Nicaragua canal.

Because of his strong expansionist sentiments, and the support he had lent to the Roosevelt-Lodge-Mahan doctrines, Morgan also had a unique kind of leverage. In most other respects he was a good Bryan Democrat and a Southerner to the core. Still he could usually count on support from the other side of the aisle when he needed it. And for several years now he had been more welcome at the White House than any Democrat in town.

Morgan wanted an American canal under American control no less than did Roosevelt. Nor had he ever been the slightest bit tentative about that, which was among the chief reasons for Roosevelt's admiration. Several of his strongest arguments for a *Nicaragua* canal were, nonetheless, avowedly provincial. An ocean passage at Nicaragua would mean a return of prosperity to the South. A Nicaragua canal would be closer to any American port than would a canal at Panama, but a Nicaragua canal would also be *seven to eight hundred miles closer* to the Gulf ports of Mobile, New Orleans, and Galveston than to New York or Boston. He foresaw his native southland fronting on one of the world's principal sea lanes and every Gulf port a major coaling station. World markets would open for southern lumber, southern iron, cotton, manufactured goods. It was a position that made him extremely popular at home.

But on top of this Morgan believed quite sincerely that Nicaragua *was* the superior choice from an engineering standpoint and in view of political considerations. His technical argument was much the same as that advanced by Grant's canal commission or by Menocal and Ammen at the Paris congress: Nicaragua offered the lowest pass anywhere on the Cordilleras from Alaska to Tierra del Fuego; Nicaragua provided fifty-odd miles of magnificent lake, perhaps as much as sixty miles of navigable river; the lake offered a limitless supply of water at the summit level of the canal. Politically, Nicaragua was a stable country in which to make so vast an investment of American capital and effort. A Nicaragua canal had already been the subject of six treaties between Nicaragua, Costa Rica, and the United States. Nicaragua was clean, fertile, relatively free of disease; it had great potential for development. And he could marshal impressive facts and figures, drawing from his prodigious memory, government reports, and such widely respected authorities as A. G. Menocal.

By the same token, his contempt for the Panama route was monumental, his utterances on the subject, if anything, even more notable. Earlier in the year, as the newly elected Vice President, Theodore Roosevelt, presided rather nervously over a Senate debate on the canal, Morgan had called the Panama plan "a job which has disgusted France . . . until she had shuddered like a sick baby at the enormity of the villainies perpetrated by her own people." The entire affair had been "gangrene with corruption." The Compagnie Nouvelle du Canal de Panama was the *so-called* New Panama Canal Company, the words spoken as though they had an unpleasant smell. The company's assets and franchises were held to be virtually worthless, its stockholders little better than common thieves. Its officers were paid schemers and to be trusted under no conditions. These people, Morgan warned, had no intention of finishing the canal; their present efforts in Panama were a thin sham; their only objective, their only reason for existence, he insisted, was to sell their poisonous junk heap to the United States.

And since this was closely in tune with the opinion of the vast majority of Americans, his position seemed impregnable. The very dark cloud that hung over Panama in the popular mind appeared immovable, while Nicaragua, by stunning contrast, was seen as a sunny, hospitable land much favored by fortune. Nicaragua would be a fresh start.

Those few Americans who had spoken out for Panama Morgan regarded as fools or, worse, stooges for the transcontinental railroads that were conspiring to defeat any waterway through Central America that they could not own or control. Morgan was a railroad fighter of long standing and the railroads, he insisted, were as much opposed to a canal at Panama as they were to one at Nicaragua. But by playing up Panama they hoped to stall a congressional decision on Nicaragua. He accused no one in particular, but there was no call to. At the heart of the "Panama Plot," the public and most of the press assumed, were E. H. Harriman, J. P. Morgan, and James J. Hill. And quite possibly the assumption was correct, or at least partly so, although neither Morgan nor the newspapers were ever able to produce substantive proof.

In the place of proof were the frequent declarations of the railroad people themselves, and since the railroads had shown no prior aversion to political wirepulling, and since their grip on the country as a whole had become a very live political issue, the specter Morgan raised of paid railroad agents scheming to wreck the canal was one nobody took lightly.

Speeches by others on the subject of the Nicaragua canal filled hundreds of pages of the *Congressional Record*. In the archives of the House and Senate were tens of thousands of pages of reports from special canal committees, testimony from explorers, engineers, sea captains, all supporting the fundamental wisdom of the Nicaragua route. (If pens were spades, remarked the Minneapolis *Times*, the canal would have been dug long since.) There were all the maps and surveys of the Grant expeditions, tabulations on weather and tides and annual rainfall gathered by still further Nicaragua expeditions in the 1880's, when the French were busy at Panama. Most of the popular magazines—*Harper's Weekly*, *Atlantic Monthly*, *Munsey's*, *Century*—had carried major articles on the Nicaragua canal. The Maritime Canal Company, before it went bankrupt, had built a magnificent scale model of its canal, complete with running water and tiny locks that actually worked, and this had been exhibited in Washington and a dozen other cities. American boards of trade, state legislatures, scores of civic groups of one kind and another, had passed solemn resolutions for a Nicaragua canal. A Nicaragua canal had been a showpiece in both the Republican and Democratic platforms. But a clinching argument for Nicaragua heard repeatedly was that if Old Morgan, knowing all he did, having given the better part of a lifetime to the subject, said it was the place, then certainly that must be so.

A Nicaragua canal bill would go before Congress, it was presumed, and Morgan would see it safely and speedily through. The one remaining piece of business was the release of a Presidential study on the "most practicable and feasible route" for the canal. The study had been ordered by William McKinley and authorized by Congress in 1899. It was the work of the Isthmian Canal Commission, the second such high-level commission established by McKinley (the first, the Nicaragua Canal Commission, had been organized in 1897), and it was to be the final word on the subject. Chairman of the commission was Rear Admiral John G. Walker, who had also headed the earlier study, and hence it was referred to as the Second Walker Commission, or more commonly as time passed as simply the Walker Commission. Besides Walker, eight others, most of them eminent civil or military engineers, composed the board. A million dollars had been appropriated. The field work had involved two years, hundreds of men–surveyors, engineers, naval officers, physicians, geologists–and it was in November, only a few days before Hay and Pauncefote met to sign their treaty, that Admiral Walker had marched up the steps to the State Department on his way to Hay's office, two men trailing a few paces to the rear carrying the long-awaited report in two large wooden boxes.

The report was supposed to have remained secret until the President had read it and sent it on to Congress, but on November 21, three days after the Hay-Pauncefote signing, William Randolph Hearst broke the results in the New York *Journal*. One of the admiral's stenographers had been bribed and Hearst had a carbon copy of the full text.

Having considered all factors of climate, health, legal rights, existing franchises, having arrived at probable figures for the cost of construction and operation of ship canals in both Panama and Nicaragua, the Walker Commission had again declared Nicaragua the preferred choice. The issue, it seemed, had been settled once and for all. The rest would be largely a matter of legislative formality.

For those few who bothered to read the commission's report, however, it was obvious that the important news was not the concluding decision for Nicaragua–a decision that had been expected all along–but the exceedingly strong case being made for Panama. There was no need to read between the lines. All one had to do was to look at the technical arguments being presented, none of which was very technical or complicated.

The deciding factor had been the price put by the French company on its Panama holdings. Nicaragua was the "most practicable and feasible" route "after considering all the facts developed by the investigations . . . and having in view the terms offered by the New Panama Canal Company," which were "so unreasonable that its acceptance cannot be recommended by this commission." Yet with amazingly few exceptions the editorial writers and politicians chose to pay no attention to that. The commission's findings were hailed as the ultimate confirmation of the American route.

The *Journal* followed its great scoop with an article on a minority report (also provided by the obliging stenographer) in which the virtues of the Panama route were stressed in further detail by the most eminent civil engineer on the commission, George Shattuck Morison. *The New York Times* and one or two other papers had also made mention of a "Panama Lobby" stepping up its "gumshoe campaign" in Washington and of a "powerful coterie" in the Senate working secretly for the Panama route, irrespective of the commission's conclusions. But the stories were generally discounted. Asked by reporters if he had any knowledge of Panama sentiment among his colleagues, John Tyler Morgan drawled, "I haven't heard a brush crack in the woods about it."

When Congress convened in the first week in December, a House bill for a Nicaragua canal was pushed through committee without a hitch. Its author, William Peters Hepburn, of Iowa, was a Republican with a large streak of vanity who had once blocked a similar bill because it was then called the Morgan Bill. He had decided that if any one individual or party was to be immortalized by the canal legislation it was to be Congressman Hepburn and the Republicans. Morgan had since assured Hepburn that he would not respond in kind, that he would be quite happy to see it be a Hepburn Bill, and so it was expected to pass quite handily.

On December 10, a formal diplomatic convention was signed in Managua "with a view to the construction of a Nicaragua canal by the United States." On December 16, to nobody's surprise, the Senate ratified the Hay-Pauncefote Treaty. Three days later the House of Representatives, by unanimous consent, placed the Hepburn Bill on the calendar for immediate consideration following the Christmas holidays.

Then just before Christmas came reports from Paris that the president of the Compagnie Nouvelle had suddenly resigned. A stockholders' meeting on December 21 had become so tumultuous that the police had to be called in. The gist of the speeches had been to get the United States to buy the canal at any price.

To date, technically speaking, the French company had never really fixed a price for its holdings. Admiral Walker had been informed only as to what the company considered the Panama property, equipment, and franchises to be worth—which was $109,000,000. Having nothing else to go by, Walker and his commissioners had taken that to be the price and had based their decision on it.

The new price, the first price actually quoted from Paris, was presented to Walker by representatives of the company early on January 4, 1902, the morning most of Washington was absorbed in accounts of Alice Roosevelt's coming-out party at the White House the night before. Walker and his eight-man commission had concluded in their report that what the French company had to sell was worth considerably less than $109,000,000. The useful portions of the French excavations they valued at $27,400,000. They were willing to include $2,000,000 for the French maps, surveys, drawings, and records. The Panama Railroad they judged to be worth nearly $7,000,000, and another $3,000,000-plus had been added to cover possible oversights. So the total estimated value came to $40,000,000, which, interestingly, was the precise figure the French were now offering to sell for.

Walker had hurried over to the State Department at noon and from there the news had been taken next door to the White House. The French had not only slashed their price, they had cut it by more than 60 percent. As Admiral Walker was to tell the Morgan Committee in his deadpan fashion, "It put things on a very different footing."

But when the House took up the Hepburn Bill, the debate, if it can be called that, lasted all of two days. On January 9, the House voted all but unanimously—308 to 2—to proceed with the Nicaragua canal. As Mark Hanna observed, probably not one congressman in four had even read the report of the Walker Commission. Morgan, who had read it, and closely, announced that he would commence hearings and see that the bill reached the Senate with all dispatch.

The Administration all this while had been keeping silent, the implicit understanding being that the choice was the prerogative of Congress and that Roosevelt remained a Nicaragua man. But no sooner had the House acted than Roosevelt called the members of the Walker Commission to the White House, one by one, for private consultation. He wished their own personal views, freely expressed, one man at a time.

A meeting of the full commission followed, a closed, secret meeting in the President's office, during which Walker and the others were told to get together and issue a supplementary report. Roosevelt wanted the French offer to be accepted. The conclusion of the commission, he said, was to be unanimous.

Morgan was incredulous when Mark Hanna confronted him with the news. "Go ahead and ask the President if you do not believe it," Hanna replied, and Morgan went down to the White House that same day. What sort of exchange he and Roosevelt had neither man ever disclosed.

That was on Thursday, the sixteenth. By Saturday the papers were saying that Roosevelt had a new canal report in his hands. Walker, intercepted by reporters between the State Department and the White House, would say only that the report was likely to be a disappointment to the public. On Monday, January 20, the story was out. On the motion of George S. Morison, the commission had reversed its decision: Panama was now declared the unanimous choice for the canal.

A general inventory of the French property was provided for the first time. There were some thirty thousand acres of land, which, along with land belonging to the Panama Railroad, comprised nearly all the ground required for the canal itself. There was the railroad. There were more than two thousand buildings (offices, living quarters, storehouses, shops, stables) in addition to the large central headquarters in Panama City and the hospitals at Panama City and Colón. There was "an immense amount of machinery" (tugs, launches, dredges, excavators, pumps, cranes, locomotives, railroad cars), as well as surveying instruments and medical supplies. The excavation already accomplished, that excavation that would be of value according to the commission's own plan, was figured to be 36,689,965 cubic yards.

Very few in Washington missed the point.

Assuming Theodore Roosevelt was as impatient to build the canal as he appeared, then his fastest, most expedient course would be to ignore the last-minute overtures of the French company and let John Tyler Morgan handle the rest. That way there would be no more time wasted. But Roosevelt quite obviously had chosen not to do that. Instead he was flying head-on against the Senator in defiance of all the old man's authority and power, not to mention the popular sentiment of the country. He was throwing all past faith in the Nicaragua route to the winds, and by so doing he was risking still further delays, more interminable debate, and very likely a personal defeat at the outset of his Presidency. Clearly something or somebody had caused him to conclude that Panama was not just the better alternative, but so much better as to be worth making a fight for.

Or possibly, it was being said, he was no less susceptible than his predecessor to the will of the Senator from Ohio. And if Mark Hanna was for Panama, there was no special mystery about that, since it was axiomatic that Mark Hanna spoke for the railroads.

Hanna could not "bamboozle" the American public like a lot of children, declared an irate press. The American people are not fools, said the New York *Herald*. National opinion was unanimous for Nicaragua and the lesson of democracy was to trust the public instinct:

> All the objections shown have been admitted by competent scientific authorities, but their weight is *nil* compared with the instinctive conviction so deeply rooted in the American nation, that the Nicaragua canal project is a purely national affair, conceived by Americans, sustained by Americans, and (if, later on, constructed) operated by Americans according to American ideas and for American needs. In one word, it is a *national* enterprise.

Sentiment, the editors insisted, must be reckoned in national as in personal affairs. The fundamental question was whether the United States Senate would prove more "permeable to foreign influence" than the House had.

The Louisville *Courier-Journal*, in an editorial that was carefully clipped and saved by John Tyler Morgan, wrote of the "bare-faced comicality of the medicated steal: twenty millions to enable the thieves on this side to pass the bill; twenty millions for the insiders on the other side; a few rusty pots and pans and an international law suit for Uncle Sam."

Morgan's frequent assertions that the title of the property was invalid, that Colombia would never willingly abandon its rights on the Isthmus, that political unrest was endemic in Panama, were all very much in evidence now. "Talk about buying a lawsuit," wrote William Randolph Hearst in the New York *Journal*, "the purchase of the Panama Canal would be buying a revolution. Apparently the only way in which we could secure a satisfactory concession from Colombia would be to go down there, take the contending statesmen by the necks, and hold a batch of them in office long enough to get a contract signed."

Meantime, the Colombian minister in Washington, Dr. Carlos Martínez Silva, assured the State Department and the press that his government was ready to deal liberally with the United States concerning

Colombia's isthmian province. The government in Bogotá would show "no mean nor grasping spirit. Everything in the way of a concession the United States needs to warrant it in undertaking to build the Panama Canal, Colombia is willing to grant."

On January 28, Senator John Coit Spooner introduced an amendment to the Hepburn Bill. It authorized the President to acquire the French company's Panama property and concessions at a cost not to exceed $40,000,000; to acquire from Colombia perpetual control of a canal zone at least six miles wide across the Isthmus of Panama; and to build a Panama canal. If a clear title or a satisfactory agreement with Colombia could not be reached within "a reasonable time," then the President was authorized to proceed with a canal at Nicaragua.

If passed, the proposal would obviously transform the House bill into an entirely new measure. It was the strongest evidence of all that Roosevelt had made up his mind that it must be a *Panama* canal. Spooner had shown no prior partiality for the Panama route. But Spooner was an able floor leader for Administration bills who would never have taken such a stand without full White House approval. So plainly the plan had emanated from the White House.

Theodore was still the spinner.

10
The Lobby

> In the course of a very active and very extended professional career . . . the firm of Sullivan & Cromwell had found itself placed in intimate relations, susceptible of being used to advantage with men possessing influence and power.
>
> —WILLIAM NELSON CROMWELL

> The first bugle-note had been heard. I hastened to settle up my business affairs and left France on the *Champagne* . . . for this crusade which was to result in the resurrection of Panama.
>
> —PHILIPPE BUNAU-VARILLA

I

As Chairman John Tyler Morgan gathered his committee for the first hearings on the Hepburn Bill, the idea of building the American canal at Panama, of buying out the French and finishing what they had begun, was altogether devoid of popular appeal and without a single spokesman of national reputation. What open support there was for the Panama proposition was just barely discernible—a few newspapers (the New York *Evening Post* being the most persistent), a few Midwestern business groups, perhaps a half-dozen prominent civil engineers. Political support appeared to be nonexistent. Extraordinary as it may seem in light of what was to transpire, by the start of 1902 not a single politician of importance had ever declared himself in favor of a Panama canal. The idea had no constituency, whereas the enthusiasm for Nicaragua, within Congress and without, appeared to be overwhelming.

Any ordinary citizen who dared even to suggest that perhaps the French had picked the best place after all, or that a Panama canal ought not be dismissed out of hand because it was a French idea or because it would be a *Panama* canal, spoke virtually alone. Old John Bigelow, for example, had become something of a curiosity for espousing such views, as well as something of a nuisance to such influential former colleagues as John Hay, who responded with due courtesy, but nothing more, to Bigelow's lengthy, reflective letters on the matter.

As things stood, there was every reason to assume that the commerce of the world, not to mention the white ships of the United States Navy, would one day be plying the waters of beautiful Lake Nicaragua. And this is doubtless what would have occurred had it not been for certain unexpected events and a mere handful of extremely determined individuals, two of whom comprised the main thrust of what the newspapers darkly referred to as the "Panama Lobby." They were William Nelson Cromwell and Philippe Bunau-Varilla. Their activities to date require some explaining.

Both Cromwell and his French counterpart were small, aggressive, fatherless men who would each be compared to Napoleon. Only Cromwell, however, had made "influence" a profession. Cromwell was something new in the legal world, a corporation lawyer, a kind of mutation sprung forth in the Wall Street jungle during the rise of the railroads. An almost pretty little man, with thick, curly, prematurely white hair and white mustache, he had large, glittering blue eyes—"as clear as a baby's," according to one account—and a smooth, pink complexion that "would not shame a maiden." In striped trousers and morning coat he looked like a clever drama student dressed for the part of elder statesman. But the look he fancied, the role he cultivated, were those of the man with all the cards, and possibly several more up his sleeve. As one young protégé would recall, Cromwell delighted in being known as a mystery man, a puller of strings. An incensed congressman was to call him "the most dangerous man the country has produced since the days of Aaron Burr," which was extravagant, but exactly the sort of remark from which Cromwell took extreme satisfaction.

He had no interest in sensational trial work, never courted publicity. He was a talker man to man. "No life insurance agent could beat him," a reporter for the *World* wrote after a long interview. "He talks fast, and when he wishes to, never to the point." His great genius was for "arranging" things, for planning every move in advance. "Accidents don't happen," he would admonish young associates, "they are permitted to happen by fools who take no thought of misadventure."

William Nelson Cromwell—he preferred the use of all three names—was the good, eager, poor diminutive boy from Brooklyn, the son of a Civil War widow, "a lad of delicate health," who had once played the organ in the Church of the Pilgrims and went to work first as an accountant in a railroad office. He was the model of Ambition Rewarded who began each day at first light and advised others: "A successful man never forgets his work. He gets up in the morning with it, he works all day with it, he takes it home with him, he lives with it." He had worked his way through Columbia Law School in his off-hours, was graduated in 1876, and three years later, with an older, well-established trial lawyer named Algernon Sullivan, founded the Wall Street firm of Sullivan & Cromwell. When Sullivan died in 1887, Cromwell became the senior partner at age thirty-three.

He hired equally promising young men (one of whom was John Foster Dulles) and busily cultivated his own legend. To his more staid peers he seemed a touch vulgar. His "training in finance and accounts," an associate would explain, had "developed in him valuable skills unusual to lawyers of that day who were generally trained in literature, logic, rhetoric, philosophy, and the classics."

Cromwell's fees for straightening out the affairs of troubled corporations or arranging giant mergers were the largest of their kind up until that time. Still in his early forties, he was already a millionaire many times over. When the New York firm of Decker, Howell & Company failed in 1891, with debts of $10,000,000, Cromwell, who had been named assignee, had the company's affairs straightened out in six weeks—creditors paid, operations resumed—and his fee was an unheard-of $250,000. By 1901 he had reorganized the Northern Pacific Railroad and assisted J. P. Morgan in founding the United States Steel Corporation. (He was also among those privileged to participate in the stock syndicate that made the giant steel combine possible, along with such "Lords of Creation" as H. H. Rogers, W. K. Vanderbilt, and John "Bet-a-Million" Gates. Cromwell's share was for $2,000,000, for which he had been required to put up a bare 12½ percent.) He was adviser to and confidant of several of the most powerful men in America, whom he admired and flattered to the skies. Once, speaking before a Wells, Fargo stockholders' meeting, he declared, "Mr. Harriman is the one man to be thanked for what this company has gained through the favor of the railroads. He cannot be replaced, for he moves in a higher world which we cannot hope to enter." Nor had he the slightest compunction about trading openly on such friendships.

In 1894, the year the New Panama Canal Company was organized, Cromwell had become general counsel for the Panama Railroad, a stockholder, and a director. This had come about because he was at the time involved with C. P. Huntington and the Southern Pacific, which by then virtually controlled the Panama Railroad as the result of a traffic agreement. Presently he had started looking after the "interests" of the New Panama Canal Company, promising its officers an "open, audacious, aggressive" campaign of "publicity, enlightenment, and opposition" all planned with "Napoleonic strategy." He was to profess most earnestly later on that his underlying purpose at all times had been to give the United States the best possible canal. But from other things he said and did it is clear that his fundamental objective was to sell the French company to the United States government, or, that failing, to some other government or combination of foreign capital. And for such efforts he expected to be well paid. His fee for services rendered when finally submitted to the Compagnie Nouvelle would be for $800,000.

Few lobbyists had ever gone about their task with such intensity or imagination. He made lobbying one of the lively arts, as someone said. No opportunity was missed. Editors and congressmen were supplied with reams of material on Panama, the French company, the drawbacks of the Nicaragua route. He was in Washington again and again, often for weeks at a time, seeing people on the Hill, negotiating with the Colombians. He had some help from a lawyer named William Curtis and a newspaperman, Roger Farnham, whom he had hired away from the *World*. But he was the spearhead. It was he who counted Hanna and Spooner among his "intimate" friends. It was he who called at the White House.

He made liberal use of his own and his client's money. He brought people together. Once he had even arranged a meeting between his client's representative and William McKinley. On the Hill his strategy was to do everything possible to dampen the Nicaragua ardor and he was as "ubiquitous and ever present" as John Tyler Morgan said he was. Indeed, the hatred he engendered in the old Senator is probably the clearest proof of his effectiveness.

His most demonstrable achievement was the establishment of the Isthmian Canal Commission, at least such was to be his lifelong claim. To bring this off he had concentrated on House Speaker Thomas B.

Reed and Congressman Joseph ("Uncle Joe") Cannon, chairman of the House Ways and Means Committee, both Republicans who quickly saw, he later said, "the wisdom, the justice, and the advantages" of one conclusive, grandiose scientific study and gave it their backing, which was all that was needed. It was an inspired delaying tactic—and a critical one, as things turned out—but it was also an enormous gamble, since a verdict by the commission in favor of Nicaragua would utterly demolish his client's already slim prospects.

Once the idea was in motion he had moved quickly to influence the selection of the nine men who were to serve on the commission. He urged McKinley not to reappoint Admiral Walker. A Corps of Engineers officer, Colonel Peter Hains, and a professor of civil engineering from the University of Pennsylvania, Lewis Haupt, were also unacceptable in his view, since, like Walker, they had served on the earlier Nicaragua Canal Commission and were therefore not without Nicaragua bias. (Professor Haupt was actually on record as saying that nothing could change his mind about the superiority of the Nicaragua route.) A fourth man, Alfred Noble, a noted Chicago engineer, had also been compromised, Cromwell argued, by service on a still-earlier, short-lived Nicaragua canal board. Among the Army engineers, nearly all of whom were strongly, if privately, behind the Nicaragua plan, Cromwell's influence with McKinley was described as "too powerful for ordinary mortals to counteract."

Cromwell failed to block the appointments of Hains, Haupt, Noble, and Walker, but the three other civil engineers chosen were from Cromwell's acceptable list, and among them was the Olympian George S. Morison, whose reputation among Cromwell's railroad friends was second to none.

Once the new commission was set up for business in the Corcoran Building, it was the "silver-tongued" Cromwell who convinced Admiral Walker that the place to commence his studies was in Paris, not Central America. And so it was to France that the nine commissioners and several of their wives had sailed in August 1899, Cromwell, meantime, having hastily departed on an earlier ship.

The final report issued by the commission contains no mention of Cromwell. It is stated only that in Paris the officers of the Compagnie Nouvelle "received the commissioners with great courtesy and were ready at all times to assist them in making a study of this [Panama] route in all its aspects." The most important and attentive of those officers, however, had been the American lawyer. It was Cromwell who turned up at the Continental Hotel to greet the commissioners the morning after their arrival. It was Cromwell who served as master of ceremonies throughout their five-week stay, and who came to bid them each farewell the day they left.

A staggering quantity of material had been gotten up for display, its value far exceeding any Panama data then available in Washington or anything the commission could possibly have assembled on its own in the time available, as Cromwell appreciated perfectly well and as his guests quickly saw for themselves.

They gathered at the company's offices at Number 7 Rue Louis-le-Grand. There were maps, engineers' reports, hydrographic studies of the Chagres River, geologic profiles, reports on test borings along projected dam and lock sites, plans for dams, plans for locks, records of tidal observations on the Pacific, reports on excavation expenses at Culebra, a detailed inventory of the company's equipment and property. Everything was beautifully arranged. Printed copies of the most important documents, a total of 340 different items, had been prepared for each member of the commission, the documents contained in fifteen neatly labeled cream-colored folders and these secured with dark-green ribbons.

At the end of August, Cromwell had arranged a special convocation of the Comité Technique International, a board of engineers established earlier by the Compagnie Nouvelle to evaluate the work accomplished on the Isthmus. It was a genuinely impressive body. The chairman was a retired inspector of the department of Ponts et Chaussées; General Henry Abbot had recently retired from the United States Army Corps of Engineers; there were a chief of the Manchester Canal, a noted Russian engineer, and a former technical director of the Kiel Canal. The year before, this same group had declared its unanimous confidence in the feasibility of completing the Panama canal.

To introduce the gentlemen of the Comité to the gentlemen from the United States, a luncheon was arranged at the sumptuous Pavillon Paillard, a restaurant in the park on the Champs Élysées directly across from the gardens of the Élysée Palace. Everything was done just so—personalized menus with an engraved view of the Pavillon, four wines, six courses—"a very fine lunch & pleasant occasion," noted George S. Morison in his diary, which for George S. Morison was a positively rhapsodic accolade.

General Abbot, who was well known to the commissioners and unquestionably able, told them he was so convinced of the soundness of the French company's overall scheme for a lock canal that he was sure some other country or some combination of foreign capital would jump at the chance to carry on, should the United States be foolish enough to proceed at Nicaragua. For six years after the de Lesseps company failed, the canal had been idle, no digging, no work at all to speak of; but for the past four years, since 1895, things had begun to stir again on the Isthmus, since progress of a kind had to be shown by the new company in order to maintain the Colombian franchise. This was no mere token effort, Abbot assured them, however modest in scale. A long-needed railroad wharf had been built at Panama City; some excavation had been resumed at Culebra. The place was being tidied up, the jungle chopped back again, equipment looked after. This was phase one, he explained; phase two would be to go to "the great moneyed people of the world and show them it would be a good investment."

The technical discussions that followed during the next several weeks were conducted as if "before a court of highest jurisdiction," according to the dapper little attorney from New York, who by his own subsequent declaration was "in attendance" every moment. And afterward, back in Washington, before the commission left for its own firsthand inspection tours at Panama and Nicaragua, he had "kept in constant and personal communication with various members of this body, adding to their information, furnishing documents . . . overcoming their hesitations . . . etc."

Yet his single most valuable service, Cromwell later avowed, was the personal conversion of one man, Mark Hanna. Just when Hanna saw the light is not clear in Cromwell's account, but it was he, and no one else, Cromwell insisted, who had led the famous Senator to the truth; it was he who had made Hanna his specialty, from the time McKinley asked Hanna to post himself on the business and technical aspects of the canal project and Hanna had dutifully taken his place on Old Morgan's committee.

II

Cromwell's counterpart in the crusade, the former acting director general of the Compagnie Universelle, was no less passionately committed than in earlier years to The Great Adventure of Panama. Now in his mid-forties, he looked a little stouter than before, the hairline had receded considerably, and what hair there was he kept cut extremely close. He had also acquired a certain fixed look of fierce pride. In photographs from the time, he focuses directly on the camera; he is flawless, stiff-necked, and unflinching, the eyes steady and grave. As in earlier days the face is dominated by a large mustache, only now it has been waxed to fine spikes and looks ornamental, overdone. It might be something pasted on in jest were it not for the eyes, which are plainly those of a man who never did anything in jest. Roosevelt called it the look of a duelist.

Philippe Bunau-Varilla was to be greatly misunderstood in another generation. The tendency among historians would be to see him as an almost comic figure, a sort of road-show French schemer who, though colorful enough in his fashion, should not be taken altogether seriously. Possibly the mustache had a bearing on that judgment. But primarily it was Bunau-Varilla's own account of all that happened, his obsession with the first person singular in everything he wrote, which to even the most tolerant modern reader seems so absurdly one-sided, so inflated by self-interest, as to be ludicrous. In his books, the most important of which is *Panama: The Creation, Destruction, and Resurrection*, his ideas are invariably brilliant, his actions invariably bold, inspired, *pivotal.* Anyone who opposed him or dared to disagree with his point of view is portrayed as stupid or villainous or mentally deranged. Those who see things as he does are gentlemen of the first magnitude, uncommonly intelligent and marked by a high sense of moral purpose.

He saw himself as the gallant crusader—"a soldier of the 'Idea of the Canal' "—going forth to battle Prejudice in the cause of Scientific Truth. He was still the central figure in a spacious romance. He would, he had resolved, restore the honor of France, an honor tarnished by Panama and by the Dreyfus case. Single-handedly, if necessary, he would salvage "The Great Idea of Panama."

Things happened to him, he writes, as if they were occurring in a work of fiction. Fortune "smiled" on him as it did on very few. "At every turn of my steps it seemed as if I were accompanied by a protecting divinity."

Yet in view of what in fact did happen, considering the romantic tradition he was a product of, there is little wonder he felt as he did. The shame is that he also felt compelled to unfold it all in such highblown fashion. He never seems to have understood how much more readily his story would have been accepted—especially in a less posturing, more skeptical age—how much more impressive it would have been, had he only told it straight. Moreover, there is ample evidence that the man himself bore little resemblance to the character he becomes in his books. In truth he was a hardheaded, practical, personable, exceptionally intelligent, almost unbelievably energetic individual who made an impression on people that they would remember all their days.

If anyone failed to take him seriously at the time, there is no evidence of it. Edward P. Mitchell, of the New York *Sun,* among the ablest newspapermen of the day, later wrote: "When I came to know him well I found him to be in mind and will one of the most surprising dualities it was ever my privilege to encounter; Napoleonic, indeed, in his practical energy and resourcefulness, yet an idealist of the first grade in disinterested devotion to a patriotic sentiment." John Hay, who was to have more direct dealings with him than anyone in Washington, was astounded by the man's diversity and by the uncommon speed with which he could accomplish things. George Morison, appraising him purely on professional grounds, declared him "brilliant," a tribute George Morison seldom conferred on any man.

Even those who were instinctively suspicious of his motives never seem to have discounted his ability. It would be a grave mistake to underrate this man, the Chicago engineer Alfred Noble warned Senator Morgan in confidence at the time Bunau-Varilla arrived in the United States to begin his campaign.

His English was excellent and spoken with marked precision. There were bows for the ladies; his table manners were impeccable. He was the cultivated, upper-class European *par excellence* and he knew exactly how to gain attention wherever he went. "He didn't just come into a room, he made an *entrance,*" recalled Alice Roosevelt Longworth admiringly. It was he, rather than the theatrical-looking Cromwell, who had the actor's timing, the intuitive feel for the dramatic gesture. The engineer was the evangelist of the pair, oddly enough, and it was he who became "the peripatetic spellbinder" (as John Tyler Morgan would say), carrying his campaign cross-country much as de Lesseps had done twenty years before.

The impact of his whirlwind tour was unmistakable. He was a novelty. American audiences had simply not encountered an authority on Panama before, let alone an engineer who had had the experience of actually attempting to dig a canal there. And the engineering argument for building at Panama rather than at Nicaragua had never been set forth publicly and with conviction.

"Every phase of the canal question was at tongue's end with this envoy of the Panama idea," wrote Mitchell. But most appealing, one gathers, was the capacity to invigorate others with his vision, to light the imagination with the possibilities of a Panama canal. It was a capacity many of his listeners thought quite remarkable in a Frenchman. He had, the newspapers said, "a sort of resourceful energy which some people are accustomed to regard as peculiarly American." He was "the Frenchman who is like an American."

When the original canal company went bankrupt in 1889, Bunau-Varilla's first impulse, he afterward explained, was to rally his countrymen to carry on with the work. To this end he had plunged into politics and campaigned for election to the Chamber of Deputies—"to lash slander with the whip of truth." Discouraged after a narrow defeat, he had come to New York to look up John Bigelow and get his advice. This friendship between the older man and the brilliant youth "ripened into almost a father-son relationship," as Bigelow's biographer would write. Bigelow told him to go home and put his case in writing, with the result that in 1892 he produced a book, *Panama: Past, Present, and Future,* in which the Panama and Nicaragua routes were compared on purely technical grounds, something that had not been done before other than in government reports of the kind produced in the 1870's.

But the idea of getting the United States to take over at Panama had either not dawned on him as yet or was still too much at odds with his vision of French destiny. So he had gone first to see the Russians.

There had been a chance meeting with a Russian prince on a train in 1894, after which Bunau-Varilla rushed to St. Petersburg to try to convince Tsar Alexander III that Russia should provide the capital to finish the canal. He never saw the Tsar, only the Tsar's powerful Minister of Finance, Count Sergei Witte. He told Witte that a Panama canal and the Trans-Siberian Railroad, then under construction, could be the perfect Franco-Russian counterpart to the Anglo-American combination of the Suez Canal and the transcontinental railroads. A lock canal at Panama, Bunau-Varilla said, could be finished in four more years if the Russian sovereign would give a guarantee of 3 percent to the necessary capital, which he put at $140,000,000. Witte promised to present the plan to the Tsar and Bunau-Varilla returned to Paris bursting with expectations. The French government was astonished by what he had to report, and highly interested, according to Bunau-Varilla, whose word is all we have to go by. The government fell shortly afterward, however, and Alexander II was assassinated. Moreover, the liquidator of the defunct Panama company, furious over Bunau-Varilla's meddling, saw to it that he would have no more say in company matters. What the consequences might have been had the Russian scheme gone any further is interesting to speculate on.

Not for five more years was his American crusade launched, in

Paris, the summer of 1899, when the Isthmian Canal Commission arrived.

"Everybody, the world over, then supposed that the Nicaragua Canal—the old American solution of the problem—would be carried out. I determined thenceforth to center my efforts toward the adoption of Panama by the United States. The task seemed impossible of achievement!"

There had been a letter from John Bigelow, an amazingly well-timed, plot-turning letter just like those in novels. What was urgently needed was "someone competent to persuade our engineers," wrote the dignified old New Yorker. "I shall be eighty-one years old the 25th of this month," Bigelow wrote, "and of course am not of much use in a fight except perhaps to beat the drum." He had sent one of the engineers on the commission, Colonel Oswald Ernst, to see Bunau-Varilla first thing on arriving in Paris, and a neat, scrubbed-looking Yale man and lawyer named Frank D. Pavey, who was in Bunau-Varilla's pay later, if not then, was also instrumental in arranging the first meetings.

The little Frenchman applied himself in the tradition of his former leader and idol, Ferdinand de Lesseps. But while Cromwell was devoting his energies to the entire commission, Bunau-Varilla, who at this stage had still to meet Cromwell, concentrated on just three of the group—Colonel Ernst, Professor William Burr, of Columbia University, and George S. Morison, the three who, with the concurrence of Admiral Walker, had agreed to make Panama their particular concern. And of the three, Morison was the primary target, Morison having the greatest professional eminence and a reputation for being a highly independent and persuasive individual in his own right.

"Our conferences were long and frequent," wrote Bunau-Varilla, among the few understatements he ever permitted himself. They met at one of his favorite restaurants or at his palatial gray stone *hôtel particulier*, on the Avenue d'Iéna, near the Arc de Triomphe, then, as later, the most fashionable of Paris addresses. "Dinner with him meant half past eight," Frank Pavey would recount, "and after dinner we settled down in his library, and he never let go of an American victim when he got one in that library until he thought he had converted him . . . the first time I dined in his house I stayed until two o'clock the next morning, listening to his picturesque and fascinating argument."

The wives of the visiting Americans were often included in such evenings and their host could not have been more charming. He had a fund of fascinating conversation on all manner of subjects, but the great dominating topic that summer was the Dreyfus trial and to their amazement they learned that he personally had played a critical part in the drama. Among his interests since returning from Panama had been the newspaper *Le Matin*, which he had purchased and put under the charge of his brother, Maurice. It was Maurice who had obtained a photograph of the incriminating letter that Captain Dreyfus had allegedly written to the German attaché. Philippe had known Dreyfus years before at the École Polytechnique, where they had been friends and classmates, and upon seeing the photograph he had hunted up an old letter from Dreyfus. The difference in the handwriting was not merely obvious, but astounding. So, convinced of Dreyfus' innocence, the brothers had published pictures of both letters in *Le Matin*, a sensational bit of journalism that led to the reopening of the case and a story that held the American guests spellbound.

Before their departure that September, the three engineers had been given a copy of *Panama: Past, Present, and Future* and instructed by the author to throw it away if a single mistake in fact or logic could be found. "When my three eminent new friends left Paris a large hole had been made in the dam of prejudice then existing against Panama in their minds—as in everybody's."

It was not quite large enough, however. The following year, in the autumn of 1900, the commission issued a preliminary report recommending Nicaragua. "The fight to a finish was now to begin," wrote Bunau-Varilla and from this point on, by his own account, he was accompanied by deep mystical feelings of Fate taking charge. It was as though everything that happened had been prearranged.

An unexpected cable arrived, an invitation from some Cincinnati business people who wanted him to come to their city as soon as possible to lecture on the comparative values of the Panama and Nicaragua routes. They too had met him during a summer sojourn in Paris. "We have not forgotten the presentation with which you favored us," they wrote in a follow-up letter, "—so vivid, so comprehensive, and so convincing—and we are anxious to have it reach our American public in the most effective manner we can devise."

The "bugle-note had been heard." He sailed on a ship called *Champagne.* There was little chance of his influencing public opinion in America, he had decided, but he was bound to "conquer for the Panama side" those who could.

Strolling the deck he struck up a friendship with a French priest who after hearing him expound on his favorite subject suggested that he look up an American whom he, the priest, had met in Rome. The man's name was Myron T. Herrick. He was a Cleveland banker, a friend of President McKinley's, explained the priest, who, as Bunau-Varilla tells the story, had now become part of the great puzzle Fate was piecing together. "Every time I was in need of a man he appeared, of an event it took place."

But a letter written by Lieutenant Commander Asher Baker, an American naval officer and another of those Americans Bunau-Varilla had managed to cultivate in recent years, suggests that Bunau-Varilla had more than Fate working for him. "Everything has been done for Philippe," Baker informed the lawyer Pavey. Baker, who met the Frenchman at the ship, was being reimbursed for his expenses and services by Pavey, who himself was serving as Bunau-Varilla's "man" in New York.

The historic whirlwind crusade began with an after-dinner speech before the Cincinnati Commercial Club in a large room bedecked with French and American flags the evening of January 16, 1901. The speech, the first Bunau-Varilla had ever attempted in English, was an unqualified success. He said approximately what he would say wherever he went thereafter and with such winsome conviction that he held everyone's attention. Included in the large collection of his papers on file in the Library of Congress is an affectionate little note from his young daughter back in Paris, who had enclosed a tiny map of Central America that she had drawn most carefully with pen and crayons. Hovering over Nicaragua is a black devil brandishing a pitchfork, while above Panama sails a winged angel. The conception was the very same as that of her *Papa adoré* and it was precisely that kind of partiality, as much as the barrage of facts he had at his command, that so held his audience. It was "the intensity of conviction which inspired all your utterances" that had the most telling effect, wrote one of his hosts, adding, "I love a man who loves a great cause."

The standard speech began with a profession of independence: he represented no private interests, which was to be taken as a guarantee that he had not come as a salesman for the new Panama company. His mission was purely to defend "a grand and noble conception which gave me many happy years of struggle and danger . . . during which I do not remember one hour of despair."

That said, he would get to particulars. He stressed basically what was to be stressed by the revised report of the Walker Commission: a Panama canal would be a third the length of a canal at Nicaragua; it would have fewer curves; it would require less excavation in total, fewer locks; it would cost less.

He talked about the railroad at Panama, the harbors at Colón and Panama City. He referred to the Chagres as "this monster of the imagination." He did not talk about the rains or the slides at Culebra. He did not talk about yellow fever or malaria. He did not mention the uprising of 1885 or how he had felt on seeing Colón burned to the ground. He did not describe the Chagres in flood. Nor did he bring in the fact that he was a stockholder in the Compagnie Nouvelle or the circumstances by which that had been brought about.

There was, however, a further element to the set speech that seemed almost incidental at first, but that quickly became its most important element.

Panama had no volcanoes. There was not a single volcano, active or inactive, within 180 miles of the Panama line, he assured his listeners. In Nicaragua this was by no means the case. In Nicaragua in 1835 the eruption of the volcano known as Coseguina had lasted nearly two days. The noise, he said, had been heard a thousand miles away and enough stone and ashes had been ejected every six minutes to fill a Nicaragua canal.

He was not the first to have raised the issue. Humboldt had cautioned that there was "no spot on the globe so full of volcanoes" as Nicaragua. John Lloyd Stephens, as will be recalled, had made much of Mount Masaya and its potential as a tourist attraction. At the Paris congress, Commander Selfridge had cautioned against Nicaragua for this very reason. But those who heard Bunau-Varilla lecture regarded it as a fascinating revelation.

Always "the force of things" had driven men to build at Panama, he would conclude; it had been the Spanish gold trail to begin with, then the American railroad, then the de Lesseps canal. At times men had thought otherwise and intended to build elsewhere, "but the force of things drives them to Panama and it will again."

It was the volcano part of the speech, however, that had the greatest impact.

Among his Cincinnati hosts were several who were personally acquainted with Myron T. Herrick. Herrick was not just a friend of McKinley's, Bunau-Varilla now learned, but of "a man far more important for my purpose," Mark Hanna. A phone call was put through, letters of introduction were prepared, and Bunau-Varilla took the night train for Cleveland.

At a private lunch at a Cleveland business club, seated with the tall, inordinately handsome Herrick and some twenty other pillars of Cleveland enterprise, he held sway for three and a half hours, popping up every so often to illustrate a point on a blackboard that had been wheeled in. "Never did a more propitious occasion offer itself, nor a completer success crown my efforts. All who listened to me, and whom I had made sincere and deeply convinced believers in Panama, formed the circle of Senator Hanna's intimate friends."

From Cleveland he went to Boston where he spoke at a banquet at the New Algonquin Club the evening of January 25, 1901. "This French engineer," observed the Boston *Herald* in a long, glowing editorial, "treated the matter [of a canal at Panama] from a distinctly professional point of view," something quite novel in Boston. He was in Chicago a week later, accompanied by Asher Baker, who handled the advance arrangements. His host was James Deering, of the Deering Harvester Company, and the lecture this time was at the Central Music Hall, where he was introduced by the illustrious civil engineer William Sooy Smith.

"He lectured before 250 representative people," Baker reported excitedly to Frank Pavey. ". . . Western Society Civil Engineers, members of the Nicaragua Canal Commission, most of the solid and very well known Chicago Clubmen were there. I introduced him to Marshall Field, Robert Lincoln and a lunch was given him at the Club . . . there was a dinner and theater every night. Coquelin and Bernhardt were in town, the whole trip was simply perfect." Best of all, Baker went on, "I have arranged through most important people TO HAVE HIM MEET SENATOR MORGAN in Washington (!!!!) . . . in an *intimate* and *friendly* way. It would take *pages* to explain how this was brought about."

Back in New York briefly, Bunau-Varilla dined with George S. Morison, who advised him to make less of the volcano issue. They made quite a pair. The stiff, tiny Frenchman with his waxed mustache and bullet head was often taken for a military attaché; Morison, a figure of vast bulk, ponderous double chins, and walrus mustache, looked like a German sausage maker. While Nicaragua was undoubtedly an area of volcanic activity, Morison did not believe that would have any serious effect on canal structures. From the engineering point of view, the issue was a phony.

Through Cyrus McCormick, of Chicago, a speech was arranged at Princeton University, and it was followed by a half-hour appearance before the New York Chamber of Commerce, this being, in Bunau-Varilla's private estimate the most important of all possible public platforms. The Chamber of Commerce audience was polite and unenthusiastic; still, the resulting publicity had great value. At Philadelphia, Bunau-Varilla told an especially large and attentive audience that to prefer Nicaragua over Panama was equivalent to preferring the stability of a pyramid resting on its point to one resting on its base; ". . . and to that stability is attached the prosperity and welfare of a whole continent."

He stopped always at the best hotels. He was extended guest privileges at the best clubs. In return he was generous with theater tickets and fine cigars ($28 worth of "Segars" are included on one hotel bill). For the wives of his hosts there were enormous bouquets of roses and, invariably, a prompt, gracious thank-you note (for "one of the most grateful remembrances of this agreeable sojourn in America"). For the dutiful Asher Baker, there was a $100 clock from Tiffany.

Busy as a day might be, there was always time for a dozen or more letters—to people he had just met, or, more often, to friends of people he had just met—asking for doors to be opened, introductions arranged, contacts. He prepared a pamphlet entitled *Panama or Nicaragua?* and had thirteen thousand copies printed and mailed. Again, disregarding what Morison had said, he hammered away at his volcano story. Let those inclined to dismiss his warnings take note:

> Open any dictionary of geography, any encyclopedia, and read the article entitled "Nicaragua." I will say also: Look at the coat of arms of the Republic of Nicaragua; look at the Nicaraguan postage stamps. Young nations like to put on their coats of arms what best symbolizes their moral domain or characterizes their native soil. What have the Nicaraguans chosen to characterize their country on their coat of arms, on their postage stamps? Volcanoes!

The mailing list for the pamphlet included every congressman, the governor of every state, a thousand bank presidents, some six hundred shipowners, two hundred merchants reputedly worth more than $100,000, the editors of four thousand newspapers and magazines, hundreds of boards of trade and chambers of commerce, plus all those names on the list he himself had compiled during his travels, a list that by now came to nearly a thousand names.

He had John Bigelow send copies to Secretary Hay and Admiral

Walker, with covering letters explaining how he had first met the distinguished French engineer in Panama. Hay, who had once served as Bigelow's aide in the Paris embassy, confided in response that of course there was "a good deal of searching of hearts" over the proper path for the canal, but reminded his old friend that the decision did not "lie in the discretion of the Executive." Walker's reply was that the Frenchman was making too much of the volcano matter.

Having been steadily on the move for close to three months, Bunau-Varilla talked privately in New York to Bigelow and others of sailing for home. The interview with Senator Morgan remained on his schedule still, and he had not met Senator Hanna, but for pressing personal reasons he thought it time to wind things up. Once again, if his story is to be believed, Fate stepped in.

> Towards midnight, as I was about to go out for a breath of fresh air before retiring, I met a party of people in evening dress entering the Waldorf Astoria. My surprise was great when I saw at the head of them Colonel Herrick with a lady on his arm, and behind them Mrs. Herrick, accompanied by a short stout gentleman who limped slightly.
>
> His characteristic face, so frequently reproduced in the papers was familiar to me. . . .

It was Hanna, and Herrick happily made the introductions. "Ah!" Hanna said (recounts Bunau-Varilla). "Monsieur Bunau-Varilla, how glad I am to meet you!" More important, the Senator wished to have M. Bunau-Varilla call on him in Washington anytime that was convenient. "The ice was broken, under the best and most cordial conditions," wrote the author of the scene.

His love of the chance encounter, of famous figures in elegant attire, of fateful exchanges between men of power made in a suitably grand setting, was very great. Perhaps this is the way it happened, perhaps it is not. But he *did* go directly to Washington "to attack the political fortress." He saw Hanna at the Arlington Hotel, then the Senator's Washington residence, and Hanna smoked and listened, his large cigar poised in a surprisingly delicate hand. According to Bunau-Varilla, the interview was decisive, which makes a mockery of Cromwell's subsequent claims. "Monsieur Bunau-Varilla, you have convinced me," Hanna is said to have exclaimed when it was over. He naturally wanted to find out what Panama specialists on the commission thought, but: "If, as you assert, they think as you do, I shall go over to your side."

A few days later Bunau-Varilla was at the White House, chatting pleasantly with William McKinley. The introduction this time had been made by Charles G. Dawes, Comptroller of the Currency, whose friendship Bunau-Varilla had acquired in New York and again as a result of another chance encounter in the Waldorf lobby. He did no more than pay his respects at the White House. As he later explained, he did not wish to "inflict" a long lecture on the President, knowing the value of his time *and* "that the opinion of Senator Hanna would be his [McKinley's]own." In other words, he had already spoken to the head man.

There was another, final encounter in Washington and it must have been a memorable one.

A little after dark he rode up Pennsylvania Avenue toward the Capitol and turned left at John Marshall Place to a tall brick row house, Number 315, the home of John Tyler Morgan—"the Lion's cage." He went convinced that the old man was a dangerous paranoid. "The fanatical and almost demented state of mind of the old Senator, after twenty years' uninterrupted efforts for Nicaragua, prompted him to see conspirators everywhere." That Morgan might be a man of keen intelligence, whose motives, by his own lights, were quite as noble and patriotic as his own, apparently never occurred to him. Again Bunau-Varilla's account is the only one available.

"My visit produced a deep impression on him. In spite of his apparent courtesy I saw he was trembling with passion." Morgan insisted on doing all the talking and this it seems was more than Bunau-Varilla could endure. "But the volcanoes of Nicaragua—" he blurted in desperation, cutting Morgan off in mid-sentence. Morgan would hear none of it. "Now, between ourselves," he thundered, "you would not put one dollar of your own money in this absurd project—in this rotten project—of Panama!"

Apparently they were both on their feet by this time and Bunau-Varilla, unable to contain himself, lifted his hand to strike Morgan across the face. But the hand stopped in midair; he had a sudden vision of giant newspaper headlines—FRENCH ADVENTURER ASSAULTS DEFENDER OF NICARAGUA DREAM. Morgan had deliberately provoked him, he now saw in a flash; the whole encounter had been arranged to trap and destroy him. "I lowered my half-raised hand, and extending it solemnly toward the Senator, I said: 'You have just inflicted upon me, sir, a gratuitous and cruel insult. But I am under your roof, and it is impossible for me to show you my resentment without violating, as you do, the laws of hospitality.' "

And having delivered that little speech, the Bonaparte of Engineers turned on his heel and strode out the door.

He sailed for France on April 11, 1901.

What had it all cost? And who had paid for it?

Philippe Bunau-Varilla would say only that he had met all his expenses himself, out of a private source that also remains something of a mystery and that had been the subject of resentful, unpleasant talk in Paris.

The situation was this. Years before, at Panama, when he resigned his position with the canal company and went to work as a private contractor at Culebra, he had been able to take only a government salary because of a rule requiring all French government engineers to remain in service, accepting no pay or fees from private sources, for a minimum of five years. However, he had seen to it that his brother, Maurice, was put into the Paris office of the contracting firm as its financial manager, and he and his brother had made a secret agreement. A salary would simply be put aside for him until the required five years were up. It was a maneuver that evoked no little disdain when revealed later, but Bunau-Varilla maintained that the money was rightfully his and, furthermore, that it enabled him to "consecrate" his life to the Panama canal, "to save the noble conception of French genius through its adoption by America." How much money was involved, how much of it he may have used, if any, has never been determined.

It is quite certain, nonetheless, that he did have a direct monetary interest in the fortunes of the new canal company, since he and his brother were what were known as "penalty stockholders."

The founding of the Compagnie Nouvelle had been arranged in a most ingenious fashion, which was the chief reason why Senator Morgan and others viewed that whole organization as no better than an assembly of crooks. The court-appointed liquidator of the old de Lesseps company, in the interests of the stockholders in the old company, had devised a very direct and effective means of capitalizing the risky new company.

Those French contractors who had worked on the canal—and who were still solvent—were simply told that they could either invest in the new company or face prosecution for fraud and breach of contract. The rush to buy stock was pronounced. Two-thirds of the new company's capital, some $8,000,000, was raised in this fashion. In plain fact there would have been no new company had the liquidator not resorted to this bit of blackmail, a point Morgan had made more than once on the floor of the Senate.

The largest of these "penalty stockholders" was Gustave Eiffel, and so ostensibly he stood to gain the most were the company's holdings sold to the United States. Threatened with an 18,000,000-franc breach-of-contract suit, Eiffel had put 10,000,000 francs ($2,000,000) into the new company. The investment of the Bunau-Varilla firm was 2,200,000 francs.

So one theory is that Bunau-Varilla had come to the United States representing not only his own and his brother's interests, but those of Eiffel and the other penalty stockholders, none of whom was permitted to have any say in the management of the company, and few of whom had much respect for the way in which the new company was being managed.

Another intriguing theory is that Bunau-Varilla had been "discovered" and subsidized by the Seligmans, the great Jewish financiers of New York, whose reputation for the strictest integrity had been badly stained by their prior role in Ferdinand de Lesseps' Comité Américain. The late Jesse Seligman had been vigorously interrogated before a congressional committee at the time of the Panama Affair. Nothing very serious had been turned up by the committee, other than the obvious fact that the Seligman firm had been paid an exorbitant amount to do no more than lend its name to the de Lesseps scheme. Nonetheless, the Seligmans were eager to see the legitimacy of the Panama idea restored and thereby justify their prior involvement. And so, the theory goes, it was they who invented "The Man Who Invented Panama," Philippe Bunau-Varilla, who had initially caught their attention through his role in the Dreyfus case.

That Bunau-Varilla knew the Seligmans quite well, and Isaac Seligman in particular, that the family took a keen interest in his lobbying activities in Washington, are matters of record. Isaac Seligman, for example, wrote letters of introduction in his behalf, including one to Mark Hanna, and went out of his way to speak to Hanna privately about Bunau-Varilla's engineering credentials. But if Bunau-Varilla was actually the creature of the Seligmans, or in their pay, there is no solid evidence of it, and to his dying day he would angrily denounce any suggestion that he had ever been anyone's agent or taken money for anything he ever said or did about Panama.*

* As late as 1939, when *Life* magazine ran an article in which he was referred to as a lobbyist and an adventurer, Bunau-Varilla, at age eighty, responded that he

But the puzzle the man presents is made still more complex by the very existence of the private, personal sources from which, by his own account, he drew his expenses, as well as still larger outlays to come. Where, how had he acquired all the money? How could he afford the enormous house on the Avenue d'Iéna, a house in which there were "servants to wait on the servants," as one member of the family would recall. No one knew, or at least no one said. The son of an unwed mother of no apparent wealth, a scholarship student at the École Polytechnique, he had gone to Panama, where theoretically he had earned only a modest government salary, then returned to Paris to dabble unsuccessfully in politics, buy a newspaper, and write books about the inherent Genius of the Idea of Panama. Yet somewhere along the line he had become an extremely wealthy man. His wife, a semi-recluse who took her meals alone in her room for fear of catching some disease, was not a wealthy woman. To his own descendants the origins of the family fortune would remain a mystery.

The only hireling, the only mercenary in the crusade, according to his version of the story, was Cromwell, whom he had come to detest and whom he customarily referred to as "the lawyer Cromwell," the word "lawyer" to be taken as an epithet. The most Bunau-Varilla could ever bring himself to say for Cromwell was to call him "an active and useful messenger between important men," but then added on another occasion: "An active go-between will easily think he is the author of the messages he has to carry."

It was not until the following year, 1902, that these two remarkable figures actually met for the first time. During Bunau-Varilla's initial campaign they had kept as clear as possible of each other despite the obvious benefits some degree of cooperation might have produced. The Frenchman never asked the attorney for favors; the attorney made no use of the Frenchman's technical expertise or his skill at persuasion. Most likely Cromwell had been so instructed by his client in Paris, who, with Bunau-Varilla's Russian episode in mind, probably regarded him as unreliable and a possible embarrassment. And Bunau-Varilla doubtless felt that any overt connection with someone known to be in the employ of the Compagnie Nouvelle, and especially someone whose allegiances were so plainly for hire, could only jeopardize his own stance as the Champion of Truth.

Whatever the explanation, each man would cast himself in the hero's

---

had been no such thing: "Unless you call adventurer a man who sacrifices his time, his money and his scientific capacities to the glory of his nation and to the service of her great friend the United States. . . ."

role when it came time to account for what happened and would pointedly belittle or ignore any constructive part claimed by the other.

Cromwell's claim that he had inspired the creation of the Walker Commission was, for example, utterly absurd, according to Bunau-Varilla. He was the one who had done that; he had convinced Asher Baker that Panama was the place for the canal and Asher Baker, during the winter of 1898–1899, had "enlightened" Speaker Reed and Congressman Cannon.

As for Cromwell's boasted influence on Hanna, that, said Bunau-Varilla, was strictly a question of business as usual for "the lawyer Cromwell." During the Presidential campaign of 1900, the chairman of the Republican National Committee, Mark Hanna, had received a donation from Cromwell of $60,000, a donation that Cromwell had charged off to the Compagnie Nouvelle. In return for the donation, said Bunau-Varilla, Hanna had seen to it that the Republican platform called for the construction of an "isthmian" canal, rather than specifying one at Nicaragua, as the Democrats had done. And to that extent only would Bunau-Varilla acknowledge a Cromwell role in Hanna's conversion.

How he learned of the donation Bunau-Varilla never said. But the idea that $60,000 would have caused Hanna to make any such change seems highly remote and suggests that Bunau-Varilla may never really have understood Mark Hanna, who was accustomed, as he himself said, to frying bigger fat and never with strings attached. Neither Hanna's vote nor his public expressions were ever for sale, whatever his faults and irrespective of his notoriety as "Dollar Mark," the brutal money-bags of the party.

Later, for the public record, Cromwell would tally up the most amazing list of accomplishments in behalf of his client, but that was mainly to justify his staggering $800,000 fee. He was the professional putting the best shine possible on services rendered. To Bunau-Varilla the client was posterity, the judgment of history, before which he wished only to appear as the unrivaled knight-errant. Neither one ever fully appreciated the contributions made by the other. Neither one was ever quite capable of telling the whole truth.

III

On April 10, 1900, Admiral Walker had addressed a letter to the president of the Compagnie Nouvelle. Did the company have a clear title to its franchises and property on the Isthmus, the admiral wished to know, and for what sum, in dollars and in cash, would the company be willing to sell these franchises and property?

On June 25, 1901, still having received no definite answers from Paris, the admiral made a special trip to New York to call on William Nelson Cromwell at his offices at 49 Wall Street. The commission was nearly finished with its studies, the admiral told Cromwell. There was, therefore, an urgent need for a firm price from the French company. Did Mr. Cromwell have an idea what figure his client had in mind?

Cromwell promised to look into the matter. His cable to Paris, sent later in the day, was so blunt about the state of things that the officers of the company not only refused to make a definite offer of sale, but they informed Cromwell by return cable that his services as attorney were no longer desired. Apparently they had had enough of his high-pressure methods and his liberal use of their money.

So that fall, following the death of McKinley, when the report of the Walker Commission was about to be released in Washington and the second Hay-Pauncefote Treaty was about to be signed, the Panama Lobby had been reduced to a party of one, Bunau-Varilla, who now came hurrying back to New York.

The assassination at Buffalo had been a terrible blow to Cromwell and Bunau-Varilla alike, both having spent so much of their time and energies cultivating Mark Hanna, whose relations with the new President were known to be far from smooth. When Roosevelt had been merely Vice President, neither Cromwell nor Bunau-Varilla had bothered to pay him any attention.

Arriving in New York on November 13, Bunau-Varilla found the situation "as bad as it could possibly be." He rushed about trying to determine which way the wind was blowing. He must meet Roosevelt face to face he told Frank Pavey and others, but nobody seemed to know how to arrange that. Within a week Hay and Pauncefote had signed their treaty and cartoons in the papers showed John Bull swinging wide the gate to Nicaragua as a jaunty Uncle Sam marched through with pick and shovel.

But then Hearst broke the Walker Commission report, and if Hearst and others missed its importance—that the French company's price tag was all that had kept the commission from naming Panama as the most advantageous route—Bunau-Varilla did not. With little delay he was on his way back to France again.

Exactly what happened in Paris in the next few weeks can only be roughly pieced together. On December 17, he received a telegram from Washington from a man named Walter Wellman, a reporter for the Chicago *Times-Herald* and another of the contacts he had established. Perhaps he was paying Wellman, perhaps Wellman was doing favors for some of the Chicago industrialists who had been caught up in the Panama campaign.

VARILLA
53 AVENUE D'IÉNA, PARIS

CONFIDENTIAL INFORMATION. COMMISSION SENATE PROBABLY ACCEPT OFFER FORTY MILLIONS. IMPERATIVE NOT HIGHER. MOVE QUICKLY.

WELLMAN

Bunau-Varilla's answer read as follows:

WELLMAN
1413 G., WASHINGTON

THANKFUL TELEGRAM AM MAKING MOST ENERGETIC EFFORTS TO MAKE PEOPLE UNDERSTAND SITUATION

VARILLA

He was present at the riotous stockholders' meeting of December 21, and he held forth immediately afterward in a private session with the new president, Marius Bô, and Henri Germain, of the Crédit Lyonnais, who, like Eiffel and Bunau-Varilla, had also been steamrollered into investing in the new company.

A price must be set at once Bunau-Varilla told them. Time had run out. Yesterday they might have done it; yesterday they might have gotten $60,000,000, perhaps $70,000,000. But yesterday was past. The price now must be $40,000,000 and they must accept that figure. Congress would convene again in two weeks. If by then the price had not been settled, all would be lost and they would have to accept the responsibility.

On New Year's Day, in a large advertisement in *Le Matin* that cost him nearly $6,000, he took the company to task for neglecting its own interests as well as the honor of France. On January 3, he sent identical cables to Senators Hanna and Lodge, to Wellman, John Bigelow, Myron T. Herrick, Professor William Burr, and George Morison:

CONSIDER ALMOST CERTAIN DEFINITIVE OFFER SALE PANAMA FORTY MILLIONS WILL BE CABLED TOMORROW AND OFFICIALLY PRESENTED MONDAY.

VARILLA

On January 4, the cable to Admiral Walker offering the sale of the entire Panama property for $40,000,000 was put on the wire at Paris.

And so, wrote Bunau-Varilla, the year 1902 "began with the wind blowing in the sails of Panama." When the Walker Commission reversed its decision on January 18, he sent off dozens of cables to Cincinnati and Chicago expressing his "heartfelt thanks" to all those who had enabled him to speak out "in the name of the Great Idea."

On January 27, Cromwell was reinstated as attorney for the company. The officials were in such despair, Cromwell later explained, that they asked him to resume his former connection, and so "leaving aside all our other business we acceded to this request." But Bunau-Varilla told a different story. It was he who fixed things for Cromwell as a favor to Senator Hanna. Cromwell meant nothing to Hanna, but Hanna's banker, Edward Simmons, who was also president of the Panama Railroad, had asked Hanna to ask Bunau-Varilla to have Cromwell reinstated, or at least so Bunau-Varilla would declare in a written statement prepared some years later for a House committee that was looking into the extent of Cromwell's influence. On January 27, he informed Cromwell that his case had been settled in Paris, but that it had not been easy.

According to the formal written directive from Paris, the company would rely on Cromwell's cooperation in concluding the sale of the Panama property; however, ". . . it must be clearly understood . . . that the result must be sought only by the most legitimate means; that is to say, that in no case could we recourse to methods as dangerous as they are unlawful which consist principally in gifts or promises . . ."

To Philippe Bunau-Varilla, Cromwell's return was "but a slight incident in the great struggle . . ." To Cromwell, the Frenchman was someone who served a useful purpose, but whose "pretense of influence is grossly exaggerated."

CULVER PICTURES, INC.

Theodore Roosevelt and Senator Mark Hanna at Buffalo at the time of McKinley's death

FROM PANAMA: THE CREATION, DESTRUCTION, AND RESURRECTION, PHILIPPE BUNAU-VARILLA, ROBERT M. MCBRIDE, 1920

George Shattuck Morison

LIBRARY OF CONGRESS

Captain (later Admiral) Alfred Thayer Mahan

LIBRARY OF CONGRESS

Senator John Tyler Morgan

"The Deliberations of Congress" (from *Harper's Weekly*)

CULVER PICTURES, INC.

COURTESY OF SULLIVAN & CROMWELL

ABOVE, William Nelson Cromwell

LEFT, Philippe Bunau-Varilla

RIGHT, the stamp used as "proof" of active volcanoes in Nicaragua

LIBRARY OF CONGRESS

FROM PANAMA: THE CREATION, DESTRUCTION, AND RESURRECTION, PHILIPPE BUNAU-VARILLA, ROBERT M. MCBRIDE, 1920

NATIONAL ARCHIVES

FROM THE MAKERS OF THE PANAMA CANAL, 1911

LEFT ABOVE, U.S.S. *Nashville*

LEFT BELOW, founding fathers of the Republic of Panama. Seated (left to right): José Agustín Arango, Dr. Manuel Amador, Federico Boyd. Standing (left to right): Nicanor de Obarrio, Carlos C. Arosemena, Manuel Espinosa, Tomás Arias, Ricardo Arias

BELOW, General Esteban Huertas among admirers

UNITED STATES MILITARY ACADEMY

NEW YORK GLOBE

© 1903 BY THE NEW YORK TIMES COMPANY.
REPRINTED BY PERMISSION

ABOVE LEFT, "Now Watch the Dirt Fly"

ABOVE RIGHT, "The Man Behind the Egg"

LEFT, Roosevelt at work in his study at Sagamore Hill

BELOW, Philippe Bunau-Varilla (left) and John Hay in Hay's office at the State Department, November 13, 1903, just prior to the formal recognition of the Republic of Panama

FROM PANAMA: THE CREATION, DESTRUCTION, AND RESURRECTION, PHILIPPE BUNAU-VARILLA, ROBERT M. MCBRIDE, 1920

THEODORE ROOSEVELT COLLECTION, HARVARD

NEW YORK EVENING MAIL

"He's Good Enough for Me!" Homer Davenport's famous 1904 cartoon (from the New York *Evening Mail*) was more representative than any others of the country's support for Roosevelt's actions in office, including the steps taken at Panama.

11
Against All Odds

"I do not want to be interrupted, for I am very tired . . ."

—MARK HANNA

I

At age sixty-six Rear Admiral John Grimes Walker was still a majestic figure. Even in his dark civilian suit and string tie he looked like The Old Man of the Sea, as he was sometimes known in Washington. Large and handsome, he carried himself, especially on public occasions, in grand military fashion. The gray hair was smartly parted in the exact center of his head. The complexion was ruddy, the brows heavy and beautifully arched, and from the sides of his face grew magnificent muttonchop whiskers that reached to his lapels and that were several decades out of style.

For more than forty years, Admiral Walker had been a special favorite in the capital, enjoying, it was said, more political influence than any officer in the Navy. He was direct, unaffected in manner, and if a bit self-important, he plainly meant well. His reputation for integrity was second to none.

Since his retirement in 1897, he had been devoting himself solely to his duties as head of the two Presidential canal commissions. And on the face of it he had been the ideal choice for the position. His one shortcoming was a lack of engineering background or experience, which until now nobody had made an issue of. He had never tried to assume the role of a technical authority. Over dinner at the Metropolitan Club, soon after the latest commission had been organized, he told its illustrious members that it was for them, the experts, to get at "the bottom facts," however long that took, however much it might cost; and not once thereafter had he said or done anything to make them doubt his sincerity or his willingness to trust their professional judgment.

As the commission's head, Walker was the first member to testify before the Morgan Committee. He appeared the morning of February 7, 1902. There had been several other witnesses to date, but except for Édouard Lampré, spokesman for the Compagnie Nouvelle, they had been Morgan's witnesses—that is, predictably pro-Nicaragua or anti-Panama. A. G. Menocal had appeared two days earlier, for example. Now in his sixties and also retired from the Navy, Menocal was supposedly Morgan's strongest technical witness, and prodded by Morgan's patient questioning, he had documented his whole long commitment to Nicaragua. But there was nothing new in anything he said; nothing for the newspapers.

The only remark thus far that the reporters had pounced on was one made by Morgan during an exchange with the Frenchman Lampré—that he would not give 37½ cents for Lampré's canal—a remark quoted out of context and that had been nowhere near as offensive as the headlines implied. Morgan had been trying to show his contempt for any purchase of the French property whereby the money would go to the new company, rather than to those original small stockholders who had sacrificed so much. Were they to be denied a just share, said Morgan, then he would want no part of the arrangement even if he could get the canal for 37½ cents.

Morgan had been enjoying himself enormously the whole while. It had been his show from the first day, when he kept Lampré under fire for three and a half hours, and pointedly reminded the witness several times that he was under oath. When S. W. Plume, the old Panama Railroad man, appeared the day before Walker, Morgan seemed not in the least disturbed that the room was virtually empty. Only one other member of the committee bothered to attend, Senator Kittredge, a Republican, who made a show of looking bored, but Morgan had gone right along in high spirits, questioning the witness as though the entire country were present and as if the hard-bitten old man's memories of

Panama's horrors far outweighed the views of high-powered engineers.

Walker's testimony took up the better part of one day and the morning following. In the record book the transcription fills seventy-five pages. The full committee was present this time. And from the moment Walker took his seat it was plain that Morgan and his allies had their knives out for him. To Morgan especially, Walker's new position on Panama seemed little less than treasonous. It might also prove calamitous to Morgan's cause, unless Walker could be made to look the fool or led to say, even by inference, that he had no real heart for the Panama plan. One rumor in the Senate corridors was that in return for his Panama support the White House had promised Walker the job of directing construction of the canal, from his office in Washington and with a large salary. It was easy for Walker to recommend Panama, the American minister in Nicaragua had written to Senator Morgan, since Walker would not have to *live* there. The admiral, it was known, liked his comforts.

The questioning was focused almost entirely on Walker's grasp of the technical issues involved in the commission's plan for Panama and it was Senator Harris, the one former civil engineer on the committee and Morgan's staunchest ally, who monopolized the first hour. Morgan, who had a way of glaring at people even under ordinary circumstances, never took his eye off the witness.

The commission's Panama scheme, the projected Panama canal upon which all cost estimates were based and against which all virtues or shortcomings in the Nicaragua plan were compared, had been based in large part on a plan devised by the Compagnie Nouvelle. The essential element in the plan, its key, was a giant dam that would check the Chagres River at Bohio and form a large inland lake reaching nearly two-thirds of the way across the Isthmus. The commission's decision had been to abandon the sea-level concept, as de Lesseps' engineers had finally done, and to build a lock canal much along the lines of the proposal made by Godin de Lépinay in 1879. The Isthmus was not to be severed by a vast trench, but bridged by an artificial lake, Lake Bohio, as it was called in the plan. Ships would leave one ocean, climb to the level of the lake by a flight of locks, cross the lake, then descend by another flight of locks to the ocean on the opposite side, just as de Lépinay had outlined. The one great task of excavation would be at Culebra, at the Pacific, or southern, end of the lake, where a channel of nine miles would have to be cut through the Cordilleras.

But while the Bohio dam was the most important structure on the line and a "vital necessity to the scheme," it also presented enormous "difficulties of construction." Numbers of prominent engineers considered it an extremely uncertain, hazardous solution if not an impossible one.

The dam was to be a man-made earthen hill a hundred feet high and it would create a lake some forty square miles in area, the largest artificial lake in the world. But the dam was also to have a masonry core that would extend farther below ground than the dam was high, and to achieve this, pneumatic caissons—for the foundations of the core—were to be sunk 128 feet below sea level, a depth far in excess of anything previously attempted.

Did the commission's entire Panama plan hang on the Bohio dam? Senator Harris asked. Yes, replied Walker. Everything depended on the Bohio dam, but the dam would not be the most difficult undertaking. The great cut through the spine of the Cordilleras would be more momentous still. It alone might take as much as eight years.

When it was Morgan's turn, he began with costs. Why was there a difference of $1,000,000 in the estimates for the two canals? The number of locks accounted for most of that, Walker said. There would be five locks in the Panama canal, eight in the Nicaragua canal. The reason for the fifth lock at Panama, he explained, was the great rise and fall of the tide in the Bay of Panama.

"We lift up from the Atlantic to the surface of Lake Bohio with two locks and then we drop down on the Pacific side with three locks, the last lock being the lock in which the lift varies very much, depending on the height of the tide. . . . The lock is intended as a method of passing ships from one level to another."

"I understand that," Morgan replied in a low, even drawl.

They were facing each other square on, the regal old Yankee sailor looking no less resolute than the small, white-haired one-time leader of Confederate cavalry.

Did the admiral recall any point of fact upon which he had so suddenly changed his mind as to where the canal should be built?

"Well, that would be pretty hard to answer. I went into the thing with my sympathies and prejudices, as far as I had any, in favor of the Nicaragua line, but I endeavored to take hold of this question with a mind open to proof."

Morgan said he had no doubt that this was so, but wished to know whether the admiral's new position was based upon any *fact.*

"I have changed it to this extent, that I know that the best line is the Panama line. . . ."

"In an engineering sense?"

"Yes, in an engineering sense."

Morgan, as he told reporters earlier, was convinced that the commission's sudden affection for Panama had nothing to do with engineering arguments, but was based on price alone, "cheapness" was his word.

"Well, you come to that conclusion without changing any facts in your former statements?" he said.

"Yes."

"Your judgment is convinced that you were in error in the first statement?"

"No, sir; not at all. I have not changed my mind a particle."

Then how, Morgan demanded, could he conclude that Panama was the best canal when he had recommended Nicaragua?

Because, said Walker, when he recommended Nicaragua it had been the most feasible route under conditions then prevailing.

"What conditions?"

"When I voted in favor of the Panama route it was under quite different conditions."

"What conditions do you refer to?"

"Very largely the unreasonable price that the Panama people asked for their property."

"Is it not exclusively that?"

Walker fumbled for words. "I don't think of anything—I do not go back . . . I think that the engineering features of the Panama route are better than those of the Nicaragua route, although both routes are feasible. . . ."

"You think so?"

"I think so."

Morgan, who had made his reputation in Selma as a trial lawyer, abruptly shifted ground. His interest now was in the sources upon which the commission had drawn its technical data. His purpose was to show that the commission's entire Panama proposal, hence its entire decision, was founded on the word of Frenchmen, or on French plans, which by Morgan's lights were as suspect as the people who drew them up.

"How many days did your commission spend on the Isthmus of Panama?"

"We were there about two weeks."

"You have not been back since?"

"No, sir."

"Has any member of the commission been back since?"

"Not to Panama."

"Well, in two weeks' time you did undertake, I suppose, to obtain accurate knowledge of the engineering and of all the conditions of that canal and the country through which it passes?"

"Yes," said Walker, taking the bait, "we had very good knowledge of the matter from having examined the French data with great care; we had our working parties on the Isthmus of Panama. . . . We had a locomotive and a special car every day to take us back and forth along the line, so that we lost no time, and we devoted ourselves to that work every day that we were there. . . ."

Morgan had a vivid picture in mind of the old admiral and his party breezing up and down the Panama Railroad in a private car, deciding the fate of the canal from the view from the window, attended by a swarm of hovering Frenchmen doing everything possible to put them at their ease. It was a grossly unfair picture, but it was one also shared by many in the room.

"Well, you say you made an examination of the French surveys before you went to the canal?"

"Yes."

"Where did you examine them?"

Like everyone on the committee, Morgan knew perfectly well that it had been in Paris; he knew of Cromwell's stage-managing, the displays gotten up by the French company, and so forth.

"In Paris. We spent about a month in Paris, working every day, usually two sessions a day with officers of the French company, who laid everything before us . . . and then we went to the Isthmus in person to supplement that information. . . . We made our plans for the building of a canal after we had examined this data and after we had personally visited the Isthmus and been over the ground with great care. . . ."

"You did not undertake to make an independent survey of that canal line?" (As everyone present also appreciated, the entire Nicaragua survey had been "independent," by which Morgan meant the work of American engineers alone.)

"No, sir," answered the admiral, who now apparently saw the trap being set for him. "For instance, we bored the site of the Bohio dam most thoroughly, much more thoroughly than the French had bored it. So far as what is ordinarily called surveying, topographical work, we did enough of it to convince ourselves that the French work was good and that we could accept their work as our own."

"Well, you did adopt it?"

"We did adopt it after convincing ourselves of its accuracy."

"But the basis of that survey and the basis of your calculations and plans was the *French* survey?"

"No; we accepted their survey after checking it enough to be sure that it was right, and then after that our work was our own."

"But based on what?"

"Well, based on their surveys, if you like."

"That is what I mean," Morgan said smoothly.

Morgan moved right along, taking up the geology of the Bohio valley, the design of the controversial dam, the intended use of levees, the silting up of the old French works. It was his characteristic approach, persistent and exasperatingly patient. The impression he seemed to be striving for was this: that Walker and his commission, by recommending Panama, were asking Congress and the country to risk everything on faith, faith in old John Grimes Walker, faith in the assumptions of one particular set of civil engineers, and, at root and worst of all, faith in the French.

Did the admiral happen to know whether the Bohio basin would hold water if a dam were built there? Would the admiral's entire plan hold water was the implicit question.

"Have we any right to suppose that it would not hold water?" Walker replied, obviously annoyed.

"I am just asking your judgment," Morgan said. "I am not an engineer or a commissioner. I am not recommending the government take your plan."

"I have never seen anything to make me suppose there was the slightest danger of its not holding water."

"You made no inquiry about that?"

"I did not make any inquiry about that," said Walker. "There are a great many things that I have not inquired about."

In the days when Walker had headed the Nicaragua Canal Commission, indeed until the latest reversal by the current commission, he and Morgan had often worked closely together. They had been colleagues in a common mission. But there was never the slightest sign of familiarity through this long session—no personal asides, no pleasantries—and when the committee convened the next morning, a Saturday, Morgan started in again. Often he was openly uncivil to Walker, or "spluttered" (as one reporter wrote) in exasperation at Walker's answers.

No, the locomotives left behind by the French were not worthless, Walker insisted. They would have to be overhauled, but they were quite serviceable, as were many of the French excavators and dredges.

He knew of nothing along the Panama line that was not within engineering precedents, except for the Bohio dam. No, he had not experienced Panama in the rainy season. No, he did not think that $2,000,000 for the French maps and surveys was excessive. Yes, he thought the engineers on the commission were quite able, as able as any in the United States.

Only when Morgan announced, after about an hour, that he had no more questions, did the tension in the air suddenly subside. It was then that Hanna, in sharp contrast, began to talk to the witness in genial, respectful tones. He asked merely for the admiral to tell the committee why—price aside—the canal should be built at Panama.

"To start with the route has better harbors," said Walker. "It is a much shorter canal, has easier curves, and we are surer by that route of what we are doing. While we made a very careful examination of the Nicaragua line, as thorough an examination as perhaps is ever made or likely to be made before undertaking a new enterprise . . ."

"Right there let me interrupt you," Hanna said. "You also had the advantage of all previous surveys made by the United States government of the Nicaragua route?"

"Yes. We know far less about the Nicaragua line than we do about the Panama line. It is impossible to know as much. The Nicaragua line is in comparatively wild country which has not been explored to anything like the same extent that the Panama line has. The Panama line has been a great thoroughfare, traveled for two or three hundred years. It has been examined with reference to a canal for many years past . . . and the country along the line is cleared up so that one can see what he is doing. In the wild parts of Nicaragua it is a jungle, where often we could not see fifty feet, and we would be much more likely to meet disagreeable surprises by the Nicaragua line than by the Panama line."

He told Hanna he had no misgivings about any aspect of the Panama plan. Further, he did not regard the Panama climate as any greater threat to health than any other place in the tropics.

"There was a great loss of life in building the railroad," he conceded, "and when they [the French] first went to work on the canal there was a good deal of sickness, but the surface material from which this sickness is supposed to come has largely been removed, and of late years it has been as healthy there as anywhere in a tropical country. . . ."

"Is it not likely," Hanna put in, "that in the construction of the Nicaragua canal, working a large force, turning up the surface of *that* soil, and in dredging, that malarial conditions conducive to fevers would arise?"

"Certainly," said the admiral. "As it stands today Nicaragua is a healthier route, because there is no work of that kind being done and very few people get sick, but when you get to turning up the ground there would be sickness there, as there would be anywhere."

The parade of witnesses continued on into March. Much of what was said was repetitious or boring. Frequently one committee member or another would lead the witness through drawn-out explanations simply to get some obscure point into the record. Still much was said that had not been said before, at least for the records, and was of considerable interest.

Lewis Haupt, the University of Pennsylvania professor, declared, for example, that he had signed the commission's decision on Panama only to make it unanimous. He did not think Panama the superior choice. Alfred Noble, another eminent member of the commission, said no self-respecting American contractor would take the French equipment at Panama as a gift. A third commission engineer, William Burr, told the committee that the French work would all have to be greatly enlarged, from a bottom width of 98 feet and a depth of 29 feet to a bottom width of 150 feet and a depth of 35 feet. This, he said, would make it the largest canal in the world, but none too large for the American Navy being projected.* Morgan, taking a dim view of such talk, drawled, "Do you expect to make a canal that will carry Noah's ark or something like that through it?"

"No, probably not," Burr answered. "We hope there will be no occasion for Noah's ark."

The name Ferdinand de Lesseps came up repeatedly. He and his young French engineers seemed to fill the big room like specters. There was talk of the ruined canal, of ruined machinery wallowing forlornly in the jungle, of the graves on Monkey Hill, of scandal and dishonor, and whether the same would happen all over again if Americans were to go into Panama. Morgan was convinced that it would. The engineers said no. De Lesseps' failure resulted from insufficient investigations on the grounds, said General Abbot, of the Comité Technique. An underlying theme in much of the testimony was that

* The largest ships being built for the Navy then were of the *Virginia* class, which had a beam of 76 feet. The largest commercial ship then on the ways was the *Kaiser Wilhelm II,* with an overall length of 706 feet and a 72-foot beam. As Burr pointed out, the Suez Canal was already insufficient for such ships.

industrious, practical, *moral* men–Americans–might succeed where others had failed. Indeed, in an inverse way, the downfall of the French, the sheer unpleasantness and difficulty of taking the Panama route, began to have a peculiar, compelling kind of attraction. The pesthole could be a proving ground, an opportunity to succeed gloriously, for all the world to see, where a less industrious, less manly, and less virtuous people had failed so ignominiously. One witness, a railroad contractor, had told Morgan in a letter, "Engineers are sometimes the least practical of men, they may be attracted by difficulties. . . ."

Another of Morgan's witnesses, a noted and most convincing engineer named Lyman Cooley, asked how anyone could possibly guarantee that Americans would prove three times as honest, three times as competent, as the French, because that, he said, was what it would take.

The head of the Maritime Canal Company made the legal transactions involved in the purchase of its Nicaragua properties and franchise sound extremely simple and tidy. General Edward Porter Alexander, a former Army engineer, assured the committee that there would be no technical difficulties involved in a Nicaragua canal, then finished with a tribute to the physical allure of the country that not even Morgan ever quite equaled. "It impressed me as one of the most attractive countries that I ever saw for a poor man to make a living in . . . if I had to be born again I would ask the angel that was bringing me down to take me to Nicaragua. . . ."

By far the greatest amount of time was spent on Panama, however, and the most impressive testimony was that of George S. Morison, whose reasons for wanting a canal there were essentially the same as those stated by Admiral Walker, only coming from him they appeared unassailable. Morison, as someone remarked, was a "force"–a huge, human bulwark, slow and deliberate in manner, slow to make up his mind and intractable once he had. The basic Panama plan was sound, he said; things that were impossible twenty years earlier were now quite possible. The dam could be built. The river could be controlled. The dam need not have a masonry core; the whole business of pneumatic caissons could be dispensed with.

The thing that must not be underestimated, he said, was the size of the job. The Culebra Cut would be the largest excavation ever attempted. "It is a piece of work that reminds me of what a teacher said to me when I was in Exeter [Phillips Exeter Academy] over forty years ago, that if he had five minutes in which to solve a problem he would spend three deciding the best way to do it." No less than two years, Morison said, ought to be spent just getting ready for such a task.

Again, as with Walker, Hanna asked about the problem of disease. "If the reports are correct," Morison said, "we can get rid of yellow fever by killing the mosquitoes." Nobody picked him up on that, nobody seemed the least interested, and nothing more was said on the subject during the whole course of the hearings.

The final witness, a fusty retired congressman from Nevada, who had spent some time in Nicaragua half a century earlier, appeared on March 10. Three days later the committee reported the Hepburn Bill favorably: the committee wanted a Nicaragua canal. The vote was seven to four, exactly what it would have been had it been taken before the hearings began.

The odds against a Panama victory in the Senate appeared now to be about 100 to 1. The impression also was that a great deal of time had been spent on a great deal of talk for nothing. And in contrast to the events that were soon to follow, the drawn-out testimony of the engineers would seem particularly colorless and unimportant. But the case for Panama had not gone unnoticed in certain quarters, and this was to prove of critical importance. An incident in Brooklyn, for example, caught the attention of the New York papers, and thus, almost certainly, the New Yorker in the White House. At a dinner on March 16, in honor of a Brooklyn engineer, C. C. Martin, one of the builders of the Brooklyn Bridge, the main speaker was Irving M. Scott, of San Francisco, head of the Union Iron Works, builder of the battleship *Oregon*. The canal was his topic. "We must have that canal. Whether it be Panama or Nicaragua, I care not. But have it we must." His listeners, nearly all civil engineers, did care, however. "Panama! Panama!" someone shouted. Then at once everyone in the room was on his feet, handkerchiefs were fluttering in the air. Everyone was shouting, "Panama! Panama!"

II

One topic that had been scarcely touched on, ironically, was that of "seismic disturbances."

The commission had pretty well dismissed seismic action of any kind as a serious threat to canals in either location, but between volcanoes and earthquakes, earthquakes were regarded as a more serious danger. There had been fourteen recorded earthquakes along the Nicaragua line, the report noted, including one in 1844 that did damage four miles from the canal line. But, as the report explained, canals were underground structures; even the proposed dams would be so broadly based as to be virtually part of the ground itself; the locks would all be founded on rock.

Of the fourteen or so volcanoes in Nicaragua, only a few had shown any signs of life since the time of the Spanish, and all but one of these, Ometepe, which rises on an island in Lake Nicaragua, were a considerable distance from the proposed canal line. Even Ometepe was thirteen miles from where the ships would cross the lake, and from Ometepe to the site for the nearest lock was a distance of twenty miles.

Late in April, however, on the Caribbean island of Martinique, 1,500 miles from Nicaragua, an enormous, long-dormant volcano, Mount Pelée, began rumbling ominously and spewing up clouds of hot ash. Then on the night of May 2, the mountain trembled to the accompaniment of thunderous, terrifying subterranean explosions. The city of St. Pierre, the "little Paris," was showered with volcanic dust and the sea for miles was littered with dead birds. After that the volcano was seldom still. At 7:52 on the morning of May 8, 1902, the whole mountain exploded. The city of St. Pierre was wiped out in approximately two minutes. It was one of the most appalling disasters of all time. Sailors on a cable ship anchored eight miles off shore felt the heat. A man watching from a distant mountainside said it looked "as if Martinique was sliding into the sea." Nearly thirty thousand people had been killed, and above the island, blotting out the sky, was a tumultuous black cloud, perhaps fifty miles across. The sole survivor of St. Pierre was a prisoner locked in a windowless underground jail cell who had no idea what had happened until he was discovered by rescue workers.

At the White House, disregarding all red tape, Roosevelt ordered the cruiser *Cincinnati* to leave for Martinique at once under full steam and the *Dixie*, a converted freighter, followed, carrying Army rations, medical supplies, and doctors. Congress was asked for an immediate appropriation and Congress quickly granted $200,000. Pelée kept erupting in subdued fashion, meantime.

For Philippe Bunau-Varilla the news was heaven-sent. "What an unexpected turn of the wheel of fortune!" he would write. "If not the strongest of my arguments against Nicaragua, at least the most easily comprehensible of them was thus made a hundred times more striking . . ."

In the past months he had been chasing about seeing newspaper editors, talking to Hanna, talking to everyone who would listen, with such fanatical zeal that word had reached Paris that he had lost his mind. Now he swung into action as never before. A letter outlining the "terrible object lesson" of Pelée was rushed to the White House and a still more elaborate version went off to Senators Hanna and Spooner. There was an impassioned appeal on Waldorf stationery to John Tyler Morgan.

In New York the ancient and reliable John Bigelow was stirred into action still one more time. Through Bigelow a meeting was arranged with editor Edward P. Mitchell, of the *Sun*, the result of which was a vigorous editorial declaring that the "volcanic menace" in Nicaragua could no longer be dismissed as a remote issue.

On May 14, incredibly, came a dispatch from New Orleans describing the eruption of Momotombo, in Nicaragua itself. Now "even the mountains of Nicaragua are enlisted in the alleged conspiracy to defeat the great purpose of Senator Morgan's life," observed Mitchell in the *Sun*. Momotombo was said to be shooting great shafts of fire into the sky, and an accompanying earthquake had supposedly sent a government dock plunging into Lake Managua. On May 20, Pelée exploded a second time, leveling what little remained of St. Pierre, and on the island of St. Vincent, just to the south of Martinique, still another volcano erupted.

At the request of John Tyler Morgan, the Nicaraguan minister in Washington cabled Managua for verification of the Momotombo story. Morgan, meantime, had also obtained assurance from John Hay that the President would remain silent on the canal matter. "He greatly prefers, as did President McKinley," Hay wrote assuringly, "that the question of the route be decided by Congress . . ." The Senate was to commence debate on June 4.

On June 3, the Nicaraguan minister reported back to Morgan with a copy of the reply from Managua, a cable signed by José Santos Zelaya, president of Nicaragua:

> THE NEWS PUBLISHED ABOUT RECENT ERUPTIONS OF VOLCANOES AND EARTHQUAKES IN NICARAGUA ENTIRELY FALSE.

"I may add also," wrote the Nicaraguan minister in a covering letter to Morgan, "that Nicaragua has not had any volcanic eruption since 1835, and at that time Coseguina discharged smoke and ashes, but no lava. No one was killed or injured and no property was destroyed by that occurrence." Momotombo was a hundred miles from the proposed canal line.

The minister's name was Luis Corea. Whether he actually received such a cable from President Zelaya or was responsible for the copy he passed on to Morgan is impossible to determine. In any event, Momotombo had definitely erupted; the cable accredited to President Zelaya was quite false.

It was Old Morgan who made the first speech when the debate on the Hepburn Bill began in the Senate, Wednesday, June 4, 1902.

"Mr. President, I do not care to approach the discussion of this important measure in a cloud of volcanic smoke and ashes which the opponents of the measure outside of the Senate have brought as a funeral pall to place over its bier, and I think it proper that I should try to clear the atmosphere. . . ."

He read Luis Corea's letter, then two others from the American minister at Managua, his old friend William Merry, who emphasized that there had been no seismic disturbances along the proposed canal line. But mainly the speech was an attack on Panama for its political violence, "its mixed and turbulent people," for *its* seismic disturbances. He read a vivid eyewitness account from the Panama *Star & Herald* of the destruction caused by the earthquake of 1882. If seismic disturbances were the only way to defeat the Nicaragua canal, then he was sorry to report that the argument would carry Panama down with it.

For quite a while the Senate was subjected to a lesson in geopolitics, as Morgan explained the peculiar relationship of Panama to the rest of Colombia because of geography. Taking Bancroft's *History of Central America* as his text, he explained the Bidlack Treaty and the touchy nature of American involvement in Panama. The one possible solution to Panama's political demoralization, the immortal Bancroft had said, was a strong government "provided from abroad." Did this mean then, the historian had asked rhetorically, that the United States—"as the power most interested in preserving the independence" of a Panama waterway—would take upon itself "the whole control for the benefit of all nations?" Only time would tell, Bancroft had written. Only time would tell, repeated the frail, white-haired Senator from Alabama. That, he said, looking about the chamber, was what concerned him above all. Should the United States decide on a canal at Panama, it would be merely a matter of time before the United States would be compelled to take Panama *by force*. And he wanted no part in that. It would "poison the minds of people against us in every Spanish-American republic in the Western Hemisphere, and set their teeth on edge against us . . . it would tarnish our national honor to enter Panama under the pledge that our purpose is to build a canal and follow it with the annexation of Panama." He was Jeremiah now, and the part came easily. "And no actual necessity for annexation, however imperative it may be, would ever excuse or palliate that result, in the opinion of the Spanish-American people. If this is to be . . . as a necessity for the protection of the canal, it would be the most dangerous national pitfall into which we could plunge."

The place for the canal, as always, was Nicaragua, "where all the people are anxiously awaiting the coming of the United States to their assistance, with eager hopes and warm welcome, to their fertile, healthy, and beautiful land."

Morgan was not much of an orator. He paused often to frame his sentences and like all his speeches this one would read better than it sounded. He closed with an appeal for an act of kindness, as he said, for his beloved South. ". . . I would brighten that land with the bloom of prosperous industry, and bring back to my brethren the consciousness that they live and move in the current of human affairs. I hope to see the waters of the Gulf of Mexico and the Caribbean Sea . . . as busy with commerce as the bay of San Francisco."

He had spoken for some two and a half hours and he had made no mention of engineering considerations.

The following day, June 5, shortly before two in the afternoon, Marcus Alonzo Hanna limped down the aisle of the Senate to deliver the most important speech of his career. Behind him hurried a secretary carrying a stack of books, papers, and pamphlets.

The chamber was nearly full and all about were hung enormous maps and plans. One map of Central America and the Caribbean islands reached from the gallery railing to the floor. It showed the location of every principal volcano, active or extinct, the active ones being marked in red, the extinct in black, with the result that an almost solid band of red dots ran from the Mexican-Guatemalan border on the Pacific shore to about midway into Costa Rica. Eight of the red dots were in Nicaragua. In the Colombian province of Panama there were none. Also, as many observers were quick to note, Pelée, on Martinique, was indicated as extinct.

The maps and the plans were merely colossal enlargements of several from the report of the Walker Commission, copies of which had long since been made available to every member of the Senate and to the press. But Hanna, who had an instinct for promotion, knew the effect such a display would have. No graphic presentation of such scale had ever been seen before in the Senate.

"Ladies and diplomats, reporters, agents of the powers, all jammed the gallery's aisles," we are told. Telephones had been ringing across town all morning with the message that Hanna was to speak.

Largely misunderstood in his own day, largely forgotten by the time the last of his generation had passed from the scene, Marcus Alonzo Hanna was an original, the plain, plain-spoken, brilliant, eminently practical man who had made a business of politics and made himself the nearest thing there had ever been to a national political boss. Coal merchant, ironmonger, owner of ore boats and newspapers, he was burly, but seemingly bland, with a bald head and large brown eyes and a bad case of rheumatism. He had put his adored friend William McKinley in the White House by amassing the biggest campaign fund on record and through keen political judgment, and when McKinley made Senator Sherman Secretary of State in 1897, Hanna was appointed to fill his place. He was still finishing out Sherman's term and until now he had had comparatively little to say in the Senate. He thought political speeches "gas."

But the talk now, increasingly, was of Hanna for President in 1904, which was a major reason for the stir over the speech he was about to make. For while he made light of any possible Presidential ambitions, the idea of big, able "Uncle Mark," friend of business, friend of labor, host of sumptuous parties, holding forth in the White House in his own style appealed to many, and especially to Wall Street, where Roosevelt was thought "unsafe." History, moreover, seemed to be on his side. Of the four Vice Presidents who had previously succeeded to the Presidency upon the death of a President not one had even been nominated by his party to run for the office in his own right.

"He has a mass of material," Cromwell informed Bunau-Varilla the day before, "but he says he will use little of it and speak only in his own simple and direct way. . . ." Cromwell, who was sitting in the gallery as Hanna made his entrance, had been personally responsible for a volume of testimony, now in the hands of Hanna's trailing secretary. Some eighty-three shipowners, shipmasters, officers, and pilots—those who would *use* the canal—had given their unanimous preference for the Panama route. The lawyer had prepared a businesslike questionnaire covering trade winds, weather conditions, canal curvature, towing, and nighttime navigation, and through his Wall Street connections he had seen that copies were sent to the ranking officers with the Cunard, White Star, American, and Red Star lines. Only two had said they would risk their ships through a Nicaragua canal at night. None thought the Nicaragua canal would offer a saving in time, because of its length, and without exception they agreed that the shorter the canal, the less time spent in the canal, the less risk to their ships, and so the better the canal.

Bunau-Varilla, no less busy than Cromwell, had contributed a number of clever diagrams of his own design, all based on the canal commission's own statistics, each pointing up Panama's essential engineering and navigational virtues. The diagrams were as simple as illustrations in a child's primer, conveying their message at a glance and easy to remember. They were an inspiration, Hanna saw instantly. The inevitable problem with technical reports, with any arguments based on technical data, was that few would read them, and the *only* advantage that Panama could claim was its technical superiority. So it had been arranged that the pamphlets containing the diagrams be delivered to the Senate in quantity at the close of Hanna's speech.

Hanna was even less an orator than John Tyler Morgan. He was also in poor health and would be forced to spread his remarks over two days, stopping suddenly, unexpectedly, on this first day after being on his feet little more than an hour. But he had a disarming manner of talking as though he had no intention of inflicting a speech on anybody. He made things sound easy and sensible. He had thought the subject through, as a business proposition. It was the voice of common sense speaking, of American enterprise, of the North, of power and "stubborn facts," as he called them. "This plain old person in a dull gray suit," wrote a biographer, "was doing something and a drama heaped itself in the warm chamber while he drawled along, explaining this investment without an eloquent phrase."

We have passed the experimental stage, Hanna began. We have passed the sentimental stage, we want the best route, we want the best canal, we want a canal to serve the needs of the entire world, we will build not just for today or next year but for all time.

"It is the great, broad, liberal American policy for which we stand in the building of a world canal. I sympathize with all those who in other days, laboring for an isthmian canal, had but one star to guide them—Nicaragua—and who must now naturally feel like giving up an old friend to pass it by. But in this age of progress and development, Mr. President, the American people are looking to Congress to answer to them on this question without regard to sentiment. . . ."

De Lesseps had been no fool. At Panama there could be a sea-level canal, at Nicaragua there would be no chance for that, never ever. If Panama was unsettled politically, all Central America was unsettled. In any event, an American canal would be a great peacemaker.

It was then that his legs gave out. He was too tired to go on, he said suddenly, causing a commotion in the gallery, and at once he sat down. But the next day he was back again and took up as though nothing had happened.

Panama was the place to build the canal for the following reasons, Hanna began, as his secretary, who sat behind him, handed up a sheaf of papers. One: A Panama canal would be 134.57 miles shorter, terminal to terminal. Two: It would have considerably less curvature. Three: The time in transit, by steam, would be less than half that at Nicaragua—twelve hours against thirty-three. Four: Panama required fewer locks. Five: Panama had better harbors. Six: Panama was "a beaten track in civilization." Seven: Panama had a railroad "perfect in every respect." Eight: A Panama canal would cost less to run. Nine: "All engineering and practical questions involved in the construction of the Panama canal are satisfactorily settled and assured. . . ."

He wanted no one to underestimate the importance of reason nine. The *engineers* wanted a Panama canal and they were the ones to listen to. "Why, Mr. President, there are now done a great many things which fifty years ago were unheard of, never dreamed of, never thought possible, as a product of human intelligence and ingenuity in engineering. It has become almost a byword today that in the hands of a skillful engineer nothing is impossible."

Morgan had called for a boon to his native southland; Hanna said neither sentiment, sectionalism, nor personalities ought ever enter into so momentous a decision. Morgan had cited the preference of past canal commissions for Nicaragua; Hanna urged his colleagues to think not of the past but of the future.

When Senator Mitchell, an ardent Nicaragua man, tried to break in as Hanna read Cromwell's survey of shipmasters, demanding to know who the author of this questionable document might be, Hanna had replied, "I do not want to be interrupted, for I am very tired . . ."

He ended on a warning. If the United States were to build a Nicaragua canal, what then was to prevent some other power—by which he meant Germany—from finishing the French canal? Our competitors then, he said, would have all the advantages.

"*Mais, il est formidable!*" the Russian envoy was heard to remark.

It was the finest speech Hanna ever made. There were no ringing phrases, but apparently it did something very rare in the Senate; it changed some votes. One Senator told Hanna that he had been undecided until then. He would vote now, he said, for the "Hannama Canal."

Even so, the Hanna forces felt the tide was running against them as the debate continued in the days after. A Nicaragua speech by Senator Harris received blazing newspaper acclaim. More serious was the spreading belief that the volcano scare in Nicaragua was something Hanna and his cohorts had manufactured. The Nicaraguan embassy stuck by its denial of any serious disturbances, and a cartoon in the Washington *Star* showed Hanna at an easel slapping out Nicaraguan volcanoes by the yard to the delight of two onlookers, Philippe Bunau-Varilla and the head of the Great Northern railroad, James J. Hill. (In Minneapolis, Hill had sounded off to reporters about asinine congressmen and the "nasty, crooked" Nicaragua canal; only if a volcano were belching beneath the seat of his pants would any congressman ever take heed, said James J. Hill.) Hanna was enraged by the cartoon and he, Cromwell, and Bunau-Varilla tried desperately to think of some response.

"It was absolutely necessary to reply with emphasis," Bunau-Varilla recalled, ". . . but it could have no weight unless official. How could I obtain such a document? Nicaragua was far away. The authorities had shown their bad faith. It seemed impossible to procure anything whatever. . . . Only six or seven days remained."

And it was then that he remembered the postage stamp.

"*Young nations* [he had written in his pamphlet of the previous year] *like to put on their coats of arms what best symbolizes their moral domain or characterizes their native soil. What have the Nicaraguans chosen to characterize their country on their coat of arms, on their postage stamps? Volcanoes!*"

He knew the exact one, a pretty little one-centavo Nicaraguan stamp showing a railroad wharf in the foreground and, in the background, Momotombo "in magnificent eruption." Rushing about to every stamp dealer in Washington he managed to purchase ninety altogether, one for each senator. He pasted the precious stamps on sheets of paper and below each typed out: "An official witness of the volcanic activity on the isthmus of Nicaragua." The stamp arrived at the office of every member of the Senate with the morning mail on Monday, June 16, three days before the deciding vote. He had, declared Philippe Bunau-Varilla, fired the last shot of the battle.

• •

But that was not so; not quite. The last shot, like the first, was fired by Old Morgan, who, the following day, rose from his seat still one more time, to enter into the debate the name of William Nelson Cromwell, unleashing, as he spoke, years of stored hatred for the lawyer's "humiliating and repulsive" intrusion into the decisions and policies of the United States government.

No talk of volcanoes, not the cleverest propaganda, could disguise the insidious course of this hired agent, Morgan said. "I trace this man back . . . to the beginning of this whole business." It was Cromwell who had fed Hanna every supposed fact that Hanna stood behind, Cromwell who had intruded the commercial interests of the nefarious French company into congressional legislation, the hearings, the deliberations of technical commissions. "He has not failed to appear anywhere in this whole affair . . ." It was all a matter of record and the record was as much as Morgan ever wished to know of such an individual. "I would not dare to follow him when he is not on the surface."

Panama, declared the old Senator at length, was "death's nursery"; those who wished "to touch that thing" might go ahead and do so.

On June 19, after fourteen days, the debate ended. The majority report—for the Nicaragua canal—had been spoken for by Morgan, Harris, Mitchell, Turner, Perkins, Stewart, and Pettus. Besides Hanna, those for the minority—for the Panama canal—included Kittredge, Cullom, Gallinger, Teller, Allison, and Spooner. There had been a full gallery most all the time; the press and the country had followed the story very closely. Hanna claimed to have forty-five votes, exactly half the Senate. Everybody knew it would be extremely close.

The test came the afternoon of the nineteenth. The vote was 42 to 34. Panama had won by eight votes. So had there been a difference of just five votes, the result would have been a Nicaragua canal.

III

They were all very busy congratulating one another—Hanna, Spooner, Cromwell, Bunau-Varilla. Telegrams went off, effusive letters of gratitude to supporters in Cincinnati and Chicago, editorials were clipped and saved. Bunau-Varilla called it a "conclusive vindication." Hanna was told that his place in history was fixed forever. Cromwell was happily confiding to almost anyone who would listen that he had written most of Hanna's speech.

As time went on, as the details of what happened emerged, Cromwell and Bunau-Varilla would both be amply credited (or scorned) for the parts they had played in the astonishing victory, which, as one able historian would write, ranks among the masterpieces of the lobbyist's art. Hanna's role would be weighted heavily by his colleagues in the Senate. Senator Orville Platt would call Hanna's speech the most effective he had ever heard in all his political career. Senator Frye said Hanna had converted him from a lifelong advocacy of Nicaragua.

Pelée and Momotombo would also figure prominently in all subsequent accounts of the "Battle of the Routes," as well they should. Had there been no such eruptions that spring, it is quite unlikely that the Senate would have voted as it did.

How much the little postage stamp really mattered, whether it actually changed any votes, is impossible to say. Probably it did not, Bunau-Varilla's assertions notwithstanding. His diagrammatic pamphlet probably had a more telling effect. Still, the stamp was an inspired bit of propaganda, perhaps even worth ten thousand senatorial words, and it would brighten after-dinner reminiscences in Washington and Paris for years to come.

But a careful study of the record and some reasonable conjecture suggest, in retrospect, that one other figure had been a great deal more influential than had met the eye, a man who deserves our recognition. That was George S. Morison.

If one traces back through the chain of events that led to the Senate vote, keeping count of who was influencing whom and when, and if it is remembered that Morison, unlike Hanna, Bunau-Varilla, or the garrulous Cromwell, made no effort to glorify his contributions, at the time or later, then Morison emerges a bit like the butler at the end of the mystery—as the ever-present, frequently unobtrusive, highly instrumental figure around whom the entire plot turned.

The significant thing about the outcome of the "Battle of the Routes" is that it was decided on technical grounds. It was the technical view, the considered judgment of the engineers, that triumphed in the Senate. The situation was the exact reverse, interestingly, of that at the Paris congress of 1879 and yet Panama was still the end result. The emotional power, the force of personality—Morgan's—had been on the side for Nicaragua this time. At Paris it had been the engineers—Menocal, Eiffel, and others—who had urged a decision for Nicaragua.

The most articulate and forceful of the engineers and by far the most stubborn Panama proponent was Morison. It was Morison whom Bunau-Varilla singled out as the leader on the commission. It was Morison, in the summer of 1901, before McKinley's death, who prevailed against the inclination to wind things up with a unanimous decision for Nicaragua because the French company refused to set a price. It was Morison who wrote the minority report in favor of Panama. It was Morison who did the most to convince others on the commission and Walker in particular that the Bohio dam could be built, and who then convinced Mark Hanna.

Indeed Morison seems to have had just about everybody's ear at one time or other, except Morgan's, of course, and most important of all he seems to have been the one who worked the conversion of Theodore Roosevelt. Someone did, it is certain. It could not have been Cromwell or Bunau-Varilla. Neither of them had entrée to the White House as yet. It was not Hanna.

The most solid evidence we have is a letter Morison wrote to the President dated December 10, 1901. The letter sets forth in the clearest, strongest terms the technical reasons why the canal should be built at Panama and Morison's own personal unwillingness to accept Nicaragua as the only choice. The date is important. For it means Roosevelt received the letter *before*—a month before—he called the commissioners in to see him one at a time, to hear each explain things in his own words.

Morison himself seems to have had little doubt of what he accomplished. Years later, his close friend and confidant, Dr. Leonard Waldo, of Peterborough, New Hampshire, the Morison family physician, wrote that Morison "practically alone" had changed official opinion regarding Panama and it is unlikely that the doctor had any other source for the claim than George S. Morison.

Nor is it without significance that Morison was exactly the sort of man Theodore Roosevelt admired, trusted, and listened to. His whole career had been built on intelligence and daring. He was at the very top of his profession because of what he had *done* and he had done it in spite of his background. He was a preacher's son, a classics major at Harvard, who had made himself an engineer on his own, through self-study, self-development, sheer will power. Bright but not distinctive as an undergraduate, he had gone to Harvard Law School and finished in the same class as Justice Holmes. But the law bored him—as it had Ferdinand de Lesseps, as it had Roosevelt—so he had decided to be an engineer, "that I may lead a good and useful life."

Unlike Roosevelt, Morison was a lifelong bachelor and a prude. The sole failing he ever admitted to was an inability to handle horses. "There is a kind of man," he once said, "that likes animals and handles them well, particularly horses, and such men are usually the type who are popular with others, and are known as 'good fellows,' but—but such men are usually fellows with lax morals." Morison was never known as a good fellow. He was arrogant, inflexible, most unpopular, a man who was easy to admire from a distance.

Probably it was his total candor, the unshakable air of authority, that appealed most to Roosevelt. The Harvard background, a mutual interest in the West, in books, also gave them common ground. But in the last analysis Morison *was* brilliant, he *did* know what he was about, and he knew how to make other intelligent men feel that in their bones. "I hate to eat my lunch with Morison, he always quarrels with the waiter," one noted engineer once remarked, "but I'd trust his judgment sooner than that of any other engineer I know."

Originally he had been for Nicaragua, Roosevelt was to say, until the engineers convinced him otherwise. How often he was exposed to Morison is not known, but it is easy to picture them being quite direct with each other. Had Morison lived, it is probable that Roosevelt would have asked him to take a major part in the building of the canal. Very possibly he would have been made chief engineer, but he died in 1903, during his first illness since childhood, at age sixty.

The important fact is that Theodore Roosevelt had been convinced that Panama was the superior choice from the strictly objective technical standpoint. And to have a fair understanding of Roosevelt's subsequent moves this must be kept in mind. "I took the Isthmus" was to be his arrogant, unfortunate claim, but in a very real and crucial sense, quietly, rationally, without fanfare, well *before* the Panama revolution, he "took the Isthmus" because the sort of men who would have to build the canal assured him that Panama was the place to put it. A momentous policy decision was determined by technical advisers here at the start of the new century.

And Panama *was* the superior choice, as George Morison said, and for the reasons he, Hanna, and the others cited. Given the sort of canal that was needed, considering the size of the ships of the day, taking into account all the advantages offered by the two routes, Panama was the place. The choice was never so clear-cut as Bunau-Varilla made it out to be, and while a Nicaragua canal would have taken longer to build and would have cost more, it would not have been a failure. Furthermore, if such nonengineering concerns as health and Central American politics are entered into the discussion, as Morgan had always insisted they must be, then the issue becomes as debatable in hindsight as it was then.

• • •

On June 26, the House passed the Spooner Bill by an overwhelming vote of 259 to 8, the Nicaragua forces in the House having received word from Senator Morgan that the game was not up quite yet, since failure to obtain a clear title to the French properties was certain and this would force the Administration to revert to Nicaragua. "Make way for the canal!" cried one congressman. "Make way for the canal!"

The President signed the Spooner Act two days later, June 28, 1902, and so it became the law.

12

Adventure by Trigonometry

The plan seems to me good.

—Manuel Amador

I

Although negotiations for the canal treaty with the Republic of Colombia had begun well before passage of the Spooner Act, it was not until January of the following year that the agreement was at last signed, and for those most directly involved, the negotiations had been the most difficult, tortuous experience of their professional lives. Dr. Carlos Martínez Silva, the first of three successive Colombian diplomats, had been retired in a state of complete exhaustion and would die a year or so after returning home, a victim apparently of the strain in Washington. His replacement, Dr. José Vicente Concha, suffered a physical and emotional collapse upon resigning his post and reportedly was put on a ship in New York in a straitjacket. Even the indomitable William Nelson Cromwell succumbed to a siege of "nervous exhaustion" at one point in October. For John Hay it was the most thankless and exasperating episode in a long career.

To begin with, the routine at the State Department had been greatly altered by what Hay's biographer would describe as "a new impelling force"—the man next door in the White House. The overriding aggravation, however, had been the Colombians, about whom, by 1903, even the temperate and very proper Hay could speak of as disparagingly nearly as did Roosevelt. Accustomed to treating with such adroit, worldly professionals as Julian Pauncefote—men much like himself—Hay had been forced to deal with a succession of edgy, inexperienced Latin Americans who were obliged to consult with Bogotá at every move. Communications with the Colombian capital were dreadful. An official exchange of letters between the minister in Washington and his government could consume three to four months, and Bogotá's shifting, frequently cryptic positions were an endless source of frustration for the Colombian ministers no less than for anyone else.

Martínez Silva had made his first call on Hay in March of 1901. His instructions from Bogotá were to do all in his power to make possible the adoption of the Panama route by the United States, and with coaching from Cromwell he had been extremely conscientious, working with Hay, checking regularly with Admiral Walker, releasing statements to the American press that his government was ready at any time to deal liberally with the State Department. But then his superiors in Bogotá decided that he had allowed himself to become too closely associated with the French canal company—that is, attorney Cromwell—and so he had been replaced by the nervous, painfully proud José Vicente Concha, a former Colombian Minister of War, no diplomat either by training or temperament, who had never been outside his own country before, and who spoke no English.

The primary issue, as stated both by Martínez Silva and Concha, was Colombian sovereignty over the proposed canal zone, and in the fall of 1902, just at the critical point in Hay's conversations with Concha, Colombia's seemingly interminable civil war had flared up anew on the Isthmus. To secure the Panama Railroad, Roosevelt sent American Marines ashore without first receiving the expressed consent of Colombian authorities—neither those on the Isthmus nor those in Washington—as had always been done before whenever American forces had been landed. The Marines were withdrawn eventually but the damage done to progress on the treaty seemed irreparable. Of particular aggravation to the Colombians was the decision of an American admiral to prevent any movement of Colombian troops on the Panama Railroad at one crucial stage. To Dr. Concha such use of American force had been not merely a violation of the 1846 treaty, but an inexcusable humiliation and the perfect expression of the underlying imperialistic ambitions of the United States. His hostility to Hay personally became such that he refused to see him for weeks. The fact that Hay cabled Bogotá his regret at the misunderstanding that had arisen, declaring there had been "no intention to infringe sovereignty or wound the dignity of Colombia," did not improve matters.

Meantime, Senator Morgan and other pro-Nicaragua figures on Capitol Hill, along with several influential newspapers, were saying that the "reasonable time" allowed by the Spooner Act for treaty negotiations was fast expiring.

The strain on Concha was severe. He grew ever more suspicious, ever more obstinate about the sovereignty issue. He also stood fast on Colombia's right to make its own bargain with the Compagnie Nouvelle before releasing the company from the provision in the Wyse Concession that explicitly prohibited the sale of the franchise to any foreign power. It was Concha's position, as it had been Martínez Silva's before him, that if the French company was to receive $40,000,000 for its Panama properties and the rights granted by the Colombian government, then in all justice Colombia ought to receive an appreciable part of that sum in return for its willingness to permit the sale. It was a question of tremendous financial importance to Colombia.

Describing the Americans he dealt with for his home office, Concha wrote, "The desire to make themselves appear, as a Nation, most respectful of the rights of others forces these gentlemen to toy a little with their prey before devouring it, although when all is said and done, they will do so in one way or other." For their part, Hay and Cromwell were undecided as to whether Concha was wholly sane, and from the American minister in Bogotá came a report that Concha was known there to be "subject to great nervous excitement."

Infuriated by what he took to be insults to the honor of his homeland, Concha resigned several times in succession, only to be instructed to stay with his responsibilities. He was certain that his messages to Bogotá were being intercepted and requested his home office to change codes. In November he was at last ordered by Bogotá to sign the treaty, whatever his feelings about it (the final decision would rest with the Colombian Congress, he was reminded), but this his "conscience" would not permit him to do and so he quit. His distaste for dealing with Hay, he wrote, amounted to a "neurosis."

His replacement was a sixty-year-old career diplomat, the Colombian chargé d'affaires, Dr. Tomás Herrán, a naturally tactful, intelligent, and rather sad-looking man, who was the son of General Pedro Alcántara Herrán, who in 1848 had campaigned so effectively in Washington for ratification of the Bidlack Treaty. It had been then, traveling with his father as a small boy, that Herrán had first come to the United States. He was a graduate of Georgetown University, the master of four languages; his English was perfect; he had innumerable friends in Washington. By Hay's standards he was a vast improvement and so it had been presumed at the State Department that the final details of the treaty would now be dispensed with smoothly and swiftly. But Herrán proved extremely cautious and burdened with apprehension. In private, in his correspondence with Bogotá, he expressed his fear that Roosevelt's "impetuous and violent disposition" might lead him to seize Panama by eminent domain, on the ground of universal public utility. It was only when Hay issued a sharp ultimatum—by command of the President, as Hay stated—that the impasse was broken. If Colombia refused any longer to agree to the treaty as it stood, then Hay would commence negotiations for a Nicaragua canal.

The ultimatum was issued on January 21. The Hay-Herrán Treaty was signed at Hay's home the afternoon of the following day. "I feel," Herrán wrote to a friend, "as if I am waking from a horrible nightmare. Gladly shall I gather up all the documents relating to that dreadful canal and put them out of sight." Among those documents, one he quietly buried in the legation archives, was a cable from Bogotá, received three days after he had signed the treaty, directing him not to sign but to await further instructions.

The reaction in Washington was immensely favorable. To nearly everyone it seemed a solid, straightforward treaty, and in spite of the fiery, often brilliant, unyielding opposition of John Tyler Morgan, who proposed no less than sixty amendments, it was ratified by the Senate on March 17, without amendment and by an overwhelming margin (73 to 5).

By the treaty the Compagnie Nouvelle was authorized to sell its "rights, privileges, properties, and concessions" to the United States, and Colombia granted the United States control of a canal zone six miles wide from Colón to Panama City, but not including either of those cities. The franchise was for a hundred years and was renewable at the option of the United States. In return the United States was to pay the Republic of Colombia the lump sum of $10,000,000 cash (gold) plus an annual rent of $250,000. Though Colombian sovereignty over the canal zone was specifically recognized in Article IV, the United States was permitted to establish its own courts of law within the zone and to enforce its own regulations concerning the canal, ports, and the railroad. Police protection for the canal and the railroad was to be provided by Colombia, but if Colombia was unable at any time to meet this obligation, the United States could act with Colombia's consent, or in an emergency, without that consent.

The response to the agreement in Bogotá was another matter, however, and the Colombian Congress had yet to grant its sanction. The Colombian government insisted still that it had the right to negotiate its own settlement with the Compagnie Nouvelle. The annual payment of $250,000 was regarded as too little, since it was no more than what was being received yearly from the Panama Railroad as things already stood. Being vastly larger, more important, more valuable, the canal ought to pay more than the little railroad, it was felt, and the payments for the canal, as the treaty presently read, were not to start until nine years after ratification. Ten million dollars was not enough for the cession of any territory in Panama, wrote a noted Colombian intellectual and political activist, Raúl Perez, in the pages of the *North American Review*. "Panama is bone of the bone and blood of the blood of Colombia, and has always been her cherished hope."

Nor did the expressed guarantee of Colombian sovereignty within the canal zone appear quite so conclusive in Bogotá as it did in Washington.

Repeated warnings of Colombian anger over the treaty were cabled to the State Department by the American minister at Bogotá, Arthur Beaupré. "Without question public opinion is strongly against its ratification," Beaupré wrote as early as March 30. What began as suspicion had quickly become outspoken hostility, he reported two weeks later. The Colombians, he reported next on May 4, believed the guarantee of sovereignty meant nothing, that "the lease is perpetual . . . the whole document is favorable to the United States and detrimental to Colombia."

Beaupré was to be the target of much criticism later. He would be blamed for his roughshod, amateurish handling of the situation, his disregard of Latin sensitivities. But the charges are at odds with the facts of his career (he had had six years' experience in Guatemala, Honduras, and Colombia), his reputed "urbane, dignified manners and courtly demeanor," and the perception apparent in his striking dispatches. That he was frequently blunt, even dictatorial, in his pronouncements to Colombian officials is also a matter of record, but in view of his orders from Washington he was left with little choice.

As of April 28, for example, Beaupré was instructed to inform the

Colombian government that the United States would consider any modification whatever of the terms of the treaty as practically a "breach of faith" on the part of the Colombian government. By June, as the Colombian Congress was about to convene in special session, John Hay had abandoned any pretense of regard for the wishes or feelings of the Colombian people. The message of June 9 to Beaupré was strikingly ominous. "If Colombia should now reject the treaty or unduly delay its ratification, the friendly understanding between the two countries would be so seriously compromised that action might be taken by the Congress next winter which every friend of Colombia would regret." Beaupré was ordered to communicate the substance of this verbally to the Colombian Minister of Foreign Affairs. It was, as one noted diplomatic historian would observe, "an aggressiveness rarely found in friendly diplomatic intercourse." The contrast in tone to that of the correspondence relating to the Hay-Pauncefote Treaty, for example, could not have been much more pointed.

The nature of the threatened "action" was never specified officially, but just four days later, William Nelson Cromwell emerged from the White House after a "long conference" with the President and immediately dispatched his press agent, Roger Farnham, to the Washington bureau of the New York *World*. The following morning, June 14, the *World* carried this remarkable item:

Washington, June 13, 1903

President Roosevelt is determined to have the Panama canal route. He has no intention of beginning negotiations for the Nicaragua route.

The view of the President is known to be that as the United States has spent millions of dollars in ascertaining which route is most feasible, as three different Ministers from Colombia have declared their Government willing to grant every concession for the construction of a canal, and as two treaties have been signed granting rights of way across the Isthmus of Panama, it would be unfair to the United States if the best route be not obtained.

Advices received here daily indicate great opposition to the canal treaty at Bogotá. Its defeat seems probable for two reasons:

1. The greed of the Colombian Government, which insists on a largely increased payment for the property and concession.

2. The fact that certain factions have worked themselves into a frenzy over the alleged relinquishment of sovereignty to lands necessary for building the canal.

Information also has reached this city that the State of Panama, which embraces all the proposed Canal Zone, stands ready to secede from Colombia and enter into a canal treaty with the United States.

The State of Panama will secede if the Colombian Congress fails to ratify the canal treaty. A republican form of government will be organized. This plan is said to be easy of execution, as not more than 100 Colombian soldiers are stationed in the State of Panama.

The citizens of Panama propose, after seceding, to make a treaty with the United States, giving this Government the equivalent of absolute sovereignty over the Canal Zone. The city of Panama alone will be excepted from this zone, and the United States will be given police and sanitary control there. The jurisdiction of this Government over the zone will be regarded as supreme. There will be no increase in price or yearly rental.

In return the President of the United States would promptly recognize the new Government, when established, and would at once appoint a minister to negotiate and sign a canal treaty. This can be done expeditiously, as all the data is already supplied.

President Roosevelt is said to strongly favor this plan, if the treaty is rejected. . . .

It is known that the Cabinet favors the President's idea of recognizing the Republic of Panama, if necessary to secure the canal territory. The President has been in consultation both personally and by wire with leading Senators, and has received unanimous encouragement. . . .

It is intended to wait a reasonable time for action by the Colombian Congress, which convenes 20 June, and then, if nothing else is done, to make the above plan operative.

The article was unsigned. The White House issued no denials.

A large part of the problem in Washington was a pervading ignorance of, indeed a chronic disinterest in, Colombia in general, Colombian politics, or the individuals with whom Beaupré was dealing. Bogotá itself was still as removed from the rest of the world as it had been when Lieutenant Wyse made his trek over the mountains in 1878. It was one of the most isolated, inaccessible cities in the world. In all Colombia in 1903, in a country as large as the combined areas of California, Oregon, Washington, and Arizona, there were less than four hundred miles of railroad. To reach Bogotá from either Buenaventura or Barranquilla still required anywhere from two weeks to a month of arduous travel. By North American standards Colombia was pathetically poor and backward and its government was both unstable and financially destitute as a result of a disastrous three-year civil war that had only ended in November 1902.

The present head of state was a Conservative, a direct descendant, politically speaking, of Rafael Núñez, whose triumphs following the uprisings of 1885 had centralized political power in Bogotá under a new constitution. He was José Manuel Marroquín, an elderly, bearded scholar and man of letters whose shyness had kept him in the background most of his career. In 1897 he had been elected first *designado*, or vice-president, on a ticket with an eighty-five-year-old presidential candidate named Manuel Sanclemente, also a scholar, who was in extremely feeble health. In 1900 Marroquín had taken power by *coup d'etat*.

At the State Department and at the White House, Marroquín was understood to be an iron-handed despot. His power was thought to be absolute and thus the whole process of treaty ratification by a specially convened Colombian Congress was regarded in Washington as a charade. Marroquín supposedly had only to say the word and the treaty would be accepted. At both the State Department and the White House the consensus was that Marroquín and a few cohorts would one day retire to some private place and divide up the $10,000,000 payment from the United States.

But in fact Marroquín's power was limited, his personal prestige was in the decline, and in his every move concerning the canal issue he was subject to savage attacks from his political opponents, the Liberals, who accused him of selling out to the North Americans. Personally he was a rather abstract and inefficient idealist and, generally speaking, those about him were politically high-minded men of character, whatever their limitations, and exceedingly sensitive about their national honor—none of which, again, was ever quite understood in Washington.

Colombian regard for the political ideals of the United States was enormous. The country's federal and state system had been modeled after that of the United States. Bolívar, the Liberator, was known as the "George Washington of South America." Wealthy and educated Colombians sent their sons to be educated in the United States. By no means did the leading political figures fit the portrayal Theodore Roosevelt was to provide. An American minister to Colombia, James T. Du Bois, was to write some years later in this connection, "An impartial investigation at Bogotá . . . convinced me that, instead of 'blackmailers' and 'bandits,' the public men of Colombia compare well with the public men of other countries in intelligence and respectability. . . ." Until 1903, Du Bois said, Colombia was the best friend the United States had south of the Rio Grande.

Another large and important part of the problem in Washington was William Nelson Cromwell, since Roosevelt's and Hay's appraisal of the situation in Bogotá, the picture they had of Marroquín and his regime, the tactics devised, even the wording of instructions to Beaupré, were strongly influenced by Cromwell.* Especially was his influence felt at the State Department. He was consulted just about daily; he was relied upon for information (Cromwell had his own paid operative in Bogotá to keep him posted); his say, his expertise, were major factors at virtually every important juncture, as is plain in the record and as Cromwell himself would later boast.

It was a highly unorthodox arrangement, to say the least, to have the attorney for the corporation most directly in line to benefit from the treaty, a man with no official title, no rightful business to be involved in any official capacity, operating at will at the highest diplomatic level, instrumental to a degree exceeded only by the Secretary and the President, and with full impunity. But such was this exceptional man's influence over Hay and such it had been since the negotiations began. When Hay had at last put his large, legible signature to the treaty in January, it was to Cromwell that he turned and presented the pen.

Most important, Cromwell had succeeded in persuading Hay—and thus Roosevelt—that the United States must not sanction or be party to any move by Colombia to deal independently with his client, the Compagnie Nouvelle. Roosevelt had earlier sent Attorney General Philander C. Knox to Paris to make a thorough examination of the French firm, which Knox reported to be "vested with good and sufficient title to the property it intended to convey." On February 17, Knox had formally notified the company that its offer of sale was hereby accepted by the United States; that is, that the United States would pay the full asking price of $40,000,000 for the Panama holdings. So as the company's attorney, Cromwell was determined that nothing should jeopardize this unprecedented transaction, this largest real-estate sale on record.

Specifically, he was determined that Colombia would be given no chance at any part of the $40,000,000.

Hay not only agreed to this in principle, but instructed Beaupré to make it known in Bogotá in the plainest terms. Thus the State Department was being used to secure the interests of the French company, its stockholders, and its American attorney, who, if the sale went through, stood to profit more personally than any other individual involved.

* Reportedly it was Senator Hanna who put Roosevelt on to Cromwell. "You want to be very careful, Theodore," Hanna is supposed to have advised in fatherly fashion, "this is very ticklish business. You had better be guided by Cromwell; he knows all about the subject and all about those people down there." Roosevelt replied that "the trouble with Cromwell is he overestimates his relation to Cosmos." "Cosmos?" said Hanna. "I don't know him—I don't know any of those South Americans; but Cromwell knows them all; you stick close to Cromwell."

And it was when Beaupré reported back that ratification could probably be secured if the French company were to agree to pay Colombia $10,000,000 that Hay responded with his ominous telegram of June 9, a telegram drafted by Cromwell.

That the United States government had no rightful authority in a dispute between a foreign power and a private corporation was lost sight of. A settlement by Colombia and the Compagnie Nouvelle would have cost the United States nothing, and in retrospect it would appear that even a comparatively modest settlement—plus a little tact—could have resolved the whole issue quite swiftly. But to Hay, to Roosevelt, talk of a Colombian lien on the French company was patent extortion, a "holdup." As men of honor they could never be "party to the gouge," as Roosevelt expressed it.

Their personal regard for Latin-American politicians of any nationality had never been particularly high, it must be emphasized, but Cromwell had succeeded in convincing them that here they were dealing with the slipperiest, most corrupt variety of Latin American and that the sovereignty issue was the purest political hypocrisy. So the report from Beaupré that $10,000,000 from the French company would settle everything served only to substantiate all their worst suspicions of "those bandits in Bogotá," as Roosevelt was to call them.

When in early August the State Department released the information that the Colombians were holding out on the treaty in the hope of getting more money, the press mistakenly reported that the increased payment was to come from the United States—a misstatement the State Department did not bother to contradict. The widespread impression, therefore, was that Colombia was trying to "hold up" the United States, which was not at all the case.

The generally arbitrary tone assumed by the State Department, the threatening cable of April 28, were all designed to call Colombia's bluff, Cromwell would later explain. The story he planted in the *World* had also been contrived, he said, purely to frighten the Colombian Congress into ratifying the treaty.

Quite possibly that is so. But it is also possible that his underlying intention right along, perhaps even the underlying hope at the White House, was for the treaty to fail at Bogotá. In other words, if the language used by Beaupré or a story leaked by Cromwell following an ostentatious exit from the White House infuriated the Colombians, it was because they were meant to, because a crisis situation was wanted.

A great deal would be written and said to refute accusations that the

White House or the State Department was ever in any way party to the kind of scheme Cromwell had prophesied for the *World* in such amazing detail. And in fact the full story of what was transpiring behind the scenes will probably never be known. Cromwell, for example, for all that he would have to say for public consumption, appears to have purposefully created one of the larger gaps in the historical record. For in the otherwise complete file of his business dealings, still in the possession of the firm of Sullivan & Cromwell, there is not a paper relating to his Panama operations; all correspondence, cables, documents, expense vouchers, and the like are mysteriously missing.

On August 14, having finished a lunch with the President at Sagamore Hill, the President's home at Oyster Bay, Long Island, Senator Shelby Cullom, chairman of the Committee on Foreign Relations, held a press conference with reporters from the New York papers. The President, he said, was fully prepared for bad news from Bogotá and the President still wanted a Panama canal. "What will be done is a matter of discussion and consideration after the Colombian Congress has finally acted." When a reporter asked how the canal could be built without the treaty, Cullom replied in a matter-of-fact way that "we might make another treaty, not with Colombia, but with Panama."

Was the United States prepared to foster a Panama revolution? he was asked.

"No, I suppose not. But this country wants to build that canal and build it now."

The interview appeared the morning of August 15, the same day the long-awaited cable from Beaupré reached the State Department. Three days before, on August 12, the Colombian Senate had rejected the Hay-Herrán Treaty by a unanimous vote.

The President was notified immediately, as was the Secretary of State, who was also away from Washington at his summer place on Lake Sunapee in Newbury, New Hampshire. All matters of consequence at the State Department were being attended to by Acting Secretary Francis B. Loomis and Second Assistant Secretary Alvey A. Adee, both conversant with the Colombian situation. Adee, a man of vast experience with the department, a friend and colleague of John Hay's for thirty years, had installed a cot in his office and was on duty virtually around the clock.

Roosevelt's mounting frustration with the entire situation had become a source of some concern to Hay. "Those contemptible little creatures in Bogotá ought to understand how much they are jeopardizing things and imperiling their own future," Roosevelt had written earlier in the summer. Hay's instinctive response, then as now, was to urge restraint, patience. "I would come at once to Oyster Bay to get your orders," he now wrote, "but I am sure there is nothing to be done, for the moment." He advised consultation with Hanna and Spooner. Then, pointedly, he referred to Nicaragua as the "simple and easy" course, adding, "If you finally conclude to close with Nicaragua, it will be quick work to get a treaty ready. But I presume you may think best to do nothing definite until our Congress meets . . ."

It was Hay at his best and a very different letter than he would have written had Cromwell been with him in New Hampshire. It crossed in the mail with another angry missive from Roosevelt saying, "We may have to give a lesson to those jack rabbits."

These must have been exciting days for anyone sorting the mail in the little New Hampshire community. On August 18, Alvey Adee cautioned the Secretary by letter against any thought of American involvement in a Panama revolution. "Such a scheme could, of course, have no countenance from us—our policy before the world should stand, like Mrs. Caesar, without suspicion." A day later Adee wrote again to suggest to Hay that maybe, after all, Nicaragua was the best way out. ". . . We are very sorry, but really we can't help it if Colombia doesn't want the Canal on our terms."

But writing to Hay that same day, Roosevelt left no doubt as to his position concerning Panama. ". . . It seems that the great bulk of the best engineers are agreed that that route is the best; and I do not think that the Bogotá lot of jack rabbits should be allowed permanently to bar one of the future highways of civilization . . . what we do now will be of consequence, not merely decades, but centuries hence, and we must be sure that we are taking the right step before we act."

Roosevelt's particular interest at the moment was a memorandum that had been sent on to him by Acting Secretary Loomis, a lengthy document prepared by a specialist in international law and diplomacy at Columbia University, Professor John Bassett Moore, which Roosevelt now forwarded to Hay. Professor Moore's thesis, in essence, was that by the tenets of the old Bidlack Treaty the United States already had sufficient legal grounds to proceed with the canal. The "right of way" at Panama was already "free and open" to the United States, as stated in the treaty of 1846.

Hay went to Sagamore Hill on August 28, and there, while numerous Roosevelt children in light summer clothes scampered over a great green sweep of lawn, the two men conferred at length. Afterward, a correspondent for the *Herald* sent a long dispatch to his office. The President and the Secretary had settled on three possible courses of action in view of the failure of the canal treaty. The first was to proceed to construct the canal under the treaty of 1846, and "fight Colombia if she objects." (This, it was felt, would be a short and inexpensive war.) The second was for the President to move in accordance with the Spooner Act and turn to the Nicaragua route. The third course was to delay the great work "until something transpires to make Colombia see the light," then negotiate another treaty.

> It will, doubtless, be a surprise to the public that a course which is sure to involve the country with war with a South American Republic is one of the methods of procedure being soberly contemplated by the United States. . . .
>
> Persons interested in getting the $40,000,000 for the Panama Canal Company are of course eager that this government shall go ahead and seize the property, even though it leads to war.

II

When Dr. Manuel Amador first landed in New York he had still to meet William Nelson Cromwell. Dr. Amador, whose full name was Manuel Amador Guerrero, was a leading physician and a popular figure in the social life of Panama City. His wife was the brilliant María de la Ossa and he himself was known as a man of "unblemished character," large property interests, and much political acumen. Born in Turbaco (near Cartagena), Colombia in 1833—which made him just seventy in 1903—he was a graduate of the University of Cartagena who had come to Panama at the time of the gold rush. His political career as a Conservative had flourished along with his medical practice until 1867, when he was designated president of the Department of Panama, but did not take office because of a revolution. Defeated, captured by the opposition, he was sent into exile, not to return again for a year, at which point he went back to medicine at the Santo Tomás Hospital, where he became superintendent. It was as chief physician of the Panama Railroad, however, that Dr. Amador had attained most of his influence and prestige, as well as his interest in the canal. He had been among those prominent Panamanians to appear at the various occasions arranged to honor Ferdinand de Lesseps. When Lieutenant Wyse made his second journey to Bogotá in 1890 to secure an extension of the Wyse Concession for the court-appointed liquidator, the doctor was the head of a delegation of Panamanians who joined Wyse there to lobby in his behalf.

He was a neat, frail-looking man of medium height with thin white hair, a shaggy white walrus mustache, large ears, and heavy black brows, and he wore small steel-rimmed glasses. In his photographs at least, he seems to have had a habit of looking at people with his head cocked slightly sideways. He was also a man of nerve and ambition and he had come to New York to help arrange a revolutionary takeover at Panama.

The first known organized meeting of the movement had been held at a country estate outside Panama City on a Sunday late in July, probably July 25, 1903, a meeting at which Dr. Amador had not been present. Those who were there included his old friend Senator José Agustín Arango, Carlos Constantino Arosemena, and an American named Herbert G. Prescott, all of whom, like the doctor, were employees of the Panama Railroad and had been in regular communication with William Nelson Cromwell. Arango, a senator from the Department of Panama, was the railroad's attorney on the Isthmus, its land agent and chief lobbyist; Arosemena was a staff civil engineer; Prescott was the assistant superintendent. Also present were the United States consul general at Panama City, Hezekiah A. Gudger, and two officers from the U.S. Corps of Engineers, the only American Army officers on the Isthmus, who had been sent by the Walker Commission, Major William Black and Lieutenant Mark Brooke. The complete guest list is said to have numbered twenty-five or twenty-six people and the hosts were Ramón and Pedro Arias. Hezekiah Gudger, who was the main speaker, would be unable to recall later exactly what the meeting established, other than that "plans for the revolution were freely discussed."

All evidence is that Senator Arango (*El Maestro*), a most distinguished-looking gentleman with a white Vandyke beard, was the inspirational force; that in May he had begun talking revolution with his sons, sons-in-law, and the "intelligent and devoted" Arosemena, men in their late twenties or early thirties, all of whom had been educated in the United States. It was also *El Maestro*, whose office was side by side with Amador's, who personally recruited Amador early in August.

At that point the group was anxiously awaiting the return from New York of another railroad employee, James Beers, freight agent and port captain, who had left on a secret mission to see Cromwell. Whether he went on Cromwell's orders, whether he was actually summoned to New York, as later charged, cannot be proved. But at least six men were to testify that Cromwell sent for Beers, and it is unlikely that Beers or any other railroad employee would leave his job to stir up a revolution unless sent for by a superior, and as the railroad company's New York attorney, Cromwell ran the railroad. Beers, "a shrewd and calculating" former sea captain, had Cromwell's confidence, in any event, and he returned to Panama with the word that Cromwell was ready to "go the limit" for them. (Cromwell, in Arango's subsequent account, *Data for the History of the Independence*, would be referred to never by name, only as "the responsible person.")

Arango, Amador, and Arosemena became the nucleus of the conspiracy, to be joined shortly by Federico Boyd, son of the founding editor of the *Star & Herald*. Amador insisted that Arango should become the first president of the projected new republic of Panama. Arango, out of courtesy, said it should be Amador, and Amador agreed.

Amador, it was further decided, should also be the one to go to New York to see "the responsible person" in order to line up the necessary arms and money, and to secure some kind of assurance from the American Secretary of State that a revolution would be given military support by the United States. Amador, it was thought, would arouse the least suspicion since he had a son in the United States, a doctor with the United States Army, who was then stationed in Massachusetts. As instructed, the son sent a cable to his father saying, "I am sick; come." So on August 26 Amador sailed for New York, taking with him a cable code that he and the others had devised to cover every possible contingency.

The code had thirty numbered expressions for Amador to use in cables to Panama—he need only cable the appropriate number or combination of numbers—and sixteen numbered expressions for those at home to use in reply. Amador's list is especially interesting in that it shows how very uncertain things were at this stage. It shows, for example, that the conspirators had not excluded the possibility that Cromwell might be a liar.

Amador's code went as follows:

1. Have not been satisfied with Hay in my first conference.
2. Have had my first conference with Hay, and I found him determined to support the movement effectively.
3. Have not been able to talk to Hay personally, only through a third

person; I believe that everything will turn out in line with our desires.

4. Hay is determined to aid us in every way, and has asked me for exact details of what we need to ensure success.
5. My agent is going with me, fully authorized to settle everything there.
6. Cromwell has behaved very well, and has facilitated my interviews with important men who are disposed to cooperate.
7. You can hurry up matters, as everything here goes well.
8. I am satisfied with the result and can assure success.
9. Minister Herrán has suspected something and is watching.
10. Have not been able to obtain assurances of support in the form in which I demanded it.
11. Delay of Cromwell in introducing me to Hay makes me suspect that all he has said has been imagination and that he knows nothing.
12. It appears that Hay will not decide anything definitely until he has received advices from the commissioner who is there [in Panama].
13. I understand that Hay does not wish to pledge himself to anything until he sees the result of the operation there.
14. The people from whom I expected support have attached little importance to my mission.
15. Those who have decided can do nothing practical for lack of necessary means.
16. I have convinced myself that Hay is in favor of the rival route, and for that reason will do nothing in support of our plan.
17. News that has arrived from there on facilitating the construction of the canal has caused opinion here to shift in regard to our plan.
18. The pretensions manifested in the new draft of an agreement [treaty] render all negotiations between the two Governments impossible, and for this reason I have again resumed conferences.
19. The new commissioner is expected here to negotiate. On this depends my future movements.
20. I consider that I can do nothing practical here now, and for this reason I have decided to take passage for home.
21. Await my letter, which I write today.
22. Here it is thought best to adopt a different plan in order to obtain a favorable result for the construction of the work.
23. Cromwell is determined to go the limit, but the means at his disposal are not sufficient to ensure success.
24. Hay, Cromwell, and myself are studying a general plan of procedure.
25. The commissioner there is an agent of Cromwell's, of which fact Hay is ignorant.
26. I wish to know if anything has been advanced there and can I fix date here to proceed.
27. Delay in getting satisfactory reply obliges me to maintain silence.
28. B. [Beaupré apparently] communicates here that the contract can be satisfactorily arranged.
29. I have considered it prudent to leave the capital [Washington] and continue negotiations from here [New York] by correspondence.
30. I await letters from there in reply to mine, in order to bring matters to a close.

Of additional interest is the fact that Amador departed from Colón with insufficient cash to meet even the most modest travel expenses. It was only as a result of several good days at the poker table during the voyage that he was able to make ends meet.

There was trouble almost immediately. Among Amador's fellow passengers–indeed, the one he had won the most from at the poker table–was a man of "large interests" in Panama, J. Gabriel Duque, a Cuban by birth and a naturalized American who owned the *Star & Herald*, an ice plant, a construction company, and the extremely lucrative Panama lottery. Though not part of the Arango-Amador inner circle, Duque was aware of all that was going on and appeared entirely sympathetic. On reaching New York he went directly to Wall Street to see Cromwell, while Amador trailed off uptown to find an inexpensive hotel.

It is unclear exactly what Duque was up to, but Cromwell said that if Duque provided $100,000 to finance the revolution, then he, Cromwell, would see that Duque was made the first president of the new republic. He told Duque, furthermore, that the Secretary of State was eager to see him, and with Duque sitting before him, he picked up the phone, put through a call to Hay at the State Department, and set up the appointment. It was further suggested that Duque go to Washington by overnight train to avoid registering in a Washington hotel, a suggestion that Duque followed the next evening.

He arrived in Washington at seven in the morning, September 3, and after a breakfast at Harvey's Restaurant, he went to the State Department. At ten o'clock Hay appeared. The conference lasted for the next two and a half hours. Hay is said to have given Duque no promise of direct American assistance in the conspiracy. But in the same breath he emphasized that the United States was determined to build a Panama canal and did not propose to let Colombia stand in the way. Then, allegedly, he went still further. Should revolutionists take possession of Colón and Panama City, he said, they could depend on the United States to stop Colombia from landing troops to put down the revolution. This, Hay said, would be done to guarantee "free and uninterrupted transit" on the railroad, which the United States was treaty bound to maintain.

Duque understood perfectly. And no sooner had he descended the front steps of the State Department than he was on his way to the Colombian legation to see Tomás Herrán and tell him everything.

Perhaps this had been his intention all along out of spite over some real or imagined insult on the part of the inner circle. Perhaps it was a sudden impulse resulting from something Hay had said, or the way he said it. Or possibly he thought such a warning, when relayed to Bogotá, would jolt the Colombian regime into apprising the seriousness of the situation. Whatever the explanation, Cromwell had been double-crossed.

Herrán immediately sounded the alarm. He cabled Bogotá that revolutionary agents were in Washington seeing Hay, and that if the treaty was not ratified, Panama in all probability would secede, and with American support. He notified the Colombian consul general in New York of the Panama Railroad Company's involvement in the plot and of Amador's activities. The plot, he wrote, had been "well received" in Washington. He put detectives on Amador's track, then wrote to Cromwell and to the Paris office of the Compagnie Nouvelle to warn that they would be held directly responsible for any secessionist movement on the Isthmus. Implicit in the warning was the threat of full abrogation of all rights and privileges possessed by the Compagnie Nouvelle—all that it was about to sell for $40,000,000—if the company or its agents were party to an act of sedition.

For Amador, meanwhile, the mission to New York had suddenly become a bewildering dead end.

Unaware of what Duque had done, oblivious of the fact that he was being trailed by detectives, he saw only that Cromwell, the model of hospitality and enthusiasm on first meeting, had turned unexplainably rude and unreceptive. Amador appears to have made his first call on Cromwell on September 2, or the day after Cromwell saw Duque. Cromwell made "a thousand offers in the direction of assisting the revolution," even promised Amador that he would finance the undertaking. "I was to go to Washington to see Mr. Hay," Amador would recall. But by the time Amador returned for a second conference with the lawyer, Duque had been to see Herrán. Herrán had fired off his warning letters to New York and Paris, and Cromwell, determined to protect himself and safeguard the interests of his client, had decided to have no further ostensible dealings with conspirators from Panama.

Amador was told that Mr. Cromwell was out. When Amador declined to leave, insisting that he be received by the attorney, Cromwell at last burst out of his office and ordered Amador to go at once and not come back. In the end the elderly physician was shoved into the hall and the door was slammed behind him.

No explanation had been given and there appeared to be nowhere else to turn. Furious, worried that his money was running out, at a loss to report what had happened by means of his code phrases, he sent off a one-word cable (in English)—"Disappointed"—to his friends in Panama and prepared to leave on the next ship.

But then through a Panamanian banker based in New York, Joshua Lindo, of Piza, Nephews & Company, at 18 Broadway, Amador received word that if he simply remained quiet in New York there would be "help from another quarter."

"We can never know too much about the personality of Theodore Roosevelt," a learned student of the President's career once remarked. And to the many who were trying to appraise what was happening in regard to Panama, or more important, what the next turn might be, that personality seemed the crux of the matter.

"The warning I gave [in a previous telegram] . . . is founded on threatening statements which he has uttered in private conversations, and which by indirect means have come to my knowledge," wrote Tomás Herrán to his home office on September 11. "Your excellency knows the vehement character of the President, and you are aware of the persistence and decision with which he pursues anything to which he may be committed. These considerations have led me to give credit and importance to the threatening expressions attributed to him."

And Roosevelt's ultimate response to the Panama situation was to become the most disputed act of his career largely because it appeared to be an act of such violent impulse, an expression of what even many of his strongest admirers saw as an arrogant, nearly infantile insistence on having things his way and plunging ahead heedless of obstacles or consequences. To some observers there seemed something unpleasantly appropriate about the fact that his recreational passion at Sagamore Hill that summer of 1903 was the so-called point-to-point "obstacle walk," the one rule, the only rule, being that the participant must go up and over, or through, every obstacle, never around it. He was invariably the leader on such escapades, followed by a band of excited children, perhaps a stout-hearted guest or two. Once his sister Corinne Robinson saw him approach "an especially unpleasant-looking little bathing-house with a very steep roof" and she hoped this time a detour would be in order. "I can still see the sturdy body of the President of the United States hurling itself at the obstruction and with singular agility chinning himself to the top and sliding down the other side."

But for all the "vehement" reputation, for all his unpleasant private remarks concerning contemptible little creatures in Bogotá, for all the deliberate misstatements or threats by Cromwell and others that he let stand, Roosevelt was still taking no action as summer ended and the record shows that he was still giving serious thought to several possible "ways around" at Panama.

The expert on international law, Professor Moore, was invited to spend a night at Sagamore Hill to elaborate on his theory. Mark Hanna was queried for his views. (Hanna recommended patience and moderation. He was sure a satisfactory settlement could be reached with Colombia.) Hay remained a steady sounding board.

Hay observed in early September that a revolution on the Isthmus was "altogether likely," but advised caution and careful consideration. "It is for you to decide whether you will (1) await the result of that movement, or (2) take a hand in rescuing the Isthmus from anarchy, or (3) treat with Nicaragua."

Roosevelt's reply of September 15 was to agree to do nothing until he returned to Washington at the end of the month. "Then we will go over the matter very carefully and decide what to do." Only to this extent had he reduced the field of choice. Henceforth, he told Hay, he wanted no further dealings with "those Bogotá people."

"No one can tell what will come out in the Isthmian Canal business," he wrote that same day to his friend "Will" Taft in the Philippines.

Once, in describing his method of executive leadership, Roosevelt remarked to a friend, "When I make up my mind to do a thing, I act. A good many . . . call me jumpy and say I go off half-cocked, when, as a matter of fact, I have really given full consideration to whatever it is that is to be done." It was his quickness in following up on a decision that misled people, his cousin Nicholas would reflect years later; the decision itself, however, was rarely ever arrived at without enormous forethought.

For the moment, Roosevelt was waiting also for response to a request made through Secretary of War Elihu Root as far back as mid-March, before the Senate approved the Hay-Herrán Treaty. He had ordered that two or three picked men from the Army be sent to Panama in civilian dress to appraise the situation from a military point of view and to report back to him personally.

On October 1, or several days after Roosevelt's return to Washington, Tomás Herrán reported the official policy to be one of "watchful waiting."

III

Manuel Amador's "help from another quarter" arrived September 22. "I had gone to New York by pure chance," Philippe Bunau-Varilla maintained later. But so "fortuitous" a coincidence would strike many as highly improbable. It had been just two weeks since Amador had sent his "Disappointed" cable, or exactly time enough for Cromwell to have wired Paris and have Bunau-Varilla catch the next steamer.

At any rate the audacious little engineer had been busy much of the summer burning up the wires to Bogotá with vigorous, costly cables to Marroquín warning of dire consequences should the Colombians reject the treaty. He had been corresponding with Loomis at the State Department and publishing lengthy paid announcements in *Le Matin*. He was sure the dark forces that had destroyed Ferdinand de Lesseps were loose again.

Once in New York, according to his subsequent testimony, he "never even saw the shadow" of Cromwell. He had come, he said, to pick up his thirteen-year-old son, Étienne, a hay-fever victim who had been spending the summer at John Bigelow's country place on the upper Hudson. "I naturally took advantage of my presence in America to visit and to question, as to the state of affairs at Panama, those who could give me any information," he wrote later. In fact, through Joshua Lindo, whom he had known in Panama years before, Bunau-Varilla was in touch with Amador by phone the day after his arrival.

He had checked in at his favored Waldorf-Astoria and it was in his room there, Room 1162, that he and Amador sat down to talk for the first time on the morning of September 24, at precisely 10:30, according to Bunau-Varilla's recollection. They too had known each other in years past on the Isthmus, but Bunau-Varilla had by no means forgotten what the difference in rank had been between a mere local physician in the employ of the railroad and the director general of the Compagnie Universelle.

Amador, "deeply moved by emotion and indignation" (according to Bunau-Varilla), unburdened himself of all that had happened. If his friends in Panama should be found out and shot as a result of Cromwell's meddling, Amador declared, then he would kill Cromwell. Bunau-Varilla called it "unpardonable folly" ever to have listened to Cromwell in the first place. "With your imprudence you have indeed brought yourselves to a pretty pass," he lectured. The situation, however, was not hopeless. To extricate themselves from their plight the doctor and his friends had only to appeal to reason and to put the matter in the hands of Philippe Bunau-Varilla. "Tell me what are your hopes and on what are based your chances of success. Tell me all calmly, methodically, precisely."

The particulars, as Amador presented them, were these:

Only a small, weak garrison of federal troops was maintained on the Isthmus. The soldiers had not been paid for months and their commanding officer, young General Huertas, was known to be sympathetic to the revolutionary movement. Colombia, however, had command of the sea and so could land more troops at will.

(Except for the part about General Huertas, it was all information that could have been obtained from a careful reading of the newspapers over the past few months—or from Cromwell, were Bunau-Varilla and Cromwell secretly in contact with each other, a side of the story that will never be known.)

The immediate need was money Amador said. Exactly what figure did he have in mind? Bunau-Varilla asked. Six million dollars replied Amador, which would cover the cost of the necessary gunboats.

Bunau-Varilla told Amador that he now understood the situation perfectly and that he would need a few days to devise a solution. In the meantime Amador was to keep out of sight and talk to no one. Amador had impressed him as a risky confederate—"a childish dreamer." If Amador wished to make contact by phone, he was to use the name Smith. "I shall take that of Jones."

According to Bunau-Varilla he was now confronted with a grave question of conscience. "Had I the moral right to take part in a revolution and to encourage its development?" The answer, he quickly decided, was yes. "Yes, because Colombia was obviously prosecuting a policy of piracy aiming at the destruction of the precious work of Frenchmen."

IV

At noon on October 10, 1903, Assistant Secretary of State Loomis escorted Philippe Bunau-Varilla across from the State Department to the White House. Loomis was a handsome man in his early forties, starched, eager, with a marvelous handlebar mustache and thin black hair plastered with brilliantine. A former press agent, Loomis planned to introduce Bunau-Varilla to Roosevelt informally as the publisher of *Le Matin*. Loomis, another of Bunau-Varilla's "personal friends," had been spotted and cultivated by the Frenchman in Paris a few years before when Loomis was en route to a post in Portugal.

According to the subsequent recollections of both Bunau-Varilla and Roosevelt, their conversation began with talk of the Dreyfus Affair and of the part played by *Le Matin,* after which Bunau-Varilla asserted, "Mr. President, Captain Dreyfus has not been the only victim of detestable political passions. Panama is another." The exchange that followed was conducted by all parties with scrupulous care. Describing the scene later, Roosevelt would remark that there might just as well have been a Dictaphone in the room. Bunau-Varilla predicted a revolution on the Isthmus and according to Bunau-Varilla the "features of the President manifested profound surprise."

"A revolution?" murmured Roosevelt (according to Bunau-Varilla's account). "Would it be possible?"

Bunau-Varilla said later that he never asked Roosevelt what the United States would do in the event of such an uprising. But in Roosevelt's version, given off the record ten years later, Bunau-Varilla asked point-blank whether the United States would prevent the landing of Colombian troops, then added, "I don't suppose you can say." To which Roosevelt replied in substance that he could not. All he could say was that Colombia by its action had forfeited any claim on the United States—and that he had no use for the Colombian government.

In a letter to Roosevelt written only a few months later, a letter very possibly written on request, Loomis stated, "Nothing was said that could be in any way construed as advising, instigating, suggesting, or encouraging a revolutionary movement."

Be that as it may, Bunau-Varilla left the President's office positive that he knew where Roosevelt stood, and Roosevelt allowed later that had Bunau-Varilla failed to grasp what he, Roosevelt, intended to do then Bunau-Varilla would not have been very bright. "Of course I have no idea what Bunau-Varilla advised the revolutionists," Roosevelt would tell John Bigelow, ". . . but I do know, of course, that he had no assurances in any way, either from Hay or myself, or from anyone authorized to speak for us. He is a very able fellow, and it was his business to find out what he thought our Government would do. I have no doubt that he was able to make a very accurate guess, and to advise his people accordingly. In fact, he would have been a very dull man had he been unable to make such a guess."

Bigelow, however, was among those who would remain unconvinced. And his private observations cast a very different light on the situation. Bunau-Varilla had come to Bigelow's country place at Highland Falls, on the Hudson—his family was staying with the Bigelows—immediately after seeing the President, and from the conversation that had passed between them there, Sunday the eleventh, Bigelow clearly understood that Roosevelt had been fully informed as to Bunau-Varilla's revolutionary plan. Bigelow, it should also be noted, was a staunch admirer of Roosevelt's and would later share none of the scruples over the role of the United States in "the Panama business," as Roosevelt called it. The following is taken from Bigelow's private journal. It was written within a week of Bunau-Varilla's White House conference:

> Bunau-Varilla was up over Sunday, has seen the President and the Ass't Secretary of State; unfolded to them his scheme for proceeding with the Isthmian Canal without much more delay. . . . It is in brief to have Isthmians revolt from the Colombian govt. declare their independence . . . issue a Proclamation to that effect, adopt the Constitution of Cuba at the same time, and give Dictatorial powers to the President [Amador] who is an old and trusty friend of B-V., have the U.S. send vessels to protect . . . the new state from any hostility that could do it any harm, etc. &c.

But according to Bunau-Varilla it was only when he was on the train back to New York from Washington that he conceived his plan. Fundamentally, it was no different from what Cromwell had outlined for the *World* months earlier, except that Bunau-Varilla saw American gunboats playing the key role. He knew it would work, he said later, because he had watched it happen during the revolution of 1885. In point of fact, however, he had seen no such thing, since no American force at Panama had prevented the landing of Colombian troops in 1885 or during any previous disturbance.

He saw Amador at the Waldorf the night of the thirteenth and told him there was no cause to buy gunboats, explaining why in only the most general terms. Amador insisted that there was still a great need for money to guarantee the support of the Colombian garrison at Panama City. Bunau-Varilla thought $100,000 ought to be sufficient and promised to provide that amount from his own pocket if necessary. Amador remained highly agitated. He had had visions of much larger sums; he felt there ought to be a commitment from someone in Washington, something beyond the Frenchman's mere say-so. He was distressed over a notion of Bunau-Varilla's that the new republic need only comprise the canal zone, not the whole of the Isthmus. If he and his friends wanted the entire Department of Panama, they could take it later, Bunau-Varilla said; with the canal treaty ratified, they would get the $10,000,000 authorized by the Spooner Act and could wage all the war they wanted.

They separated "coldly," but Amador was back again first thing the next morning, pale and haggard after a sleepless night, and declared himself prepared to go along with whatever Bunau-Varilla wished.

"This is what I call a sensible speech," responded Bunau-Varilla. He would be leaving again that morning for Washington. The doctor meantime was to prepare himself to sail for Panama. On Bunau-Varilla's return from Washington, he would be given the precise program of action.

At his Washington hotel over the next few days, prior to seeing John Hay, Bunau-Varilla prepared everything he thought Amador would need—a ready-made revolution kit, including a proclamation of independence, a basic military plan, a scheme for the defense of Colón and Panama City, the draft of a constitution, a code by which he and the rebels could correspond. The one element lacking for the moment was a flag for the new republic.

Things began to move rapidly. On October 15, Cromwell, who had been staying under cover this whole time, sailed for France, removing himself thereby from any possible association with activities in Washington or New York. As a parting gesture, in a letter to Roosevelt dated the fourteenth, he advised that his associates in New York would be on call at the President's command. "Never before was this problem of the ages so near solution as at this moment," Cromwell wrote, "and, if the opportunity be lost, it probably will be lost for centuries to come."

It was also on the fifteenth that a dispatch went out from the Navy Department to Admiral Henry Glass, commander in chief of the Pacific Squadron. One week hence, on the twenty-second, he was to proceed with his squadron on an "exercise cruise" to Acapulco. Further instructions would follow.

The next day, October 16, with Loomis again serving as his entrée, Bunau-Varilla saw John Hay at Hay's house on Lafayette Square—this at Hay's suggestion. The meeting occurred in the afternoon and it was a most curious and ultimately critical confrontation.

Bunau-Varilla had pictured Hay as cold and severe, an American Bismarck, as he later wrote, but instead he found a man of "delicate and refined mind" whose ideas "coincided rigorously with my own." Together they deplored "the blindness" of Colombia. The entire state of affairs, declared the Frenchman, would end in a revolution, and Hay agreed that this, unfortunately, was the most probable hypothesis. "But we shall not be caught napping," Hay said. "Orders have been given to naval forces on the Pacific to sail towards the Isthmus."

To anyone with a personal interest in a revolution, this was, as both men appreciated, a momentous, an invaluable, piece of information. But as in Bunau-Varilla's exchange with the President, the tone remained one of perfect propriety. Bunau-Varilla appears neither to have registered any response to the news nor to have given Hay any indication as to his own intentions.

What followed instead was a long diversion by Hay, a lot of small talk seemingly about a novel the Secretary had just read and that he happened to have at hand. Giving Bunau-Varilla the book, he told him to take it and read it at first opportunity. The title was *Captain Macklin*, and the author was Hay's friend Richard Harding Davis. Bunau-Varilla was to write that it was "the subtle symbol, the password exchanged between Mr. Hay and myself."

The story, as Hay briefly recounted for Bunau-Varilla, concerns an idealistic young West Point cadet, Royal Macklin, who sails to Central America (on a ship called *Panama*) to seek his fortune after being expelled from the Academy for a minor infraction of the rules. He casts his lot with an older French officer and together they bring off a revolution in Honduras. There is much of the author's customary zest for manly combat, and at one point, the hero, a figure very much like the author, kisses the locket given to him by his sweetheart and thereby recovers his confidence and determination. But it is also a book in which the local political leader is a figure of ridicule, the revolution is a "comic opera," and Central America is seen as a frontier of untold opportunity if only the white man were to take charge.

" 'I know all of Central America, and it is a wonderful country,' " observes one character, a North American.

> "There is not a fruit nor a grain nor a plant that you cannot dig out of it with your bare fingers. It has great forests, great pasture-lands, and buried treasures of silver and iron and gold. But it is cursed with the laziest of God's creatures, and the men who rule them are the most corrupt and the most vicious. . . . They are a menace and an insult to civilization, and it is time that they stepped down and out, and made way for their betters, or that they were kicked out."

The book's most memorable figure, however, is the French freebooter, a tragic, sad, honorable, chivalrous knight-errant who fights on at the head of Latin soldiers, but who is plainly too good for them. To John Hay, as he told Bunau-Varilla, the ambitious young Macklin and the French officer were brothers in spirit, "searchers after the Ideal."

By his own account Bunau-Varilla came away from the interview in something like a spiritual daze. He had met "one of the most noble characters it has ever been given me to know"; he would "cherish . . . the memory of Mr. Hay [with] an almost religious admiration." Was Hay saying that they too were "searchers after the Ideal"? Did Hay see this Frenchman's own soldierly part in the epic of Panama as akin to that of the Frenchman in the story? Plainly the American Secretary of State was trying to tell him something. "Did he not wish to tell me symbolically that he had understood that the revolution in preparation for the victory of the Idea, was taking shape under my direction? . . . It only remained for me to act."

The same evening, the evening of October 16, two Army officers were ushered into the President's office at the White House to report on their confidential mission to Panama. Captain Chauncey B. Humphrey was an instructor of drawing at West Point; Lieutenant Grayson M.-P. Murphy had graduated from the Academy only the year before. Either might have been picked to portray Captain Macklin had anyone been casting a stage production.

From all they had seen, the officers told the President, a revolution on the Isthmus could be expected at any moment. Rifles and ammunition were being smuggled into Colón in piano boxes, a fire brigade recently organized in Panama City was intended as a revolutionary military unit, a man named Arango was the ringleader (they had picked up the name of the wrong Arango, as it happens), and the people of Panama seemed unanimous in their low regard for the government at Bogotá.

As Lieutenant Murphy would later confide, the prospects for a swift, neat, potentially lucrative revolution had struck them as so very certain that they were thinking of resigning their commissions forthwith and "assisting in its consummation." Their plan was to approach J. P. Morgan for the necessary financing. For bringing the revolution off, their fee was to be $100,000 each, a fair cut they believed of the $10,000,000 the United States was to turn over to Panama.

That their reconnaissance had been no chance or casual assignment is borne out by a subsequent written report, which includes such vital details for a military campaign as the best positions for artillery to command Colón and Panama City and the estimated number of mules that could be procured in remote interior villages. Nor is there much question that the word of such men would have carried great weight with Roosevelt. Probably theirs was the first account of the situation in Panama that he felt he could regard as trustworthy.

However, the response they saw was such that they left the White House convinced the United States would play no part in any such revolution. They had been astonished at how much Roosevelt seemed already to know of the topography of the Isthmus and other such details, but their immediate mutual reaction was to scrap any further thought of seeing J. P. Morgan. "There goes our revolution," Murphy said. "I sail for the Philippines."

Philippe Bunau-Varilla was now in high gear. He had gone directly from Hay's house to the station, had taken the next train to New York, and during the stop at Baltimore had sent a telegram telling "Smith" that "Jones" expected to see him at the usual place at 9:30 in the morning. At the stated hour Amador knocked at Room 1162 at the Waldorf, which, according to Bunau-Varilla, deserved to be regarded as "the cradle of the Panama Republic."

He had Amador sit down. There was no time left to quibble over details. Amador was to question neither his assertions nor his sources. He could now assure Amador that he and his junta would be protected by American forces within forty-eight hours after they proclaimed their new republic. Only one prior commitment was required. They must agree to entrust him, Philippe Bunau-Varilla, with the diplomatic representation of the new nation at Washington. He must be the one to draw up the canal treaty with the American Secretary of State.

Amador objected. To have a foreigner serve as their first representative abroad would be a blow to Panamanian pride.

"I can easily see that," Bunau-Varilla answered, "but a supreme law must dictate our resolution. It commands us to assemble every element which may ensure final success. A battle royal will be fought at Washington. Let him wage it who is best equipped to win the victory."

Amador promised to see what he could do.

The following day, October 18, was a Sunday, a sparkling autumn Sunday along the Hudson River, as Bunau-Varilla went north by train again to John Bigelow's place at Highland Falls, adjacent to West Point. On Monday he returned, looking like any other man of affairs newly refreshed by country air and vistas, but bringing in his suitcase a strange silk "flag of liberation" that Madame Bunau-Varilla and Bigelow's daughter Grace had spent nearly all Sunday stitching together "in the greatest secrecy."

Amador appeared again at Room 1162 for a last briefing. Bunau-Varilla displayed the flag, which, according to Bunau-Varilla, Amador "found perfect." The design was very like that of the flag of the United States, the differences being that the white stripes were yellow and in place of white stars were two yellow suns (symbolizing the two continents) joined by a yellow band (for the canal). As one Roosevelt biographer would note, there is a certain injustice to the fact that Roosevelt was unaware of these latest preparations. With his sense of humor and boyish love of adventure he would have savored every detail.

Amador said it would take him at least fifteen days after reaching Panama to get everything ready. His ship was due at Colón on the twenty-seventh. Bunau-Varilla said he could not wait that long. He wanted the revolution to occur on November 3—election day in the United States—which gave Amador exactly seven days once he reached home. If he and his friends could not do what had to be done in that time, if the revolution did not occur on the third, then they were on their own and he would take no responsibility for the consequences.

Amador received all the documents Bunau-Varilla had prepared, the code, the flag, on the following morning just before his ship sailed. He was also given the text for a telegram that he was to send to Bunau-Varilla the moment the new republic was proclaimed.

> The government has just been formed by popular acclamation. Its authority extends from Colón inclusive to Panama inclusive. I request you to accept the mission of Minister Plenipotentiary in order to obtain the recognition of the Republic and signature of Canal Treaty. You have full powers to appoint a banker for the Republic at New York, and to open credit for immediate urgent expenses.

Upon receipt of this message, and *only* upon its receipt, he would send Amador the promised $100,000 and the guaranteed military protection would arrive within forty-eight hours. According to Bunau-Varilla, Amador then departed from the Waldorf, having solemnly affirmed his complete agreement with all "conditions thus stipulated."

Amador sailed on the steamer *Yucatan*, the same ship that had once carried Theodore Roosevelt and his Rough Riders to the war in Cuba. To the purser, he entrusted a package, telling him to put it in the ship's safe and to guard it carefully, as it was vital to the future of Panama. The purser, young George Beers, was the son of Captain James Beers.

"The plan seems to me good," Amador had written to his son, the Army doctor, in a letter mailed just before sailing.

For Bunau-Varilla the week that followed was interminable. He busied himself with cables to Paris banks to arrange for a loan of $100,000 borrowed against personal securities and saw to the transfer of the money by cable to a New York bank (Heidelbach, Ickelheimer & Company). But mainly he worried over the possible movement of Colombian troops from Cartagena to Panama, a turn of fate that, if it came too soon, could wreck everything. He watched the New York papers and on October 26 read with "indescribable joy" a small dispatch saying that General Tobar, commander of the Colombian troops at Cartagena, though expected to leave soon for the Isthmus would probably not do so until November.

On the twenty-seventh, the day the *Yucatan* was due at Colón, there was no word from Amador. And there was nothing on the twenty-eighth. But on the morning of the twenty-ninth, Bunau-Varilla received the following signed "Smith":

> FATE NEWS BAD POWERFUL TIGER. URGE VAPOR COLÓN.

The first part of the message was in Bunau-Varilla's code and was perfectly clear.

Fate—This cable is for Bunau-Varilla
News—Colombian troops arriving
Bad—Atlantic
Powerful—Five days
Tiger—More than 200

Though the rest—*Urge vapor Colón*—did not conform to the prearranged code, he took it to mean *Send steamer Colón.*

These were puzzling and annoying words and he was at a loss to understand what they implied. But then it dawned that Amador wanted him to send an American man-of-war to Colón at *his*, Amador's, request. He was asking Bunau-Varilla to prove to the others at Panama, to the rest of the junta, that he could deliver on notice and exactly what was needed. "It was not information which was transmitted to me, it was a test to which I was being submitted."

The little Frenchman was at once in a grand state of agitation. An

American ship must be sent to Colón at once; everything depended on it. But how? "If I succeeded in this task the Canal was saved. If I failed it was lost." He could think better on a train he decided, and so out of the Waldorf he hurried, on his way to Washington again.

He saw Francis Loomis at his home that evening and told him to keep in mind the date November 3. There would be a repetition of what had happened at Colón in 1885, he said, and it would be a terrible shame if no American ship were on hand, or if her commander were to behave as had the commander of the *Galena* during the Prestan Uprising. Apparently Loomis said he could not and would not commit himself. But the following morning, as Bunau-Varilla was walking about Lafayette Square wondering whether to call on Hay directly, he ran into Loomis and this time Loomis declared in a notably formal manner that the situation at Colón was indeed "fraught with peril" and that it would be "deplorable if the catastrophe of 1885" were to be repeated.

And this, according to Bunau-Varilla's subsequent account, is all Loomis said. Still the message was clear: "The words I had heard could have but one interpretation: 'A cruiser has been sent to Colón.' "

He was at that moment like the character in *King Solomon's Mines* who, recalling that a solar eclipse is imminent, tells his savage captors that he will show his powers by blotting out the sun. He now had only to inform Amador that the ship was on its way and the first sight of it on the horizon at Colón would have exactly the desired effect.

A very great many people, however, were to find this explanation extremely difficult to believe. The Frenchman had gone to Washington, it would be charged, not to clear his thoughts or to stroll idly about Lafayette Square, but to tell Loomis to send a ship at once and that Loomis had assured him the next morning that the ship was on its way.

The answer given to this charge is vintage Bunau-Varilla:

> My only reply to such critics is that they have not the slightest idea of scientific methods.
>
> I built all this subtle diplomatic structure as a bridge is built: that is, by calculating its various elements, and not by trying to obtain direct information which it would have been impossible to obtain.
>
> The abstract operations of trigonometry lead to results more certain than physical measurements, when both operations are possible, but in the majority of cases trigonometry alone can be used. I have made diplomacy as it were by trigonometry.
>
> Such a method will without doubt seem incomprehensible to many minds.

He had noticed in the New York papers the reported movements of certain American naval vessels. The *Dixie* had been reported on its way to Guantanamo a few days earlier in *The New York Times*. The *Nashville* was at Kingston, Jamaica. If a ship had been ordered to Colón, it would be the *Nashville*, the one stationed nearest Colón, which, as he knew, had been in Colón earlier in the month. Figuring the ship's speed to be ten knots, he decided that she could cover the five hundred miles from Kingston to Colón in two days. He then added twelve hours for preparations before sailing, which, he reckoned, would bring her over the horizon at Colón on the morning of November 2.

Having talked to Loomis, he took the morning (eleven o'clock) train back to New York and again at Baltimore got off to send a wire to Amador. The wire went off a little after noon on October 30. Decoded it read as follows:

ALL RIGHT WILL REACH TWO DAYS AND HALF.

In *The New York Times* delivered to his room the morning of Sunday, November 1, on page 4 in the bottom right-hand corner, Bunau-Varilla found a small dispatch datelined Kingston, Jamaica, October 31:

> The United States gunboat *Nashville* sailed from here this morning under sealed orders. Her destination is believed to be Colombia.

13

Remarkable Revolution

> It was a remarkable revolution—I think the most remarkable I ever read of in history.
>
> —SENATOR SHELBY M. CULLOM

I

Manuel Amador's arrival at Colón had been without incident. By prior agreement none of his fellow conspirators were waiting to greet him when his ship docked. He was met only by Herbert G. Prescott, assistant superintendent of the Panama Railroad, who came on board with the port captain, as appeared perfectly routine, to carry off any papers or documents Amador would not wish to have found in his possession. The doctor came ashore looking innocent enough, and there seemed nothing unusual about the fact that he and Prescott departed together on the next train to Panama City. Had the doctor been searched, however, he would have been found to have an odd-looking flag wrapped about his waist.

The trouble started that evening when the flag and the rest of Philippe Bunau-Varilla's revolutionary paraphernalia were presented to the others at a secret gathering in a house on Cathedral Plaza, the home of Federico Boyd. Amador's report evoked only disappointment or harsh disapproval. The mere promises of an unknown Frenchman impressed no one. The idea of an independence movement that did not include the whole of the Isthmus was viewed as asinine, since several of those present owned extensive properties outside the Frenchman's proposed zone. The expectation had been that Amador would return with an actual agreement signed by John Hay or possibly even Roosevelt himself. Nobody liked the flag, which was thought to look too much like the flag of the United States.

It was nearly midnight when the meeting disbanded and the only agreement reached was that emissaries should be sent to the interior to drum up revolutionary support. Amador went home thoroughly dejected.

The following day it appeared that the whole game was up. Tomás Arias, one of the wealthiest, most influential members of the junta, came to tell Amador that he was backing out. "You are an old man," Arias said, "Arango is an old man, and you don't care if you are hung. I do not like to be hung."

Within hours Amador further learned from José de Obaldía, governor of the Department of Panama, that a force of picked Colombian troops—a detachment of *tiradores*, or sharpshooters—was on its way to Colón from Barranquilla. Obaldía, another wealthy landowner, had been appointed governor by José Marroquín only the summer before and because Obaldía was known to favor separation from Colombia in the event the canal treaty fell through, the appointment had caused a great stir in Bogotá. He was not involved in the conspiracy, only sympathetic and a close personal friend of Amador's. In fact, he happened to be living temporarily in Amador's house.*

Amador decided, first, that he himself had gone too far to pull back and, second, that for the time being he would confide this latest piece of information to no one except Herbert Prescott, who would be described as a "very energetic and typical railroad man, one who does not do things halfway." Together they agreed to "bluff it out."

That was October 28. Early the following day, Amador made his move. He would demonstrate to the others what could be achieved by merely saying the word. The crucial telegram—"Fate news bad powerful tiger. Urge vapor Colón."—went off to New York. Conferences were hastily arranged with Porfirio Meléndez, a stout, highly political police chief and part-time straw boss for the railroad at Colón,

* According to John Bigelow's private journal, Bunau-Varilla had actually received a letter from Obaldía leaving no doubt as to his sympathy with the planned revolt.

who agreed to manage the uprising on the Atlantic side, and with General Ruben Varón, commander of the *Padilla*, one of two Colombian gunboats presently in the Bay of Panama. For the promise of $35,000 in silver, Varón agreed to turn his ship over to the junta the moment their revolution commenced. The uprising was scheduled for November 4.

By Sunday, November 1, Amador had his answer from New York, which "had the effect of putting fresh life into the conspirators." Tomás Arias instantly regained his faith in the scheme. Dr. Carlos Mendoza, leader of the Liberals, and two of his prominent compatriots, Eusebio Morales and Juan Henríquez, agreed to prepare a proper manifesto and to improve upon Bunau-Varilla's declaration of independence. A new flag was designed by Amador's son Manuel and was sewn together by Señorita María Amelia de la Ossa, the fiancée of Herbert Prescott's brother, who was the railroad's chief telegrapher. The flag was composed of four rectangles, the lower left of blue, the upper right of red, the upper left of white with a blue star in the center, the lower right of white with a red star in the center. And it was rapidly duplicated by other ladies, including Señora Amador and her daughter Elmira (who was married to the nephew of United States Vice-Consul Felix Ehrman), and various members of the Arango and Arosemena households.

Even J. Gabriel Duque lent his unqualified support, promising that the city's fire brigade, of which he was the leader and major financial support—some 280 men—could be counted on when the time came.

None of the so-called inner circle, not even Arango, was as yet aware that Colombian troops were en route. Only Amador knew, probably his wife, Herbert Prescott, and, as of now, Colonel James Shaler, superintendent of the railroad. Shaler had to be included: the railroad was not only the one means of moving men from Colón (where the troops would land) to Panama City (where the revolution was to begin), the railroad had the only telephone and telegraph system between Colón and Panama City.

Prescott had gone over to Colón to confer with Shaler as soon as Amador apprised him of the situation; and with Shaler's blessing, Prescott had shifted all idle rolling stock, every car that might be used to transport troops, out of the yards and back to the Panama City end of the line. The railroad, these two men saw immediately, was the key. Shaler remained where he was; Prescott returned to Panama City to "wait until something turns up."

Which is how things stood on Monday, November 2, when the *Nashville* came over the horizon.

The urgent cable to Commander Hubbard of the *Nashville*, a cable classified as secret and confidential, had been sent on October 30, the day of Philippe Bunau-Varilla's chance encounter with Assistant Secretary Loomis in Lafayette Square. The *Nashville* was to proceed at once for Colón; Hubbard was to telegraph in cipher the situation there once he had consulted with the United States consul; and he was to keep his destination secret. Nothing was said of an expected revolution on the Isthmus or of any action to be taken in such event. So to Hubbard and his crew as they steamed out of Kingston, it had seemed a relatively routine matter. The long, white two-stacked gunboat had called at Colón on other occasions and as recently as two weeks before.

Nor was the ship's arrival at Colón taken as any particular cause for alarm by those Colombian or local officials who knew nothing of the schemes afoot. It only surprised them that the ship had returned so soon.

To Amador's fellow conspirators, however, it was the long-awaited decisive moment, the irrefutable sign that the United States stood prepared to guarantee their success, that Amador's Frenchman was truly their deliverer. "Have just wired you that the *Nashville* has been sighted," James Shaler wrote in a quick letter to Prescott at about four the afternoon of November 2. "This, I presume, settles the question."

The ship dropped anchor in the harbor at 5:30, or only about eight hours later than Bunau-Varilla had specified. Hubbard went ashore and found that "everything on the Isthmus was quiet." But he also talked to Shaler and there is no reason to believe that Shaler kept anything from him. So Hubbard undoubtedly appreciated exactly what the situation was when the Colombian gunboat next arrived in the harbor.

According to Hubbard's log it was nearly midnight when the *Cartagena* steamed in, her lights all aglow. Whether anyone was on duty at the railroad office at that hour or could determine what ship she was, what message Shaler may have put on the wire to Panama City that night, if any, are not known. But at daybreak, November 3, Hubbard took a launch to the Colombian ship, went aboard, and was informed by General Juan Tobar that she was carrying nearly five hundred troops and that he, General Tobar, intended to put them ashore at once.

Hubbard made no protest, despite what he knew. He had no orders to prevent such a landing and as yet there was not a sign of disturbance of any kind by which he might have justified his own intervention.

News of the landing was immediately telephoned to Panama City, and to those conspirators who had been kept in the dark this whole time, it was a crushing revelation. Word of a Colombian warship standing off Colón would in itself have had a devastating effect; but far worse was the realization that the American ship had made no move to prevent the Colombian troops—and assuredly a Colombian firing squad—from coming ashore. All the bravado engendered by the arrival of the *Nashville* the evening before was undone in an instant. The conspirators saw themselves as the victims of a diabolic Yankee betrayal. Even Amador, by all accounts, was having his own bleak second thoughts and might have called the whole thing off right then, early on the morning of the third, had it not been for the stately Arango, who declared himself ready to stand by his old friend, and for Señora Amador, a woman "of courage and snap" (as William Howard Taft would later describe her) who was considerably younger than her husband and who declared that it was time to get on with the fight, soldiers or no soldiers.

A plan was hurriedly improvised, an extremely neat stratagem that appears also to have been the inspiration of Señora Amador, and the details were quickly communicated by telephone to Colonel Shaler, he being the one chosen to bell the cat.

Colonel James Shaler was seventy-seven years old, older even than Amador, and he was widely regarded as the most important and popular North American on the Isthmus. A New York reporter who met Shaler a few months later perceived that "the impress of his personality" could be felt everywhere. In a land where most of mankind was short or medium-sized, brown-skinned and black-haired, Shaler was tall and lean and was made especially conspicuous by a huge white mustache and a great bushy crown of pure white hair. In a society where prolific families counted above all else, he was also a bachelor, a quiet man of quiet, contemplative pleasures (books, billiards). But he openly adored Panama and both his physical and mental vitality belied all traditional accounts of the torpor engendered by permanent residence in such a climate. To the junta he was suddenly indispensable. As several of them were to acknowledge later, without him there would never have been an independent Republic of Panama.

General Tobar and the Tiradores Battalion (plus perhaps a dozen wives) landed at the old Panama Railroad wharf, Tobar and his aides "glittering in elaborate uniforms and bristling with all the arms it was permissible for officers to bear." They were being received with customary deference by various local officials when Shaler approached from the railroad office. Shaler introduced himself, bid the officers welcome, and calmly recommended that they depart at once for Panama City on a special train, a single car and a locomotive, which had been arranged for their convenience, he said, at the personal request of Governor Obaldía. The troops could not be transported immediately because of a temporary shortage of equipment, he explained, but they would follow shortly. Tobar hesitated; Shaler was insistent, saying that the time fixed for departure had already passed and that there was no reason in the world why the officers should have to stand about in the killing heat a moment longer.

"I pointed out to him," Tobar explained afterward in his own defense, "that it was necessary for me to take the proper measures for the disembarkation of the troops . . . [but] as he insisted in his efforts, and as I was able to satisfy myself, even by the assurance of the prefect himself, that the troops could and would go over in a special train . . . I found no justifiable reason to persist in my refusal. . . ."

A young officer was picked to remain in command of the battalion, a Colonel Eliseo Torres.

But just as Tobar and his aides—fifteen men in all—were being comfortably settled in their special car, Tobar's second-in-command, General Ramón Amaya, grew suddenly uneasy about the arrangement, saying he must get off at once. Tobar objected, claiming it would be unseemly if the two of them were not to arrive in Panama City together. The issue was resolved only when Shaler stepped quietly to the rear of the car, pulled the signal cord, and hopped off the train. He was smiling broadly and waving as the train steamed away.

The railroad office now became a kind of command post. Shaler telephoned Prescott and told him to expect the generals at about eleven. He would do all he could to hold the troops in Colón, Shaler said, but he did not know how long it would be before they became suspicious and decided to take things into their own hands.

Sometime between 10:30 and 11:00, Commander Hubbard appeared at the office eager to know the situation, as he had just received a most important cable from Washington. His specific, secret orders now—orders issued the day before, November 2—were to prevent the landing of Colombian troops. The cable, a document of particular interest in time to come, read as follows:

NASHVILLE, CARE AMERICAN CONSUL, COLÓN:

SECRET AND CONFIDENTIAL. MAINTAIN FREE AND UNINTERRUPTED TRANSIT. IF INTERRUPTION THREATENED BY ARMED FORCE, OCCUPY THE LINE OF THE RAILROAD. PREVENT LANDING OF ANY ARMED FORCE WITH HOSTILE INTENT, EITHER GOVERNMENT OR INSURGENT, EITHER AT COLÓN, PORTO BELLO, OR OTHER POINT. SEND COPY OF INSTRUCTIONS TO THE SENIOR OFFICER PRESENT AT PANAMA UPON ARRIVAL OF *BOSTON*. HAVE SENT COPY OF INSTRUCTIONS AND HAVE TELEGRAPHED *DIXIE* TO PROCEED WITH ALL POSSIBLE DISPATCH FROM KINGSTON TO COLÓN. GOVERNMENT FORCE REPORTED APPROACHING COLÓN IN VESSELS. PREVENT THEIR LANDING IF IN YOUR JUDGMENT THIS WOULD PRECIPITATE A CONFLICT. ACKNOWLEDGMENT IS REQUIRED.

DARLING, ACTING.

Shaler told Hubbard what he had done with the generals, and Hubbard left to send a return cable to Washington. Colombian troops were already ashore, he reported; however, no revolution had been declared (Washington had said nothing of a revolution) and there had been no disturbances. Still: "Situation is most critical if revolutionary leaders act."

Hubbard was being scrupulously careful. Nothing would be done out of line, nothing, that is, without specific instructions from Washington. That Shaler had decided to "act," that things were also moving swiftly on the other side of the Isthmus, were perfectly obvious. Hubbard, in fact, was probably present when Porfirio Meléndez came into the office and Shaler and Meléndez began concocting a plan in the event that the soldiers demanded a train at gunpoint. The plan, as Shaler later explained, was to put all their rifles and ammunition in the rear car. When the train reached Lion Hill, one of Meléndez' men would pull the rear coupling pin and leave the arms stalled in the jungle. The engineer would then run the train full steam to Culebra, where he would abandon his engine and let the stranded, unarmed soldiers walk out whichever direction they chose.

To inform the others in Panama City of this scheme, and of a plan to hijack the *Cartagena*, it was decided to make up an unscheduled train and send Meléndez' daughter across. Aminta Meléndez, a tiny, cheerful eighteen-year-old who appeared considerably younger than her age, made the journey as asked, an act of considerable courage, which she would modestly discount afterward. She was neither stopped nor questioned by anyone. She simply found Arango, whom her father regarded as the real leader of the movement, and delivered the message. And as things turned out, the information had no effect one way or the other, since neither scheme was to be necessary, but Aminta Meléndez and her "ride" would become an essential element in the story of the revolution. In a favorite version she would be pictured as riding in the cab with the engineer, when in fact she sat quite comfortably in a coach.

The trap for Tobar and Amaya was being neatly set, meantime.

As soon as Herbert Prescott received Shaler's message that the generals were on their way, he went to Amador's house and told him it would have to be "now or never." Some very fast thinking was called for, as they had about two hours to get things ready. Amador was also convinced, from what he had learned during his trip to New York, that excessive bloodshed would seriously jeopardize American sympathy for their cause. The revolution, it was decided, would take place that afternoon.

Amador at once ordered his carriage and drove to the Cuartel de Chiriquí, the barracks of the Colombian garrison, a large pale building by the seawall, facing onto the Plaza Chiriquí. In command of the garrison was General Esteban Huertas, small, smooth-faced, impeccable, young, and very ambitious, as Amador well knew. According to the recollection of one of Huertas' own men, who was standing nearby when Huertas received the white-haired doctor, Amador said that he himself was old and tired but that Panama and the general had a great future ahead.

"If you will aid us, we shall reach immortality in the history of the new republic." An American ship had arrived, more were coming, Amador added. "You and your battalion can accomplish nothing against the superior force of the cruisers, which have their orders. Choose here, glory and riches; in Bogotá, misery and ingratitude."

Huertas is said to have remained "impassive" for a moment, then put out his hand. "I accept."

But since this appears to have been the only time the two met more or less privately that morning, an agreement must also have been reached regarding the sums Huertas and his men were to receive for their part, unless, of course, the bargain had already been worked out in secret in the days preceding, which is perfectly possible. In any event, payment to the soldiers was to be $50 per man, while Huertas was to be compensated for his revolutionary fervor with $65,000, an absolute fortune in Panama in the year 1903.

At 10:30, in full uniform, Huertas marched at the head of his regiment down the Avenida Central to receive the generals at the railroad station.

At 11:30 the train pulled in and Tobar, Amaya, and their aides stepped down to an amazing welcome. Governor Obaldía was there, accompanied by all his official family; General Francisco Castro, military commander of Panama, with his aides; United States Vice-Consul Ehrman, who was also head of the important Ehrman bank in Panama City; and Huertas with his troops, drawn up on the dusty little plaza across from the station. There was much saluting, much cheering, Obaldía was full of mellifluous words of welcome, and a line of sleek carriages stood waiting.

"There was," Tobar said later, "nothing that did not show the greatest cordiality and give me the most complete assurance that peace reigned throughout the department."

An elaborate luncheon followed at the Government House. But as the afternoon wore on, with still no sign of his troops, Tobar grew increasingly suspicious and finally demanded to be taken to military headquarters at the Cuartel, where he promptly assumed command. An officer confided that rumors of an uprising were sweeping the city; a cryptic note from a prominent citizen warned Tobar to trust no one.

Sometime near two o'clock the anxious general sent several of his aides to Obaldía to inform him of these rumors and to request that Obaldía order the immediate dispatch of the troops from Colón. The aides returned saying the governor had assured them that everything would be taken care of.

Apparently satisfied by this, Tobar and a number of his officers crossed to the barracks, where, joined by Huertas and Huertas' own retinue of officers, they inspected the local troops. The seawall was next, Tobar showing Huertas where he wanted the best marksmen placed to command the streets running from the harbor to Cathedral Plaza.

All this time Amador had been extremely busy completing his arrangements. He had met with Arango, who was to tell Carlos Mendoza to have the declaration of independence ready. He had met with J. Gabriel Duque to tell him the uprising would begin promptly at five, that the fire brigade must be at Cathedral Plaza, ready to march on the barracks. He had met Huertas on a street corner near the plaza, just before Tobar went to the barracks, and had listened rather impatiently as Huertas argued for a different plan. (Huertas wanted to strike later in the evening when there was to be a band concert and it would be easier to take the generals separately.)

Tobar and Huertas were still on the seawall when a secretary to Governor Obaldía appeared to tell Tobar that unfortunately the railroad superintendent, Colonel Shaler, was placing difficulties in the way. The troops could not be moved, Shaler had insisted, until their fares had been paid in full and in cash; it was a company regulation. Tobar told the man to go straight back and inform Obaldía that he was prepared to pay and that the troops were to be dispatched at once.*

Reports reached the barracks that things were getting out of hand elsewhere in the city. The head of the Panama Treasury, Eduardo de la Guardia, arrived to inform the generals that an uprising was certain and that Obaldía would do nothing to suppress it. By now it was nearly 4:30.

At about five o'clock, as Tobar, his officers, and Huertas sat conferring on a bench outside the barracks near the gate to the seawall, Tobar was informed that a crowd had begun gathering at the front of the building. General Amaya went out and returned to confirm the report. Huertas asked if it was not time to order out the first patrol. Tobar assented and Huertas, excusing himself to change out of his dress coat, went inside, followed by General Castro.

When a company of soldiers marched out with fixed bayonets, the generals were still sitting in the same place. The soldiers wheeled to the right of the seawall gate, as if to pass in front of the generals, but then suddenly opened into two files, one going in front of the seated men, the other behind. At a command the soldiers stopped and swung about with bayonets lowered at the astonished generals.

"Generals, you are my prisoners," said the officer in command, a young captain named Salazar.

"I am the commander in chief," Tobar declared.

"You and your aides," answered Salazar.

"By whose orders?"

"General Huertas'."

Tobar lunged at the nearest soldier in an effort to escape, but was instantly hemmed in by bayonets. He appealed to Salazar, begging him not to be a traitor. He called on sentinels along the wall, the other

* Tobar, as it happens, was well supplied with cash. Knowing that the national treasury at Panama was virtually empty, he had had the foresight to bring some $65,000 in American money to meet his own payroll and that of the local garrison.

soldiers, to come to the defense of their country, all to no avail. But neither he nor any of his companions had attempted to draw a sword or reach for a side arm.

Disarmed, they were marched out the seawall gate, through a crowd of several thousand people, and on to Cathedral Plaza, across the plaza and up Avenida Central to the jail, the crowd shouting "Viva Huertas! . . . Viva Amador! . . . Viva el Istmo Libre!" Those in the crowd who were armed began firing shots in the air.

Minutes after the generals were locked up, at 5:49 by the wall clock in the railroad office, Herbert Prescott was on the phone to tell Shaler and Meléndez. It was "the hour of freedom."

Amador ordered that Obaldía be taken into custody—as a matter of form—then went himself to see the American vice-consul, Felix Ehrman, who dictated a cable to Washington:

> UPRISING OCCURRED TONIGHT, SIX; NO BLOODSHED. ARMY AND NAVY OFFICIALS TAKEN PRISONERS. GOVERNMENT WILL BE ORGANIZED TONIGHT, CONSISTING THREE CONSULS, ALSO CABINET. SOLDIERS CHANGED. SUPPOSED SAME MOVEMENT WILL BE EFFECTED IN COLÓN. ORDER PREVAILS SO FAR. SITUATION SERIOUS. FOUR HUNDRED SOLDIERS LANDED TODAY, BARRANQUILLA.

Then, immediately, Amador, Ehrman, Arango, Federico Boyd, and Tomás Arias repaired to Cathedral Plaza to be acclaimed by the crowd.

At dusk, as the municipal council met to give the junta its formal recognition, the Colombian gunboat *Bogotá* opened fire, throwing five or six shells into the city, killing one man—a Chinese shopkeeper who had been asleep in bed—and a donkey. These were the day's only casualties. When a shore battery responded, the ship withdrew behind an island in the bay and was heard from no more. The *Padilla*, meantime, had kept perfectly silent.

So by nightfall there remained only the problem of the troops at Colón.

II

It was very early on the morning of November 4 that Commander Hubbard of the *Nashville* issued the order addressed to Superintendent Shaler forbidding the movement of "troops of either party [Colombian or insurgent] or in either direction by your railroad." So when the young Colombian colonel, Eliseo Torres, who had been left in charge, appeared again at Shaler's office that same morning to resume his effort to get transportation for his men, Shaler had only to tell him that his hands were tied. The troops and the number of women who accompanied them were camped in the streets and were the cause of much curiosity. There had been no friction with the local populace; not the slightest sign of trouble. And Torres, having no means of communication with Panama City, knew nothing of what had transpired there the day before and had yet to grasp the extreme gravity of his own situation.

Not until noon was he told—by Porfirio Meléndez, who, after conferring with Shaler, escorted Torres across Front Street to the saloon at the Astor Hotel. Over a drink, Meléndez explained what had happened to the generals, warned the young officer that more American help was on the way, and offered him a handsome honorarium in cash if he would be so sensible as to order his men back onto the *Cartagena* and quietly sail away.

The response of the young officer was explosive. He "flew into a violent passion" and, like Pedro Prestan, announced that he would burn the town and kill every American in it unless the generals were released by two that afternoon.

So there followed two extremely critical hours.

Hubbard had all American women and children put on board a German steamer then in port and on another ship belonging to the railroad. He gathered the men inside the railroad's stone warehouse and landed a detachment of forty sailors with an extra supply of arms. Cleared for action, the *Nashville* weighed anchor and moved in closer to shore, her guns trained on the railroad wharf and on the *Cartagena*, which to the surprise of everyone got up steam and departed at full speed.

The Colombians had the railroad building surrounded almost immediately, their purpose being, in Hubbard's view at least, to provoke an attack. It was a situation ripe for catastrophe. Yet for all the tension on both sides, no shots were fired and at about 3:13 Torres walked up to the barricaded building and told Hubbard that in fact he was "well disposed toward the Americans" and wished only to make contact with General Tobar to find out what he was supposed to do. He proposed that he withdraw his own troops to Monkey Hill, that Hubbard and his force return to the *Nashville*, and that he be permitted to dispatch an emissary to Tobar to explain the gravity of the situation and to bring back Tobar's answer.

After a hurried conference with Shaler, Meléndez, and the American consul at Colón, a man named Oscar Malmros, all of whom were impressed by Torres and convinced of his "good faith," Hubbard agreed to the proposition. Two emissaries were chosen, one of Torres' men and a local policeman. Shaler at once produced a special train, then put through a call to apprise Panama City of what had happened.

A murderous showdown had thus been averted for the moment. A number of people had kept their heads. However, with the *Cartagena* no longer standing by to evacuate Torres and his troops, the problem of their departure had also been compounded, and their quickest possible dispatch from Colón—from the Isthmus entirely—was of paramount importance to the success of the revolution. For as long as loyal troops remained where they were, Bogotá's claim to *de facto* sovereignty over Panama was quite as valid as that of the junta. Colonel Torres was in a comparatively strong position, furthermore. No insurgent force had as yet made itself known in Colón, and if the American commander stood by his own order that neither loyal nor insurgent forces could be transported on the railroad, then no insurgent force could be brought over from Panama City to challenge him. With nearly five hundred well-armed veteran troops at his command, he was unquestionably a force to reckon with, and he certainly had it within his power to lay waste to Colón as threatened, and to much of the railroad and its property. Most important, as he wrote in a note for Tobar, he and his men were fully prepared to "resist any attack rather than be traitors."

Torres was, in fact, the trump card and everything depended on how Tobar chose to respond.

At Panama City it was decided that a personal appeal by Amador (*El Presidente*, as the crowds were now calling him) might do the trick. The day at Panama City had been a very different one from that at Colón. The junta was riding high; the whole city was celebrating; the new flag had been raised at the Government House and at Cathedral Plaza. "The world is astounded at our heroism," Amador had told the troops at the barracks that morning. "Yesterday we were but the slaves of Colombia; today we are free. . . . President Roosevelt has made good. . . . Long live President Roosevelt! Long live the American Government!" He and Huertas had stood beside eight large wooden boxes filled with Colombian silver delivered for the troops from the Ehrman Bank. (Huertas and his officers, as they were informed privately, would receive their share in another five days, with checks drawn on another local bank, Isaac Brandon & Brothers.) "We have the money! We are free!" exclaimed Huertas, who was picked up in a chair and borne in triumph through the streets at the head of an enormous crowd. When a sudden downpour struck just as the parade reached the plaza, all who could crowded into the Central Hotel, where for another jubilant hour bottles of champagne were poured over Huertas' head.

Amador got to the police station about five o'clock, or roughly half an hour before the arrival of the emissaries from Colón. He talked to Amaya first, then to Tobar, and his point was the same with both: that further resistance on their part was useless since the United States was involved. "You must understand that we who started this movement are not insane," he told Amaya, whom he had known for years but had yet to face in quite this way. They were seated in the guardroom alone. "We fully appreciated the fact that in no case could we withstand all the rest of the nation, and in consequence we had to resort to means that, although painful, were indispensable. The United States has fully entered into this movement . . . and our independence is guaranteed by that colossus."

With Tobar he had been more explicit, saying that the plan had been sanctioned in Washington, that the United States had already supplied him with $250,000 to meet the expenses of establishing the new republic, which was quite untrue.

The generals refused to be swayed. "I answered Señor Amador," Tobar later related, "that I would take no account of what he had just told me, as my duty and the duty of the army I commanded was sufficiently clear, and that in consequence no human force could drag from me the order he desired."

Apparently Amador was no more out the door when the two messengers from Colón were brought in. Tobar read the note from Torres and said he positively refused to order the evacuation. Yet neither would he order Torres *not* to depart. Colonel Torres knew his duty, the general insisted.

Nor had he anything different to say the following morning, November 5, when it was reported from Colón that Torres and his men had marched back into town from Monkey Hill—claiming the mosquitoes had driven them out—that Hubbard had landed his force once again, and that therefore the situation was fully as serious as before.

Urgent meetings were called on both sides of the Isthmus. Shaler was on the phone to Prescott perhaps five or six times. Then at about five o'clock Tobar, Amaya, and the other Colombian officers were told to get ready to leave for the railroad depot. It had been decided to take them back to Colón.

They left the jail, surrounded by a large, well-armed escort. But at the station Herbert Prescott refused to put Tobar aboard until Tobar gave his word that he would make no attempt to escape. An extended "altercation" took place, Prescott insisting that Tobar must go as a voluntary prisoner because orders from the American government prohibited the transportation of soldiers to guard him.

Tobar, unfortunately for his cause, stood on his dignity. As an officer he could give no such guarantee; they could either transport him as a prisoner in fact, he said, or they could return him to prison. The argument dragged on, more time passed. Prescott called Shaler on the phone to ask what to do. Commander Hubbard, who was in the Colón office at the moment, told Shaler to tell Prescott to put the generals under an armed civilian escort and send them across.

And this was what was about to be done, the generals were actually seated on the train, when Shaler called again, great excitement in his voice. He and Porfirio Meléndez had just succeeded in getting Colonel Torres to agree to embark on the *Orinoco*, a Royal Mail steamer that had come in the day before. The price, Shaler told Prescott, would be $8,000. Prescott had only to get the money from the junta and he, Shaler, would pay Torres out of the railroad's safe and the troops could start their evacuation immediately.

Ordering one of the others to hold the generals until he got back, Prescott rushed out of the depot and took a carriage back to Cathedral Plaza where he found Amador, Boyd, and Arango. The only available cash, the three said, had been given out already to Huertas' troops, but the Brandon bank would vouch for whatever was necessary. Not wanting to lose a minute more than necessary, Prescott raced back to the phone, called Shaler, and told him he had the money. Fifteen minutes later Shaler called back and said the Colombian troops were just beginning to go on board the *Orinoco*.

Commander Hubbard, as he later testified in Washington, had had no part in the bargain struck with Torres. Shaler and Porfirio Meléndez had done all the talking, and Torres, as Shaler acknowledged, had agreed to their offer only after Shaler assured him that five thousand American troops were about to arrive. Then, at 6:20, as if on cue, the *Dixie* had been sighted on the horizon.

The $8,000 for Torres was carefully counted out in the railroad office by the company's cashier, a Mr. Wardlaw, and by Joseph Lefevre, a local resident who was later to become Minister of Public Works for the new Republic of Panama. The money, all in American twenty-dollar gold pieces, was put in two sacks and was carried out the door.

The only snag had occurred just as the troops were crowding onto the wharf beside the *Orinoco*. The local agent for the Royal Mail Steam Packet Company had suddenly specified cash in advance for their passage to Cartagena. Shaler told him there was not money enough left in the railroad safe, but not to worry, that whatever the cost it would be covered soon enough. Citing various regulations, the agent at first refused, then said he would clear the ship if Shaler and Hubbard put their signatures on a voucher for the passage money, which came to something over £1,000. Both men signed their names, and as a final gesture, Shaler sent Torres two cases of champagne.

At 7:05, while the troops were still going aboard, the *Dixie* anchored in the harbor. It was pitch dark by this time and raining very hard. At 7:35 the *Orinoco* cast off and steamed away, and in less than an hour four hundred Marines under Captain John A. Lejeune had landed.

The formal proclamations were read the following morning in front of the Colón prefecture. "We separate ourselves from our Colombian brothers without hatred and without joy," Porfirio Meléndez read from the declaration of independence, but the joy of the crowd was unmistakable. As a gesture of gratitude, Meléndez then asked Major William Black, the Walker Commission officer, to raise the new flag.

In Panama City that same morning, Señor Don Eduardo Ycaza, who had been appointed paymaster by the junta, began writing checks drawn on the Brandon bank—$30,000 to Huertas, who was to get another $50,000 later on (why he wound up with $80,000 all told, rather than the $65,000 originally promised, has never been explained); $35,000 for General Varón of the *Padilla*, $10,000 each for Captain Salazar, who had handled the actual arrest of the generals, and several other of Huertas' officers whose loyalty was deemed important.

Tobar and his generals, who had been returned to police headquarters, were again released and transported by train to Colón to await passage on the next ship to Cartagena.

Cables to Secretary of State Hay were composed and sent in the meantime, one from Arango, Arias, and Boyd, the other from Vice-Consul Ehrman. The authority of the new republic, the cables said, had been established and enthusiastically received throughout the entire Isthmus (in fact no news of the uprising had as yet reached several important parts of the interior), and Philippe Bunau-Varilla had been appointed "confidential agent" in Washington.

The reply came that afternoon. It was dated November 6–12:51 P.M. The United States government had formally recognized the new Republic of Panama.

III

It had been fifty-seven years since Benjamin Bidlack had signed the treaty at Bogotá, fifty-five years since the United States Senate had confirmed the treaty, fifty-one years since the first trains had begun rolling on the Panama Railroad. And in all that time, throughout the entire second half of the nineteenth century, there had been no serious misunderstandings as to the critical agreements of the treaty contained in Article XXXV. In no way was the arrangement to impair Colombian sovereignty over the Isthmus; Colombia was to remain the sole protector of the Isthmus and of the isthmian transit against domestic obstruction. The clear specific intent was to safeguard for Colombia its sovereignty in perpetuity, a guarantee for which Colombia had been willing to grant to the United States the right to create an isthmian transit–rail or canal. The United States was obligated to maintain order only when requested by Colombia and, as President Cleveland once stated, "always in maintenance of the sovereignty of Colombia."

"The purpose of the stipulation [Article XXXV]," Abraham Lincoln's Secretary of State, William Seward, had declared in 1865, "was to guarantee the Isthmus against seizure or invasion by a foreign power only. It could not have been contemplated that we were to become a party to any civil war in that country by defending the Isthmus against another party." Concerning Colombia, the United States desired nothing more, Seward wrote, than the enjoyment of "complete and absolute" sovereignty, and if that were "assailed by any power at home or abroad," the United States would be ready to cooperate with Colombia to "maintain and defend" its sovereignty.

The same or similar policy had held under subsequent administrations in Washington, including three illustrious Republican secretaries of state–Hamilton Fish, William Evarts, and James G. Blaine. Secretary Fish, for example, had on one occasion notified the American minister at Bogotá that the treaty of 1846 "has never been acknowledged to embrace the duty of protecting the road [the Panama Railroad] . . . from the violence of local factions . . . it is . . . the undoubted duty of the Colombian Government to protect it [the railroad] against attacks from local insurgents."

Thus the secret orders cabled to Commander Hubbard from the Navy Department, November 2—to prevent the landing of any armed force "either government or insurgent"—had been contrary not only to the spirit and intent of the treaty but to long-established policy and precedent. Colombia, the sovereign, was to be denied the right to land its own troops on the pretext that the United States was obligated to maintain "free and uninterrupted transit" on the railroad. In addition, the orders had been issued when there was not a sign of disturbance as yet anywhere on the Isthmus, when no revolution had even been declared, let alone physically set in motion.

Since General Tobar and his troops had already landed by the time Hubbard actually received the November 2 order, the United States, of course, had still done nothing out of line up to the moment when Hubbard took charge of the railroad. It was at that point, early on the morning of November 4, that American armed power had become an actual, rather than symbolic, factor in the plot, and at that point there was still no sign of trouble in Colón—no mobs gathered, no guns brandished—and nothing whatever had put the railroad or its operations in jeopardy. Neither had there been the least sign of an uprising in Colón even as Hubbard and his small force faced Colonel Torres from within their barricaded warehouse. No local patriots had rushed to help Hubbard, it should be further noted; the only violence threatening at Colón was between the Colombian troops and the American sailors.

Hubbard had taken command of the railroad because those were his orders. Indeed, as would be revealed afterward, Washington had been so anxious that he understand this that on November 3 two cables ordering Hubbard to take the railroad and keep the Colombian troops bottled in at Colón were sent to Consul Malmros from Washington, and another to Vice-Consul Ehrman at Panama City. The first of these cables, signed by Acting Secretary of State Francis B. Loomis, had been sent at 8:45 in the morning—or nine hours before the uprising took place at Panama City. (In fact, Acting Secretary Loomis was so overly anxious about things in general that a little later that same morning he cabled Ehrman, "Uprising on Isthmus reported. Keep Department promptly and fully informed." Ehrman replied, "No uprising yet. Reported will be in the night. Situation critical.")

What settled the fate of the infant republic, however, was the arrival of the *Dixie* followed, all within a week or so, by the *Atlanta*, *Maine*, *Mayflower*, and *Prairie* (at Colón), and the *Boston*, *Marblehead*, *Concord*, and *Wyoming* (at Panama City). The ships had come from Acapulco and Kingston; the *Maine*, among the last to arrive, had been on maneuvers at Martha's Vineyard. In several public appearances Theodore Roosevelt by now had mentioned "an old adage which runs, 'Speak softly and carry a big stick; you will go far.' " By the big stick he meant a strong Navy and he was wielding it for the first time. The latest orders from Washington were to prevent the landing of Colombian troops anywhere within the Department of Panama, not merely in the vicinity of the railroad. On the Pacific side the *Boston* and the *Concord* patroled as far east as the Gulf of San Miguel. More American troops were landed, some were sent into the interior. Rarely had there ever been so neat and effective a practice of all that Captain Mahan had preached.

Without the military presence of the United States—had there been no American gunboats standing off shore at Colón and Panama City—the Republic of Panama probably would not have lasted a week. Rear Admiral Henry Glass, for example, would conclude after a careful appraisal of the republic's capacity to defend itself that at the very most six hundred men might have been furnished with adequate arms. Taft, on his first visit to Panama a year later, would describe its army as "not much larger than the army on an opera stage." Colombia, had it had free access from the sea, could have landed several thousand veteran troops on both sides of the Isthmus, just as the conspirators themselves had appreciated from the beginning. As it was, a Colombian force of some two thousand men did attempt an overland march through the Darien wilderness, but ravaged by fever, they gave up and turned back.

The orders that sent Hubbard ashore at Colón, that secured the railroad, that started ten warships converging on Panama from points several thousands of miles off, had all emanated from the State, War, and Navy Building and were accredited to the Secretary of the Navy William H. Moody or to Acting Secretary Charles Darling and to Secretary of State Hay or to Acting Secretary Loomis. But the responsibility for "the dynamic solution of the Panama Question" (in the words of John Hay's biographer) rested entirely with Theodore Roosevelt, as Roosevelt himself would proudly acknowledge. "I did not consult Hay, or Root [Secretary of War Elihu Root], or anyone else as to what I did, because a council of war does not fight; and I intended to do the job once for all."

The American flag would "bring civilization into the waste places of the earth," he had declared in one of his speeches earlier in the year. The burden of empire was to advance liberty and order and material progress. "We have no choice as to whether or not we shall play a great part in the world," he had told another cheering crowd at San Francisco. "That has been determined for us by fate. . . ." They were popular words and very like those in a novel that was to appear less than a year after Panama became a republic—*Nostromo*, Joseph Conrad's tale of a Latin-American revolution and of the self-deceptions men work with the words they summon to deceive others. " 'We shall run the world's business whether the world likes it or not,' " a San Francisco financier remarks early in the story. " 'The world can't help it—and neither can we, I guess.' "

On the morning of November 3, the morning General Tobar and his *tiradores* came ashore at Colón, Roosevelt, as expected, had been at Oyster Bay, having taken the night train from Washington in order to vote in his hometown. On the second floor of Fisher's Hall on Main Street, over a Chinese laundry, with reporters and well-wishers crowding about, he had cast his ballot for two New York state judges and an assemblyman. He was back at the White House shortly after eight that night and from then on was caught up in "the Panama business."

Yet even as the crisis was still unfolding he had begun to plead his case, searching for exactly the right phrase or expression. In a letter to his fourteen-year-old son written the following night he explained that the United States had been policing the Isthmus for too long, that he had no intention "any longer to do for her work which is not merely profitless but brings no gratitude." Two days later, the afternoon of the sixth, he was happily talking of a "covenant running with the land" on the Isthmus, an expression his friend Oscar Straus, author, lawyer, diplomat, had used over lunch to suggest a basis for an American claim on the canal zone. Vice-Consul Ehrman's cable from Panama declaring the apparent success of the "Isthmian movement" had been delivered to the White House at 11:31 that morning; at 12:51, just seventy minutes later, Panama had been recognized by John Hay; and at virtually the same moment, Straus had produced what seemed the perfect legal ground. "Why that is splendid—just the idea," Roosevelt exclaimed and he sent Straus straightaway from the table to "explain that to Hay."

On Roosevelt's orders the following day, Hay assured reporters that "the action of the President is not only in the strictest accordance with the principles of justice . . . but it was the only course he could have taken in compliance with our treaty rights and obligations."

"It is reported we have made a revolution, it is not so," Roosevelt confided to the French ambassador, Jules Jusserand. ". . . it is idle folly to speak of there having been a conspiracy with us," he assured Dr. Albert Shaw, one of his kitchen cabinet.

The faculty of Yale University was up in arms, meantime. The head of the American Bar Association spoke angrily of the "crime" committed. A torrent of outrage was unleashed in editorial columns.

The first news of the revolt had been given very little play in the papers because of the election news. The front pages of the New York papers, for example, the morning the story broke, were taken up almost entirely by the triumph of George B. McClellan, Jr., the Democratic candidate for mayor. But Panama was the lead story everywhere in the days that followed, and many powerful papers immediately commenced a blistering attack on the Administration, holding Roosevelt strictly responsible for what had happened. *The New York Times* scarcely let a day pass without some new assault on the President and his "act of sordid conquest." Cartoons in the *World* by the brilliant Charles Green Bush showed a brutish Rough Rider, armed to the teeth, pouncing on Panama or glowering down the barrel of an enormous cannon at a helpless little Colombia.

But to the editor of *The Northwestern Christian Adovcate* Roosevelt wrote of the "oppression habitual" suffered by the people of Panama and insisted that "our Government was bound by every consideration of honor and humanity . . . to take exactly the steps that it took."

The explaining, the affirmations of high purpose, would continue for weeks, months, indeed for years—in a special message to Congress, in private conversation and correspondence, in magazine articles, speeches, his memoirs.

The United States had a mandate from civilization to build the canal, he told Congress on January 4, 1904, in a message devoted entirely to the subject. "The time . . . for permitting any government of anti-social and of imperfect development to bar the work, was past." The fundamental purpose—"the great design"—of the treaty of 1846, he claimed, had been to secure the construction of an isthmian canal; so therefore Colombia was violating the treaty, "the full benefits of which she had enjoyed for over fifty years." No American warships had been present, no American troops or sailors, when the revolution took place at Panama City. At Colón, Commander Hubbard had acted with "entire impartiality toward both sides, preventing any movement, whether by the Colombians or the Panamanians, which would tend to produce bloodshed. . . . Our action was for the peace both of Colombia and of Panama."

The people of the Isthmus, he said, "rose literally as one man." ("Yes, and the one man was Roosevelt," remarked Senator Edward Carmack, of Tennessee.) "I think proper to say, therefore, that no one connected with this Government had any part in preparing, inciting, or encouraging the late revolution on the Isthmus of Panama, and that save from the reports of our military officers . . . no one connected with this Government had any previous knowledge of the revolution except such as was accessible to any person of ordinary intelligence who reads the newspapers. . . ."

"We did our duty, we did our duty by the people of Panama, we did our duty by ourselves," he wrote in one of his several magazine pieces. "We did harm to no one save as harm is done to a bandit by a policeman who deprives him of his chance of blackmail." To talk of Colombia as a responsible power—"to be dealt with as we would deal with Holland or Belgium or Switzerland or Denmark"—was a mere absurdity, he informed a correspondent. "If they [the people of Panama] had not revolted, I should have recommended Congress to take possession of the Isthmus by force of arms . . ."

His action had been the farthest thing from impulsive, he would stress in a long chapter in his *Autobiography*. Nine-tenths of wisdom was to be wise at the right time; his whole foreign policy, he wrote, had been based on "the exercise of intelligent forethought and of decisive action sufficiently far in advance of any likely crisis" and Panama was "by far the most important action I took in foreign affairs." Colombia had proved itself utterly incapable of keeping order on the Isthmus; Colombia had no right to block a passageway so vital to the interests of civilization. For reasons of national defense no further delays could be tolerated. He had been prepared to act; no bloodshed had resulted. "From the beginning to the end our course was straight-forward and in absolute accord with the highest standards of international morality. Criticism of it can come only from misinformation, or else from a sentimentality which represents both mental weakness and a moral twist."

John Hay lent his support, sounding more and more like Theodore Roosevelt. "Some of our greatest scholars, in their criticisms of public life, suffer from the defect of arguing from pure reason, and taking no account of circumstances," he wrote to a member of the Yale faculty. "It was a time to act and not to theorize . . ." An attempt would be made by Hay's admirers to establish that he had been "disgusted" with all that went on and that on the pretense of poor health he had taken no active part; but the claim was without support and Hay himself, publicly and in private, remained "as emphatic and free from doubt about our Government's course" as the President.

Others in the Cabinet fell into line, without apparent qualm, nor with anything approaching Roosevelt's solemn air of righteousness. Attorney General Knox, having been asked by Roosevelt to construct a defense, is said to have remarked, "Oh, Mr. President, do not let so great an achievement suffer from any taint of legality." At another point, during a Cabinet meeting, Roosevelt talked of the bitter denunciations in the press, then entered into a long, formal statement of his position. When he had finished, the story goes, he looked about the table, finally fixing his eye on Elihu Root. "Well," he demanded, "have I answered the charges? Have I defended myself?"

"You certainly have, Mr. President," replied Root, who was known for his wit. "You have shown that you were accused of seduction and you have conclusively proved that you were guilty of rape."

But years later, on March 23, 1911, at Berkeley, California, at the climax of a speech before eight thousand people in the Greek Theater at the University of California, Roosevelt, in academic gown, was to make the remark that would undo virtually all of his other utterances concerning his "most important action" and that would be remembered afterward, by critic and admirer alike, as the simplest and best explanation of what the Panama revolution came down to. The speech, until that point, had been a heartfelt call to the youth of the Pacific slope to carry on with the high courage and purpose of the vanishing pioneers. And the audience had been profoundly stirred. Then his mood had shifted:

> The Panama Canal I naturally take special interest in because I started it. [*Laughter and applause.*]
>
> There are plenty of other things I started merely because the time had come that whoever was in power would have started them.
>
> But the Panama Canal would not have been started if I had not taken hold of it, because if I had followed the traditional or conservative method I should have submitted an admirable state paper occupying a couple of hundred pages detailing all of the facts to Congress and asking Congress' consideration of it.

> In that case there would have been a number of excellent speeches made on the subject in Congress; the debate would be proceeding at this moment with great spirit and the beginning of work on the canal would be fifty years in the future. [*Laughter and applause.*]
>
> Fortunately the crisis came at a period when I could act unhampered. Accordingly I took the Isthmus, started the canal and then left Congress not to debate the canal, but to debate me. [*Laughter and applause.*]

"I took the Isthmus" was an expression of the kind that came naturally to him, "the kind of exaggeration that he liked to make," as Root observed. It was what Hay called "a concise impropriety," like "We want either Perdicaris alive or Raisuli dead," the famous declaration made during the Moroccan kidnapping incident in 1904 (a statement Hay himself had actually written). It was also, in its fashion, as misleading and as self-congratulatory as some of the other things he said in his defense, since it seemed to dismiss out of hand the contributions of Amador and his fellow revolutionaries, not to mention the railroad personnel, or Cromwell, or General Esteban Huertas and his garrison, or Philippe Bunau-Varilla. "I took Panama because Bunau-Varilla brought it to me on a silver platter," Roosevelt is supposed to have remarked privately, which would be a more accurate summation.

But primarily, questions of morality aside, it was a mistake to have implied a deliberate, master strategy conceived and directed from the Oval Office. Tremendous effort would be made by newspaper reporters and latter-day historians to prove that Roosevelt had told Bunau-Varilla what to do, that Amador too had actually gone to Washington in secret and had been briefed by both Roosevelt and John Hay, that the money for the junta had been supplied, as Amador believed to be the case, from Washington. But no solid evidence, no evidence of any kind, was ever found to support these charges. And in fact one need only review the steps by which the plot unfolded to see how very tenuous it all had been and how many critical turns were determined by the individual responses of people about whom Theodore Roosevelt knew nothing.

Had James Shaler not pulled the signal cord when he did, had Señora Amador failed to fire her husband's flagging resolve, had the Colombian general Tobar been less concerned over his injured dignity, had he gone peacefully to Colón and merely remained there quietly with Torres and the troops, had any of a dozen small but critical developments gone differently, Theodore Roosevelt's ships would have arrived to find a wholly different situation and in all probability there would have been no new Republic of Panama either to proclaim or to protect.

If American sea power had settled the issue on the instant, made Panama an immediate *fait accompli*, it is equally obvious that belief in an American involvement far in excess of reality was for the actual conspirators the vital sustaining force: what Amador and his compatriots *believed* the situation to be—their mistaken impressions as a result of the arrival of the *Nashville*—was far more important than were the facts of the situation. An enormous gamble with far-reaching, immensely vital consequences was made by a variety of participants, and by all ordinary rules of chance the story should never have come out as it did. But as Conrad also observed in *Nostromo*, "Men of affairs venture sometimes on acts that the common judgment of the world would pronounce absurd; they take their decisions on apparently impulsive and human grounds."

Roosevelt's haste, his refusal—his inability—to see the Colombian position on the treaty as anything other than a "holdup," were tragically mistaken and inexcusable. It seems certain that with a modest amount of good will and patience the issue with Bogotá could have been resolved to the satisfaction of both sides; another six months' delay would have mattered little. In truth he was doing no more than to guarantee that the Compagnie Nouvelle received its full $40,000,000—which would lead to the charge that he was protecting the French investment because certain of his friends and relatives were secret stockholders, a charge that would later precipitate a sensational lawsuit. In 1908 Roosevelt had the government prosecute Joseph Pulitzer, owner of the *World*, for libel, but the court found that though there were "many very peculiar circumstances about . . . this Panama Canal business," Roosevelt had no case against the *World*. And since there was not a shred of evidence to support the charge against Roosevelt, the whole furor came to nothing.

For Colombia, already crippled by a costly civil war, Roosevelt's "most important action" meant the loss of what since the days of Bolívar had appeared to be its most valuable natural treasure, the Isthmus, with its unique geographic position "between two oceans." It meant also the loss of the $10,000,000 lump sum that was to be paid by the United States, the $250,000 annual payment by the Panama Railroad (for decades a crucial part of the national income), and the $250,000 annual payment that was to be forthcoming from the United States as part of the canal agreement. There were riots in Bogotá; desperate offers were to be made by special Colombian emissaries dispatched to Washington, including an offer to accept the treaty as it stood, which served only to satisfy the Administration conclusively that the earlier rejection of the treaty had been an outrageous act of extortion.

The damage done to American relations with Colombia, indeed with all of Latin America, was enormous, just as John Tyler Morgan had prophesied. As an American minister at Bogotá, James T. Du Bois, would write in 1912, the breach worsened as time passed.

> By refusing to allow Colombia to uphold her sovereign rights over a territory where she had held dominion for eighty years, the friendship of nearly a century disappeared, the indignation of every Colombian, and millions of other Latin-Americans, was aroused and is still most intensely active. The confidence and trust in the justice and fairness of the United States, so long manifested, has completely vanished, and the maleficent influence of this condition is permeating public opinion in all Latin-American countries, a condition which, if remedial measures are not invoked, will work inestimable harm throughout the Western Hemisphere.

"I fear," declared a much embittered John Tyler Morgan on the floor of the Senate, "that we have got too large to be just and the people of the country fear it." But in fact the people of the country were generally well satisfied with what had happened, with the results —and with Theodore Roosevelt.

14

Envoy Extraordinary

I had fulfilled my mission . . . I had safeguarded the work of the French genius; I had avenged its honor; I had served France.

—Philippe Bunau-Varilla

I

The newly designated "confidential agent" of the Republic of Panama —a citizen of France who had not laid eyes on Panama for eighteen years—had waited out the birth of the nation in the privacy of his room at the Waldorf-Astoria Hotel. On first word of success from Dr. Amador he had cabled an emotional reply hailing the infant republic ("small in extent but great in the part she will play") and like proud fathers he and the banker Lindo had celebrated with a bottle of champagne in the Waldorf dining room.

Some anxious days followed, during which Amador made the expected request for $100,000 but made no mention of diplomatic powers for Bunau-Varilla, who grew "very suspicious" and refused to honor his side of their bargain. Cables went back and forth; for a time something went wrong with the code. Then at last came official word from Arango, Boyd, and Tomás Arias designating him "Envoy Extraordinary and Minister Plenipotentiary near the Government of the United States of America." So by Sunday, November 8, he was installed in the new Willard Hotel in Washington, ready to embark on another chapter of "The Great Adventure," and the next day at a private lunch at Lafayette Square, he was received by the Secretary of State, whose only concern was a dispatch in the morning papers saying that a special commission was about to leave Panama for Washington to make the canal treaty.

The situation was indeed perilous, declared Bunau-Varilla. "Mr. Secretary of State, the situation harbors the same fatal germs—perhaps even more virulent ones—as those which caused at Bogotá the rejection of the Hay-Herrán Treaty." Before, there had been only the intrigues of the Colombians to contend with. Now, he said, to the intrigues of the Colombians would be added the intrigues of the Panamanians. ("Against my work formidable interests were up in arms," he would confide in a memorable, melodramatic aside in his *Panama: The Creation, Destruction, and Resurrection.* "Fortunately the firm basis of clearness and straightforwardness, which I had, throughout my life, taken for my acts, defied the most desperate assault." The departure of a special commission from the Isthmus could only conceal a "maneuver" and Amador was a party to it. "I knew his childish desire to sign the Treaty." It was the start of "a plot against me.")

"So long as I am here, Mr. Secretary," he said, "you will have to deal exclusively with me." It was all John Hay wished to know.

The special commission sailed on the mail steamer *City of Washington* the following day, November 10. On the eleventh, Vice-Consul Felix Ehrman cabled the news of their departure to the State Department, explaining that they were coming only to assist Envoy Bunau-Varilla. "I am officially informed," the cable said, "that Bunau-Varilla is the authorized party to make treaties."

Who had thus informed the American vice-consul remains obscure, but in actuality the written instructions for Bunau-Varilla being carried north from Colón were quite to the contrary. It was quite clearly specified that the envoy extraordinary was to "adjust" a canal treaty, that all clauses in the treaty were to be discussed in advance with Amador and Boyd, that he was "to proceed in everything strictly in accord with them."

Just seven days later, on November 18, 1903, came the signing of the Hay–Bunau-Varilla Treaty. What had transpired within those seven days, the document that resulted, would evoke a whirlwind of controversy. The treaty would be the subject of one of the angriest debates in the history of the United States Senate and would remain a bone of contention between Panama and the United States for generations to come. This is what happened.

Between his Monday luncheon with John Hay and a ceremonial first call at the White House on Friday the thirteenth—the call that marked the formal diplomatic recognition of Panama by the United States—Envoy Bunau-Varilla made a flying visit to 23 Wall Street, the six-story, white-marble headquarters of J. P. Morgan & Company. He saw the regal head of the firm in a glass-enclosed ground-floor office and in less than an hour an arrangement was agreed to. Morgan would serve as financial agent for the new republic. Morgan would supply an immediate loan of $100,000, which, for the moment, Bunau-Varilla would cover through the Lindo Bank. Morgan & Company was also to have the "exclusive faculty" (granted by Envoy Bunau-Varilla) of cashing the $10,000,000 indemnity to be paid by the United States. If Morgan's friend and legal adviser Cromwell or Senator Hanna had played any part in this affair, as is quite possible, Bunau-Varilla never said so. He stressed only the wonder of a fledgling republic coming into its own not merely under the guardianship of the United States of America, but of the house of Morgan.

Returning to Washington with all speed, and accompanied now by his thirteen-year-old son, who he hoped might witness some of the history about to transpire, he had himself photographed in full diplomatic regalia—morning coat, striped trousers—and at 9:30 sharp the morning of Friday the thirteenth was at the State Department, ready to be escorted to the White House. Hay, however, insisted that first they pose together for an official portrait in his office. Chairs were drawn up to the end of a heavy, polished table. Thin patches of sunlight fell on a Brussels carpet. Hay, impeccable, stiff as a tailor's dummy, his small white hands in his lap, sat in profile, gazing vacantly toward the windows; but Bunau-Varilla faced directly at the camera, his expression deadly serious, and cocked an elbow on the table. He was a man in perfect accord with his surroundings it would appear, and in perfect command of the situation.

They went down to the street, to Bunau-Varilla's waiting carriage, where young Étienne Bunau-Varilla sat idly absorbed in the passing scene. Hay, as Bunau-Varilla would recall, "instantly had the charming idea of taking him to the White House," so moments later the boy too was ushered into the Blue Room and was provided a chair as his father and the President read their formal declarations and then shook hands.

"What do you think, Mr. Minister," Roosevelt asked, "of those people who print that we have made the Revolution of Panama together?"

"I think, Mr. President, that calumny never loses its opportunity even in the New World. It is necessary patiently to wait until the spring of the imagination of the wicked is dried up, and until truth dissipates the mist of mendacity."

He then introduced his son to the President, who was plainly pleased by the boy's presence.

The ceremony was over. Panama, in the formal sense, had attained legal status in the family of nations. Not a Panamanian had been present; not a word had been spoken in Spanish. And as was understood by all who had participated, there remained only four days until the special commission from Panama was due in New York.

To much of the press it had been a bit of barefaced comic opera. "Doubtless M. Bunau-Varilla, whirled along on the torrent of his own tropical eloquence, came almost to believe in it, and was too impassioned to wink," wrote *The New York Times*. "Neither, we may be sure, did the President yield to his human impulse to drop the eyelid." A cartoon in the Sunday *Times*, titled "The Man Behind the Egg," portrayed the envoy extraordinary in top hat and spats clutching French canal stock in one hand and with the other applying a candle marked "intrigue" to hatch the Panama chick for Theodore Roosevelt.

Having seen Roosevelt, Bunau-Varilla concentrated next on Jusserand, the French ambassador, writing formally that same Friday of Panama's abiding love for France and requesting that he be received officially by the ambassador (that is, that France recognize Panama) at the soonest possible moment. Then—"in a purely private character"—he went to see Jusserand to assure him that all former contracts and concessions between Colombia and the French canal companies would be honored by the new government of Panama, that there would be no "intrigue against French interests."

Two days later, on Sunday, a large Manila envelope was hand-delivered to the Willard Hotel for M. Philippe Bunau-Varilla, Envoy Extraordinary for the Republic of Panama. It contained a copy of the Hay-Herrán Treaty with minor penciled modifications and a covering note marked "Most Confidential" from John Hay requesting that the document be returned with the envoy's own suggestions at his earliest convenience.

Bunau-Varilla recalled later that he devoted the full day to appraising the treaty and the entire night as well, except for two hours' sleep between midnight and two in the morning. By dawn he had decided it would not do. The "indispensable condition of success" was to write a new treaty "so well adapted to American exigencies" that it would be certain to pass in the Senate by the required two-thirds majority. A failure to obtain that two-thirds majority, he was convinced, could still mean a Nicaragua canal after all. The major difference in the new treaty must be in the share of sovereignty attributed to the United States within the canal zone.

Frank Pavey, his lawyer and long-time colleague in the Panama crusade, had already been summoned from New York and appeared at the Willard suite shortly after breakfast. A stenographer was installed in a room down the hall and the two men got to work. Bunau-Varilla drafted all the articles of the new treaty in longhand and in English. Pavey "corrected the literary imperfections, polished the formulas, and gave them an irreproachable academic form."

Once during the day Bunau-Varilla hurried across town to tell Hay what was going on, and presumably Hay had no objections. By ten that night two finished copies of the treaty were ready. Despite the hour, Bunau-Varilla left directly to see Hay, but on reaching Lafayette Square found the big house in darkness. So it was not until the following morning, November 17, the same day the steamer from Colón was due at New York, that the treaty was delivered.

"It was with anxiety that I awaited a summons from the Department of State during the day . . ." Bunau-Varilla would remember. "It did not come. Mr. Hay made me no sign."

To the Panamanian delegates, meantime, he sent a telegram of welcome through Joshua Lindo, instructing them to remain in New York, "to observe the greatest secrecy with regard to their mission," and to say nothing to the newspapers.

By nightfall he and Pavey were still at the Willard waiting for word from Hay, not daring to leave their room. At ten, unable to stand the suspense any longer, Bunau-Varilla sent his own message. "I cannot refrain from respectfully submitting to you that I would like very much to terminate the negotiations . . ." he told Hay. "I feel the presence of a good deal of intrigues round the coming Commission and people hustling towards them who will find great profit in delaying and palavering and none in going straight to the end.

"I beg, therefore, dear Mr. Secretary, that we should fulfill our plan, as originally laid, to end the negotiations now."

The answer, by return messenger, was immediate: He could come over that night if he preferred.

Bunau-Varilla left directly for Lafayette Square, where a "long conference" ensued, at the conclusion of which Hay congratulated Bunau-Varilla on his work. But whether Hay was actually ready to accept the treaty remained unclear to Bunau-Varilla, who ended the evening with what he intended as the plainest possible inducement to action.

"So long as the delegation has not arrived in Washington," he declared, "I shall be free to deal with you alone, provided with complete and absolute powers. When they arrive, I shall no longer be alone. In fact, I may perhaps soon no longer be here at all."

"As for your poor old dad, they are working him nights and Sundays," John Hay was to tell his daughter in a letter. "I have never, I think, been so constantly and actively employed as during the last fortnight." Nor, in truth, had he ever enjoyed himself quite so much. The breathless sense of motion, the hurried messages after dark, the extraordinary Frenchman with the waxed mustache, were all the stuff of an adventure tale, and so to Hay, a devotee of the form, a novelist himself, all wonderfully therapeutic. In the morning mail, that Tuesday the seventeenth, was a light-hearted note from Richard Harding Davis saying he had been about to write a novel telling how a foreign adventurer robbed Colombia of Panama. "The day I started to write the story," said Davis, "Panama became a republic, and somebody owes me the money I lost on the story."

Hay was also thoroughly satisfied—overjoyed one must imagine—at the treaty he had been presented. Though much of the wording was identical to that of his own Hay-Herrán Treaty, the privileges now granted to the United States were far more sweeping and advantageous to the United States than in the earlier pact. As Hay himself was to confide in a letter to Senator Spooner, the new treaty was "very satisfactory, vastly advantageous to the United States, and we must confess, with what face we can muster, not so advantageous to Panama. . . . You and I know too well how many points there are in this treaty to which a Panamanian patriot could object."

The basic tenets were these:

The United States was empowered to construct a canal through a zone ten miles in width (in contrast to a zone of six miles in the Hay-Herrán pact). Colón and Panama City were not to be part of the zone, but the sanitation, sewerage, water supply, and maintenance of public order in these terminal cities were placed under United States control. Further, four little islands in the Bay of Panama—Perico, Naos, Culebra, and Flamenco—were granted to the United States and the United States had the right to expropriate any additional land or water areas "necessary and convenient" for the construction, operation, sanitation, or defense of the canal. In return the United States guaranteed the independence of Panama.

The French canal company was granted the right to transfer its concessions and property (including the Panama Railroad) to the United States and the compensation to Panama was to be the same as offered earlier to Colombia—$10,000,000 on exchange of ratifications and an annual annuity of $250,000 that would commence nine years later.

The most significant difference, however, between this and the earlier treaty was contained in Article III, which specified that Panama granted to the United States within the canal zone "all the rights, power and authority . . . which the United States would possess and exercise if it were the sovereign of the territory . . . to the entire exclusion of the exercise by the Republic of Panama of any such sovereign rights, power or authority." Though not the sovereign within the canal zone, the United States was to be able to *act* like the sovereign.

And instead of a one-hundred-year lease of the zone that would be indefinitely renewable, as in the Hay-Herrán pact, the zone was to be held by the United States "in perpetuity."

His one remaining obligation, Hay felt, was to have Root and Knox look the document over, and Leslie Shaw, the Secretary of the Treasury, all of whom he convened for lunch at Lafayette Square on Wednesday the eighteenth—that is, the very next day after Bunau-Varilla's night visit.

Hay was "putting on all steam." The lunch went smoothly, all were in agreement, and he hurried back to the State Department and "set everybody at work" drawing up final drafts. To Bunau-Varilla he sent a one-sentence note requesting that he call at his house at six that evening and either Hay or his stenographer was sufficiently excited to put the wrong date on the note.

Bunau-Varilla arrived at Hay's front door at the hour stated and it was not a moment too soon, as he had just learned from Joshua Lindo that Amador, Boyd, and Arosemena had departed from New York and were due in Washington in a matter of hours.

When reporters outside Hay's house asked the envoy if he was there to sign the treaty, the reply was that he did not know.

He was received with inordinate solemnity, the Secretary now addressing him as "Excellency." It was his wish, Hay said, providing his

Excellency was in agreement, to sign the treaty to permit the construction of the interocean canal. Bunau-Varilla, responding in similar tone, said he also was fully prepared to sign—"at the orders of Your Excellency."

They then went over the document, the text of which, with the exception of only a few small alterations, was that submitted by Bunau-Varilla. In Article II, he had used the expression "leases in perpetuity" to describe American control over the canal zone. Hay preferred to have it read "grants to the United States in perpetuity the use, occupation and control . . ."

"You see," said Hay, "that from a practical standpoint it is absolutely synonymous." No other substantive change had been suggested.

"If Your Excellency agrees to it the Treaty will now be read and we will then sign it."

Bunau-Varilla without hesitation agreed and suggested that the reading be "abridged as far as possible." Time was very much on his mind. They had been together now for approximately forty minutes.

According to Hay's letter to his daughter, it was exactly seven o'clock when they signed "the momentous document." They were in a small drawing room with blue walls overlooking the square and the lights of the White House. It was two years to the day, Hay noted, since the signing of the second Hay-Pauncefote Treaty. Bunau-Varilla had brought no seal with him, so Hay gave him a choice either of his own signet ring (a ring, as Hay explained, that Lord Byron had worn the day he died) or a ring with the Hay family arms. For Bunau-Varilla it was the one difficult decision of the hour and, as he later stressed, "I had not a long time to think it over." He chose the family seal.

The pen was dipped in an inkwell that had once belonged to Abraham Lincoln. Bunau-Varilla signed first, in a small, rapid, controlled hand, just beneath the final line of the pact: "Done at the City of Washington the 18th day of November in the year of our Lord nineteen hundred and three." Then Hay affixed his own large, clear signature, and with a few final words presented the pen to the envoy extraordinary.

"We separated not without emotion," Bunau-Varilla would recall, "and I hastened back to my hotel . . ." In another two hours he was at the railroad station to greet "the travelers with the happy news!"

He was standing on the platform as the Panamanians stepped off the train and his first words were these: "The Republic of Panama is henceforth under the protection of the United States. I have just signed the Canal Treaty."

According to Bunau-Varilla's own account, Amador looked as though he was about to faint. Federico Boyd, again according to Bunau-Varilla, was no less able to mask his "consternation." But as the story would be told later in Panama, considerably more than consternation was expressed. The Panamanians had been by stages incredulous, indignant, then livid with rage. Federico Boyd is said to have hit Bunau-Varilla across the face.

In the days that followed, Boyd kept insisting that Bunau-Varilla had acted without authority, illegally, specifically contrary to his written instructions. Bunau-Varilla, with marked impatience, assured him that any such protestations were quite pointless, "as everything is finished."

The treaty had only to be ratified by the United States Senate and by the government of Panama, and Bunau-Varilla was determined to obtain immediate ratification from Panama, before the treaty went to the Senate, a move the three Panamanians absolutely refused to be party to. He warned that a special mission from Bogotá was expected in Washington at any time to treat with the Americans and that the surest guarantee against any possible trouble was to confront that mission with a ratified treaty. He took Amador and Boyd to see Hay, trusting to the conciliatory effect of the old diplomat's grace and courtesy, and Hay even went so far as to promise a supplementary treaty to correct any possible defects. But it was all to no avail. They were without authority to act on the treaty, the Panamanians told the American Secretary of State; nor would they retreat from that position when, after leaving the State Department, Bunau-Varilla accused them of bad manners and warned that they had made a "decidedly bad impression" upon the Secretary.

So once again the kinetic Frenchman decided he must remove matters from "inexpert hands" and place them in his own, and it was this next phase of his activities, more even than the hurried signing of the treaty, that would be regarded in Panama as his great act of treachery.

Though he and the three Panamanians were barely speaking by this time, it was agreed among them that the actual treaty would be sent to Panama on the *City of Washington* when the ship sailed the following Tuesday, and that they would all convene in New York to officiate over the placing of the document safely on board.

Bunau-Varilla, however, left Washington before the others and on Saturday, November 21, without consulting any of them, he sent the entire text of the treaty by cablegram to Panama's new Minister of

Foreign Affairs, Dr. F. V. de la Espriella. He informed the minister that the commission had behaved very poorly before American officials and that Panama's sudden caution and unwillingness to cooperate were proving an embarrassment to the government in Washington, which was being condemned by its political enemies for having acted with undignified haste in recognizing Panama. He wanted immediate ratification of the treaty by Panama and "the immedate expedition of telegraphic instructions" so that he might inform Washington. The treaty, he warned, would not be sent to the Senate until Panama did its part.

The reply from Minister de la Espriella did not reach Bunau-Varilla at the Waldorf until Monday night and it was negative. Amador and Boyd, he was certain, had intervened, had cabled Panama to ignore his advice and to give no sanction to his treaty.

Nonetheless, they all convened as agreed the following morning in Bunau-Varilla's usual room, Number 1162, where, he reminded them, their liberation had been prepared. The treaty was placed in an envelope, the envelope was wrapped in the flag of Panama and was placed in a small safe filled with cotton. The safe was sealed; then in a body they went out the door with it.

When the *City of Washington* sailed that afternoon, the safe was on board and Bunau-Varilla was on his way back to Washington. He had decided, he later explained, "to shake off the web which I felt was being woven about me."

The evening of the twenty-fifth, he sent a 370-word cable to Minister de la Espriella that struck an entirely new note of fear. If the government of Panama failed to ratify the treaty immediately upon the treaty's arrival at Colón, then the almost certain consequence would be an immediate suspension of American protection over the new republic and the signing of a canal treaty with Bogotá.

Possibly, as would be suggested, John Hay had had something to do with this astonishing message. Bunau-Varilla never implied as much, but Hay was to remark later in a letter to Senator Cullom that "we" insisted on immediate ratification.

In any event, it was the ultimate knife at the throat and wholly spurious. The notion that Roosevelt would abandon Panama at this point, that he would leave the junta to the vengeance of Colombia, that he would now suddenly turn around and treat with Bogotá, was not simply without foundation, but ridiculous to anyone the least familiar with the man or the prevailing temper in Washington. Nothing of the kind was ever even remotely contemplated at the White House or the State Department.

"This time I hit the mark," Bunau-Varilla was to exclaim. "The Government of Panama was at last liberated from the morbid influence of its delegation." The following day, November 26, he received cabled instructions from Panama—from Arango, Tomás Arias, Manuel Espinosa, and Minister de la Espriella—to notify Washington officially that the treaty would be ratified and signed as soon as it arrived at the Isthmus.

The provisional government of Panama kept its pledge. The treaty was formally approved, unanimously and without modification, on December 2, less than one month after the revolution and just five days before Congress was due to convene in Washington.

The debate in the Senate began at once, with John Tyler Morgan and Hoar, of Massachusetts, a dogged anti-imperialist, leading the assault not simply on the treaty, but on the means by which Panamanian independence had been obtained and on the President's conduct. As it was, however, for all the acrimony that filled the pages of the *Congressional Record*, comparatively little fault could be found in the treaty, even among those most inclined to oppose anything put forth by the Administration. In one long speech, for example, Senator Hernando de Soto Money, of Mississippi, an old ally of Senator Morgan's and die-hard champion of the Nicaragua route, conceded that the treaty "comes to us more liberal in its concessions to us and giving us more than anybody in this Chamber ever dreamed of having . . . we have never had a concession so extraordinary in its character as this. In fact, it sounds very much as if we wrote it ourselves . . ."

Support for the measure, moreover, was massive and led by Spooner, Cullom, and Lodge. The forthcoming Presidential election helped solidify Republican ranks. So to all practiced observers the outcome seemed foreordained. "The debates will be long and heated," wrote Dr. Herrán, "but there is no doubt that the treaty will be finally approved." The one missing element this time was Mark Hanna. Ill, exhausted from a successful campaign to make his friend Myron Herrick governor of Ohio, Hanna remained in his rooms at the Arlington Hotel, slowly dying.

In a letter written February 10, Roosevelt told his son Theodore, Jr., that Panamanian opposition seemed "pretty well over." On the same day, Dr. Herrán closed the Colombian legation and notified John Hay that he was departing for home—"with crushed spirits and broken health," as Herrán wrote to a friend. He would die soon afterward.

On February 15 Mark Hanna died at age sixty-six, and soon thereafter, on February 23, 1904, the Senate ratified the treaty by a vote of 66 to 14.

II

In Panama a National Constitutional Convention had convened in the capital city on January 15 and completed its labors in less than a month. The new constitution was generally in accord with that of the United States, but with some notable added provisions. Only the government could import or manufacture arms, for example. No president could succeed himself, nor could he be succeeded by any member of his family. Citizenship could be suspended for habitual intoxication.

The government was divided into three parts, the legislative, executive, and judicial, the delegates in the legislature and the president being elected by popular vote. And with the work of framing the constitution completed, the convention at once resolved itself into a National Assembly and elected Manuel Amador the first president. His inauguration took place in Cathedral Plaza on February 20, 1904, three days before Senate ratification of the Hay–Bunau-Varilla Treaty.

The nation over which the new government presided extended from the Costa Rican border on the west to Colombia on the east, where the precise location of the border line would remain undetermined and a rankling issue for years to come.

Since no complete survey of the country had ever been attempted, its total area could only be approximated. It was thought to be from 30,000 to 35,000 square miles. In fact, with the Colombian border resolved, Panama was something less than 30,000 square miles (29, 208, not counting the Canal Zone). So it was about the size of South Carolina, smaller than Portugal or Scotland, larger than Ireland or Sierra Leone. Its population in 1904 was estimated at 350,000, which was roughly the population of the District of Columbia.

From a strictly economic perspective, the future appeared grand beyond imagining. The long night at Panama, the years of economic stagnation since the demise of the Compagnie Universelle, was ended. The one significant, hopeful development in the local economy during the 1890's had been in banana production, as a result of the systematic efforts of an American entrepreneur, Minor C. Keith. Though the origins of the entire banana trade dated back to the 1860's when, as an experiment, small shipments were made from Colón to New York by a man named C. A. Frank, it was not until Keith began buying up land on Chiriquí Lagoon, west of Colón and established enormous banana plantations there, that production of the fruit on the Isthmus became a serious, large-scale enterprise. At Bocas del Toro, on Chiriquí Lagoon, once a tiny Negro village, thousands of acres were under cultivation and the population by the early 1900's was perhaps nine thousand. Keith talked of eventual rail connections from Panama all the way to New York. He "saw into the banana future and builded for it," in the words of one account. In 1899 he had joined forces with the Boston Fruit Company (which had plantations on Jamaica and Cuba) to form the United Fruit Company, which at its inception was the world's largest agricultural enterprise. To many on the Isthmus it had seemed that the banana was to be Panama's economic salvation.

But now construction of the canal promised the return of boom times, prosperity that would surely surpass even the French era. There would be all the commensurate demands for local goods and services, and payment this time in dollars. With the United States committed to the task, no one appeared in the least doubtful that it would be completed, that not very far off in the future, Panama was to be what Simon Bolívar had once prophesied, "the emporium of the universe."

As a physician, Amador foresaw sweeping advances in sanitation and an end to centuries of plague. Panama could become a model for all the tropics.

On the more immediate level, his country was also commencing operations from the unique position of being wholly debt free. Instead of a national debt, Panama had what amounted to a national endowment in the form of the stipulated $10,000,000 payment from the United States. Of this some $750,000 was kept in hand as working capital; roughly $2,000,000 was applied at once to much-needed public works; while the major portion, approximately $6,000,000, was very profitably invested in first mortgages on New York City real estate, the arrangements for this being handled by Panama's designated fiscal representative in the United States, J. P. Morgan & Company, and by Panama's appointed New York counsel, William Nelson Cromwell.

First payment of the $10,000,000 was made on May 2, when Morgan & Company received a Treasury draft for $1,000,000, of which $200,000 was promptly shipped to the Isthmus. On May 19, two weeks after Panama formally transferred control of the Canal Zone to the

United States, Secretary of the Treasury Leslie Shaw paid Morgan & Company the remaining $9,000,000, the greater part of which was to be retained in New York.

It was, however, the far larger financial transaction between Washington, New York, and Paris that most concerned the Roosevelt Administration, attorney Cromwell, J. P. Morgan, Philippe Bunau-Varilla, and the scores of others involved on two sides of the North Atlantic.

The purchase of the French holdings at Panama was the largest real-estate transaction in history until then. The Treasury warrant for $40,000,000 made out to "J. Pierpont Morgan & Company, New York City, Special Dispensing Agent," was the largest yet issued by the government of the United States, the largest previous warrant having been for the $7,200,000 paid to Russia for Alaska in 1867. Participation by the house of Morgan had been agreed to by both the buyer and the seller and in late April, prior to receipt of the Treasury warrant, J. P. Morgan sailed for France to oversee the transaction personally. His bank shipped $18,000,000 in gold bullion to Paris, bought exchange on Paris for the balance, and paid the full sum into the Banque de France for the account of the Compagnie Nouvelle and the liquidator of the Compagnie Universelle. On May 2, at the offices of the Compagnie Nouvelle on the narrow, little Rue Louis-le-Grand, the deeds and bills of the sale were executed. On May 9 in New York the United States repaid the $40,000,000 to the house of Morgan. Morgan's fee for services, charged to the Compagnie Nouvelle, was $35,000.

With the $10,000,000 paid to Panama and the $40,000,000 to the Compagnie Nouvelle, the United States had spent more for the rights, privileges, and properties that went with the Canal Zone—an area roughly a third the size of Long Island—than for any actual territorial acquisition in its history, more than for the Louisiana Territory ($15,000,000), Alaska ($7,200,000), and the Philippines ($20,000,000) combined.

At the then exchange rate, $40,000,000 came to 206,000,000 francs. By an earlier private contract the directors of the Compagnie Nouvelle and the liquidator of the Compagnie Universelle had agreed that approximately 38 percent of any future canal profits would go to the Compagnie Nouvelle and 62 percent to the liquidator. Thus the Compagnie Nouvelle now received 77,400,000 francs; the liquidator, 128,600,000. So as it worked out, shareholders in the new company, the largest of whom were the penalty shareholders such as Eiffel and the brothers Bunau-Varilla, received 130 francs per share, which was the equivalent roughly to a 3 percent return on their investment over the ten uneasy years since the company had been organized. The Bunau-Varilla firm, for example, recovered all of its 2,200,000-franc stake ($440,000) in the Compagnie Nouvelle, plus a profit of about 66,000 francs, or $13,200.

When the liquidator distributed his 128,600,000 francs among some 200,000 claimants—all bondholders in the old Compagnie Universelle—the return on their investment came to approximately ten cents on the dollar. Stockholders in the old company received nothing.

Cromwell, who placed the value of his services to the Compagnie Nouvelle at $800,000, was forced to wait another several years as the fee was arbitrated in Paris. In 1907 the firm of Sullivan & Cromwell was awarded the sum of $200,000.

Philippe Bunau-Varilla, meantime, had returned to Paris and resumed his duties as publisher of *Le Matin*. Three days after ratification of the treaty by the Senate he had submitted his resignation as envoy extraordinary, asking that in lieu of salary for his services the money be withheld by the new republic for the erection of a monument at Panama City to Ferdinand de Lesseps. His replacement, the new Panamanian ambassador to Washington, was former Governor Obaldía.

The Frenchman's final official act had been performed on a bright February morning in Washington when he and John Hay formally exchanged the ratified treaties and bid each other farewell. It was, he wrote later, a deeply moving moment for both of them. And in his vaulting literary fashion he recalled a rush of private thoughts—

> of all those heroes, my comrades in the deadly battle, worthy grandsons of those Gauls who conquered the Ancient World, worthy sons of those Frenchmen who conquered the Modern World, who fell in the struggle against Nature . . . of the shameful league of all the passions, of all the hatreds, of all the jealousies, of all the cowardices, of all the ignorances, to crucify this great Idea . . . of my solitary work, when I went preaching the Truth on the highways . . . of the untold number of stupidities I had had to destroy, of prejudices I had had to disarm, of insults I had had to submit to, of interests I had had to frustrate, of conspiracies I had had to thwart, in order to celebrate the Victory of Truth over Error and mark at last the hour of the *Resurrection of the Panama Canal.*

At the moment, however, looking at Hay, he had said only, "It seems to me as if we had together made something great."

• •

The actual delivery of the canal works at Panama occurred early on the morning of May 4, 1904, and to the Panamanians, who adored ceremony and celebration, who remembered Cathedral Plaza festooned with palm branches and French flags, who remembered parades and banquets and Ferdinand de Lesseps prancing on horseback, it was a terrible disappointment and most unbecoming to the occasion. At 7:30 A.M. Lieutenant Mark Brooke met with half a dozen American officials and a duly authorized representative of the Compagnie Nouvelle at the company headquarters on the plaza, the old Grand Hotel. On being handed the keys to the storehouses and to the Ancon hospital, Lieutenant Brooke swiftly signed the receipt of the property and read aloud a brief proclamation. The transaction occupied no more than a few minutes. Scarcely anyone other than those present was aware of the event. Lieutenant Brooke had not even thought to invite President Amador.

Having shaken hands with the Panamanians and the French officials, the young officer raised the Stars and Stripes to the top of the hotel flagpole.

BOOK THREE

The Builders
1904–1914

15
The Imperturbable Dr. Gorgas

The world requires at least ten years to understand a new idea, however important or simple it may be.

—Sir Ronald Ross

I

"It is all unspeakably loathsome," concluded a New York reporter who was among the earliest to arrive at Colón. That his countrymen could and would build the mighty ship canal, he, like his countrymen, took as a matter of course. Any contrary view would have placed him among an all but indistinguishable minority. Having tamed a continent, having achieved industrial supremacy, having embarked upon the great adventure of world leadership, the American people—some 80,000,000 strong—would now triumph where the French had failed so ignobly. "There is nothing in the nature of the work . . . to daunt an American," the reporter insisted. "I have made three excursions over the canal route . . . and while I do not pretend to speak expertly of the engineering aspects of the problem, I should say that the building of the canal will be a comparatively easy task for knowing, enterprising and energetic Americans."

Still, Colón was troubling. The Negroes lived in the most appalling fashion in rotting shanties propped on stilts in a swamp, "a morass, a vast expanse of black water covered with green scum." There was no plumbing, not one sewer. The stench was like nothing in his experience; the nights were made "hideous" by the interminable din of thousands of frogs.

He described the poisonous mists rising over the Chagres River, mists quite visible from Colón in the early morning; and not wishing to appear ignorant of advanced medical theory, he wrote also of the mosquitoes. What conceivable chance, he asked, was there to make so vile a place safe for white men?

The article appeared in the New York *Tribune* the first week in February 1904, three months before Lieutenant Brooke raised the flag over the old French administration building, and three weeks before the President emphasized comparable concern to the chairman of his new Isthmian Canal Commission. "As you know, I feel that the sanitary and hygienic problems . . . on the Isthmus are those which are literally of the first importance, coming even before the engineering . . ." Roosevelt declared.

The tragic experience of the French was never far from mind. But more immediate and vivid was the memory of Cuba in 1898, when thirteen times the number of American troops killed by the enemy had died of yellow fever, malaria, and typhoid fever. For Roosevelt personally, Cuba had been an unforgettable lesson in the havoc disease could bring down on an expeditionary force. Cuba had been primarily the fault of bad or indifferent leadership in the field and "not too many gleams of good sense" in Washington. It was Roosevelt, following the capture of Santiago, who with Leonard Wood wrote the famous round-robin letter to General Shafter, saying that the army must be moved at once or else perish of malaria. His own brigade, he said in a second letter, was "ripe for dying like rotten sheep."

To head the new commission, he had turned again to old John G. Walker, who, as time would tell, was an unfortunate choice; he was as ill-suited for his responsibilities as had been the Secretary of War in 1898, Russell Alger, the cause of much of the anguish in Cuba. But the difficulties of mounting the task at Panama were compounded still more by the unwieldly composition of the commission itself, a matter over which Roosevelt had no say. The Spooner Act required that there be seven members, at least four of whom must be "learned and skilled in the science of engineering," and of those four, two must be military officers (one Army, one Navy). All seven were to have equal authority.

Walker was the Navy man; for the Army there was General George W. Davis, "a fine old plains warrior" who had been a vice-president of the defunct Nicaragua Canal Construction Company. The civilian engineers were Professor William H. Burr (from the prior Walker Commission), Benjamin M. Harrod, Carl E. Grunsky, and William Barclay Parsons. The seventh member, Frank J. Hecker, was a businessman. On the surface it was a distinguished body. But none of them had ever organized a gigantic construction project. None was accustomed to handling problems of supply, labor, or overall planning on a scale even approaching what was now called for.

Nor had any of them had the least medical training. Congress had looked upon the canal as a problem of engineering construction exclusively. The presence on the commission of a physician or of someone experienced in sanitation had not been deemed essential, so none had been named.

That a sanitary officer was assigned to serve under the commission was due primarily to the insistence of Dr. William Henry Welch, of Johns Hopkins, who, during a personal call at the White House, had urged Roosevelt to tackle the sources of disease prior to every other effort on the Isthmus. Welch, one of the celebrated "big four" (Welch, Osler, Kelley, and Halsted) at the Johns Hopkins Medical School, went to the White House with a delegation of prominent physicians, plus, significantly, the chief of the Bureau of Entomology at the Department of Agriculture, Leland O. Howard. "We passed through a room crowded with persons waiting to see the President," Welch recalled, "and I felt that he must begrudge every minute we occupied . . ." But Roosevelt afterward told Walker to find the very best medical man in the country to take charge of the hospitals and sanitary work at Panama, and although he did not tell Walker who that man should be, Walker was instructed not to make the appointment before consulting with Welch.

The result was the naming of an Army doctor, a former student of Welch's, Colonel William Crawford Gorgas, who since the death of Walter Reed (of appendicitis in 1902) was known in professional circles as the outstanding authority on tropical disease. Gorgas was forty-nine years old, a courtly, white-haired man whose humorous eyes and "sunny" Alabama manner concealed a marvelous tenacity. Everyone had a good word for him. He was the son of the Confederate general Josiah Gorgas, the career soldier from Pennsylvania who became Jefferson Davis' chief of ordnance; and so, for political purposes, the appointment helped offset a preponderance of Yankees on the commission.

The seven had been chosen rather hastily—too hastily, Roosevelt later conceded—during the last week of February. On March 4, he sent seven telegrams announcing the appointments and giving each recipient four days to drop what he was doing and convene at the White House. (Walker and General Davis were already in Washington; Burr and Parsons in New York, Hecker in Detroit, Harrod in New Orleans, Grunsky in San Francisco.) He was not interested in their politics, Roosevelt said at their first meeting. They had been picked solely on their reputations for integrity and ability. He had nothing to offer concerning the details of the work. "What this nation will insist upon is that the results be achieved."

The commission set up headquarters in the Star Building on 14th Street and the position of chief engineer was offered to a Chicago railroad official named John Findley Wallace, an offer he accepted with the understanding that the job required residence on the Isthmus. His salary was to be $25,000 a year, which was more than that of any other government employee in 1904, with the exception of the President, and $10,000 more than he, Wallace, was receiving as general manager of the Illinois Central. William Gorgas' annual income as an Army colonel was $4,000.

On May 9 came the executive order placing the commission under the direct supervision of the new Secretary of War, William Howard Taft.* The commission thus had full authority to proceed with the canal.

So six months following the Panama revolution, everything seemed set to go. To the Yale professors who still challenged the legality of his part in the revolution, the Harvard man in the White House responded with another of those spontaneous, eminently quotable retorts for which he had such a gift, wholly avoiding the issue but greatly pleasing the country: "Tell them that I am going to make the dirt fly!"

To William Gorgas there was no real problem about what had to be done at Panama. Nor does he appear to have had any doubt that he

* This was done "inasmuch as the War Department is the department which has always supervised the construction of the great civil works for improving the Rivers and Harbors of the country and the extended military works of public defense . . ."

could succeed, his assumption being that full support would be forthcoming from the office in the Star Building. He would concentrate on yellow fever first. Malaria was the larger, more serious threat in his judgment, but an outbreak of yellow fever could result in panic and yellow fever was his specialty. Though the means by which both diseases are transmitted had become known, only yellow fever had been eradicated in any one plague spot as a result of such knowledge—in Havana in 1901—and he had been the man chiefly responsible.

The carrier of yellow fever, it had been determined, was the small, quiet, silvery household mosquito known as *Stegomyia fasciata*, exactly as the Cuban physician Carlos Finlay had announced years earlier. As is presently known there are no fewer than 2,500 different species of mosquito (rather than 800-odd, as Finlay believed), and these belong to three important genera: *Culex*, *Anopheles*, and *Aëdes*. The *Culex* group includes the ordinary gray household mosquito found in northern latitudes (*Culex pipiens pipiens*). The *Anopheles* are the only known carriers of malaria and also transmit encephalitis, or sleeping sickness. The *Aëdes aegypti* is *Stegomyia fasciata*, as it was then called, the yellow-fever mosquito.

Credit for finding the cause of malaria, one of the greatest medical discoveries of all time, belonged to the English physician Ronald Ross, who had addressed himself to the problem alone in a remote field hospital in Secunderabad, India. Though malaria was a worldwide killer, flourishing in a broad zone on both sides of the equator, its greatest toll in human life was in Asia, and in India it was the arch destroyer, taking possibly a million lives a year. Ross, once the indifferent student, amateur poet and musician, had figured out the pattern by which the disease is spread, a pattern that seemed simple enough only after he had explained it. In the summer of 1897 he dissected under a microscope an *Anopheles* mosquito after it had fed on a malaria patient. In the insect's stomach he saw the same circular cells that the French physician Laveran had discovered in Algeria in 1880, the malaria parasite *Plasmodium falciparum*. Besides, he determined that the cells were growing.

In his notebook that night he wrote a poem describing the moment. It was to be his only published verse, the last stanza of which became famous:

> I know this little thing
> A myriad men will save.
> O Death, where is thy sting?
> Thy victory, O Grave?

Not until the following year, however, was he able at last to prove the mosquito theory by locating the mosquito's salivary gland and determining that the expanding parasite within the mosquito's stomach eventually penetrates all parts of the mosquito's body, including that gland. "The door is unlocked," he wrote in an exultant letter to England, "and I am walking in and collecting the treasures."

The solution was this: *Anopheles*, comparatively large, brown mosquitoes with little black dots on their wings, transmitted malaria only after having bitten someone already infected with the disease. The mosquito drew the blood containing the parasite, the parasite multiplied in the stomach of the mosquito, then moved to the salivary gland whence the parasite was delivered to the bloodstream of whomever the mosquito bit next. The insect was not the source, in other words, only the agent of conveyance. But this particular insect was the only means of conveyance. Furthermore, it could cause damage only when there was infected blood to feed on, when there were people about who were sick with malaria. So the way to stamp out malaria was not simply to get rid of the *Anopheles* mosquito, but to make it as difficult as possible for the *Anopheles* mosquito to get at anyone who had the disease.

The idea of preventing malarial epidemics by exterminating the *Anopheles* was first put forth in a letter addressed by Ross to the government of India on February 18, 1901. The following year Ross published a small book, *Mosquito Brigades*, explaining how such a campaign should be organized. By then, however, Gorgas and his Army doctors had demonstrated at Havana what almost no one had believed possible. In 1901, in one of the worst fever cities on earth, they had eliminated yellow fever in less than eight months and very nearly got rid of malaria. There had been nothing comparable in medical or military history to their war on mosquitoes.

Gorgas had played no part in determining the cause of yellow fever. As it happens, he had been one of the last of the Army doctors in Cuba to accept the mosquito theory. Like Ross he had even come to the practice of medicine itself without noticeable interest or enthusiasm and most of his professional life had been spent at "hitching post" forts, as far removed as Secunderabad from the salient discoveries and innovations that had so dramatically transformed medical science during the last part of the nineteenth century. In 1880, the year Gorgas finished his training at Bellevue, the germ theory of disease was still a subject of debate; from 1880 to 1900—the years that marked the emergence of Pasteur, Koch, Lister, the founding of Johns Hopkins—Gorgas was at such places as Fort Brown, Texas, or Fort Randall, South Dakota, attending to routine duty. Fort Randall, set in a boundless prairie, was seventy-five miles from the nearest railroad.

Gorgas became a doctor because of a boyhood determination to have a military career. As a child in Richmond he had seen Lee and Jackson confer with his father in the front parlor, and in the final winter of the war, as ragged Confederate troops filled the streets, he insisted, at age ten, on going barefoot. He received a bachelor of arts degree from the struggling little University of the South at Sewanee, Tennessee, where his father had been made president, but still determined to be a soldier, much against his father's wishes, he tried for an appointment to West Point. When West Point turned him down, he chose medicine because it was the only way left to get into the Army.

Classmates at the Bellevue Medical College in New York would remember "Billy" Gorgas as a devout Christian, a careless speller, too poor to go home for vacations, the most likable man in the class, and "imperturbable." It was at Bellevue that he encountered Dr. Welch, then in his late twenties, and after receiving his degree in June 1879 Gorgas spent another year as an intern at Bellevue Hospital.

His father, meantime, learning that his ambition remained unchanged, protested that the life of an Army doctor "would not be a life to look forward to as a permanent thing. It is not in the army that the sphere of a doctor is ennobling."

To Gorgas later it would seem that the pattern of his life had a large, definable purpose, perhaps God-willed. It was as if every important turn had been designed to prepare him for one historic task. He was not a person to dwell overly on questions of cosmic destiny. Rather, genuine modesty was an important part of his considerable charm. "I am not much of a doctor," he once remarked to a gathering of prominent physicians, "that is, I am not experienced in the care of the sick, and I am not very much of a military man, although I have been in the army service practically all my professional life." But while he seemed incapable of taking himself seriously, he was intensely serious about his work and what it could mean to his fellow mortals.

To him the years at frontier outposts had been supremely invigorating. He had grown physically rugged, accustomed to riding out hardships and boredom of a kind to defeat other men. He learned discipline, acquired an unequivocating devotion to duty. His wife would recall nights when, wrapped in buffalo robes, he would set off in a sleigh in the midst of a North Dakota blizzard to deliver an Indian baby in a cabin sometimes thirty miles away. "Life like this was more than an education in medicine," she would write.

The degree to which yellow fever had affected his career, his personal life, even his very existence, was quite uncanny. It was because of a yellow-fever epidemic at Mobile in 1853 that his mother and father first met. She was Amelia Gayle, daughter of a former governor of Alabama, who for her safety had been sent to stay with an older brother at Mount Vernon, a small town on high, dry ground to the north of Mobile and the site of a federal arsenal. Her brother, the company surgeon at the arsenal, lived next door to the young ordnance officer who was in command, Josiah Gorgas. The wedding was in December that same year, and William, the first of six children, was born the following year, October 3, 1854. The attending physician, as noted earlier, was Alabama's pioneer yellow-fever specialist, Josiah Nott.

Gorgas' own first test as a young officer in the Army Medical Corps was a yellow-fever outbreak in Texas. In 1882 he was assigned to Fort Brown, on the Rio Grande, close to the border settlement of Brownsville where the epidemic raged. "I am sending you the most progressive young surgeon under my command," the surgeon general informed the post commander, and having set himself to work, Gorgas proved so very persistent, ignoring strict orders to keep away from the infected wards, that he was put temporarily under arrest.

Of particular concern as the epidemic worsened was Miss Marie Doughty, sister-in-law of the post commander, for whom little hope was given. A grave had been dug for her in the yellow-fever burial ground on an island in the Rio Grande and Gorgas had volunteered to go with the body and read the last rites. But then he too fell ill, nearly died, but recovered, as did Miss Doughty, with the result that they were convalescent at the same time and fell in love. She became the doctor's wife following a visit to Tuscaloosa to meet his widowed mother, and since they were now permanently immune to yellow fever, he would be summoned repeatedly for special duty wherever the disease broke out.

He advanced in rank, to major by the time of the Cuban war, but preferred then, as later, to be called Dr. Gorgas. When the American forces occupied Havana in 1898, he was put in charge of the yellow-fever camp at the village of Siboney. One patient, recalling Gorgas' efficiency and kindness, wrote that had it not been for him the death rate among the troops might have been twice what it was. But Gorgas was no closer to understanding the cause of yellow fever than he had ever been. When the situation at Siboney became critical, his only solution was to burn the town, along with all the medical supplies.

The great yellow-fever discoveries at Cuba, those later dramatized in Sidney Howard's play *Yellow Jack* and in a movie made from the play, were the work of Gorgas' superior officer, Dr. Walter Reed, who had taken his lead from Carlos Finlay. At the time the Americans moved into Havana, Finlay was still laboring to prove that *Stegomyia fasciata* —and only *Stegomyia fasciata*—was the yellow-fever carrier. And for all his persistence he had gotten nowhere. His carefully tended laboratory mosquitoes drew blood from diseased patients and were then permitted to attack the bare arms of willing victims. Yet no yellow-fever cases had ever resulted for Finlay. His notion that the mosquito had to be the transmitting agent was perfectly correct and like Ross in India he had hit upon it alone. What he did not know, what Reed and his staff were to demonstrate, was that *Stegomyia fasciata* transmits the yellow-fever parasite only according to a very particular time pattern, that the development of the parasite within the insect requires what is known as the period of "extrinsic incubation."

For the mosquito to become infected it must suck the blood of the yellow-fever patient within the first three days after the patient has contracted the disease. Then once the mosquito has taken the blood, another twelve to twenty days must pass before the mosquito can transmit the infection. Finlay had simply been applying his infected mosquitoes too soon.

Determination of this critical time factor, a phenomenon that evaded not just Finlay but all previous investigators, had been made in Mississippi the year of the Cuban war by an aristocratic Virginian who was with the Public Health Service. Dr. Henry Rose Carter, like Gorgas, had been immunized by a previous attack of yellow fever and consequently made the disease a specialty. He had noted that often when a man developed yellow fever on board ship none of his shipmates became ill. He wondered if somehow there might be a period of delay, or incubation, at the source of the disease. So during an epidemic in Mississippi, he made a statistical study among isolated rural families, meticulously recording how many people—relatives, visitors—came near a yellow-fever patient, when, and with what result. What he found was that a yellow-fever patient could be visited without hazard so long as the visit was made within ten to twelve days after the patient became ill. But beyond that period, even after the patient had died and the body had been removed from the house, family or visitors were in mortal danger. Hence the sick man could not possibly be the source of contamination. Hence there had to be a period of "extrinsic incubation," as Carter named it.

Carter's findings were published two years later, in 1900, just as the Army Yellow Fever Commission under Reed arrived in Havana.* Impressed by Finlay, who volunteered to help in any way he could, and by Carter, who had been sent to Havana as a quarantine officer, Reed decided to concentrate on *Stegomyia fasciata*, to prove that yellow fever was not a filth disease. Gorgas, who remained positive it was, had been reassigned meanwhile to clean up the city, street by street, house by house.

Associated with Reed were Dr. James Carroll, Dr. Aristides Agramonte, and Dr. Jesse W. Lazear. Late in August, a few days after he had allowed a mosquito to bite him, Dr. Carroll fell ill with yellow fever, and though he recovered, his health was so damaged that he died a few years later. In mid-September, while placing mosquitoes on patients in a fever ward, Lazear saw a free mosquito of undetermined species land on his hand and he also purposefully allowed the insect to take its feed of blood. Five days later Lazear had what Gorgas described as one of the most violent cases of yellow fever he had ever attended. On September 25, the day Lazear died, he was in such wild delirium that it took two men to hold him in bed.

Convinced now of the truth of Finlay's theory, Reed pressed on with further experiments proving conclusively that *Stegomyia fasciata* was the carrier, *and* that neither filth nor "fomites," the term used for the soiled clothes or bedding of yellow-fever patients, had anything whatever to do with spreading the disease. For twenty nights, as part of one experiment, a doctor and three volunteer soldiers, confined to a one-room shack, slept in the soiled pajamas of yellow-fever patients, on beds reeking of black vomit and other excreta; and for all the discomfort of the experience, none of them suffered the least sign of illness.

Gorgas had been at the bedside of Carroll and Lazear. He had become a close friend of Finlay's and knew Reed's work to be of paramount importance. Still, he remained "unconvinced." He did not believe mosquitoes to be even the principal cause, much less the only cause. There was only one way to determine whether *Stegomyia fasciata* was the carrier, he insisted, and that was to rid the city of the insect and see if yellow fever disappeared. Reed agreed in theory but is

* Publication of Carter's vitally important observations had been delayed because the editor of a medical journal returned his paper saying it was too long.

said to have told Gorgas, "It can't be done." Gorgas, no less doubtful of success, proceeded because "it was our duty to take precautions in this direction."

His historic Havana campaign commenced in February 1901. The year before, there had been 1,400 known cases of yellow fever in the city. In 1901 there were 37 cases. But starting in October 1901 there were none. "For the first time since English occupation in 1762," Gorgas wrote to Ronald Ross, "we have an October free from yellow fever, and malaria decreased more than one half." In 1902 there was no yellow fever at all and deaths from malaria totaled 77, in contrast to 325 in 1900.

The techniques developed at Havana were what Gorgas intended to use on the Isthmus. However, as he had explained to Surgeon General Sternberg, there were considerable differences between the task at Havana and what had to be faced in a place such as Panama. Sternberg had been one of those in Washington who so badly underestimated the problems of disease when fighting the Cuban war. But he had also been responsible, at the war's end, for establishing the Yellow Fever Commission. He was one of the very few to appreciate the vast significance of what Gorgas achieved at Havana, and it was he who decided that Gorgas must become the Army's man on tropical disease. In particular, Gorgas was to prepare himself for a role in building the interoceanic canal, which both men had then assumed would be in Nicaragua. In 1902 Gorgas was sent to attend a world conference on tropical medicine at Cairo and to inspect the work being done on malaria at the Suez Canal according to precepts laid down by Ross. With Sternberg's blessing, Gorgas also spent several months in Paris going over the medical reports and records on file at the offices of the Compagnie Nouvelle in an effort to determine as nearly as possible what the French experience had been at Panama. But Sternberg had since retired from active duty; it was John G. Walker to whom Gorgas had now to address his explanations and upon whom he became dependent for support.

II

Before leaving for Panama, Gorgas made an ardent appeal for proper supplies and sufficient experienced personnel to carry out his program. The commission deliberated and decided that for the time being considerably fewer personnel would suffice than what he had in mind and that the question of supplies required further study. So the advance party that landed at Colón in June 1904 consisted of a mere seven men, including Dr. Carter and one woman, an English nurse named Eugenie Hibbard, and they were forced to begin their campaign without benefit of even such essential supplies as wire screening and disinfectants. "It is hardly an exaggeration to say . . . when they landed at Panama to engage in the mighty task of ridding this jungle of disease, [they] had little more than their own hands and their own determined spirit to work with," Marie Gorgas would recall.

The whole of the Isthmus, they found, was a "mosquito paradise." The temperature, scarcely changing the year around, allowed constant breeding of the insects. A first inspection of Panama City revealed an abundance of *Stegomyia* in practically every building. *Anopheles* were still more numerous. By local custom, household drinking water was kept indoors in red earthenware jars, *tinajas*, within which *Stegomyia* larvae abounded. Mosquito larvae, or "wrigglers," swarmed in the open cisterns and rain barrels beside nearly every building. It was the rainy season and innumerable pockets of water, perfect for mosquito breeding, collected everywhere—in large quantities along the old French diggings, in small quantities in the impressions left where railroad ties had been pulled up or in the hoofprints of cattle that grazed in an open field on Ancon Hill.

At Ancon Hospital, where Gorgas established his headquarters, there were no screens in the windows; plants in the surrounding gardens created by the French were still protected from umbrella ants by crockery rings filled with water. In the wards it was still the practice to set the legs of the beds in shallow pans of water. There was no yellow fever among the few patients in the wards, only malaria; but most of the hospital's small staff—French Sisters of Charity, two French doctors, one priest—were also infected with malaria. After dark the mosquitoes were so thick in the wards that work had to be done in relays, one set of doctors or nurses using fans to protect those working. A nurse named Jessie Murdock, who arrived in July with a small contingent of American nurses, would tell how during night duty they wrapped themselves in bandages soaked in citronella.

At Havana the task had been confined to a comparatively small area and the success with malaria had been largely a side effect of the main campaign: the assault on *Stegomyia* had simply resulted in the destruction of most *Anopheles* as well. But *Stegomyia* and *Anopheles* are quite different creatures, just as yellow fever and malaria are quite different diseases, and here, at Panama, there was not one city but two, and fifty miles of jungle between.

It was by considering the mosquitoes as predators more deadly than the most savage beasts of the jungle that Gorgas intended to solve the problem. Only by understanding the exact nature of the particular mosquitoes in question—their reproductive processes, feeding habits, flight range, and so forth—could he hope to destroy them.

Until the Cuban war comparatively little had been known about mosquitoes. It was not until 1895, for example, that a full account was published of even the common North American variety. The general impression was that all mosquitoes were more or less alike. At the time Reed and his co-workers identified *Stegomyia fasciata* as the yellow-fever mosquito, no studies had ever been made of the insect's natural life history. So this too had been part of Gorgas' task at Havana and consequently he and his associates had discovered astonishing peculiarities that were of enormous value.

Seen under the microscope, *Stegomyia* is a creature of striking beauty. Its general color is dark gray, but the thorax is marked with a silvery-white lyre-shaped pattern; the abdomen is banded with silvery-white stripes and the six-jointed legs are striped alternately with black and pure white. Among mosquitoes *Stegomyia* is the height of elegance.

Stegomyia is also, like the rat, a creature of human society. It survives by maintaining a close proximity to human beings. As among all mosquitoes it is only the female that bites—that is, only the female feeds on blood, while the male gets by on other liquids such as fruit juices and is quite harmless. For the female, blood is essential to mature her eggs. Though the female *Stegomyia* can feed on any warm-blooded animal, her decided preference is for human blood, and thus the whole life cycle of the insect must be maintained in close association with human society.

While all mosquitoes lay their eggs in water, the yellow-fever mosquito is extremely particular about where the water is located and its condition. The female *Stegomyia* will deposit her eggs only in or near a building occupied by human beings and only in water held in some sort of artificial container such as an earthenware jar or a rain barrel. In addition, it is essential that the water be clean.

With such information available, all acquired during the work at Havana, the problem of destroying the yellow-fever carrier became infinitely more manageable. "Men who achieve greatness," the brothers Mayo were to write in an essay about Gorgas, "do not work more complexly than the average man, but more simply. . . . In dealing with complex problems, with the simplicity natural to him he went directly to the point, unaffected by the confusion of details in which a smaller man would have lost himself." At Havana the hopeless task of destroying all mosquitoes was reduced to destroying a particular mosquito; then, once the natural peculiarities of that species were recognized, it was possible to reduce the task further still. The campaign would center on the insect's method of propagation. The task, very simply, was to eliminate every possible opportunity for the female *Stegomyia* to deposit her eggs. Yellow fever was eradicated chiefly by ridding the city of all standing fresh water, or by sealing it off with wire screening or wooden lids, or with a skim of oil or kerosene, an idea first suggested for mosquito control in 1892 by the entomologist Leland O. Howard. (The oil not only discouraged the mosquito from depositing her eggs, but killed any larvae already in the water, since the larvae require air to survive.)

Thousands of adult mosquitoes had been destroyed in Havana by systematic fumigation of houses wherein yellow-fever cases had been found. Doors and windows had been sealed off with newspaper, room by room, and pans full of sulphur or powdered pyrethrum (a dried flower used as an insecticide) had been burned for an hour or more. But the main attack had been on water jars, barrels, cisterns, any stray bucket, tin can, or broken dish in which rainwater might collect. A card file had been made on every house and building within city limits; the city had been divided into sections and inspectors had been sent out daily to report on the *Stegomyia*-producing status of households within their districts. Water kept indoors for household use had to be covered. It had been a laborious, often thankless task, yet extremely simple in concept, and the results had been amazingly rapid.

The female *Stegomyia* lays anywhere from 35 to 120 minute black eggs and the maturation cycle from egg to larva to pupa to mosquito takes less than ten days. So with the campaign fully organized and in effect, Havana's *Stegomyia* population diminished quite suddenly. Adult mosquitoes died off, of fumigation or of old age, after three or four weeks, and because *Stegomyia* has an extremely limited flight range (another crucial characteristic discovered by Gorgas and his people), few replacements migrated into the city from outlying villages. It was thus that victory over yellow fever had come so quickly and decisively.

Anopheles, the malaria mosquitoes, were quite different creatures and thus a wholly different kind of problem. *Anopheles*, to begin with, are not purely house-bred insects. The female, unlike her *Stegomyia* counterpart, will deposit her eggs in still water of any kind—any stagnant swamp, marsh, any clogged drain or ditch or mud puddle. *Anopheles*, therefore, are as much creatures of field or jungle as of the backyard. So while *Stegomyia* mosquitoes were always readily within range, their breeding grounds closely, neatly defined, *Anopheles* were literally everywhere, *and* in fantastic numbers, since the female deposits as many as two hundred eggs every ten days.

The sanitary measures taken at Havana—the clearing away of garbage and refuse, the installation of proper drainage systems plus the campaign on *Stegomyia*'s breeding grounds had had the effect of giving no mosquito, *Anopheles* included, much chance to propagate within the city limits. But what chance would there be at Colón with its swamps or along the canal line?

Malaria, not yellow fever, Gorgas stressed to his associates, was the problem upon which their success would ride or fall. Malaria, he emphasized, had accounted for the greatest loss of life during the French years. "If we can control malaria, I feel very little anxiety about other diseases. If we do not control malaria our mortality is going to be heavy." Knowledge of the kind gathered at Havana on *Stegomyia* would now have to be gathered on *Anopheles.*

The initial question was which particular species of *Anopheles* to go after. "It was not known how many different species of *Anopheles* existed," wrote Joseph Le Prince, one of Gorgas' advance guard, "nor was it definitely known which of them were the important malaria carriers." To Le Prince, who had also been Gorgas' right hand at Havana, it was evident that "much investigative or pioneer work" was still called for.

> We had no means of determining how seasonal changes would affect propagation, and the available data were unreliable. It was generally believed at that time that all mosquitoes traveled more or less with gentle air currents, but there was no positive knowledge of habits of flight, and the length of flight of *Anopheles* . . . was yet to be determined. It was not known if or how topography affected the distribution of species, whether *Anopheles* larvae thriving in small collections of water held by plants were of . . . importance, or whether certain species were confined to fixed geographical limits.

So while detailed information was being gathered and put on file concerning the whereabouts of *Stegomyia* larvae in Panama City, *Anopheles* larvae and pupae were being carefully taken from puddles and swamps along the canal line, scooped up in white-enameled dippers, poured into wide-mouthed jars, and carried back to a makeshift laboratory at Ancon Hospital. Live adult *Anopheles* collected in villages along the railroad could not survive the return trip, it was found, unless carefully protected from direct sunlight, rain, and strong air currents—an observation that was to have considerable subsequent value.

This preliminary survey disclosed the presence of *Anopheles* breeding grounds in or near every existing settlement, every abandoned camp built by the French. At a work camp at Culebra, a village of roughly five hundred Jamaicans, Gorgas found that every child he examined had a greatly enlarged spleen, a sign of chronic malarial infection. Every adult he talked to spoke of attacks of chills and fever within the preceding six months. On a hill above this same village a detachment of United States Marines was encamped, half of whom already had malaria. "The condition," he wrote, "is very much the same as if these four or five hundred natives had smallpox, and our Marines had never been vaccinated."

No one even tried to approximate the numbers of *Anopheles* present at any given point. Within the hospital compound itself their presence was phenomenal. On the panel of a single doorway one dutiful assistant counted fifty-four. Like *Stegomyia*, the *Anopheles* were easily recognized by their resting stance. In contrast to the common northern mosquito, which stands with proboscis and head crooked at right angles to its body, *Stegomyia* and *Anopheles* kept proboscis, head, and body on a straight line, but at an angle to the resting surface. When feeding on an arm or wrist, an *Anopheles* looked as though it were standing on its head.

To determine the time of day or night when the *Anopheles* would take blood, the men stretched out on cots in one of the wards, each man with a supply of pillboxes and a pocket watch. Every time a mosquito bit, or tried to, it was captured, put in a pillbox, and the date and hour were recorded on the box. The *Anopheles*, it was learned, would attack a human at rest at any hour, though the night hours were by far the worst. The life span of the insect seemed to be about a month and in that time the female required a meal of blood every two to three nights. Her bite did not cause any appreciable swelling, nor was the itch especially bothersome. Often a person was not even aware of being bitten by an *Anopheles*.

After a month or so, with only a few exceptions, all the small American force, Gorgas as well, had been down with malaria.

• •

Time was the pressing concern. For although there were but one or two yellow-fever cases, and none serious, at the moment, that condition would change rapidly as soon as new human material became available for the *Stegomyia fasciata*—and thus the disease—to feed on. Gorgas' analogy to explain the violent wave effect of yellow fever—the apparent absence of the disease followed by a sudden, vicious outbreak—was the exhausted fire wherein concealed embers lay in wait for fresh supplies of fuel. The arrival of several thousand nonimmunes would be equivalent to heaping on dry kindling: nothing much would happen at first; then the disease would catch; the carrier mosquitoes would infect ever more victims with the deadly parasite, thereby creating more diseased blood for still more mosquitoes to feed upon. Unchecked, the disease would flare into a monstrous geometrical progression of death, taking hundreds, possibly thousands, of lives.

Were conditions on the Isthmus to remain as they were, and were upwards of twenty to thirty thousand men to be brought to Panama, as planned, then, Gorgas calculated, the annual death toll from yellow fever alone could run to three or four thousand.

The build-up of men and equipment was beginning. Every arriving steamer had its contingent of prospective carpenters, mechanics, file clerks, assistant engineers, all eager to be "in at the start at Panama." General Davis, who had been named the first Governor of the Canal Zone, and Chief Engineer Wallace had arrived and had taken up residence in Panama City. Gorgas, still working with the same small staff, tried to explain the situation, the need for immediate decisions, for men and supplies, and he got nowhere.

In August Admiral Walker and several of the commission came for an inspection tour and Gorgas again made his case as explicit as he knew how. The admiral and his party departed, weeks passed, nothing happened. Gorgas' cabled requests were answered evasively, if at all. Presently he was reminded by return cable that cables were costly and henceforth to use the mails.

III

The problem in essence was that Admiral Walker, Governor Davis, and several others on the Isthmian Canal Commission, as well as a very large part of the populace and its political leadership, did not seriously entertain the notion that mosquitoes could be the cause of yellow fever or malaria. To spend time and money chasing after mosquitoes in

Panama would be to squander time and money in a most irresponsible fashion.

That the minds of men in such positions could be so closed in the face of all that had been learned and demonstrated in Cuba, not to mention the insistent warnings from Roosevelt and Welch, may seem inconceivable. In the conventional understanding of history, human advancement is marked by specific momentous steps: on December 17, 1903, at Kitty Hawk, the Wright brothers fly in a heavier-than-air machine and at once a new age dawns; in a hospital ward outside Havana Dr. Jesse Lazear dies a martyr's death and the baffling horror of yellow jack is at last resolved. But seldom does it happen that way. Ideas too have their period of extrinsic incubation, and particularly if they run contrary to what has always seemed common sense. In the case of the Wright brothers, it was five years after Kitty Hawk before the world accepted the idea that their machine could fly.

During the long hearings of the Morgan Committee in 1902, prior to the vote on the Spooner Bill, despite all the concern expressed over disease in Panama, the recounting of the French tragedy, the mosquito theory had not even been discussed. George Morison mentioned it once in passing, but without evoking the slightest interest among the others, who had been content to dwell on miasmatic fumes emanating from the rank isthmian landscape. No reference was made to the breakthroughs achieved by either Ross or Reed, nor was anything ever said of Gorgas' demonstrated success at Havana. Yet all the efforts of the Yellow Fever Commission, all of Gorgas' work, had been initiated by the Army, all the resulting reports had been published at government expense.

In the autumn of 1904, with the situation on the Isthmus unimproved, Gorgas returned to Washington to plead his case. It had been nearly four years since the epochal report of the Yellow Fever Commission. Ross had won the Nobel Prize in 1902 for his discoveries. A scientific congress held in Paris in 1903 had thoroughly reviewed Reed's work and declared that the mosquito transmission of yellow fever was a "scientifically determined fact." But to Walker, Davis, and their fellow commissioners, Gorgas was wasting their time. Walker's word for the theory of mosquito infection was "balderdash." Gorgas spent hours waiting in Walker's anteroom, hours more with Walker going over the evidence in support of the theory. The correct procedure, Walker insisted, was to get rid of the garbage and the dead cats, paint the houses, pave the streets. In his zeal he even offered to give Gorgas a detailed set of rules to guide him in the work.

Walker remained a bastion of integrity. The years he had devoted to the canal question, his study of the French disaster in particular, had convinced him that material wastefulness and graft were the gravest threats to American success at Panama. To Gorgas he remarked, ". . . whether we build the canal or not we will leave things so fixed that those fellows up on the Hill can't find anything in the shape of graft after us."

With Davis, who had returned briefly to Washington, Gorgas had no better success. "What's that got to do with digging the canal?" was his rejoinder to Gorgas' plan. Davis professed great friendship for Gorgas, then in tones that Gorgas later described as kindly and patient, Davis tried to "set him right." "On the mosquito you are simply wild," Davis said. "All who agree with you are wild. Get the idea out of your head."

Before Gorgas had left for Washington, several of his staff had urged him to resign rather than face such continuing ignorance and obstructionism. But having gotten nowhere with the commission—indeed, having been shown how very little actual authority he possessed—he sailed again for Colón within a few weeks. Old Walker's attitude, however galling, was of a kind he had experienced before in the military. He had kept his temper. He made no attempt to undermine or dislodge anyone on the commission, nor did he resort to political or social by-paths to the War Department or to the White House. The doctor was neither fighter nor schemer. But then neither would he give up. "That persistence which had always been his chief asset . . . forced him to the task," according to his wife. He was not merely returning, moreover; he was returning to stay, for this time he took her with him.

Of all those who had been named to positions of importance that year, only Gorgas would stay on. Not only was he the one perfectly qualified man for his particular role and the one really solid appointment made in 1904, he was also to be the only major official of importance to stay with the work on the Isthmus from start to finish.

It is in Marie Gorgas' published reminiscences that we find some of the earliest first impressions of Panama as recorded by one of the American canal force. Colón was "unspeakably dirty," swarming with naked children, ugly, dilapidated, and terribly depressing. Yet the two-hour ride on the Panama Railroad more than compensated. In places the jungle swept the sides of their car and the jungle itself she found astonishingly beautiful. From the depot at Panama City they rode to

Ancon in an open victoria through "a sea of mud reaching at times almost to the hub . . ." Their quarters were to be on the second floor of the hospital, in Ward Eleven, once the officers' ward for the French.

> . . . Dr. Carter was on hand to greet us . . . A flight of stairs led to the gallery of the second floor. Although it was only a little after five o'clock, the short twilight gave a somber though refreshingly cool aspect. My spirits rose. On the right, facing the stairs, was the large living room, comfortably furnished . . . with wicker chairs and small tables—a room of many windows. Following the gallery to the right was a bedroom running across the back porch . . . Two small rooms across the back porch were fitted up as bath and servants' rooms. There was no running water, and, as I found out afterward, water was exceedingly scarce, being delivered daily in small quantities.

During the years of French occupation, she was told, more men had died of yellow fever in this building—in these very rooms—than in any other building on the Isthmus.

After dinner with the staff—served "Spanish style" in six or seven courses, in a screened, candlelit room on the ground floor—she, her husband, Dr. Carter, and several others returned to the upper gallery.

> There is an alluring something about a night in the tropics. Dr. Gorgas experienced a melancholy pleasure listening to the sighing of the royal palms . . . in imagination visioning through the haze of his cigar the ghosts that haunted the old building. . . .
>
> It was a beautiful and starry evening. Beyond stretched the great Pacific, the dotted islands in the distance dimly seen . . . every place teeming with the history of departed glory and vast enterprise.

The night, with its "creeping noises on the roof and on the floor," had considerably less charm. In the morning she had her first full view of the city and bay spread below.

> We were on a high point, with only the road separating our ward from the sheer descent to the valley below, a descent so steep that a retaining wall had been built as a protection. The road was bordered by a row of stately royal palms, planted by the French . . . Beyond a stretch of green valley the hills and mountains were seen emerging from the heavy mist . . . The sun was rising from the Pacific, a strange phenomenon, and the rays gave a jeweled appearance to the dew-soaked plants and the leaves of the trees. . . .
>
> The city of Panama lay tantalizingly near . . . From our high point . . . the pastel shades of the Spanish tiled roofs were easily discernible; also the animation and movement of the streets. . . .

She remained "content," however, to confine her excursions to walks about Ancon Hill in company with the French nuns in their blue gowns and vast white headdresses, or with Laura Carter, wife of Dr. Carter, who arrived in midsummer.

Miss Hibbard, the head nurse, Jessie Murdock, and the other American nurses (Margaret Magurk, Mary Markham, Eleanor Smith, Anna Turner) were quartered in still another ward. "Old rusted French beds, with mildewed mattresses and pillows lined the walls," Bessie Murdock remembered. "Each [bed] had a candle, but it was soon found that it was not wise to keep these burning, as they attracted moths and all sorts of flying insects. Yet, in spite of these many difficulties, we were not disheartened, but thoroughly enjoyed the novel experience."

Accommodations in the ward for the unmarried male members of the staff were no less "deluxe," as one of them, W. C. Haskins, would recall for his fellow townsmen in Oelwein, Iowa:

> One straight-backed chair was made to do for the entire bunch. . . . We had but one lamp. . . . There were no mirrors, and the fortunate possessor of an individual looking glass was to be envied. Some combed their hair and shaved with the aid of the swinging glass windows backed up against the wall. There were but two washstands for all of us. . . . We lived in constant dread of the alacrán, or scorpion, who seems to have a penchant for buildings long unused, and for going to sleep in your clothes or shoe. . . .

Gorgas, who spoke a little Spanish, had already done more to win the trust and friendship of the Panamanians than any other American on the Isthmus. The name Gorgas was Spanish. According to family tradition, he was descended from a Spaniard who settled in Holland in the sixteenth century, when Spain ruled the Low Countries. And from his experience in Cuba he was appreciative of Latin pride and humor. In Manuel Amador, a fellow physician, he had an especially important ally. His tact, his sensitivity to the feelings of others, were unfailing. An American engineer would remember him as "a grand, quiet, lovable man." Dr. Victor Heiser, a young American physician who was passing through en route to the Philippines, saw Gorgas stopped in the street by a beggar. "He bowed to the man, shook his hand, even inquired for his name, then gave him a single penny—the only coin he had I suppose—but it was all done so perfectly naturally, with such dignity that the man walked away very pleased."

From Gorgas' private secretary it would also be learned years later that the kindly doctor, seemingly as imperturbable as ever, could also become so incensed over red tape and bureaucrats in Washington that he would sweep the papers from his desk, lock them in a drawer, and storm out of the office, not to be seen again for a day or more.

He rose early, cared little about his clothes, his customary ensemble a rumpled three-piece civilian suit, stiff detachable collar, black tie with stickpin. His main pleasures were food—virtually anything set before him—horseback riding, a glass of beer, conversation, and books, his reading being done according to a lifelong routine. He always kept three books at hand—one scientific, one of classical literature or history, one light fiction—which he took up in turn, giving each exactly twenty minutes according to a pocket watch placed on the table beside his chair. In this fashion, he said, he was able to remember what he read.

"He loved especially the adventure stories of H. Rider Haggard," his daughter would recall; "*King Solomon's Mines!* I think those books had a great deal to do with his enthusiasm for the adventure of Panama, for being there in the jungle then."

It was in the autumn of 1904, during the time Gorgas was on home leave, that Ronald Ross made his brief, almost universally ignored, visit to Panama as the result of an invitation issued by Gorgas. Ross crossed the Atlantic to attend the world's fair in St. Louis, where, among other things, he noted surprising numbers of both *Stegomyia* and *Anopheles* mosquitoes nicely thriving. Then, following a pleasant few days with William Osler at Baltimore, he sailed for Colón on the steamer *Advance.* Gorgas came down to the New York pier to see him off. For an hour before the ship sailed, they sat chatting on the fantail, the two men who had done more than any others alive to rid the world of tropical plague and whose lonely, unending battle with official indifference or official enmity had long since made them brothers of the same blood.

After a week on the Isthmus, Ross described Gorgas' projected campaign as sound in every particular. Panama, Ross declared, could be made an example for the entire world.

John F. Wallace

John Stevens

NATIONAL ARCHIVES

Fumigation brigade, Panama City

PANAMA CANAL COMPANY

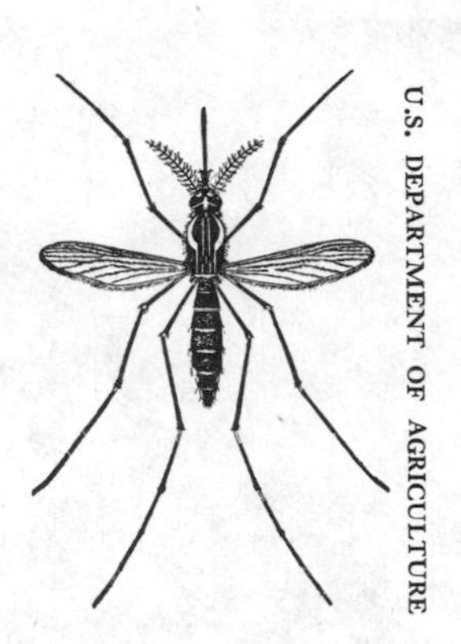

U.S. DEPARTMENT OF AGRICULTURE

Stegomyia fasciata, an adult female (much enlarged)

Yellow-fever patient inside a portable isolation cage at Ancon Hospital

NATIONAL ARCHIVES

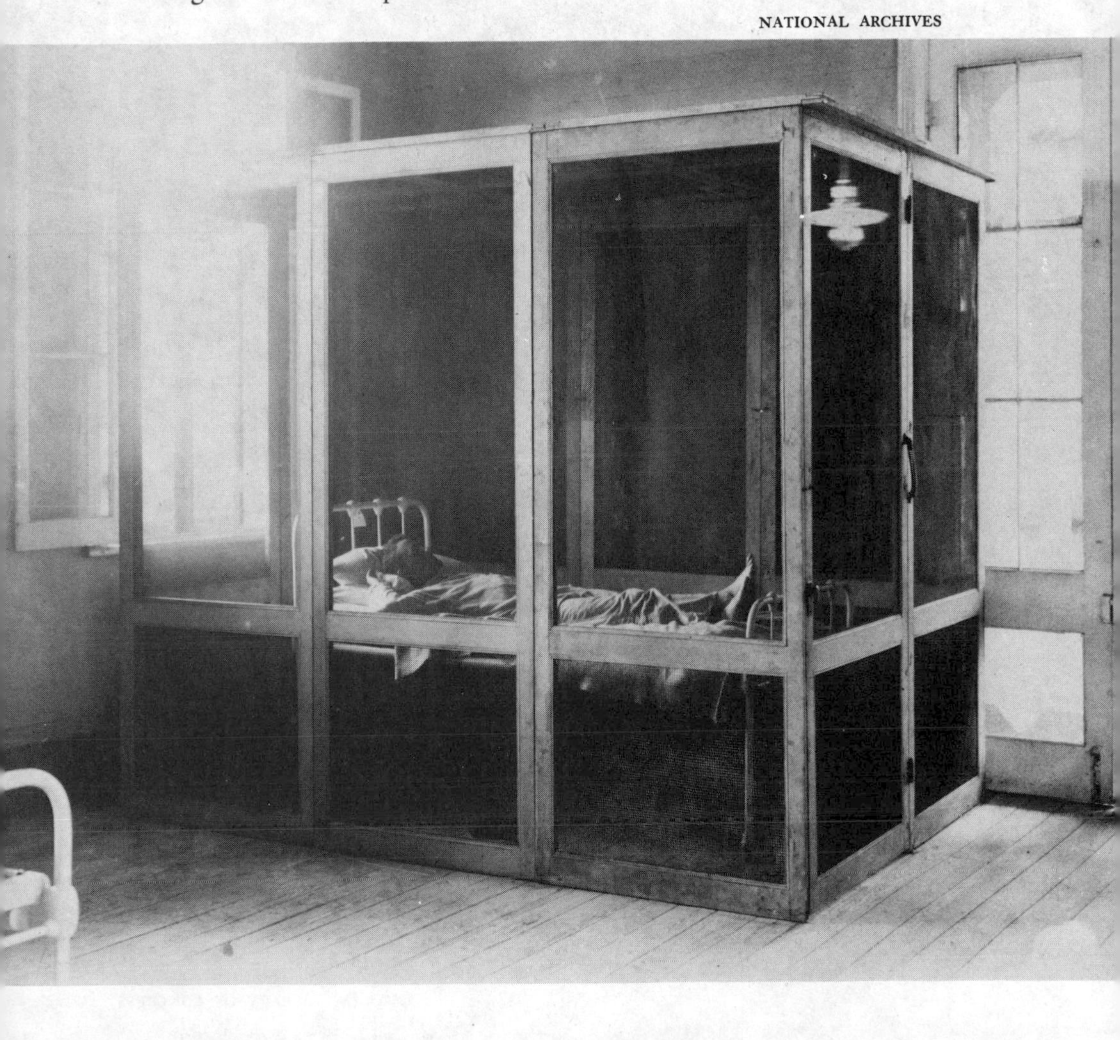

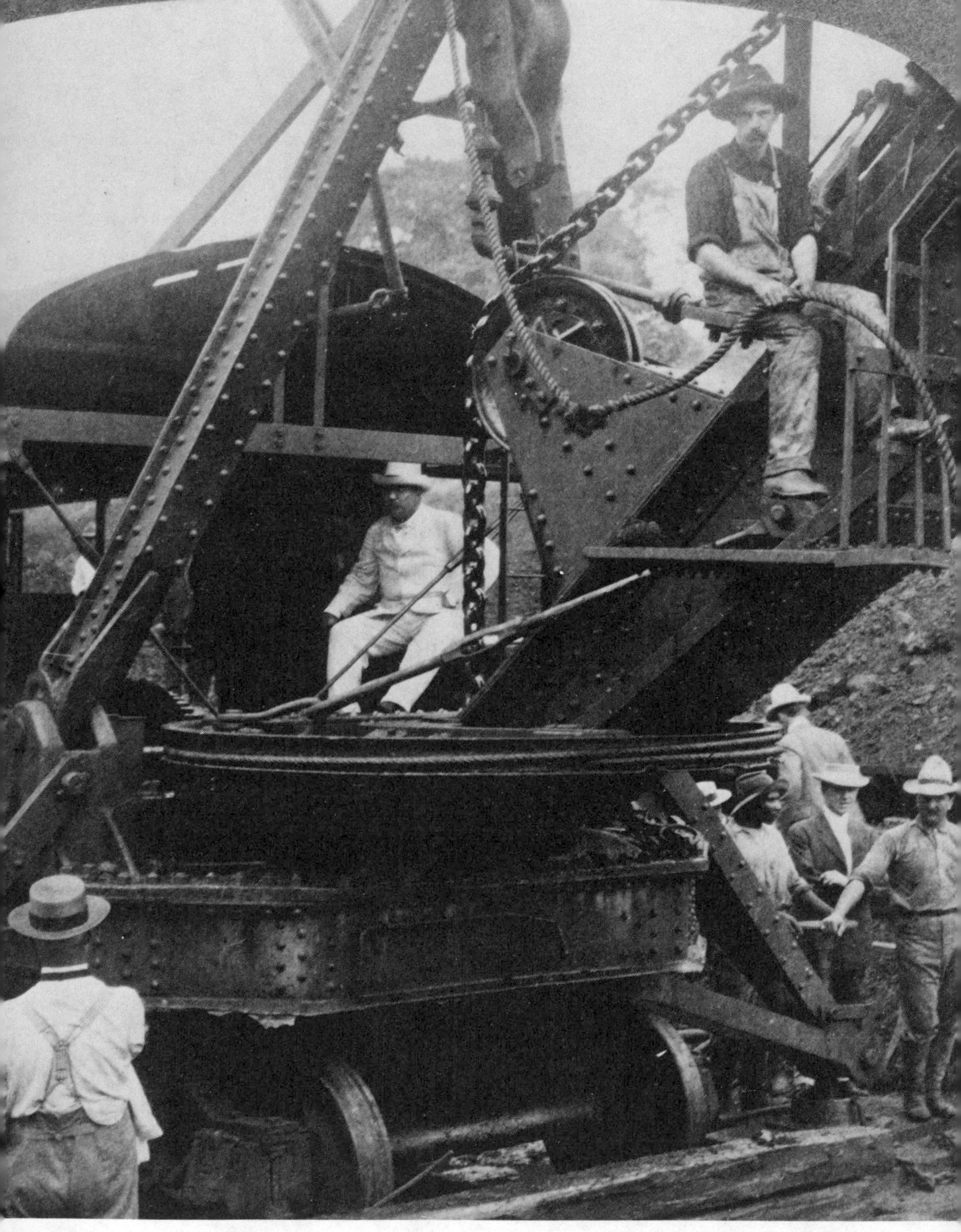

LIBRARY OF CONGRESS

LEFT, "The Man and the Machine": Theodore Roosevelt strikes his famous pose at the controls of a ninety-five-ton Bucyrus shovel at Pedro Miguel.

BELOW, five unidentified American workers, one of whom holds an issue of the *Canal Record*, the weekly paper that began publication in 1907

COLLECTION OF J. W. D. COLLINS

FROM THE PANAMA GATEWAY, JOSEPH BUCKLIN BISHOP, CHARLES SCRIBNER'S SONS, 1913

Goethals and his high command on the steps of the Administration Building at Culebra. Left to right, Lieutenant Colonel William L. Sibert, Joseph C. S. Blackburn, Rear Admiral Harry H. Rousseau, Joseph Bucklin Bishop, Colonel George W. Goethals, Lieutenant Colonel Harry F. Hodges, Colonel William C. Gorgas, and Lieutenant Colonel David D. Gaillard

ON PAGES 436–437, Culebra Cut
PANAMA CANAL COMPANY

RIGHT ABOVE, West Indian dynamite crew

RIGHT BELOW, motion-picture still of a mechanical track shifter in operation

BELOW, Spanish track gang

PANAMA CANAL COMPANY

PANAMA CANAL COMPANY

NATIONAL ARCHIVES

16
Panic

You are going to have the fever,
Yellow eyes!

—James Stanley Gilbert

I

Later, in his own defense, Chief Engineer John Findley Wallace would say that he was denied the free hand promised at the time he accepted the job. He would complain of red tape—"System gone to seed"—and of the mad clamor to "make the dirt fly."

How much autonomy, if any, he may have been assured is impossible to know. But the red tape was quite as horrendous as he said it was and in that regard he was wholly blameless. What began in Washington as a conscientious concern over possible misuse of funds, anything that might nurture graft, rapidly became an obsessive fear of the least extravagance. Each member of the seven-headed Isthmian Canal Commission considered himself personally responsible for every step taken, every dollar expended. An elaborate, insanely deliberate system of forms and regulations was handed down and every detail of procedure had to be cleared by the seven who sat two thousand miles from the scene. The well-meaning but intractable Walker and his commissioners had to pass with due formality on virtually every purchase voucher, irrespective of importance, with the inevitable result that delivery of equipment and material took months instead of weeks to reach Colón. One shipment of urgently needed water pipe ordered in August would not arrive until January, and then by sailing schooner. When Wallace, like Gorgas, cabled Washington in despair, he too was "delicately informed" of the high cost of telegraphic communication.

The appointment of any employee at a salary exceeding $1,800 required the approval of the full commission. The commissioners very often had trouble agreeing with one another, while Walker's insistence on his prerogatives as chairman had an increasingly stultifying effect. When he departed from office later, no less than 160 requisitions would be found unopened in his desk, many of them months old.

On the Isthmus, to hire a single handcart for an hour required six separate vouchers. Carpenters were forbidden to saw boards over ten feet in length without a signed permit. The clerical work required for each fortnightly payroll was amazing: by September, with 1,800 workers on the books, payment took six and a half hours and involved the filling out of 7,500 separate sheets of paper weighing in all 103 pounds.

In anticipation of long delays in Washington, department heads would order material in excessive quantity and well in advance of need, only to see more shipments arrive than there were men enough to unload, or a staggering oversupply of some odd item for which there was no real or immediate demand. Wallace's chief architect, as an illustration, determined that fifteen thousand new doors would be required eventually, for which he would need fifteen thousand pairs of hinges. Consequently, twelve thousand doors were shipped to Colón without delay, but because someone in the Washington office decided that fifteen thousand pairs of hinges would be insufficient, and because the architect's original order appears also to have been inadvertently duplicated by someone else in Washington, the order of doors was accompanied by 240,000 pairs of hinges.

If Philippe Bunau-Varilla, as he told his admiring American audiences, never knew a single day of despair in Panama despite the most crushing setbacks, John Findley Wallace seems to have known little else but despair. From the time of his initial reconnaissance in July, he had been openly incredulous and discouraged. He had seen only "jungle and chaos from one end of the Isthmus to the other." Yet in fact there was comparatively little jungle to face along the actual canal line; in contrast to the French in 1880, he and his engineers began with the decided advantage of being able to *see* the problems before them along the length of the fifty-mile corridor. The only chaos, when Wallace first arrived, was in the forlorn wreckage left in the path—the millions of dollars' worth of French equipment lying in huge scrap heaps, the silent lines of rusted locomotives overgrown with vines and brush. Within a mile radius of Cristobal (the former Christophe-Colomb) there were eighty French dredging machines toppled over or sunk in the shallows. Shiploads of beautifully machined and tooled castings for the Eiffel locks had been dumped in the same vicinity, and for miles along the canal line the discarded rails and pipes, the enormous gears and axles and hundreds of nameless parts and pieces strewn everywhere, gave the look of bitterly contested ground in some titanic battle of machines.

Most of the French buildings, unoccupied for years, were in sad disrepair. Floor joists and roof beams had rotted; mold, rats, and termites had all taken their toll. Some of the work camps had become so overgrown by vines and bamboo scrub as to be nearly impossible to find. Several years later, while studying one of the old French maps, George Goethals would note a camp marked Caimito Mulato that did not appear on the American maps. He sent some men to look into it and they found an entire village built by the French completely buried in the jungle.

The Panama Railroad was also in deplorable condition, service slow and unpredictable, equipment worn out, coaches ramshackle and filthy, freight cars in short supply and ridiculously undersized. The entire line was virtually without signals or siding. Bridges were in dangerously poor repair.

But the French had left their successors a canal, a navigable water passage upward of twenty-five feet deep and seventy feet wide, running inland from Colón to Bohio, a distance of eleven miles. At Bohio, across the river from the railroad, was a vast excavation in solid rock, where the Eiffel locks were to have been built. For the next thirty miles, as far as Miraflores, there was evidence of excavation along the entire route, except for one seven-mile stretch just before the summit at Culebra. At Culebra itself the ground had been cut to a depth of 163 feet below its original surface; and beyond Miraflores, through the salt marshes of the lower Rio Grande on out to the Bay of Panama, ran another open channel. Indeed, to most new arrivals "the first surprise" of Panama was the "magnificence of the French failure." "One cannot spend much time on the Isthmus without discovering in himself a mighty respect for the French" was the grudging conclusion of a writer for *Everybody's* magazine. "They showed skill in every part of their work," wrote a correspondent in *The Outlook*, "and the excellence of all their material is the wonder of every practical man who tests it." A third man declared, "One appreciated more and more the wonderful amount those French had really accomplished. It is vastly more than the popular impression . . . It touches from ocean to ocean."

To be sure, there was much that could deceive the novice. The channel at the Pacific end was scarcely a third as deep as it would have to be; even the terraced cut at Culebra, by far the most dramatic evidence of the French assault, was a bare beginning compared to the canyon that would have to be created before a ship passed through.

What is more, a large part, something over half of the French work, would be of little or no use to a canal of the kind planned by the Walker Commission, as had been pointed out in the commission's own report. For instance, the huge network of diversion channels built by the French as a last-resort answer to the problem of the Chagres River would be of only marginal value to a lock-and-lake canal, and those channels alone ran to more than thirty-three miles.

Still, the useful portions of the French work amounted to approximately 30,000,000 cubic yards of excavation; that was 30,000,000 cubic yards of Panama that no longer stood in the path of the canal, a volume equal to about a third of the excavation of the Suez Canal.

Locomotives and dump cars and two French excavators were still in service at Culebra, where the Compagnie Nouvelle had been scratching away, however slowly, since 1895. Quantities of French tools, machines, stationary engines, carloads of spare parts, were still safely under cover. Six large machine shops and a power plant were in working order. Furthermore, a surprising percentage of the equipment abandoned by the wayside, despite its appearance, could be put back in running order, as Wallace would soon determine. By December of 1904 there would be six of the old French excavators in use in Culebra Cut. By 1905 a hundred of the boxy little Belgian locomotives would be in service and two thousand French dump cars. Wrecked dredges and sunken tugboats would be raised, floated, rebuilt. Of the 2,149 buildings left by the French, a total of 1,500 would be refurbished as time went on.

The real problem, as much nearly as with the bureaucracy in Washington, was with Wallace himself.

Wallace was a competent enough technician and someone who worked well with men of large affairs. The son of a Presbyterian clergyman, equable and intelligent-looking, he had built railroads, and a number of impressive terminals for the Illinois Central (at Chicago,

New Orleans, Memphis); he had devised the system for transporting the crowds in and out of the Columbian Exposition at Chicago in 1893 and few American engineers had attained such professional honors. At fifty-one, he was a past president of the Western Society of Engineers, past president of the American Railway Engineering and Maintenance-of-Way Association, and a member of the Institute of Civil Engineers of Great Britain. But at Panama he never displayed the least enthusiasm for the work. He was tentative, withdrawn, wholly uninspirational. Men who served under him would have nothing particularly derogatory to say about him afterward, nor anything very complimentary. Most seriously he appeared to have no clear idea how he would build the canal.

He said he wanted a year at least to experiment with the French equipment—to test it against modern American equipment—and to experiment with various kinds of American equipment. And though he would insist later on that he had a "regular system" in mind the whole while, his assistants were never given the slightest hint as to what it might be.

He had started digging at Culebra without delay, or rather he simply continued on with the machines and token force of the Compagnie Nouvelle. By early November he had managed to install a new American steam shovel—a ninety-five-ton Bucyrus, a machine three times the size of the American shovels used by the French—and on November 12 it clanked into motion and began gouging away at the brick-red slope of Gold Hill, on the east bank of the Cut. The impression at home was that the American canal was under way. It was one of the year's outstanding developments, in company with the Russo-Japanese War and Theodore Roosevelt's victory over the hapless Democratic candidate for President, Judge Alton B. Parker. But the effort at Culebra was entirely random. For all practical purposes it meant nothing. The laborers who had to put down the track for the new shovel and dirt trains did not even have the right tools. Railroad spikes were being driven with axes.

Like those in Washington with whom he felt at such cross-purposes, Wallace insisted on putting second things first. Had George Morison been alive and serving as head of the Isthmian Canal Commission or as chief engineer, this would not have happened. ("It is a piece of work that reminds me of what a teacher said . . . that if he had five minutes in which to solve a problem he would spend three deciding the best way to do it.") Wallace failed to see that his primary responsibility, the priority task, was not to dig but to prepare the way—to formulate a comprehensive plan, to assemble the necessary equipment, to provide facilities for feeding and housing the army of laborers who would do the work, to put the railroad in order, to settle the problems of yellow fever and malaria in the quickest, most effective manner possible. *And* he failed to see that it was his duty to make all this plain to the commission. The call to make the dirt fly was as naïve and misguided as the cry of "On to Richmond" before the First Battle of Bull Run, and would look equally silly in the light of subsequent events.

A charitable assessment of Wallace by his successor, John Stevens, was that he failed because he was not sufficiently aggressive, that a more demanding approach with his superiors would have straightened things out. (There were times, Stevens wrote, "when fighting becomes a righteous duty.") When the commissioners came on their inspection tour in August, daily conferences were held at Ancon for nearly a month. In September Wallace returned to Washington to confer at still greater length. So it was not that he had no opportunity to make his case.

The men on the line seldom ever saw him. To his office staff in the old French headquarters on Cathedral Plaza, it seemed his strongest views too often centered on trifles. When an assistant, William Karner, kept him posted with weekly reports during his time in Washington, reports that Karner wrote in longhand by lamplight after hours and amid swarms of flying insects, Wallace, in one reply dated October 16, commended him for the penmanship but added peevishly that he did not like to read reports written in longhand.

Instructions were constantly changed. Men were abruptly shifted from one job to another, seldom with explanation.

"*Send forth the best ye breed*," Kipling wrote in his ironic poem "The White Man's Burden." But those being sent forth by Washington were seldom even second-best. "One young man came down with an appointment as a rodman," William Karner recalled. "On the supposition that he was a graduate of some technical school, I asked him where he graduated. He said he was not a graduate. I then asked him where he got his engineering education. He replied he had none. I then asked him if he knew the difference between a level and a transit and he frankly replied he did not. . . . He lived in one of the southern states and said his member of Congress wrote him if he wanted a position on the work he could get him appointed."

Of some two dozen supposedly experienced track hands recruited by the Washington office none had ever worked on a railroad. Lieutenant Robert E. Wood, one of the few Army officers sent by the

Isthmian Canal Commission, would recall, "The beginnings of the force recruited in 1904 . . . were largely Americans who had left the United States for this country's good—railroad men who were blacklisted on the American railroads, drunks, and what we called tropical tramps, American drifters in Latin America."

Even the best of the new men were young and inexperienced as a rule, and yet often found themselves thrust into positions of critical importance knowing little or nothing of what was expected of them. Frank B. Maltby, an engineer from Pittsburgh, was told on arrival to report to Wallace, who, after the usual observations concerning the heat and rain, plunged into the subject of what was being done and what should be done in the immediate future.

> Nothing was said as to how I came to be there [Maltby wrote afterward], my appointment, my rank or salary, and I do not recall that anything was said about dredging, except in a very general way. He then said, "I want you to take charge of the Atlantic and Pacific divisions. You will make your headquarters at Cristobal, as that is the much more important end of the canal at present. You can live in house No. 1 [the old de Lesseps' Palace]. There is a resident engineer in charge there now, who will give you information as to what is to be done. I want you to build up an organization so complete and efficient that you won't have to do anything but sit on the veranda and smoke good cigars."

By November there were 3,500 men at work. New recruits from the States who had been promised clean, furnished quarters were fortunate to find space to put a canvas cot in an unfurnished, often miserably small room with five or six others. Many of the old French quarters were put in service before repairs had been made. While two new "hotels" were under construction, the majority of the men were left to find what they could in Colón or Panama City, where decent rooms rented for three to four times what they would have at home. For unskilled black workers—that is, for about two out of every three—there was practically nothing to choose from, with the result that they crowded into the foul native side of Colón or whatever shacks could be found or improvised in villages beside the railroad.

The food available was meager, monotonous, high-priced, and as detrimental to morale as nearly all other troubles combined. There was no ice, no fresh milk, rarely a fresh vegetable; the local bread was tasteless and dirty. No one dared trust the water. Men went for weeks on a diet of canned sardines, canned Danish butter, and crackers. Nearly everyone was disheartened; a few had already packed up and left for home.

• •

Wallace was not oblivious to the situation. He had decided at the outset, for example, to provide Panama City and Colón with their own water system (which was the reason for the order of pipe) and to install sewage facilities in both cities. There is even in his organizational headings a basic appreciation of the diversity of tasks to be met other than digging dirt: Supplies, Personnel and Quarters, Buildings and Architecture, Machinery, Maps and Printing, Climatic Conditions and River Hydraulics, Communications. His thoughts, nonetheless, were tied up with the work at Culebra. Only by digging could he train the men, only by digging could he determine his unit cost—"the cost of explosives, cost of loosening and excavating material, cost of loading, cost of transportation, cost of disposition, and the cost of all the various elements of supervision and the maintenance of equipment, track and appliances, on the basis of the cubic yard." Only by digging could he satisfy Washington with monthly progress reports.

To his credit it can also be said that, as a result of his experiments, he had settled on the Bucyrus shovel as the machine to dig the canal. It was altogether a sound decision of far-reaching consequences. When in October, specifications for eleven shovels were put out for bids in Washington, they were those of the Bucyrus machine and Bucyrus, which was also low bidder, got the entire order. In the spring Wallace would order a dozen more.*

But it happens that Wallace was also stalling for time. He was in no rush to present a comprehensive plan because he had privately concluded that the canal as conceived by the earlier Walker Commission was a vast mistake. A series of recent test borings made along the site of the proposed Bohio dam indicated bedrock at not less than 168 feet below sea level, a revelation he kept to himself prior to the visit of Secretary Taft at the end of November.

Taft had been sent to straighten out certain complications with the new Republic of Panama, in particular the tariff policy and postal rates within the American Zone, and to convey the personal assurance of the President (now that he was embarking on his first full term in office) that the United States had no imperialistic designs on Panama, that the Americans were there for no purpose other than to build the canal. Taft, who weighed approximately three hundred pounds, was credited

* Chief sales representative for the Bucyrus Company, interestingly, was a young man named George A. Morison, who had left New England to join the Milwaukee firm on the advice of his Uncle George. For a young man interested in a business future, George S. Morison had said not long before his death, the then small steam-shovel manufacturer clearly had "possibilities."

by one admiring reporter with "dominating the whole scene." With the official entourage also was William Nelson Cromwell, the American counsel for the new republic, who kept very close to Taft the entire time, which had the effect of making Taft look even larger.

Taft stood with Mrs. Taft in receiving lines; he waltzed "light as a feather" with Señora Amador; he perspired mightily; and in the open, judicious manner that had made him so effective in the Philippines, he conferred at length with Amador and Arango. Panamanian goods, it was agreed, would enter the Zone duty-free; postage rates in the Zone and in Panama would be the same.

But Taft also returned to Washington convinced that the Isthmian Canal Commission must not continue as constituted, that virtually all problems with the work could be traced directly to the office of Admiral Walker, that the canal was a far larger, more bewildering task than anyone at home yet grasped, and that it must be a canal at sea level.

The futility of mounting the largest overseas effort in the country's history, the largest public work ever attempted anywhere, by placing its fate in the hands of seven men in Washington had already occurred to Taft, as to Roosevelt. But not until Taft had been to Panama, not until he had listened to Wallace expound on his problems, was he sufficiently convinced to make a move. Like many men of decisive mind, Taft prided himself in recognizing the same quality in others. During their ten days on the Isthmus, he and Mrs. Taft stayed with Wallace and his wife. Wallace had impressed Taft with his "earnestness and interest in the work, his ability, his facility of expression. . . ." (Wallace, the engineer, had also had the forethought to have one large dining-room chair taken to a blacksmith shop and thoroughly bolted and braced to bear the weight of the distinguished visitor, a courtesy Taft particularly appreciated, since, as he told Wallace, it gave him a great feeling of security throughout his stay.)

Wallace had become a frequent correspondent thereafter and because of Roosevelt's boundless respect for Taft the repercussions were not long in coming. On January 13, Roosevelt asked Congress to reduce the seven-man commission to a group of three, a solution specifically urged by Wallace. A sea-level passage became a common topic once again and an issue of controversy in the newspapers.

So while Congress debated what to do about the canal commission and Roosevelt toyed with the idea of still another blue-ribbon technical board to tell him which kind of canal it ought to be, Wallace continued with his "experiments" at Culebra. By January of the new year there were two Bucyrus shovels at work in the Cut and 1,500 men; but the shovels could not operate at even a quarter of their theoretical efficiency because there were too few trains to haul the spoil away and because what trains there were kept running off the tracks. And without an overall plan the whole effort was no less pointless than before.

The possibility of replacing Wallace had been considered, but not seriously. He was, Governor Davis reported, a "very superior man, and he ought to be retained," thus confirming what Taft already knew.

John Wallace had one further problem. He lived in mortal terror of disease.

Detesting Panama—this "God-forsaken country" he called it in one letter to Taft—he appears to have been haunted by the fate of his French predecessors. The official residence of the chief engineer was the same as it had been for the French, the old Casa Dingler on the Avenida Central, where for the first several months, until the arrival of Mrs. Wallace, he lived with his assistant, William Karner. Even the servants were holdovers from the French regime—a French butler named Benoit, a French-speaking black cook from Martinique, a Panamanian houseboy, and a personal valet who apparently was descended from one of the Irish "navvies" who built the Panama Railroad. So Wallace had heard soon enough and in detail of all that had befallen the Dingler family. To the mental burden of unit costs and endless vouchers in triplicate, to the strange night sounds of the crumbling old city, was added a vision of stark tragedy within his own walls. Under such circumstances even a trained technical mind might begin to imagine things.

Before September and his return to Washington, he saw Karner taken ill with malaria and removed to Ancon Hospital. His valet was stricken and carried from the house, only to be followed by the cook. Wallace departed filled with morbid premonitions. When he returned in November, his wife came with him, a sign of confidence it appeared, but in no time the story was all over the Isthmus that he had also taken the precaution to bring back two expensive metal caskets.

II

The first case of yellow fever was brought into the Santo Tomás Hospital in Panama City on November 21. The patient, an unemployed Italian laborer, had been found in a restaurant near the center of town. He was isolated at once in a screened ward and he eventually recovered. Little was said of the matter.

Of six more cases in December none was a canal employee and none was fatal. In January eight more cases were reported, including three on board a steamship at Colón, of whom two died. But again the victims, members of a touring Italian opera company, had no connection with the canal enterprise. It was not until later in the month when the disease broke out on the cruiser *Boston*, anchored in Panama Bay, that official notice could no longer be avoided. The ship was immediately ordered north to Puget Sound and there was but one fatality, the ship's doctor. The story, however, was out: YELLOW JACK IN PANAMA.

The mosquito specialists, meantime, had divided Panama City into eleven districts. Inspectors were assigned and a record was kept on every house, exactly as at Havana. The objective was an inventory of every essential well, water tank, cistern, water barrel, or water jar in the city. All that were unessential would be disposed of. But progress was maddeningly slow and the influx of new population increased steadily. The inspectors, mostly local men, had to be taught what to do. Frequently careless or indifferent, they had to be checked and double-checked, and a large segment of the native populace saw it all as some kind of nonsensical Yankee game played chiefly for their inconvenience.

Anyone taken suddenly ill, whatever the cause, was rushed to the closest hospital, put immediately in isolation and watched. Frank Maltby, the division head at Cristobal, became violently sick in the middle of one night, but of common diarrhea only, with the result that he spent the next week in the old French hospital in Colón. "People came and looked at me through the screen as if I were a wild animal of some sort."

A retired Army engineer, a Colonel Philip G. Eastwick, a popular figure in his hometown of Portland, Oregon, arrived in Panama City to visit his son, a member of Wallace's staff. He died of yellow fever the second week. A carpenter named Thomas Clark had been on the Isthmus only ten days when he died.

Most of Wallace's people, some three hundred technicians and office help, worked in the headquarters building on the plaza where, during an earlier inspection, Gorgas had found mosquitoes breeding in almost every office, in certain small glass receptacles in which were kept the brushes used for copying letters. Yet even here his efforts were met with little or no cooperation. Wallace was "distrustful" of the mosquito program. He and his aides regarded Gorgas' work as being largely experimental, like their own efforts at Culebra. The necessity for screening windows and doors in the building was brought repeatedly to Wallace's attention and without effect. His chief architect, M. O. Johnson (the one who had ordered the doors and hinges), declared himself too busy with serious problems to start worrying about window screens. When Joseph Le Prince went to talk to him in his third-floor office, Johnson even joked about the fuss being made. He had, wrote Le Prince, "little faith in modern ideas pertaining to yellow fever transmission."

To quell rumors about his own health, Wallace took time out to ride around town with his wife in an open carriage. Governor Davis cabled Washington to discount the stories in the press, which he categorized as "cruelly exaggerated." Conditions were actually improving, Davis insisted. But then he also was suddenly so ill with malaria that he was unable to carry on with his duties.

Meantime, Congress having failed to do anything about the Spooner Act so that a more effective commission could be organized, Roosevelt simply asked for the resignations of the present commission, Admiral Walker as well, and the chief engineer was summoned to Washington to assist in organizing a new commission along the lines he had sketched for Secretary Taft.

Again seven members were appointed, since the Spooner Act still applied, but this time Roosevelt ingeniously arranged to have just three members serve as an executive committee. Control of the work was thus placed in the hands of three, not seven, and two of the three made a quorum. The other four, three of whom fulfilled the necessary military representation, were figureheads only. The new arrangement went into effect April 1, 1905.

So the first commission had lasted less than a year. Its members retired or returned to whatever they had been doing before, with the exception of Benjamin Harrod, who stayed on as one of the figureheads.*

Under the new system the real power was not simply vested in three men, but each of the three was to head a particular administrative department. The chairman, based in Washington, would look after the purchase of supplies and serve as liaison between the commission and

* The three others were Rear Admiral Mordecai T. Endicott and two Army officers from former commissions, Brigadier General Peter C. Hains and Colonel Oswald Ernst.

the government. The chief engineer, for the first time, was to have full charge of the actual work on the Isthmus, while the governor would oversee health and sanitary conditions, besides the political administration of the American Zone.

Roosevelt's first choice for chairman was former Secretary of War Elihu Root, who had succeeded brilliantly—and against bitter opposition—in giving the Army the greatest shake-up in its history. Privately Roosevelt expressed a willingness to pay Root almost any salary to take charge of the canal. If not Root, then he wanted the steel baron Henry Clay Frick. But neither man was at all interested in Panama, which Root referred to as "that graveyard of reputations."

So the man chosen to head the second Isthmian Canal Commission, or I.C.C., was Theodore Perry Shonts, an Iowa lawyer turned railroad executive. Wallace had been among those who recommended Shonts for the job. Wallace would stay on as chief engineer. The new governor was Charles E. Magoon, a portly War Department lawyer whose specialty was colonial administration.

Taft announced the changes on April 3, but it was not until May 24 that Magoon and Wallace reached Colón. Taft, greatly concerned over increasingly grim accounts of sickness and chaos on the Isthmus, had urged Wallace to leave at once, not to wait for Magoon, who had other War Department affairs to wind up. But Wallace had wanted a vacation, and so spent another several weeks at his home in Illinois. His absence from the work was again stretched to nearly two months.

Newspapers around the country were carrying letters from embittered local citizens who had gone to Panama to build the canal. A file clerk from Cincinnati, Taft's hometown, informed readers of the *Enquirer* that at the rate things were going the canal would not be finished for fifty years. A young man named Will Schaefer, declaring that he spoke for all the Americans in Panama, wrote to the New York *Herald*, "There is not a bit of amusement or pleasure of the remotest kind here. . . . It is a case of work, work, work, all day long, and infrequently all night long, with no reward in view." Things were "altogether different" from what he had been led to believe.

"Tell the boys to stay home if they get only a dollar a day," wrote Charles Carroll to his mother in McKeesport, Pennsylvania. He was sick to death of the Panama Canal, he said in a letter picked up by the Pittsburgh papers. "Everybody is afflicted with running sores. . . . The meals would sicken a dog."

Wallace, in an interview in New York, assured reporters that conditions were no worse than to be expected. "Everything is now proceeding in harmony, with a well-defined general plan." In an article for *Harper's Weekly*, sounding very like Jules Dingler, he further observed that good health on the Isthmus was nothing more than a question of personal deportment. There were no climatic effects that a "clean, healthy, moral American" could not readily withstand.

In fact, the crisis was both real and very apparent. Charles Magoon, soon after his arrival, made no effort to conceal his astonishment at the situation, reporting to Shonts of unrest and insecurity everywhere he turned, of employees who were "ill-paid, over-worked, ill-housed, ill-fed, and subjected to the hazards yellow fever, malaria."

In Wallace's absence, yellow fever had broken out in the Administration Building and among the first to be stricken and to die was the architect Johnson. Gorgas had personally attended to the case and could do little or nothing. The only known treatment for yellow fever was the same as it had always been—to keep the patient as quiet and as comfortable as possible and hope for the best. It had been over in a few days. Johnson, who was twenty-nine, had given up his job with the Illinois Central and had come to Panama at Wallace's personal bidding. It was, wrote Governor Davis, like the "ending of many a bright young man I have seen on the battlefield." Since Wallace had been away, Johnson was buried in Wallace's metal casket.

Wallace's auditor, Robert West, had died, leaving a wife and five children back in the States. J. J. Slattery, an executive secretary in the building, had suffered the same fate. Mrs. John Seager, wife of Wallace's secretary, died. She had come to Panama as a bride only months before.

Young engineers and their families, some sixty people, had been hurriedly moved out of the city to a vacant building at Ancon Hospital. But in less than two weeks some two hundred employees had resigned, including several of the hospital staff. One nurse, upon reaching New York, told reporters that contrary to all declarations of the chief engineer—and much to her own amazement—yellow fever was taking the lives of "well set-up, clean boys with good principles." One could be neither decadent nor French, apparently, and still succumb. "A white man's a fool to go there and a bigger fool to stay," declared another returning worker, Harry Brainard, of Albany.

"A feeling of alarm, almost amounting to panic, spread among the Americans on the Isthmus," Magoon would write with total candor in his official report. "Many resigned their positions to return to the United States, while those who remained became possessed with a feeling of lethargy or fatalism resulting from a conviction that no remedy existed for the peril. There was a disposition to partly ignore or openly condemn and abandon all preventive measures. . . . The gravity of the crisis was apparent to all . . ." The "rank and file of the men began to believe that they were doomed just as had been the French before them," Gorgas would recall.

Orders were issued that all unscreened windows in the Administration Building be kept closed. The building was fumigated repeatedly. Houses in the immediate area were fumigated. Gorgas even had the holy water in the font at the cathedral changed daily after it was found that mosquitoes were breeding there, a gesture many Panamanians looked upon as possibly some subtle new form of religious persecution. But with his limited manpower and desperately limited supplies there was only so much that he could do. One critical shortage, for example, was ordinary newspaper—sufficient paper of any kind—for sealing buildings prior to fumigation. When he cabled Washington to send two tons of old newspapers on the next ship, Washington cabled back to question the request. It was thought that he was asking for reading matter for his hospital patients and that two tons seemed excessive. As a result, the ship sailed without the paper and further fumigation was delayed for ten days.

The fever wards were filling rapidly. The *Star & Herald* had introduced a regular yellow-fever report that included both obituaries and a listing of all new cases. In Colón, local undertakers, familiar with previous epidemics, had stacks of coffins standing ready in plain view at the depot.

It was the time of "The Great Scare," during which fully three-quarters of the Americans on the Isthmus fled for home. It was an episode that need never have happened and that most people, whether they stayed or left, would prefer to forget when things quieted down again, since the level of panic was so out of proportion to the actual seriousness of the epidemic. Compared to past experiences with yellow fever on the Isthmus, this was really only a mild flare-up.

Older hands among the railroad staff, veterans of the French effort, urged others to stay calm, but then it was they who also provided lurid renditions of 1885 and '86, or of the epidemic of just three years earlier, in 1902, when more than two hundred had died. In one instance in June, thirty new recruits who landed at Colón heard enough of such talk in their first few hours ashore to get back on board ship and return to New York.

You are going to have the fever,
Yellow eyes!
In about ten days from now
Iron bands will clamp your brow;
Your tongue resemble curdled cream,
A rusty streak the center seam;
Your mouth will taste of untold things,
With claws and horns and fins and wings;
Your head will weigh a ton or more,
And forty gales within it roar!

The poem "Yellow Eyes" and others of much the same vein had just appeared in a little volume with a burgundy-colored binding, *Panama Pathwork*, published by the *Star & Herald*. The author, James Stanley Gilbert, an American resident of Colón for many years, had done for Panama something comparable to what Robert Service was to do for the Yukon. His themes—in "Funeral Train," "The Isthmian Way," "King Fever," "Beyond the Chagres"—were disease, alcoholism, ever-encroaching death, and the futility of any and all human endeavor associated with the valley of the Chagres River.

Beyond the Chagres River
'Tis said—the story's old—
Are paths that lead to mountains
Of purest virgin gold;
But 'tis my firm conviction,
Whatever tales they tell,
That beyond the Chagres River
All paths lead straight to hell!

Funeral processions were continually passing through the streets. Funeral trains ran daily to Monkey Hill, their bells clanging the length of Front Street in Colón—or "*Coal*-on," as the Americans pronounced it. People feeling the least sign of illness were immediately certain that the end had come. Governor Magoon, seized with chills one afternoon, stood at his window at Ancon watching what he felt sure was his last sunset.

Then, at the very height of the panic, in mid-June, only three weeks after his return, Chief Engineer Wallace and wife packed and sailed for New York. To an editor of the *Star & Herald* and the others who came to see him off, Wallace would say only that he had "matters of importance" to take up with Secretary Taft. But almost from the moment the ship cast off, rumors were flying to the effect that even Wallace had now fled and that he had no intention of ever returning. The *Star & Herald* carried a prominent, if tentative, denial: "To say the least it does not seem plausible that a man of the type of Mr. Wallace would give up a position like the one he was occupying in the United States . . . to come to the Isthmus to engineer the canal . . . then leave . . . at this stage of the game, when the work was scarcely begun."

On July 3, with as yet nothing further in the paper concerning Wallace, came the stunning news that a man had died at Ancon Hospital of bubonic plague. A Barbadian, a stevedore on the wharf at La Boca, had been brought to the hospital on June 20 with what looked alarmingly like plague, a diagnosis that was confirmed by the autopsy. Keeping the cause of death secret, Gorgas and his staff moved with all possible speed. The La Boca wharf was put under immediate quarantine. Hundreds of rat traps were set, poison was put out; the barracks where the man had taken ill was fumigated; two ships tied up at the wharf were towed into the bay and were fumigated—all in less than twenty-four hours. The following day La Boca was declared officially off-limits and a police cordon was positioned around the entire area, on land and water. Buildings were fumigated a second time. Walls, floors, and ceilings were sprayed with a solution of bichloride of mercury. Soiled clothes and dirty bedding were soaked in the same thing. Garbage was carried away. Chicken coops, animal pens, latrines, were torn down and burned, while in Panama City a bounty of ten cents was offered for all rats and mice. Among the hundreds of rats killed at La Boca—rats the size of guinea pigs, as one Panamanian recalled—several were found to be infected with plague.

In June, despite all Gorgas' efforts, the incidence of yellow fever had been double what it was in May—62 known cases, 19 of which were fatal. There had been two or three cases of smallpox about which very little was said. For months malaria and pneumonia, tuberculosis and dysentery, had been taking a much heavier toll in life than had yellow fever, and especially among black workers, a fact no one had as yet faced up to. In all more than a thousand people had been admitted to the canal hospitals alone. But the news of one case of bubonic plague was like a scream in the night. In the language of one official report, the number of those who left the Isthmus was now "limited only by the ability of the outgoing boats to carry them away."

• •

Chief Engineer Wallace had secured permission for his return to the United States, his third such trip in a year, by cabling Taft that he had "complicated business" to discuss which could not be handled by correspondence. He gave no further explanation, which greatly irritated the Secretary, but after conferring with Roosevelt and Shonts, Taft had reluctantly cabled his consent.

From Magoon, however, Taft heard what was on Wallace's mind. Wallace had unburdened himself to Magoon in the course of two lengthy conversations, saying that he had been offered a high-paying position with a private engineering firm and that he intended to quit. He could be induced to stay on, Wallace also said, but only if he were put in as chairman (instead of Shonts), given authority over the entire work, and granted a salary of perhaps $60,000. He would insist also, Wallace said, that he be free to come to the Isthmus "when he sees fit and depart as his discretion determines."

> He made a further statement [Magoon continued], to which I attached grave significance—that he left the Illinois Central twice without telling them, directly, what he wanted, and was sent for and given three times as much as would have induced him to remain at the time he left.
>
> Evidently he considers himself essential to this enterprise, and, for the immediate present, he is. He has never secured an assistant engineer competent to take his place or keep the work going at a decent pace for sufficient time to enable a new chief engineer to master the situation. . . .
>
> Speaking of his desire to be the head of the enterprise, he told me that he figured from the first that Admiral Walker would not last more than two years and he had intended to have things in such shape by that time that he would be made chairman; but the old commission went to pieces too quickly for him. . . . I cannot escape the conviction that he is trying to "pull off" a carefully contrived *coup d'état*. . . . I hope I am doing him a grave injustice, for personally I like him . . . I can readily understand that from his point of view the action and motive I attribute to him are entirely justifiable. In railroad circles, as on the stock exchange, it is entirely justifiable and even commendable to "squeeze" friend or foe when you have the chance and can profit by it.

This letter from Magoon, dated June 11, was followed two days later by another, less harsh appraisal. As a result of talks with Wallace, Magoon now thought better of Wallace's motives. "He seems to be fully prepared to quit, but willing to remain upon terms that seem to him justifiable . . . There is no difference in its effect on the public service, but there may be considerable difference between the two mental attitudes."

Taft cared not the slightest about Wallace's mental attitudes. The genial Taft, the man known for "the most infectious chuckle in the history of politics," was in a towering rage. The new commission had been tailored specifically to Wallace's wishes. Wallace had expressed unequivocal approval of the arrangement on several occasions since April and in writing. Wallace had been profuse in his praise of Shonts, full of gratitude for being able to serve under such a man.

Taft felt betrayed. To Taft so self-serving an ethic as Magoon depicted was neither understandable nor tolerable. If Wallace deemed himself essential to the enterprise, Taft now quite emphatically did not. Though Magoon's letters reached the War Department on the eve of an important trip to the Philippines, Taft went directly to New York to confront Wallace in person three days after Wallace arrived from Panama.

Wallace was told to be at Taft's room at the Manhattan Hotel at ten o'clock the morning of June 25. To Wallace's immediate annoyance, he was ushered into the room by William Nelson Cromwell, who showed no signs of leaving. Wallace said he preferred to speak with the Secretary alone, but Taft, in what Wallace later described as a "rather peremptory manner," told Cromwell to stay where he was. He wanted Cromwell present as a witness. "This action, of course, caused irritation and apprehension on my part that the interview would be unpleasant," Wallace would recall, ". . . and the irritation under which the Secretary was evidently laboring had a tendency to prevent that calm and dignified consideration of the question in all its bearings which should have been given it."

Taft, who had spent most of his career as a judge, wished first, he said, to know what possible "complicated business" could warrant Wallace's absence from Panama at so inopportune a moment? Wallace replied that he had received an attractive offer from a private firm and explained at some length why he could "not afford" to turn it down. The income, with salary and various opportunities for investments, would be equivalent to $60,000 a year Wallace said. Life at Panama, he remarked, was lonely and "accompanied by risk" to his health and to that of his wife. He proposed that he spend the summer winding up his duties from his home in Illinois. In the future he would be glad to serve on the commission in an advisory capacity.

"Mr. Wallace," Taft began after a pause, "I am inexpressibly disappointed, not only because you have taken this step, but because you seem so utterly insensible of the significance of your conduct."

There followed an angry but measured summation lasting half an hour. He reminded Wallace that when appointed by the commission the year before, his salary had been increased by $10,000 over what it had been under his former employer. He reminded Wallace that he had known of the risks when he accepted the position, that he had both recommended and approved the latest organizational changes made by the President. "For mere lucre you change your position overnight . . . You are influenced solely by your personal advantage. Great fame attached to your office, but also equal responsibility, and now you desert them in an hour."

He demanded Wallace's immediate resignation. When Wallace said that he would prefer to discuss the matter further, that perhaps some new and different arrangement might be worked out, Taft told him there could be no further talk.

"Mr. Secretary," said Wallace, "while there is a difference between us as to the point of view we take concerning my duty, I consider that there can be no question that I have performed my full duty up to this hour."

"Mr. Wallace," Taft replied slowly, "I do not consider that any man can divide such a duty up to any one point where it suits him to stop . . . In my view a duty is an entirety, and is not fulfilled unless it is wholly fulfilled."

The following day Taft took Wallace's resignation to Cambridge, Massachusetts, where Roosevelt was attending a Harvard reunion. On June 28 the resignation was accepted by the President, to take effect immediately. On the twenty-ninth, back in Washington, Taft, as directed by Roosevelt, released to the press the full transcription of his exchange with Wallace, the text apparently being that recalled by Cromwell. John Wallace was finished.

The immediate response in the press was one of dire concern. Though a few papers suggested that it was now merely a question of finding somebody else, the majority saw the reputation of Roosevelt's Administration suddenly at stake, not to mention that of the country, and found the outlook, in the words of the Louisville *Courier-Journal*, "not cheering." In the London papers it was said that Roosevelt was paying the price for his rash "land-piracy" in Panama.

Wallace, who ultimately became president and chairman of the board of Westinghouse, Church, Kerr & Company, as well as a board member of several other industrial firms, would spend the rest of his life trying to repair the damage done to his reputation. Monetary considerations, he would insist, had never entered into the decision to "disconnect" himself from Panama. He denied any personal fear of disease, claiming he had actually had a light attack of yellow fever while on the Isthmus and so thereafter had considered himself immune.

Privately he said that he heartily disliked Shonts and knew they could not work together. And appearing before the Senate canal committee he would delight John Tyler Morgan with the further admission that it was really Cromwell whom he hated most of all. The tragedy, said Wallace again and again, was that Taft never gave him a chance to express his real reason for resigning, which was that he had "become convinced some other men in my place could render better service to the enterprise."

John Stevens, reflecting on the episode, would remark simply and without scorn that Wallace had had a "thorough case of fright."

17

John Stevens

What we needed was a fighter. And we got one.

—Frank Maltby

I

John Stevens was picked to build the Panama Canal on the recommendation of James J. Hill—Hill, who had never had any use for the project and whose personal distaste for Theodore Roosevelt was monumental. Roosevelt, the "Empire Builder" once complained, had never done anything but "pose and draw a salary."

Hill considered Stevens the best construction engineer in the country, if not the world, and apparently he told Roosevelt as much during a visit to Washington in June 1905. "Mr. Hill told the President that he knew a man who could build the Panama Canal," Stevens' son would recall, "and when the President showed much interest he spoke of my father . . . and said he would see him in Chicago and report the President's interest in him. He did so, talking the matter over in detail. . . ."

As it also happens, Stevens was about to leave with Secretary Taft for the Philippines, as a new special adviser on railroad construction. So Taft too called Stevens in Chicago about taking the Panama post instead, just as soon as he, Taft, discovered why Wallace was coming on from Colón. The Monday following the Sunday of Taft's confrontation with Wallace, Stevens was told the job was his if he wanted it. The salary was $30,000.

His immediate impulse, he later wrote, was to say no. Wallace by then was back at his home outside Chicago. Stevens sent a note asking him to meet him at the Union League Club and discuss the matter, but Wallace refused. So Stevens boarded a night train to New York, where he met with Cromwell, who persuaded him to accept. And all things considered, this was probably the most valuable service yet rendered by the clever, "silver-tongued" attorney. More active than ever as an all-purpose troubleshooter for the Republican Party, Cromwell had become greatly concerned—as had others—over what a failure at Panama might mean to the party's fortunes.

At home again in Chicago, Stevens talked things over with his wife, who told him his whole career had been in preparation for this greatest of engineering projects and that of course he should go. ". . . I allowed arguments as to what was my duty to override my own feelings, and . . . my better judgment," Stevens remembered. He wired his acceptance on June 30. Roosevelt and Taft had little time to give the matter further attention. John Hay was ill and Taft, in addition to everything else, was looking after the State Department. Roosevelt was absorbed in preparations for the historic meeting at Portsmouth, New Hampshire, at which he would bring an end to the Russo-Japanese War. On July 1, John Hay suddenly died; there was a funeral to be attended, a new Secretary of State (Elihu Root) to be installed; and the Russian and Japanese delegations were arriving meantime. When Stevens and Shonts went to Oyster Bay so that Roosevelt could meet his new chief engineer, one of the other guests at lunch was the Japanese minister.

Stevens was fifty-two years old, powerfully built, heavy-shouldered, five foot ten, with black hair and black mustache and a hard, weather-beaten, handsome face. Roosevelt wrote of him admiringly as a "backwoods boy," "a rough and tumble westerner," "a big fellow, a man of daring and good sense, and burly power."

Like Gorgas—and unlike Wallace—he had spent most of his life surviving frontier conditions. Born on a small farm in Maine, he had had little formal education, but learned surveying, and in 1873, at the age of twenty, went west to work on surveys for the new city of Minneapolis. He had come up the hard way since—as a track hand in Texas, as a junior engineer locating and building railroads in New

Mexico, Minnesota, and British Columbia. In 1886, at thirty-three, as principal assistant engineer for the Duluth, South Shore and Atlantic Railway, he had been charged with building a line of nearly four hundred miles through the swamp and pine forests of Michigan's Upper Peninsula, from Duluth to Sault Ste. Marie. By the time he went to work for James J. Hill in 1889, he had survived Mexican fevers, Indian attack, Upper Michigan mosquitoes, and Canadian blizzards. He had been treed by wolves on one occasion; he had learned to sleep sitting up while crossing the prairie in a buckboard; with surveying gear, tent, and provisions packed on his back he had traveled hundreds of miles into the Rockies on snowshoes. ". . . I became tough and hard physically," he would write. "I learned to sleep under wet skies . . . rolled only in a single blanket . . . to adapt myself . . . under the most primitive conditions. And I loved it!"

His personal faith was in the strides ordinary men might achieve. "With respect to supermen, it has probably been my misfortune, but I have never chanced to meet any of them." Hard work, he often said, was the only "open sesame" he had had any experience with. By studying on his own at night, he learned mathematics, physics, chemistry. And for all his rather rough, often profane, frontier manner, his exceptional ability was unmistakable.

Working for Hill was the turning point. In 1889 he had started as one of Hill's locating engineers, assigned to explore a route west from Havre, Montana. Hill had decided to build a railroad to the Pacific; the Great Northern was to be his own personal path of empire; and in the dead of winter, 1889, Stevens found the Marias Pass, Hill's passage over the Continental Divide. The Marias Pass saved more than a hundred miles and, at an elevation of 5,215 feet, gave the Great Northern the lowest grade of any railroad to the Pacific. Stevens became something of a legend in the Northwest, "The Hero of Marias Pass." He had made the discovery on foot and alone, his Indian guide having given up. At night, with no wood for a fire, and the temperature at 40° below zero, he had kept from freezing to death by tramping back and forth in the snow until dawn.

In the Cascades, on the western slope, he found another key pass (later named Stevens Pass). Hill made him his chief engineer in 1895 and ultimately his general manager. Stevens built bridges, tunnels (including the two-and-a-half-mile Cascade Tunnel), and more than a thousand miles of railroad, as much as had been built by any one man in the world. He built exceedingly well and Hill never intervened.

Hill was fiercely independent (the Great Northern was the only western road built without government subsidy); he was blunt, tough, a pioneer and a fighter. He had the common touch; with his physical bearing alone he exuded power. And much of what Hill was had rubbed off on Stevens, who for the rest of his life would talk of Hill as the finest man he had ever known. Of Stevens, Hill once remarked, "He is always in the right place at the right time and does the right thing without asking about it."

But Stevens was also restless, often temperamental, and in 1903 he had left Hill to become chief engineer (and eventually vice-president) of the Chicago, Rock Island and Pacific. His reason, he later explained, was that Hill had had a still higher position in mind for him, one that called for "a diplomacy which I was temperamentally unfit to exercise." Then in 1905 he turned around and accepted the Philippines assignment, because, he said, he needed a change.

In the course of conversation during the lunch at Oyster Bay, Roosevelt told Stevens to take charge and report to him directly if need be. Things at Panama were in a "devil of a mess," Roosevelt conceded. He was reminded, he continued, of the man who engaged a butler and set him to work by saying, "I don't know in the least what you are to do, but . . . you get busy and buttle like hell!"

Afterward, when Shonts and Stevens met with reporters at the Oyster Bay railroad station, it was Shonts who did the talking. Stevens would have "no one to blame if the work is not done right," Shonts said, "for he will be supreme in the engineering department." Stevens, sitting on a stone post off to one side, merely nodded in agreement.

Shonts and Stevens landed at Colón without ceremony on July 26, 1905. Neither man had ever been to Panama before.

"Shonts and Stevens will soon be with you, and the mountains will move," Taft had cabled Magoon. But whether dirt would fly or mountains move was no longer the question. The question was whether the entire American venture in Panama could be rescued from humiliating defeat. The unavoidable fact was that it had been a wretched beginning and already $128,000,000 had been spent, as *The New York Times* emphasized. For all the deprecating talk of French inefficiency, French failure, for all the proud claims of American know-how and resolve, the United States had performed with less efficiency, less purpose, and markedly less courage than had the French at any time during their ordeal. A whole year had been lost and the situation on the Isthmus was an utter shambles.

Recalling his own first impressions, Stevens wrote, "I believe I faced about as discouraging a proposition as was ever presented to a construction engineer." Accommodations on returning ships were still at a premium. Upwards of five hundred white technicians and skilled workers had evacuated. Those left behind were frightened—"scared out of their boots," Stevens said—and had every reason to despair. Some of the freight piled up at Colón had been there for more than a year. "I found no organization . . . no answerable head who could delegate authority . . . no cooperation existing between what might charitably be called the departments. . . ."

The review board appointed by Roosevelt to decide whether it should be a canal at sea level, as urged by Wallace, had not even convened as yet. So there was still no final plan to go by and everyone was waiting to be told what to do. Most of the Americans Stevens encountered seemed to believe that Shonts and he had come to order them to abandon everything and sail for home.

Theodore Shonts, with his pince-nez, mustache, and bulldog expression, looked like an older, more sedate version of Theodore Roosevelt. He was also very much the man in charge so long as he remained on the Isthmus, which was not long; with his abrupt, authoritative manner he rubbed nearly everybody the wrong way.

"Governor, what's the matter down here?" Shonts demanded over cigars the first night, as he, Magoon, Gorgas, and Stevens sat on Magoon's veranda at Ancon. Magoon was easy in manner, but very proper and highly polished—a friend likened him to a Roman cardinal—and he believed that keeping peace with the Panamanians was chief among his duties. The most pressing problem, he explained to Shonts, was food for the work force. The local merchants kept pushing prices higher and higher, until it had become nearly impossible for the men to survive on what they earned. Some workers, verging on starvation, had taken to foraging like buccaneers.

Shonts said commissaries must be established immediately, and when Magoon explained that this would be in violation of an agreement with Panama whereby all foodstuffs were to be bought from local merchants, Shonts responded, ". . . it's evident that you haven't heard the news. . . . I've come down here to build the Canal . . ." The Isthmian Canal Commission would feed the men at cost beginning immediately.

It was also Magoon's ambition to establish a model government within the American Zone, as an "edifying" example for Panama and the rest of Central America. Shonts told him to forget that. There would be no more government than was necessary to preserve order.

"Our sole purpose . . . is to build the Canal, so 'keep your eye on the ball.' "

Turning to Gorgas he remarked, "We are not here to demonstrate any theories in medicine, either." He was wholly unimpressed by what Gorgas had accomplished and put him on notice that he had four months to rid Panama of yellow fever.

Stevens, as at Oyster Bay, said very little, and when Shonts departed a few weeks later, Stevens continued to say very little. Yet in manner, appearance, in the way he treated people, he was plainly a different sort than his predecessor. He was seen out on the line daily, whatever the weather, hiking about in rubber boots and overalls, wearing a battered old hat and puffing steadily on a black cigar. "He was [seen] climbing about in the mud of the ditch, catching a switch engine, pausing among machinery," wrote a magazine correspondent. ". . . He had very little to say except to ask questions. He was very quiet, very business-like. The men were not certain at first, mostly because they could detect no pose."

Anybody could talk to him, it was discovered, and with a few terse observations he began putting spirit into the work for the first time.

"There are three diseases in Panama," he told the men. "They are yellow fever, malaria, and cold feet; and the greatest of these is cold feet." There was lost time to be made up for; there was much to learn. When it was pointed out to him that no collisions had occurred on the Panama Railroad in more than a year, he remarked, "A collision has its good points as well as its bad ones–it indicates there is something moving on the railroad."

He wanted the engineering offices moved from the Administration Building in Panama City to Culebra Cut as quickly as possible. Shown plans for an elaborate new residence for the chief engineer scheduled to be built at Ancon, he cancelled the plans. He would live at Culebra and specified an inexpensive one-story bungalow with a corrugated-iron roof. Until the house was ready he would be content with a small place on the hospital grounds.

Privately he was appalled by what he saw of Wallace's work and by the antiquated equipment in use. Once, standing on a point overlooking Culebra Cut, he counted seven trains off their tracks. Every steam shovel in view was standing idle because the crews, along with the entire labor force, were struggling to get the trains back on the tracks–"an unwise proceeding," he noted acidly, "for they [the trains] were of more value where they were."

While the public and the press at home speculated on what progress the new chief might effect, Stevens, on August 1, ordered a complete stop of all work in Culebra Cut. Excavation would not be resumed, he informed his staff, until he had everything ready. Steam-shovel engineers and cranemen were sent back to the United States. They would hear from him later.

"The digging is the least thing of all," he declared. Starting at once, Dr. Gorgas was to have whatever men and supplies he needed. Panama City and Colón were to be cleaned up and paved. Warehouses, machine shops, and piers were to be built. Entire communities were to be planned and built from scratch—houses, mess halls, barracks, more hospitals, a visitors' hotel, schools, churches, clubhouses, cold-storage facilities, laundries, sewage systems, reservoirs. He was determined, as he said in a letter to Taft, to prepare well before beginning construction, "regardless of clamor of criticism . . . as long as I am in charge of the work . . . and I am confident that if this policy is adhered to, the future will show its absolute wisdom."

Working twelve, sometimes eighteen hours a day, he saw no reason why others should not too. In the first few months, as he made his daily rounds, he walked the line from Colón to Panama. The melting heat, the rain, the terrible mud, appeared neither to discomfort nor distress him. His own health remained perfect.

He and Gorgas got on extremely well from the start. With his backing, Gorgas' real work began in earnest. Through the summer, disease of all kinds continued to cut through the ranks of the labor force. Yellow fever had abated somewhat, but only somewhat—forty-two cases in July, thirteen deaths; twenty-seven cases in August, nine deaths. But malaria, pneumonia, tuberculosis, and intestinal diseases were still rampant and taking many more lives than yellow fever, while at the same time debilitating two or three times as many as were killed, facts the public at home would never quite comprehend. Between May 1 and August 31, 1905, the time of the so-called yellow-fever epidemic, yellow fever took forty-seven lives. In this same period nearly twice as many people died of malaria, forty-nine of pneumonia, fifty-seven of chronic diarrhea, forty-six of dysentery. Black workers were hardest hit by malaria and pneumonia. When bubonic plague struck a second time at La Boca, the victim again was a Barbadian.

But while a drastic reduction of all disease was considered essential in the long run, yellow fever had to be the immediate objective. To rid the Isthmus of yellow fever, Gorgas remarked, would be to rid it of fear.

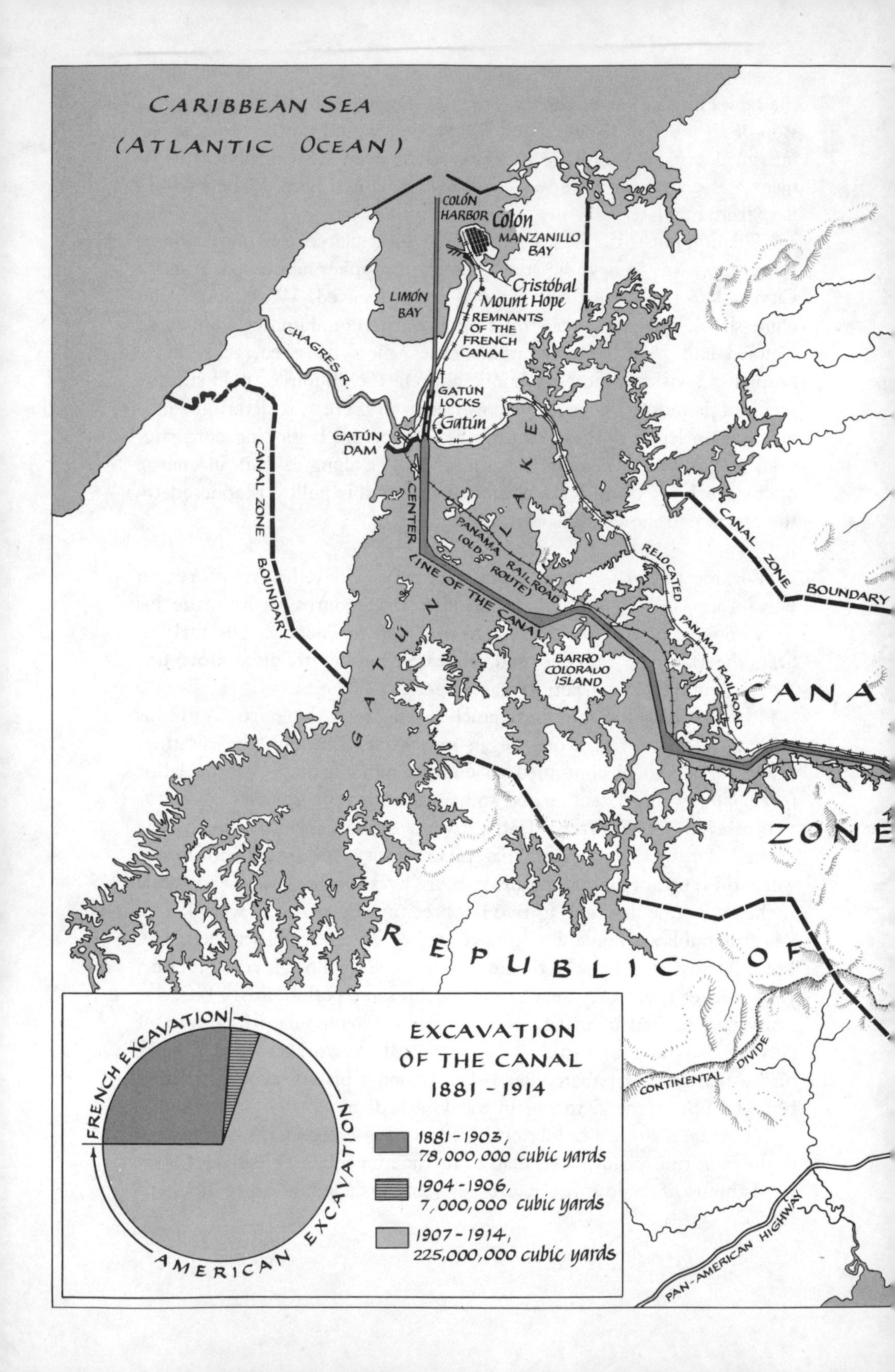
CARIBBEAN SEA
(ATLANTIC OCEAN)
COLÓN HARBOR
Colón
MANZANILLO BAY
Cristóbal
Mount Hope
LIMÓN BAY
REMNANTS OF THE FRENCH CANAL
CHAGRES R.
GATÚN LOCKS
Gatún
GATÚN DAM
CANAL ZONE BOUNDARY
CENTER LINE OF THE CANAL
PANAMA RAILROAD (OLD ROUTE)
RELOCATED PANAMA RAILROAD
GATUN LAKE
BARRO COLORADO ISLAND
CANAL ZONE BOUNDARY
CANA
ZONE
REPUBLIC OF
CONTINENTAL DIVIDE
PAN-AMERICAN HIGHWAY
EXCAVATION OF THE CANAL
1881 - 1914
FRENCH EXCAVATION
AMERICAN EXCAVATION
1881-1903, 78,000,000 cubic yards
1904-1906, 7,000,000 cubic yards
1907-1914, 225,000,000 cubic yards

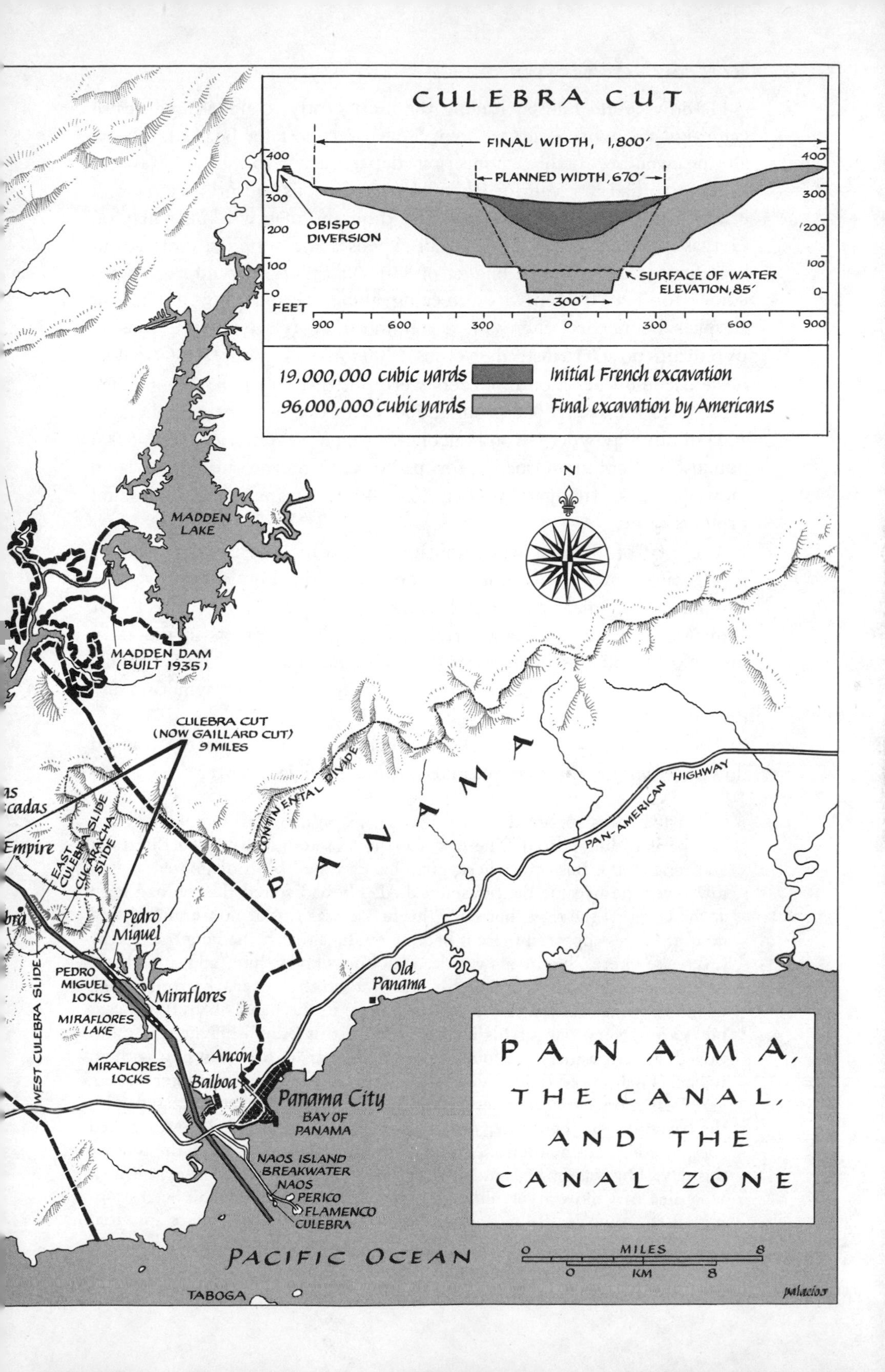

CULEBRA CUT
FINAL WIDTH, 1,800′
PLANNED WIDTH, 670′
OBISPO DIVERSION
SURFACE OF WATER ELEVATION, 85′
300′
FEET
400
300
200
100
0
900
600
300
0
300
600
900
19,000,000 cubic yards
Initial French excavation
96,000,000 cubic yards
Final excavation by Americans
N
MADDEN LAKE
MADDEN DAM (BUILT 1935)
CULEBRA CUT (NOW GAILLARD CUT) 9 MILES
CONTINENTAL DIVIDE
PANAMA
PAN-AMERICAN HIGHWAY
Empire
EAST CULEBRA SLIDE
CUCARACHA SLIDE
WEST CULEBRA SLIDE
Pedro Miguel
PEDRO MIGUEL LOCKS
Miraflores
MIRAFLORES LAKE
MIRAFLORES LOCKS
Old Panama
Ancón
Balboa
Panama City
BAY OF PANAMA
NAOS ISLAND BREAKWATER
NAOS
PERICO
FLAMENCO
CULEBRA
PACIFIC OCEAN
TABOGA
MILES
KM
0
8
0
8
PANAMA, THE CANAL, AND THE CANAL ZONE
palacios

He now found himself leading the most costly, concentrated health campaign the world had yet seen. Stevens, as he later boasted, "threw all the weight of the engineering department" to his aid. Gorgas henceforth had first call for labor. His requisitions had priority over all others. By November there were four thousand men working solely on Gorgas' projects. Until then, for all expenses and supplies, Gorgas had been limited to an annual budget of $50,000. Stevens would sign requisitions for $90,000 for wire screening alone. Gorgas now got all the supplies he needed and with a minimum of red tape—120 tons of pyrethrum powder (instead of 8 tons), 300 tons of sulphur, 50,000 gallons of kerosene oil per month. Orders were put through for 3,000 garbage cans, 4,000 buckets, 1,000 brooms, 500 scrub brushes; for carbolic acid and sulphur powder, wood alcohol, mercurial chloride; for 5,000 pounds of "common soap"; for padlocks, lanterns, machetes, lawn mowers, 1,200 fumigation pots; for 240 rat traps for the hospital grounds alone.

The city of Panama was fumigated house by house, some sections several times over. The same was done at Colón. Fumigation brigades —hundreds of men carrying ladders, paste pots, buckets, rolls of brown paper, old newspaper—trailed through the streets in the early morning like some strange ragtag army of occupation. And by nightfall, when they had gone, strips of paper fluttered from windows and doorways on hundreds of houses.

The persistency with which new yellow-fever cases were tracked down is shown in this example taken from an official report:

> . . . a man was reported ill at a hotel. . . . When search was made for him he had disappeared. The next day he was found drunk on the street and sent to the hospital, where, after his case had been diagnosed as yellow fever, he became delirious and died. He had stated that he had been at the hotel all the time, but since this house was full of non-immunes and no other cases appeared there it became evident that he had contracted the fever elsewhere. The man was dead, nobody knew him, and apparently no information was obtainable. It was known, however, that other men of the same nationality as the deceased were in the habit of visiting a certain cafe. Every one of his countrymen in this establishment was questioned. At last a man was found who stated that he had seen him with an Italian. Then every Italian who could be found in town was interviewed, and finally one was discovered who said he had seen the deceased with the bartender of the theater on two occasions. The bartender was looked for and could not be found. After a hard search he was located the following day. He was in bed and had yellow fever. He stated that the man who died of yellow fever, although registered at the hotel, had been sleeping all the time in the same room as himself in the theater. It appeared

probable that the theater had been the center of infection, and it was accordingly fumigated. A few days later a third case was discovered, that of a little girl, who had been in the theater every evening with her mother, thus confirming the indications which had already been acted upon.

Cisterns and cesspools were oiled once a week. Most critically, Panama City, Colón, Cristobal, Ancon, La Boca, Empire, Culebra, were all provided with running water, thus dispensing—after centuries—with the need for domestic water containers.

Stevens made no public declaration of faith in Gorgas or the mosquito theory. "Like probably many others I had gained some little idea of the mosquito theory," he would recall, ". . . but, like most laymen, I had little faith in its effectiveness, nor even dreamed of its tremendous importance." Still, Gorgas' presence seemed "simply an act of Providence" and Stevens' own instinct was that the only way to back Gorgas was to back him to the fullest. When a movement began in Washington to have Gorgas removed, a movement initiated by Shonts and supported by Taft, Stevens fought back. Taft thought Gorgas had "no executive ability at all." Shonts, who seems to have liked Gorgas well enough, had little confidence in the mosquito theory and was no less insistent than his predecessor, Admiral Walker, that "cleaning up" Panama must be made the priority task. Shonts had found a replacement for Gorgas, moreover, a Johns Hopkins man named Hamilton Wright, and went to Oyster Bay to tell Roosevelt what he intended to do. But Stevens in correspondence from Panama insisted that Gorgas be kept on, and so it became a test issue, the decision being left ultimately to Roosevelt, who again consulted Dr. Welch as well as a friend and hunting companion, Dr. Alexander Lambert.

Welch was actually asked for a recommendation for his Hopkins colleague, Wright, rather than for a comment on Gorgas and his relative progress at Panama. But while testifying to Wright's ability, Welch insisted that no one was better equipped for the work than Gorgas. The best man was already on the job, that was the implicit message. "Would to God," wrote Roosevelt to Welch in reply, "there were more men in America who had the moral courage to write honest letters of recommendation such as yours . . ."

Dr. Lambert expressed his views in private conversation in the study of Sagamore Hill. "Smells and filth, Mr. President, have nothing to do with either the malaria or the yellow fever," Lambert said. "You are facing one of the greatest decisions of your career. You must choose between Shonts and Gorgas. If you fall back upon the old methods of sanitation, you will fail, just as the French failed. If you back up

Gorgas and his ideas and let him pursue his campaign against the mosquitoes, you will get your canal."

Roosevelt, according to Lambert's version of the conversation, decided then and there that Gorgas would stay. Shonts was called to the White House soon afterward and was told to "get back of Gorgas." And to his great credit, Shonts accepted the decision and saw to it that the Sanitary Department became what it should have been from the start, an independent bureau reporting directly to the chairman. Shonts, as Gorgas later wrote, was a man "who thought and acted in millions [of dollars] where we army and navy officers did in thousands . . . I would never have dared even to make an application for the immense amounts of money he authorized me to spend . . ."

But the real hero, in Gorgas' view, was Stevens. "The moral effect of so high an official taking such a stand at this period . . . was very great," Gorgas wrote, "and it is hard to estimate how much sanitation on the Isthmus owes to this gentleman for its subsequent success." To Stevens, years later, he wrote privately, "The fact is that you are the only one of the higher officials on the Isthmus who always supported the Sanitary Department . . . both before and after your time. So you can understand that our relations, yours and mine, stand out in my memory . . . as a green and pleasant oasis."

The eradication of yellow fever at Havana had taken eight months. At Panama it took nearly a year and a half. But had it not been for Stevens it would have taken considerably longer. Had Stevens been chief engineer from the beginning, doubtless many lives would have been spared; there would have been far less grief and no panic. Once Gorgas' program was under way, incidence of yellow fever fell off with the same dramatic suddenness as at Havana. The epidemic was over by September, when there were only seven cases and four deaths. On an afternoon some weeks later, Gorgas and several of his staff gathered in the dissecting room at Ancon to perform an autopsy. Gorgas told them to "take a good look at this man," for he was the last yellow-fever cadaver they would see. By December the disease had disappeared from the Isthmus.

II

John Stevens ushered in what was to be known on the Isthmus as the Railroad Era. And it is one of the ironies of the story that the unseen guiding spirit as the canal got under way was James J. Hill. Stevens was not only Hill's man, but he would run the work the way Hill ran the Great Northern. Indeed, the building of the Panama Canal was among other things one of the greatest of all triumphs in American railroad engineering.

At the Great Northern the "best-fitted" men were given tremendous authority, then held strictly accountable for results. Familiarity with details was stressed at every level, but obligatory for operating officers. "Intelligent management," according to the familiar Hill dictum, ". . . must be based on exact knowledge of facts. Guesswork will not do." How many best-fitted men might be found among the holdovers from the Wallace fiasco seemed questionable at first. Losses from disease, the pell-mell rush to get away, the very serious difficulty in recruiting replacements, had inevitably meant the advancement to key positions of young men who under normal conditions would probably never have been considered. "Personally, I have always felt grateful to the yellow fever for my first great opportunity in life," wrote Robert E. Wood nearly sixty years later. As a twenty-five-year-old lieutenant he had had "no idea of getting the fever, and did not . . . Anyone who stayed was promoted." Straight, clean-shaven, as square-jawed and forthright as a Charles Dana Gibson hero, Wood was among the first spotted by Stevens as part of his reorganization. Assigned to the Department of Labor and Quarters, Wood was later to become Chief Quartermaster of the Zone; later still he would become General Wood and ultimately the commanding genius of Sears, Roebuck and Company.

Frank Maltby, Wallace's division head at Colón, was summoned to Panama City soon after Stevens got settled. "We sat on the veranda under a full tropical moon. . . . Everyone else disappeared," Maltby would write, recalling the interview. "Mr. Stevens did not talk much but asked questions–and could that man ask questions! He found out everything I knew. He turned me inside out and shook out the last drop of information. . . . At 1 A.M. we retired." The following day Stevens cabled Washington to say that the man he had in mind for the position at Colón was no longer needed; Maltby would do. Maltby, long-legged and sallow, was given the following guideline: "You won't get fired if you do something, you will if you don't do anything. Do something if it is wrong, for you can correct that, but there is no way to correct nothing."

Wallace's ranking engineer at Culebra, a glum, scowling man named W. E. Dauchy, was also kept on, but he would last only a short time longer and with few exceptions all the rest of the new regime would be composed of experienced railroad men brought in by Stevens.

Whereas Ferdinand de Lesseps had failed to see the project as fundamentally a railroad problem and neglected to send a single railroad specialist to Panama, Stevens never saw it as anything other than that, and he recruited railroad men only. Jackson Smith, young, ill-mannered, efficient, had been in railroad construction for five years in Mexico and Ecuador; he was put over Lieutenant Wood as head of Labor and Quarters. William Belding, the new chief of building construction, had been in charge of much the same thing on the Illinois Central. Edward J. Williams, the new disbursing officer, had been paymaster of the Chicago and North Western. A former general auditor of the Oregon Railway and Navigation Company became the new head of accounting. A general storekeeper for the Great Northern became the new chief of Materials and Supplies. William Grant Bierd, brought in to run the Panama Railroad, had been with Stevens at the Chicago, Rock Island and Pacific. He replaced the illustrious Colonel Shaler, who was quietly banished to permanent retirement.

The one man ever urged upon Stevens by Washington—by both Taft and Shonts—was of no interest to Stevens because he was not a railroad man. Taft thought he had found the perfect assistant engineer, someone "exceedingly able," as he told Roosevelt, who could in fact replace Stevens, should Stevens ever take it into his head to do what Wallace had. This was Major George Goethals. Shonts described him in a letter to Stevens as the top construction engineer in the Army, suitable in every way—"direct, resourceful, energetic, and a worker of the most pleasant personality." Taft even brought Goethals to the Isthmus on another of his inspection tours. But Stevens would have none of it, and so Major Goethals returned to Washington.

Stevens saw at once, as the French had not, that the Panama Railroad was the lifeline along which not only men, food, supplies, everything needed to sustain the work, would have to move freely and efficiently, but the Culebra dirt trains as well. He also saw that there was no sense in working with anything less than the biggest, heaviest equipment possible. The French had tried to improve their output by continuously modifying their "plant," using different kinds of equipment in different combinations; but generally it was all too small, too light for the size of the task. "Now I would liken that [French] plant to a modern one as baby carriages to automobiles," Stevens observed. "This is no reflection on the French, but I cannot conceive how they did the work they did with the plant they had."

The track itself, to begin with, was too light. By his standards the railroad as it stood was a pathetic toy. Equipment on the Great Northern was four times the size of that used on the little jungle line. (Hill had been the first railroad baron to equip his road with large-capacity freight cars and monster locomotives.) So within a year the line was completely overhauled and double-tracked with heavier rails. Bridges were strengthened, signals and sidings were improved upon, equipment was rehabilitated or replaced. A new telegraph and telephone system was installed, using old rails for poles. Warehouses and repair shops were built and enormous locomotive sheds were put up at Matachín. Orders were placed for freight cars, dump cars, refrigerator cars, more than a hundred locomotives, all to be shipped "knocked down," then reassembled on arrival in the new shops.

To run the line an entirely new force was brought in—yard and train masters, superintendents, dispatchers, master mechanics, and what Stevens described as "an army of conductors, engineers, and switchmen."

Until Colón's new water system was completed, he used the railroad to run trainloads of clean water into the city night and day. The railroad fed the work force, it ran the commissaries and it ran the Panama Steamship Company by which the food was shipped from New York. A tremendous cold-storage plant was built in conjunction with new terminal facilities at Cristobal. Perishable foods were soon being delivered on a regular schedule along the line every morning.

The men rebuilding the railroad, those building the new towns beside the railroad, began enjoying such luxuries as fresh eggs, lettuce, dressed meats—*ice*. A bakery was built capable of producing forty thousand loaves of bread per day.

There was no building construction, no construction enterprise of any kind not associated with the railroad. It was as if all the activity of the usual large-scale railroad project, activity normally strung across vast open space, had been compressed into this one narrow fifty-mile corridor, with the result that everything seemed tremendously intensified. The size of the labor force was tripled in six months after Stevens took over. By the end of 1906 there were nearly twenty-four thousand men at work, more than there had been working on the Union Pacific in the final race to finish at Promontory, Utah, more than there had been at any time during the French years at Panama. For several months he had twelve thousand men doing nothing but putting up buildings.

Again, as during the French effort, the labor force came from every part of the world—ninety-seven countries according to the records—

but again the unskilled pick-and-shovel workers were nearly all black men and this time it was Barbados, rather than Jamaica, that supplied the majority. Because of the suffering experienced by those Jamaicans left stranded on the Isthmus when the de Lesseps venture collapsed, the Jamaican government refused to allow any recruiting on the island and imposed a tax on anyone desiring to leave to work on the canal. As a result those Jamaicans who immigrated to Panama did so of their own volition and were mostly skilled artisans—those who could afford the tax.

On those islands where recruiting was permitted, all workers were given a contract by which they received free passage to Colón and were guaranteed free repatriation, if they so chose, after five hundred working days (roughly a year and eight months). Martinique and Guadeloupe accounted for some 7,500 men all told, but the total from Barbados was to be nearly 20,000.

Wages were ten cents an hour, ten hours a day, six days a week. Segregation by color, long an unwritten rule on the railroad, as well as in Panamanian society in general, became established policy. There were separate mess halls for blacks. Housing, schools, hospitalization, were separate and by no means equal. And it remained a "Jim Crow" railroad, though restrictions were never hard-and-fast or enforced. Travel on the line was either first or second class, and while most whites rode first class and most blacks second, low-paid white laborers frequently chose second class, just as higher-paid skilled blacks sometimes traveled in the first-class cars.

In all official rules and documents, on signs in post offices and other public places, the color line was expressed in "gold" and "silver" rather than black and white, these designations having been derived from the pay system. Pay for the unskilled work force was in Panamanian silver —balboas, as the standard coins were called. Pay for Americans, on the other hand, was in gold, that still being the monetary standard of the United States. And since nearly all the unskilled workers were black and since virtually all the Americans recruited were skilled workers and white, the terms "gold roll" (gold payroll) and "silver roll" came to be used more or less synonymously for skilled whites from the United States (who might be anything from a steam-shovel engineer to a postal clerk to a nurse to a division engineer) and for black unskilled British subjects (most likely from Barbados and most likely illiterate; but who could also be an educated, French-speaking artisan from Martinique, a Swiss surveyor, or an illiterate Spaniard or Italian).

• •

Recruiting offices were opened in New York and New Orleans and recruiting agents were sent out from Washington to rove the country in search of men. The variety of skills and trades suddenly in demand was enormous. Stevens' own estimate for the upcoming year was for 4,892 American workers skilled in some forty different specialties—bricklayers, blacksmiths, boilermakers, conductors, cooks, car inspectors, car repairers, ship captains, carpenters (1,710 carpenters was the specific request), coppersmiths, calkers, dredge operators, hand-drill operators, steam-drill operators, helpers for the steam-drill operators, engine dispatchers and their helpers, firemen, ironworkers, lithographers, locomotive engineers, locomotive foremen, molders, masons, marine engineers, machinists, plumbers, plasterers, pattern-makers, painters, pipe fitters, riggers, shipwrights, steam-shovel engineers, steam-shovel cranemen, steam-shovel firemen, stationary engineers, timers, watchmen, waiters.

Free transportation to the Isthmus was offered; free housing and medical treatment were part of the enticement. The average pay per month was $87.

The results, however, were disappointing. Too often applicants were those unable to hold a job. These were prosperous years at home and the bad publicity attending the yellow-fever scare had greatly undermined whatever patriotic or romantic appeal Panama might otherwise have had. Instead of 4,892 skilled workers, Stevens got 3,243, and within a year or so, more than half of these would find life in Panama more than they had bargained for and would quit and go home.

Though the idea of building the canal entirely with American labor, unskilled as well as skilled, received some consideration in the early stages and was described by the press as the proper course idealistically, it was never taken seriously. Unskilled white workers, even those at the very bottom end of the pay scale, had no desire to go to Panama. Union leaders strenuously opposed any wholesale shipment of men to "that deathtrap" and particularly after an inspection team from Japan, representatives of large contractors of Japanese labor, reported the Isthmus too unsafe to risk the lives of their men.

What was needed for the heavy physical work, according to the accepted doctrine, were battalions of men who by nature and habit could withstand the punishing climate: black men from the West Indies. That black North Americans might also serve—as General Ben Butler once proposed to Abraham Lincoln—was taken into account, but this too met with strenuous opposition from southern congressmen who foresaw their home states suddenly drained of their natural supply of cheap labor.

The comparative inefficiency and technical ignorance of the West Indian became a source of terrible aggravation for the American engineers and foremen, not a few of whom were naturally prone to scoff at any black man and particularly if he had a singsong British accent. By Stevens' own estimate the efficiency of the average West Indian was about one-third that of an American laborer, white or black.

Reporters were told of the West Indian's "childish irresponsibility," that he was "wasteful . . . stupid . . . possessed with unutterable hatred of exertion other than conversation." And reporters from their own observations reached much the same conclusion. A writer for the popular *Outlook* magazine declared that in all his weeks on the Isthmus he had never once seen a West Indian swing a pick properly, that "their dullness is almost beyond belief."

> It does not matter whether they are digging a drainage trench in Colon, or laying tracks at the very bottom of the Great Cut, or breaking up the ancient cobblestone pavements of Panama. Watch them work for but a single day and you are puzzling over the worst problem that faces our engineers. The only labor they can find in the Western Hemisphere for building the canal has less than one third the efficiency of our labor of the North. The West Indian's every movement is slow and bungling; every small object a subject for debate; anything at all a sufficient excuse for all hands to stop work. A slow upward look from one or two of a gang is usually the only sign that they have heard the foreman's yell, for there is no change in pace or manner of work.

Still, the same writer could see a "certain and unjustified cruelty" in forcing "poor half-fed fellows" to work eight to ten hours in such heat. "Until you have tried to do a good fifteen minutes' work with a pick and shovel during the rainy season . . . you can have no idea of the exhaustion that tropical heat brings even to the laborer who is used to it."

From his experience in the West, Stevens preferred contract Chinese labor gangs above all other choices and he wanted to bring Chinese to Panama in the shortest time possible. Consequently, bids were invited on contracts to furnish up to fifteen thousand Chinese at a pay scale the same as that of the West Indians. But the prospect of wholesale shipments of coolie labor into the Canal Zone by a government that had excluded the importation of such labor since 1882 was received at home with what Stevens called "the customary outcry." The Panamanians protested even more strenuously. In reaction to the success of Chinese merchants in Colón and Panama City, many of them descendants of Chinese laborers left over from earlier projects, the new republic had enacted its own Chinese exclusion law. Moreover, the government of China protested as well, its view being the same as that of the Japanese contractors.

So the matter was dropped. Stevens, infuriated by the politics of the incident, failed to comprehend, he said, what difference it made whether a laborer was white or black or yellow; or why some people would rather spend millions on one variety rather than another, when the performance of the other was so plainly superior.

As an experiment, he had several hundred unskilled workers brought over from the Basque Provinces of Spain. The physical endurance of these men, their effectiveness whenever gang labor was called for (such as in moving railroad track), proved so exceptional that he imported nearly eight thousand more and paid them twice what the West Indians were getting, a policy he justified on the grounds that they worked three times as hard.

But recruitment at Barbados went ahead as rapidly as possible, under the direction of William Karner, Wallace's former assistant. Work of any kind was extremely scarce on the vastly overpopulated island. The mass of the populace, black and desperately poor, survived primarily on a few months of planting and harvest on the sugar plantations, when an able-bodied man could earn about twenty cents a day, the same as he could earn in Panama in two hours. So for every man who was picked to go to Panama there were five or more others eager for the chance. On the days Karner opened his recruiting office off Trafalgar Square in Bridgetown, the police had to be on hand to keep the crowds in order.

Examinations were conducted in a large, bare loft. The men, in batches of a hundred-odd at a time, were formed in a line around the wall. Any who looked too old, too young, or too feeble were told to leave. The others were checked first for trachoma, then were told to strip, after which they were gone over for tuberculosis, heart trouble, and rupture. By the time the process was finished, only about twenty men would have passed. A correspondent who watched one such session wrote that he had never seen a more serious-looking body of men until the doctor told the remaining twenty they had been chosen to go. The change was immediate; they started to shout and dance about, clapping one another on the shoulders.

> A flood of light came in through the window at the end, and many streaks shot down through the broken shingles on their naked bodies. It was a weird sight—something like a war dance—as they expressed their relief . . . It meant semi-starvation for themselves and their families if they were rejected, and untold wealth—a dollar a day—if they passed. They were all vaccinated . . . their contracts signed, and they went prancing down-stairs to spread the good news among their friends in the square.

Sailing days for Panama were occasions remembered for years afterward, with thousands of women gathered at the wharf to bid the men farewell. "I never saw so many Negro women in my life," wrote the correspondent. "All of them in their gayest Sunday clothes, and all wailing at the top of their voices." Royal Mail steamers sailed with every inch of space occupied, the number on board generally averaging seven or eight hundred. The nearly twenty thousand men recruited at Barbados during the years of construction represented 10 percent of the island's population and approximately 40 percent of all the adult males. Virtually every able-bodied man went off to build the Panama Canal and the money they sent home to the island was something over $300,000 a year.

For the average West Indian the initial weeks at Panama—the constant movement of men and equipment, the rules, schedules, the confusion and noise—were unlike anything in his experience, often frightening, often highly unpleasant. Most of these men, it must be remembered, had never before seen or heard a locomotive. They were cane-field workers, wholly unfamiliar with modern machinery of any kind. Once, in October 1905, several hundred men inbound from Martinique were so terrified by the prospect of being vaccinated that they refused to leave the ship when it docked at Colón and so had to be put ashore by force. Fifty years later in Barbados, Jamaica, St. Lucia, Trinidad, Antigua, Martinique, St. Vincent, old men would remember being herded aboard their first labor train (". . . when saying train, I don't mean passenger train; it was a boxcar train"); or being put directly to work while still wearing their best suit or a new derby purchased especially for the trip.

They were marched out by the hundreds to dig ditches, to cut brush, to carry lumber, to unload boxcars of dynamite. "I load cement, I unload cement," remembered one of them. "I carry lumber until my shoulder peel." Previous training or trade skills were generally ignored, former schoolteachers, and skilled craftsmen were made messengers and waiters, experienced carpenters were put to work cutting points on the ends of stakes for the engineers. Rarely did a black man ever rise to a supervisory level and never over white men.

At the camps each man was assigned to a tent or to one of the new barracks and was given a tin plate, tin cup and spoon, and a brass number tag. Fifty years later most of them would still be able to recite their number. The food provided would be recalled by some as sumptuous (". . . corn beef, bread, coffee which we enjoyed . . . bread, sardines, and ice cream . . . and never forget our ice cream, I am saying here it was refreshing"). But most of such declarations date from a later time. In fact, the food from the mess kitchens in the labor camps appears to have been quite dreadful to begin with, or at least bad enough that in 1906 some sixteen thousand of the labor force preferred to fend for themselves, cooking their own meals in iron pots. Indeed, so little did the West Indians care for the food and the housing provided, so great was their distaste for the regimentation of barracks life, that not more than one in five would stay on in the camps, while the rest crowded into the slums of Colón and Panama City or put up their own ramshackle huts in the bush, exactly as in the days of the French canal.

Shortly after his own arrival on the Isthmus, John Stevens had watched three West Indians at work with a wheelbarrow. When the wheelbarrow was full, two of them hoisted it onto the head of the third man who carried it away. The scene was one he would often use to depict the variety of problems he faced. But it was Stevens who also said the West Indian would learn rapidly if given the chance and who suggested that the West Indian's diet might explain his comparative lassitude. On both counts Stevens was to prove correct. The West Indians did become increasingly proficient with tools and at working in unison and in association with heavy machinery, as many of them would recount afterward with pride. The replacement of their traditional high-starch, low-protein diet (chiefly rice and yams) with more nourishing meals did have an effect. Furthermore, while the output of the West Indians improved, that of the Spanish workers declined; they became gradually less industrious, less able to withstand the climate. In time there would be no appreciable difference in the efficiency of one group as compared to another. "The West Indian, while slow, has learned many of the trades and many of them have developed into first-class construction men," Robert Wood was to write afterward in his final official report. "The bulk of the building work on the Canal has been done by West Indian carpenters, masons and painters . . . and toward the end of the construction period the West Indian remained on the job as steadily as the Spaniard or even the American."

The racket of hammers and saws could be heard from one end of the line to the other. Stevens remarked that there was not a half mile between Colón and Panama City that did not show signs of the "tremendous activities" of his building department, a claim that if not literally true certainly agreed with everyone's impression. Under the previous chief engineer, in the one year John Wallace was in charge, a total of 336 old French structures had been renovated, 150 new buildings put up. In one year under Stevens, 1,200 structures were renovated, 1,250 new ones built.

Mountains of supplies were gathered; the worst of the old French wreckage was hauled off for scrap or dumped into swamps for fill. And while much of what went on seemed perfectly bewildering, even to the trained eye—to Major Goethals, for example, it had all looked like utter chaos—one new community after another gradually took shape, and Stevens behaved always as if everything was progressing in the smoothest and most orderly fashion.

"In his office, his desk was always clear," Frank Maltby wrote, "and apparently he had nothing to do." Reporters described him as "the type who . . . always has time on his hands," who was never a day behind in his correspondence.

Concerned greatly about the morale of the skilled American workers, disturbed by the continuing turnover among them as men gave up and left, he had clubhouses built, arranged for weekly band concerts, established a baseball league, with each settlement along the line organizing its own team. When a young clerk informed him that no funds had been allocated for building baseball fields, Stevens said to charge them to sanitary expenses.

Married men on the gold roll were encouraged to send for their wives and families as soon as housing became available, or if not married, to find a wife at the earliest opportunity. To avoid disputes or rivalry over accommodations, it was decided that each man should get one square foot for every dollar of his monthly pay. The rule applied to bachelor and married quarters alike, but wives were also entitled to a square foot per dollar earned by their husbands.* Devised and enforced by Jackson Smith, who was hence known thereafter as "Square-

* For dependents other than wives the rule became a bit more complicated: each child qualified for 5 percent of the father's base allotment for each year of the child's age (a ten-year-old thus rated 50 percent of the base allotment), while all adult members of the family other than the wife rated 75 percent.

foot" Smith, the rule established a standard understood by everybody. It also provided a strong incentive for advancement and especially with the arrival of increasing numbers of wives.

Most of the new houses were two stories, with two or four apartments each, and were enclosed completely with screened verandas. They were big, plain, pine-clapboard buildings that stood well up off the ground and were painted gray with white trim. Roofs were of corrugated iron. They were not the least fancy, but with their high ceilings and long windows on all sides, they were suited perfectly to the climate.

Each apartment was equipped with modern plumbing and was furnished at government expense. Coal for cooking, ice for the "icebox," water, electricity, garbage disposal, maintenance, grounds keeping, were all provided free of cost. The bachelor "hotels," big, rambling affairs that looked like any medium-priced summer hotel on the New Jersey shore, were kept clean by full-time janitors, but married employees were "obliged" to look after themselves.

In the eyes of a professional engineer such as Stevens, the canal in certain respects was a simpler undertaking than other less conspicuous engineering projects of the era. There was plenty of space within which to work. There were no property rights to worry about along the line of construction, no possibility of damage to existing buildings, no outside traffic to contend with. The labor force was at hand; only the steam-shovel and locomotive engineers were unionized; there were no contracts to live up to, and never any question about the money supply.

Nor were any radical or untried technical concepts necessary to handle the excavation. Most of what needed to be done had been done before.

What made the undertaking so exceptional was its overwhelming scale. "There is no element of mystery involved in it," Stevens reported to Washington, ". . . the problem is one of magnitude and not miracles."

Greatly compounding this problem of magnitude was, of course, the enormous primary task of approximating the conditions of a modern industrial community in an equatorial wilderness two thousand miles from the base of all supplies. When various of Stevens' subordinates wrote afterward that he laid the foundations for the work, it was the startling advances in housing, health, supply, his dramatic marshaling of men and machines, that they had in mind.

In more abstract terms, in terms of pure professional problem solving, Stevens' greatest contribution was the basic vision of the excavation of the canal as a large-scale problem in railroad freight. As conceived by Stevens, the Panama project was simply one of moving unprecedented tonnage—dirt—by railroad with the least possible wasted motion.

The "overshadowing" challenge would be Culebra Cut. In a letter to Shonts, with his own kind of blunt eloquence, Stevens said what no one had, but what had needed to be said for a very long time:

> Yet we must reflect that at best, even with the backing and sentiment and finances of the most powerful nation on earth, that we are contending with Nature's forces, and that while our wishes and ambition are of great assistance in a work of this magnitude, neither the inspiration of genius nor our optimism will build this canal. Nothing but dogged determination and steady, persistent, intelligent work will ever accomplish the result; and when we speak of a hundred million yards of a single cut not to exceed nine miles in length, we are facing a proposition greater than was ever undertaken in the engineering history of the world.

One recent proposal was to wash that whole section of the divide into the Pacific, using tremendous blasts of water, as in hydraulic mining. Another, equally fantastic, was to build a huge compressed-air plant in the Cut and blow all the spoil to the sea through vast pipes. Stevens' objective was to create a system of dirt trains that would function like a colossal conveyor belt, rolling endlessly beside steam shovels working at several levels at once. And his success, he knew, would depend on how well things could be managed at the disposal end of the system.

He would haul the dirt to either coast, or to both, or to wherever it was needed for fill. If a high-lake lock canal was decided on, then Culebra could supply the material to build the necessary dams. By double-tracking the railroad he had provided open access in both directions without interrupting regular traffic on the line. The distance from the point of excavation to the dumping grounds was immaterial. It made no difference whether the dirt had to be moved ten feet or ten miles. The trick was to keep the dirt trains in constant motion in and out of the Cut, to and from the dumps.

As possibly no other engineer could have, he devised an elaborate, yet ingeniously elastic system of trackage within the Cut whereby loaded trains would roll out on a downgrade and trains of empty cars would be constantly available to serve the steam shovels. For a shovel to perform at maximum efficiency, the boom had to be swinging every possible minute; and this, as he stressed, could be accomplished only by maintaining a steady supply of empty cars.

By early 1906 he had his plans far enough along and had sufficient equipment in line to resume excavation. Day after day he trudged about among the men and machines, asking questions, observing, smoking cigars like Grant at the Wilderness, as a reporter noted. The men called him "Big Smoke."

III

The summer before, when he first arrived in Panama, Stevens had assumed that he would be building a sea-level canal. He had come to the job, he later wrote, with a picture in mind of a "wide expanse of blue, rippling water and great ships plowing their way through it like the Straits of Magellan." It was the age-old preconception, the dream of Columbus, the vision that had dominated at Paris in 1879 and that persisted still in both the popular and official imagination, irrespective of the French experience. Authorized by the previous commission to design an official seal for the Canal Zone, Tiffany & Company, after much historical research, had prepared a shield upon which a Spanish galleon under full sail could be seen traversing an open strait between steep embankments, on into the Pacific, the sky aglow with a tropical sunset. Beneath the shield was inscribed the motto: "The Land Divided—The World United." And both design and inscription had been approved.

That the proper canal to build even remained an issue at so late a date was in itself a serious and immensely bothersome handicap for Stevens. In his position the French directors general had at least known what was wanted of them. There was of course much that he could do in the way of preparation, work that would be applicable regardless of the final decision, but only up to a point. As he would explain to John Tyler Morgan, it was "as though I had been told to build a house without being informed whether it was a tollhouse or a capitol."

Now, in addition to everything else, he would be required to play a political role, a role he claimed to detest and for which he felt ill equipped. Yet in restrospect, it is hard to imagine anyone doing better than he did, and at the close of a long life, he himself would look back upon it as his greatest single service.

The special international board appointed by Roosevelt to consider the problem was composed of eight Americans and five Europeans.

Chairman of the group was General Davis, the former Governor of the Canal Zone. Others included Professor Burr and William B. Parsons, from the first Isthmian Canal Commission; a former member of the Walker Commission, Alfred Noble; and General Henry Abbot, from the old Comité Technique. Joseph Ripley was chief engineer of the St. Marys Falls Ship Canal—the "Soo" Canal, as it was better known—at Sault Ste. Marie, Michigan. Frederic P. Stearns was chief engineer of Boston's Metropolitan Water Board; Isham Randolph was chief engineer of the Sanitary District of Chicago.

The foreign members were a *chef des Ponts et Chaussées*, Adolphe Guérard; William Henry Hunter, chief engineer of the Manchester Canal, who also had served on the Comité Technique; a Prussian state engineer named Eugen Tincauzer; E. Quellenac, consulting engineer of the Suez Canal; and J. W. Welcker, director of all Dutch waterways.

It was another distinguished panel drawn from an international, professional upper crust, which for Stevens, with his lack of education, was something of a world apart. At its first meeting in the I.C.C. offices in Washington, September 1, 1905, the board was presented with numerous past reports, volumes of current data, as well as assorted proposals deemed deserving of attention (including one from Philippe Bunau-Varilla, who insisted still that the surest approach was to build a lock canal and by means of dredges take it down to sea level). Roosevelt gave a lunch for the group at Oyster Bay and told them he hoped it might be a sea-level passage, but warned that time and practicality must be kept in mind. It was his vital interest to "secure a Panama waterway" in the shortest time possible. The plan must be one that would work.

John Findley Wallace returned to Washington to argue that no plan should be approved that might prevent the ultimate creation of the "Straits of Panama." And at the end of September the board went to the Isthmus to tour the line under abnormally sunny skies. In Limon Bay, in a stateroom on their ship, they interviewed Stevens, Maltby, and others from Stevens' staff. By the time of their twenty-fifth meeting in Washington, November 18, they were prepared to cast their vote. The decision, by a margin of eight to five, was for a sea-level canal. Without exception the European members wanted it that way and they were joined by Chairman Davis, Professor Burr, and William B. Parsons.

In language and logic the case as presented by the majority had a very familiar ring: the setting might have been the *grande salle* of the

Société de Géographie in 1879. Speaking for the majority, Chairman Davis declared that he had known since boyhood that Suez and Panama would be "overcome" one day; a passage at Suez had been declared impossible, a passage like that at Suez must be built in Panama. Again the issue was one of national pride and honor. Only in this instance the spokesman was an American official who knew Panama from personal experience and who no less than any of the others, presumably, should have had little difficulty understanding what had happened to the French and why. But the model was not the French canal at Panama; the model—again—was the French canal at Suez. To those in opposition, to anyone familiar with the history of the French experience on the Isthmus, the views of Chairman Davis seemed like the return of a bad dream.

> The task that confronted the private company [at Suez] . . . measured by the difficulties they had to encounter, was many times greater, it seems to me, than the task which, measured by the standard of engineering methods and capabilities that exist today, confronts the United States at Panama; but the French company carried that work to completion at Suez thirty-six years ago, and it has yielded enormous profits. The many difficulties were overcome and an open waterway was made 100 miles long, affording unobstructed navigation. . . .
>
> We know that at Panama to make a waterway similar to the Suez Canal we must construct a channel less than half as long. . . . Should the United States withdraw from the attempt to make at the American Isthmus a channel as open, free, and safe as already existing at Suez? Should they climb over the hill or remove it? . . . I think the dignity and power of this great nation . . . require that we should treat this matter not in a provisional but in a final, masterly way.

Control of the Chagres River—"the lion in the path," Davis called it—could be easily attained.

A mountainous *Report of the Consulting Engineers for the Panama Canal* was delivered to the White House on January 10, 1906. The recommended sea-level canal was to cost $247,000,000 and to be completed in twelve to thirteen years, which was approximately $100,000,000 more and three to four years longer than required for a lock canal of the kind proposed by the dissenting members.

This minority proposal was for a canal much like the one recommended by the Walker Commission; it was, that is, essentially the same canal for which the Spooner Act had been passed, a canal that would not divide the land, but bridge it with a high-level lake reached by flights of locks at either end. It had, however, one major difference.

The site of the Chagres dam had been moved from Bohio downstream to Gatun, to within four miles of Limon Bay. What had been Lake Bohio in the earlier plan now became a much larger Gatun Lake. The span of the water bridge had been extended nine miles.

The elevation of the lake was to be eighty-five feet. At Gatun there would be a single flight of three locks built into the eastern end of the dam. A ship entering the locks would be lifted to the level of the lake, then proceed twenty-three miles across the lake, south to Culebra Cut, which, like the neck of a bottle, extended for nine miles through the divide and was capped by another small dam and one lock at Pedro Miguel. There the ship would be lowered thirty-one feet to a small terminal lake, another body of fresh water, this one being contained by a dam at La Boca, beside Sosa Hill, at the edge of the Pacific. Descending through two more locks, the ship would return to sea level and thus complete the ocean-to-ocean transit.

The model for the plan, its proponents stressed, was the Soo Canal, which for fifty years had been the gateway between Lake Superior and Lake Huron. There was no more heavily traveled canal in the world. By contrast to the 105-mile-long sea-level passage at Suez, the Soo was all of a mile and half from end to end. Yet the annual tonnage through the locks—44,000,000 tons in 1905—was more than three times that of the Suez Canal, even though the Soo was closed by ice during the winter. In season, huge Great Lakes ore boats moved through with an efficiency and safety that belied all the customary arguments against lock canals. No vessel had ever been seriously injured in the locks of the Soo, not in fifty years of constant traffic. "Danger to ships in a canal is not at the locks, where they are moving slowly and under control, but in the excavated channels . . . through which they pass at speed, and where if the width is insufficient, groundings are likely to happen." The experience gained at the Soo was not only applicable to navigation at Panama, but of more value than any or all experience related to any other canal, according to the minority report, most of which was written by Alfred Noble, who in his earlier years had helped build the so-called Weitzel Lock on the Soo and who presently, at age sixty-one, was one of the two or three leading engineers in the country. (As chief engineer of the East River division of the Pennsylvania Railroad from 1902 to 1909, Noble was responsible for tunnel construction under the river and for the foundations of Pennsylvania Station in New York City.)

Yet in its essentials this latest high-level lake plan for Panama was no different from that proposed by Godin de Lépinay in Paris twenty-seven years before. Gatun, it will be recalled, was the site specified by de Lépinay for the Chagres dam.

There had been others in the interval who had also seen Gatun as the most suitable place to check the river. Two interested Americans, C. D. Ward and Ashbel Welch, had each presented papers on the subject before the American Society of Civil Engineers. But because the breadth of the valley was far greater at Gatun than at Bohio, a much larger dam would have to be built at Gatun, thus making it an even more controversial project than the one at Bohio had been. This "controlling feature" in the new proposal was to be a mountain of earth nearly a mile and a half long (7,700 feet) and more than 100 feet high. And while earth dams of nearly the same size had been built with success elsewhere, this would be the highest on record and the Gatun site appeared to offer little if any bedrock upon which to found such a structure.

Stevens was chief among those who now backed the Gatun plan. Recalled to Washington to give his views, he met with Shonts, Taft, and Roosevelt and made a memorable appearance before the Senate canal committee. Stevens, however, was not the "architect" of the lock plan, as later claimed by some of his more ardent admirers. As recently as October, when appearing before the advisory board in Colón, he had in fact quite stubbornly refused to endorse any plan for the canal, saying he was too new to the work. The present lock plan was the work of Alfred Noble and Joseph Ripley, and was based largely on their experience on the Soo. Stevens, whose background was in railroad construction only, had no knowledge of lock construction or hydraulics. The real "architect" of the present plan, if such is an appropriate designation, remained Godin de Lépinay.

Still, Stevens had experienced a revelation since October. He had seen the effect of the rains; he had seen the Chagres in flood. In conversations with Maltby and others who had served under Wallace, he had found none who favored a sea-level canal. Stevens had once believed like others that a sea-level canal "meant simply digging a little more dirt." Now he saw that the issue was one of the most momentous consequence and he could not have been more partisan. To his mind any sea-level plan for Panama was "an entirely untenable proposition," "an impracticable futility." The sea-level passage advocated by the majority of the board was to be only 150 feet wide for nearly half its length—"a narrow, tortuous ditch." He foresaw endless landslides, a precarious transit under the best of conditions. Whenever two ships passed in so narrow a channel, one would have to make fast to mooring posts, as at Suez. Even if there were no difference in cost or time of construction, he would still prefer the lock plan.

> It will provide a safer and quicker passage for ships. . . . It will provide, beyond question, the best solution to the vital problem of how safely to care for the floodwaters of the Chagres. . . . Its cost of operation, maintenance and fixed charges will be much less than any sea-level canal.

The estimated time of completion for a lock canal was nine years. He thought it could be done in eight years, by January 1914. He doubted that a sea-level canal could be built in anything less than eighteen years, or not before 1924.

As a witness on Capitol Hill, he was particularly impressive, answering all questions with characteristic confidence. Predictably, much anxiety had been voiced in the press and in Congress over the prospect of risking everything on an earth dam. Still fresh was the memory of the Johnstown Flood of 1889, when an entire city had been wiped out and more than two thousand lives lost as a result of the failure of an earth dam.

Stevens assured the committee that if properly engineered an earth dam would serve perfectly and that such suggested reinforcements as a masonry core could be dispensed with.

"Yes, if it is absolutely safe," one senator replied. "Here I suggest that that is a very positive opinion or conviction that you have."

"Well, I am a positive man," Stevens said.

His most persistent interrogator, and easily the most intelligent in his private opinion, was John Tyler Morgan, who at age eighty-one had a little more than a year to live, but who at this juncture was still going strong. At the close of the hearings, Morgan came up to Stevens and told him, "If we had [had] you on our side, the canal would be built at Nicaragua."

In November, before Stevens reached Washington, the New York *Tribune* made front-page news with an unauthorized report that Roosevelt wanted a lock canal because that was Stevens' choice. According to Stevens, however, it was only because he "talked to Teddy like a Dutch uncle" after arriving in Washington that Roosevelt swung around to favor the lock plan.

On February 5, in response to Stevens' views, the Isthmian Canal Commission overrode the majority opinion of the advisory board and chose the lock canal. Two weeks later, when submitting the reports to Congress, Roosevelt gave the lock canal Presidential sanction. It was, he said, the canal the chief engineer wanted and of all men the chief engineer had "a peculiar personal interest in judging aright."

No sooner had Stevens returned to Panama than he learned that he might be needed again to lobby further on Capitol Hill. He protested to Shonts by cable declaring that he had said all he could. As Shonts also knew, Stevens suffered severely from seasickness and dreaded every trip to or from Panama. By April the issue was still tied up in committee and Stevens, infuriated over the "vexatious manner" in which things were being handled in Washington, kept cabling Taft to do something. Professor Burr, William B. Parsons, and John Wallace had appeared before the Senate committee and denounced with notable conviction virtually every feature in the lock proposal. So Stevens was called back to Washington.

On May 17, by a margin of one vote, the Senate committee reported for the sea-level plan. Stevens, who was never known to complain of the heat at Panama, would remember the rest of his life how "for two blistering hot days" he "withstood the severest" questioning before the House Committee on Interstate and Foreign Commerce. Determined to make Congress and the country understand the nature of the problem, he kept hammering at the same fundamental idea that de Lépinay had failed to put across before the gathering in Paris. "The one great problem in the construction of any canal down there is the control of the Chagres River," he insisted. "That overshadows everything else."

He talked to congressmen, assembled statistics, and prepared a large map for display in the Senate. Most important, he drafted large sections of what was to be the major speech in the Senate, an address by the bantam-sized Philander Knox, Roosevelt's former Attorney General, who had since become senator from Pennsylvania.

In the speech, as in the lock plan itself, Gatun Dam was the focal point. Knox assured the Senate that the dam was safe. Interestingly, Knox had once been part-owner of the ill-fated dam at Johnstown, a connection that somehow eluded notice in 1906. Knox happened also to live in Pittsburgh and his personal fortune, as well as the influence of his law firm, had been built on legal services in behalf of the Pittsburgh steel empire and its leaders. In the plan as it presently stood, a total of six double locks was called for and these would require gigantic gates—gates that would be built of steel. This was a point neither Knox nor anyone else happened to raise publicly, and how strenuously the steel interests may also have been lobbying for the lock plan is impossible to determine. But the great lock gates for the Panama Canal would in fact be fabricated in Pittsburgh one day and they would be erected by a Pittsburgh contractor.

Again, as in the "Battle of the Routes," the issue was resolved in the Senate and by the narrowest of margins.

Knox spoke on June 19. Two days later the Senate voted 36 to 31 for a lock canal. A difference of three votes would have caused the United States to attempt a sea-level canal, which in all probability would have ended in terrible failure. As further experience would demonstrate, the advisory board had been no less naïve than de Lesseps concerning the true cost of so vast an excavation in terms of both money and time. George Goethals was to remark at one critical point that there was not money enough in the world to construct a sea-level canal at Panama.

Still in the opinion of a very large number of people, including a great many technical and military specialists, a dreadful error had been committed. And it was a view that would persist for years to come. Roosevelt was to be told to his face by no less a figure of valor and resolution than Lord Kitchener that he had blundered shamefully, Kitchener's contention—expressed in a very loud voice—being that a sea-level canal was the only proper canal, as any sensible person could perceive. When Roosevelt countered that there were too many technical difficulties involved, Kitchener answered, "I never regard difficulties, or pay heed to protests like that; all I would do in such a case would be to say, 'I order that a sea-level canal be dug, and I wish to hear nothing more about it.' " Roosevelt responded by saying, "If you say so, I have no doubt you would have given such an order; but I wonder if you remember the conversation between Glendower and Hotspur, when Glendower says, 'I can call spirits from the vasty deep,' and Hotspur answers, 'So can I, and so can any man; but will they come?' "

For the engineers on the Isthmus the decision was a great parting of the clouds. They knew at last the canal they were to build. Plans prepared in expectation of the decision could immediately be put into effect. Stevens had left specific instructions. Though he did not get back to Colón again until July 4, construction of a new town at Gatun was started within twenty-four hours after the Senate vote. Clearing the site for Gatun Dam was begun; tracks were laid for the dirt trains from Culebra.

Residents of the old village of Gatun, hearing that the dam was to be built where the village stood, refused to concern themselves, let alone move to the new site that had been provided for them. Such impossible things had been spoken of by the French some twenty-five years before, they said. So it was not until later, when actual construction began on the dam—and rock being dumped for the foundations crashed over several houses—that they agreed to move out.

On Stevens' return, after careful study of core samples, after much tramping back and forth, he and Maltby fixed the center line of the Gatun locks. Responsibility for the design of the locks and of the dam was put in the hands of Joseph Ripley, the chief engineer of the Soo Canal, whom Stevens had managed to recruit before leaving Washington. As Stevens wrote, "Things began to quicken everywhere."

Surveying parties were sent into the jungle to map and locate the contour line—the perimeter—of what was to be the largest artificial lake in the world. Eventually five such parties were in the field and their work was carried on almost entirely through unbroken jungle. Nearly every foot of the way had to be cleared by heavy cutting by hand. Progress was extremely slow; some parties were out for a year.

The creation of Gatun Lake would mean that approximately 164 square miles of jungle, an area as large as the island of Barbados, would vanish under water. Every village between Gatun and Matachín would be covered over, a prospect that the native populace found impossible to imagine. Mile after mile of the Chagres River, the Panama Railroad, nearly everything along the path of the French, not to mention most of the new towns being built, would be lost beneath the lake. A new railroad would have to be built on higher ground to skirt the eastern shore of the projected lake.

The preparatory period was over. Stevens' headquarters, the entire engineering department, had been moved from Panama City to Culebra, to a steep green bluff looking directly down into the Cut. It had been a year and three months since John Stevens had taken charge.

18

The Man with the Sun in His Eyes

And never did a President before so reflect the quality of his time.

—H. G. Wells

I

Theodore Roosevelt had taken a great liking to John Stevens. Stevens, in addition to his other attributes, was a reader of books, Roosevelt had discovered—"and . . . he has the same trick that I have of reading over and over again books for which he really cares." Stevens' favorite of all was *Huckleberry Finn*, which he read "continually," and this to Roosevelt was the mark of the finest literary discernment.

Roosevelt's one reservation about Stevens was his obvious insensitivity to the fact that the canal was an undertaking of the United States government. Stevens, Roosevelt complained to Taft, seemed incapable of understanding that he was no longer working for James J. Hill. Stevens had painfully little patience with congressmen, or with the Panamanians, and none at all with labor unions. When a delegation of steam-shovel engineers came to his office threatening to strike unless paid more, Stevens reportedly told them: "You all know damn well that strikes do not get you anywhere. Now, get the hell out of this office and back to work . . ." The men had returned to their machines and the story, when it got around, raised Stevens' standing even higher in the eyes of most other Americans on the Isthmus. But the steam-shovel men had sent angry protests to their union leaders at home, who very quickly took the matter to Washington where it wound up on Roosevelt's desk.

But it was the technical executive ability of Stevens and Shonts that Roosevelt valued most. They were "the very best men we could get for actually digging the canal," he told Taft; their administrative abilities were "phenomenal." And so it was in that spirit in the summer of 1906 that Roosevelt addressed himself to two propositions that would greatly change the execution of the work.

Stevens and Shonts wanted to build the canal by contract, as the French had tried, and as the transcontinental railroads had been built. It was the system to which they, as railroad men, were accustomed. Stevens envisioned a powerful syndicate of railroad contractors who "by combining their strength and influence" could bring to Panama the best people in the world to do the job. Taft was certain the plan meant trouble, and especially if selection of the contractors was to be determined by Stevens, which was what Stevens wanted, rather than by open bidding. Stevens was insistent, but Taft held his ground, and in the end Roosevelt agreed to put the work out to contract on a trial basis but only on the condition that there be open bidding.

The second and more important proposition concerned the manner in which the work was being administered. Shonts and Stevens wanted things greatly streamlined. To Shonts it would be "suicidal" to continue with the work without a "clear-cut organization with centralized power." Stevens, in a long letter from Panama dated August 5, 1906, told Roosevelt that "from now on, everything should be made subordinate to construction . . ."

"I believe that the power and responsibility should be concentrated," Stevens wrote, ". . . that the commission, constituted in whatever way it may be . . . must resolve itself into what will amount to a one-man proposition."

More correspondence followed and several meetings, and consequently a new executive order was prepared. The present three-man commission was to be abolished. Authority was to be vested in a single head, the chairman, and he in turn would report to the Secretary of War. On the Isthmus, heads of the major departments—engineering, sanitation, labor—would become members of the commission, reporting to and receiving instructions from the chairman. But since 90 percent of the employees on the Isthmus worked for the engineering department, it was really the chief engineer who had the power.

The chain of command had thus become a straight line, from Roosevelt to Taft to Shonts to Stevens. The office of governor of the Zone no longer existed. Stevens, the chief beneficiary of the change, was to be supreme commander in the field. Governor Magoon, never a favorite of either Shonts or Stevens, happened to be absent from the Isthmus on an emergency legal mission in Cuba. He would be quietly informed that he was not to return to Panama.

Since the administration of so large, complex, and distant an undertaking was a new experience for the United States government, or more specifically for the executive branch, this refinement of the commission was in itself something of a pioneering process, for which Roosevelt was ultimately responsible, although Stevens, perhaps justifiably, would later say he was the one who had mapped it all out. In less than three years' time the commission had gone from a seven-headed board that oversaw all decisions, to the three-man executive body (wherein chairman, governor, and chief engineer each had his own specific responsibilities), to the present arrangement. Had Elihu Root or even Henry Clay Frick agreed to run things earlier, possibly something of the kind would have been arrived at in much less time. Shonts and Stevens had been given exactly the power they wanted.

The order, it was agreed, would be signed when Roosevelt got to Panama.

The trip to Panama to see the canal was one of those small, luminous events that light up an era. No President had ever before left the country during his time in office and so from the day of the first advance announcement in June the journey became the talk of the country. In much of the press, serious apprehensions were expressed, even though it had been stressed that he would be in constant communication with Washington by wireless and that every possible precaution would be taken to insure his physical safety. But by and large the idea of Teddy Roosevelt going personally to Panama, like a general to the front, had tremendous appeal, and on the eve of his departure in November, even the cautious Washington *Star* lent its support. Perhaps it was a good thing after all for a President to get out and see something of the world, the paper declared; conceivably future occupants of the office might even undertake European journeys.

He sailed on November 9, 1906, on the new 16,000-ton *Louisiana*, largest battleship in the fleet, escorted by two cruisers, *Tennessee* and *Washington*. He took Mrs. Roosevelt with him, and Navy doctor Presley Rixey, his personal physician, and three Secret Service men, but no reporters. The ships traveled south at fourteen knots through quiet seas and by the time they reached the Caribbean the weather was ideal.

> Mother and I walk briskly up and down the deck together or else sit aft under the awning or in the aftercabin, with the gun ports open and ready . . . [he wrote to their son Kermit]. Mother, very pretty and dainty in white summer clothes, came up on Sunday morning to see inspection and review, or whatever they call it . . . I usually spend half an hour on deck before Mother is dressed. Then we breakfast together alone . . .

Their quarters were those of the admiral of the fleet, only somewhat enlarged, several walls having been removed for the occasion. Pictures of the rooms, with their wicker chairs and big brass beds and Oriental rugs, had already appeared in the illustrated magazines.

> It is a beautiful sight, these three great war vessels steaming southward in close column [his letter continued], and almost as beautiful at night when we see not only the lights but the loom through the darkness of the ships astern. . . . I have thought a good deal of the time over eight years ago when I was sailing to Santiago in the fleet of warships and transports. It seems a strange thing to think of my now being President, going to visit the work of the Panama Canal which I have made possible.

All together, with the voyage down and back, he was away two weeks. The most memorable part of the visit was the rain. He had picked November because it was the height of the rainy season. He wished to see Panama at its absolute worst, he said, and he was not disappointed. "It would have been impossible to see the work going on under more unfavorable weather conditions," he would report enthusiastically to Congress. It was raining the morning he landed. It was raining as he and President Amador rode through the streets of Panama City in an open carriage, Roosevelt waving a top hat to sodden but exuberant crowds. The deluge the second day was the worst in fifteen years. Three inches of rain fell in less than two hours. He saw the Chagres surge a hundred yards beyond its banks. The railroad was under water in several places. Villages were "knee-deep in water." There was even a small landslide on the railroad cut at Paraíso. The contrast between the Panama he saw and the sunny, benign land toured by Ferdinand de Lesseps could not have been much greater. "I tramped everywhere through the mud," he wrote with satisfaction.

Advance preparations had involved the efforts of thousands of people. As in de Lesseps' day, streets were scrubbed, houses were painted or whitewashed, flags were hung from windows and balconies. Programs were printed; schoolchildren were rehearsed in patriotic airs. The Republic of Panama declared his day of arrival a national day of "joy and exalted enthusiasm" and instructed the populace to behave, since "all thinkers, sociologists and philosophers of the universe [will] have their eyes upon us in penetrating scrutiny."

At Ancon, construction of a big three-story frame hotel called the Tivoli, a structure begun the year before but still far from finished, was rushed ahead with all speed as soon as Stevens learned of the visit. One wing of the building was finished and furnished in six weeks, because Roosevelt insisted upon living on shore and in the Zone.

Predictably, perhaps inevitably, Roosevelt did almost nothing by the comparatively relaxed schedule planned by Stevens. (In a cable from Washington, Taft had advised a full tour but without "overdoing matters.") The white battleship appeared in Limon Bay on November 14, a day before she was supposed to. When Amador, Shonts, Stevens, and their wives rushed by train to Cristobal that afternoon, it was only to learn that Roosevelt would stay on board through the night so as not to disrupt any of their arrangements. At 7:30 the next morning, the appointed hour, as the official welcoming party stood at the end of the pier, all eyes searching for signs of life on the big ship, an amazing figure called "Good morning" from shore. He advanced into their midst. He was wearing a white suit and a seaman's sou'wester, the brim of which reached his shoulders. The pince-nez glistened with fine raindrops. He had been rowed ashore two hours earlier, he explained, and had been having a grand time "exploring" the waterfront.

At Tivoli Crossing, a station stop built especially for his arrival at the new hotel, his immediate move was to disappear. Perhaps a hundred Zone police had been waiting to protect him. Their captain, a big, picturesque figure named George Shanton, was a former Rough Rider whom Roosevelt had personally recruited to organize the Panama force and the uniform Shanton had chosen was the same as that of the Rough Riders. So with Shanton and his men prancing about on horseback, the reception, when the train pulled in, looked very much like those staged during political campaigns.

In the confusion of the rain and the crowds, Roosevelt spotted William Gorgas and pulled him into a closed carriage. But when the carriage arrived at the hotel, escorted by the galloping Shanton, neither Roosevelt nor Gorgas was inside. Before leaving the station, they had slipped out the other door of the carriage and Roosevelt had Gorgas take him directly to Ancon Hospital for an inspection tour, two hours before he was expected.

By noon he had toured the bay in a seagoing tug and had walked unannounced into one of the employees' mess halls at La Boca, where, with several hundred "gold roll" men, he and Mrs. Roosevelt sat down to a 30-cent lunch of soup, beef, mashed potatoes, peas, beets, chili con carne, plum pudding, and coffee. According to the official schedule, he was supposed to have attended a large luncheon in his honor at the Tivoli Hotel.

The American President, said Manuel Amador in a speech from the steps of the cathedral that afternoon, was the commander in chief in the great struggle for progress. "To harmonize the various elements that had to be united . . . to reorganize the great work, to grasp, in a word, its immense magnitude, a superior man was necessary, and you were this man," said Amador. Panama and the United States, Roosevelt responded, were partners in the "giant engineering feat of the ages."

Meantime, some two hundred prominent, rain-soaked Panamanians had paraded by on horseback, all dressed as Rough Riders.

The visit lasted all of three days, which, he later stressed to Congress, was insufficient time for an "exhaustive investigation of the minutiae of the work . . . still less to pass judgment on the engineering problems." But according to the *Star & Herald*, no one in the four hundred years of Panama's history had ever seen so much in so little time.

"He seemed obsessed with the idea that someone was trying to hide something from him," Frank Maltby would recall. ". . . He was continually pointing to some feature and asking, 'What's that? . . . Well, I want to see it.' . . . he was continuously stopping some black man and asking if he had any complaint or grievance."

Everyone who tried to maintain his pace wound up exhausted and half-drowned.

He walked railroad ties in Culebra Cut, leaped ditches, splashed through work camps, made impromptu speeches in the driving rain. "You are doing the biggest thing of the kind that has ever been done," he said, "and I wanted to see how you are doing it."

He inspected the quarters for both white and black workers, poked about in kitchens and meat lockers. On the morning of the third day,

John Stevens told Maltby that it was his turn to lead the procession. "I have blisters on both feet and am worn out," Stevens said. At Gatun, Roosevelt said he needed an overall view of the dam site and pointed to a nearby hill. In Maltby's recollection, ". . . we, together with three or four secret service men, charged up the hill as if we were taking a fort by storm."

At home the papers reported his every move. "ROOSEVELT IS THERE" proclaimed the Washington *Post.* "A STRENUOUS EXHIBITION ON THE ISTHMUS" read another headline. "THE PRESIDENT CLIMBS A CANAL STEAM-SHOVEL" *The New York Times* announced on its front page.

The famous moment on the steam shovel occurred early on his second day, en route to the Cut. It was about eight in the morning and again the rain was coming down hard. At the site of the Pedro Miguel locks, Roosevelt spotted several shovels at work and ordered that the train be stopped. He jumped down, marched through the mud, and was soon sitting up in the driver's seat, engineer A. H. Grey having happily moved over to make a place for him.

He was fascinated by the huge machine and insisted on knowing exactly how it worked; he asked that it be moved back and forth on its tracks. He had to see how everything was done. "All his questions, like his movements, were deliberate and emphatic to a noticeable degree," a reporter noted; "he would stand for no ceremony. . . ."

He was at the controls for perhaps twenty minutes, during which a small crowd gathered and the photographers were extremely busy. Presidents of the United States had been photographed at their desks and on the rear platforms of Pullman cars; Chester A. Arthur had consented once to pose in a canoe. But not in 117 years had a President posed on a steam shovel. He was wearing a big Panama hat and another of his white suits. And the marvelous incongruity of the outfit, the huge, homely machine and the rain pouring down, not to mention his own open delight in the moment, made it at once an *event,* an obvious and inevitable peak for the man who so adored having his picture taken and who so plainly intended to see success at Panama. One of the photographs would quickly become part of American folklore, and as an expression of a man and his era, there are few that can surpass it.

The shovel was a ninety-five-ton Bucyrus, mainstay of the work. Going at full capacity it could dig three to five times as much as one of the old French excavators, none of which was any longer in use. It could take up five cubic yards—roughly eight tons of rock and earth—

with a single scoop. Under ideal conditions it could load a dirt car in about eight minutes.

Ten men were needed to run such a machine. In addition to the engineer, there was a craneman, who handled the dumping, two coal stokers, and a "move-up" crew of six whose job it was to level the ground and place the track so that the shovel could be advanced as it worked, always keeping its nose to the bank. The engineer earned $210 a month, which was as much as the best-paid office workers received, more even than some doctors. But unlike the locomotive engineers, they got no overtime, as engineer Grey told Roosevelt in no uncertain terms.

The rain was descending in wild silver sheets when Roosevelt entered Culebra Cut for the first time, riding along the bottom of the Cut in a special train. Water was pouring from the red clay slopes in "regular rivers." But there was a great blowing of locomotive whistles and cheering as he came into view. On the side of one shovel was stretched a big, hand-lettered banner that pleased him enormously: "WE'LL HELP YOU DIG IT."

The shovels were working along the sides of the Cut on extended terraces, or benches, as in surface mining. They were advancing from either end of the Cut toward the middle, or summit, all of them digging on the upgrade, and it was thus that the loaded dirt trains rolled out of the Cut—north and south—on the downgrade. The spur tracks for the shovels ran side by side with those for the trains, the shovels working on the lower level. The area to be excavated was drilled and blasted, then the shovel moved up to begin the heavy work of swinging the debris, much of it rock, into the dirt cars. As each shovel progressed, it made a cut approximately fifty feet wide by twelve feet deep.

Very few of the old French dump cars were in use in the Cut any longer. The spoil was being hauled out on long trains of much larger American-built cars pulled by full-sized American-built locomotives. Most of the cars Roosevelt saw were wooden flat cars that were used in conjunction with a rather crude but amazingly effective unloading device, the Lidgerwood system, as it was called. The cars had only one side and steel aprons bridged the spaces between them. The dirt was piled on, high up against the one side; then at the dumping grounds a three-ton steel plow was brought up to the last car and hitched by a long cable to a huge winchlike device mounted on a flatcar at the head of the train. The winch took its power from the locomotive. At a signal the plow was hauled rapidly forward and the whole twenty-car train was unloaded with a single sweep, all in about ten minutes. One such machine, Stevens told Roosevelt, could do the work of three hundred men under the old method of unloading by hand.

In another letter to his son Kermit, written on the *Louisiana* on the way home, Roosevelt would give this description of Culebra Cut:

> Now we have taken hold of the job. . . . There the huge steam shovels are hard at it; scooping huge masses of rock and gravel and dirt previously loosened by the drillers and dynamite blasters, loading it on trains which take it away to some dump, either in the jungle or where the dams are to be built. They are eating steadily into the mountain cutting it down and down. Little tracks are laid on the side hills, rocks blasted out, and the great ninety-five ton steam shovels work up like mountain howitzers until they come to where they can with advantage begin their work of eating into and destroying the mountainside. With intense energy men and machines do their task, the white men supervising matters and handling the machines, while the tens of thousands of black men do the rough manual labor where it is not worthwhile to have machines do it. It is an epic feat, and one of immense significance.

He had seen the Cut from above, from the rim, following lunch and a change of clothes at John Stevens' house. The excavation was still only in its early stages and because of the rain there were only about twenty-five shovels at work. Even so, it was the largest cavern yet made in the earth's surface and the noise and commotion from below were like nothing to be experienced anywhere. It was a scene, we are told in other accounts, that might only have come from the mind of H. G. Wells.

Once, earlier in the year, H. G. Wells had called at the White House. It was a bright spring afternoon and he and Roosevelt had talked at length in the garden, much as Jules Verne and Ferdinand de Lesseps had conversed in the library of the Société de Géographie. Wells was in America, he said, to search for the future and "question the certitudes of progress," for unlike Verne, he had grave misgivings about the long-range human consequences of science and technology.

Whether Roosevelt had any such thoughts as he looked down into Culebra Cut for the first and only time in his life is impossible to say. More likely it was a supreme and ineffable moment. Wells, in his travels, had seen a hall of dynamos at the Niagara Falls Power Company that evoked something verging on religious awe. They were, he wrote, the creations of "serene and speculative, foreseeing and endeavoring minds." The hall itself was a sanctuary; there had been no clatter, no dirt, no tumult, still the outer rim of the big generators traveled at the speed of 100,000 miles an hour. He had been moved to the depths of his soul by the vision of such vast power in the hands of man.

For Roosevelt at Culebra, with the rain hammering down, there had to have been something of the same sensation, though for him the noise and tumult would be the better part of it.

In their talk in the White House garden Wells had asked if the creative energies of modern civilization had any permanent value, and Roosevelt's answer had been immediate. He had no way of disproving a pessimistic interpretation of the future, Roosevelt declared. But he chose not to live as if that was so. He referred specifically to *The Time Machine*, Wells's most despairing vision of the future.

"He became gesticulatory," Wells recalled, "and his straining voice a note higher in denying the pessimism of that book. . . ." Gripping the back of a garden chair with his left hand, Roosevelt had stabbed the air with his right, the familiar platform gesture.

"Suppose after all that should prove to be right, and it all ends in your butterflies and morlocks. *That doesn't matter now.* The effort's real. It's worth going on with. It's worth it—even then."

"I can see him now," Wells remembered, ". . . and the gesture of the clenched hand and the—how can I describe it? the friendly peering snarl of his face, like a man with the sun in his eyes. He sticks in my mind as that, as a very symbol of the creative will in man, in its limitations, its doubtful adequacy, its valiant persistence. . . ."

In the long letter to Kermit written on the homeward voyage, Roosevelt said the Panama wilderness had made him wish he had more time. "It is a real tropic forest, palms and bananas, breadfruit trees, bamboos, lofty ceibas, and gorgeous butterflies and brilliant colored birds fluttering among the orchids. . . . All my old enthusiasm for natural history seemed to revive, and I would have given a good deal to have stayed and tried to collect specimens." But there was no apparent conflict between such splendors and what went on in Culebra Cut, between orchids and steam shovels. "Panama was a great sight," he told his son Ted, by which he meant everything in Panama.

To the majority of those on the job his presence had been magical. Years afterward, the wife of one of the steam-shovel engineers, Mrs. Rose van Hardeveld, would recall, "We saw him . . . on the end of the train. Jan got small flags for the children, and told us about when the train would pass . . . Mr. Roosevelt flashed us one of his well-known toothy smiles and waved his hat at the children . . ." In an instant, she said, she understood her husband's faith in the man. "And I was more certain than ever that we ourselves would not leave until it [the canal] was finished." Two years before, they had been living in Wyoming on a lonely stop on the Union Pacific. When her husband heard of the work at Panama, he had immediately wanted to go, because, he told her, "With Teddy Roosevelt, anything is possible." At the time neither of them had known quite where Panama was located.

II

His "Special Message Concerning the Panama Canal," the first message to Congress to be illustrated with photographs, was released on December 17, 1906. He sketched the progress being made. He praised the French for what they had achieved; he praised Congress for having had the sense to refuse to attempt a passage at sea level. He described the hospitals, living quarters, his meal in the mess hall at La Boca. He wrote of the rain. Only on the last morning had he caught a glimpse of the sun, and then only for a few minutes.

He did his best to depict the size of the work and urged Congress and the nation to take notice. "It is a stupendous work upon which our fellow countrymen are engaged in down there on the Isthmus," he declared. At present, he could report, there were nearly six thousand Americans on the job. "No man can see these young, vigorous men energetically doing their duty without a thrill of pride. . . ."

A very large part of the message was given over to the progress made in health and sanitation and in praise for Gorgas. The message, Gorgas wrote privately, was "indeed a corker. I had not expected anything of the kind. I do not think that an army medical officer ever had such recognition in a Presidential message. It probably marks the acme of my career."

Roosevelt called the medical progress astounding in view of Panama's past; and yet, oddly, the statistical tables included at the conclusion of the report, transcriptions from the actual hospital records, gave a very different picture and a disquieting one. The specific strides he cited were quite unprecedented and indisputable: yellow fever had disappeared, there was no more cholera, there was no plague. Among the Americans, including dependents, there had not been a single death from disease in three months, an almost unbelievable record for Panama and very impressive, as Roosevelt stressed, even by North American standards.*

Medical care and services on the Isthmus were in fact "as good as that which could be obtained in our first-class hospitals at home." The Sanitary Department was currently spending $2,000,000 a year; Ancon Hospital had a staff of 470. More than a dozen new hospitals and dispensaries had been built along the line. All hospital care was free for all employees, white and black.

Nearly a thousand laborers were kept constantly at work digging drainage ditches, cutting grass, burning brush, hauling garbage, pouring or spraying oil on streams and swamps.

But to anyone who bothered to study the records at the back of the report it was at once apparent that the success of the health crusade was really quite relative. It depended on which segment of the work force one was talking about. The white worker and his family were indeed faring extremely well; otherwise, for the vast black majority, the picture was alarming.

For the first ten months of 1906 the actual death rate among white employees was seventeen per thousand. But among the black West Indians it was fifty-nine per thousand! Black laborers, those understood to be so ideally suited to withstand the poisonous climate, were dying three times as fast as the white workers. If Panama was no longer a white man's graveyard, it was little less deadly than it had ever been for the black man. And since the black workers outnumbered the white workers by three to one, the disparity in the numbers of fatalities among the black workers was even more shocking.

In the previous ten months a total of thirty-four Americans had died, whereas the toll among men and women from Barbados alone was 362, ten times greater; 197 Jamaicans had died, 68 from Martinique, 29 from St. Lucia, 27 from Grenada.

The causes of death as listed—among all workers, irrespective of color—included everything from railroad accidents to alcoholism to dysentery, suicide, syphilis, and tuberculosis. The chief killer among black people, however, and therefore the most fatal disease on the Isthmus at the moment, was pneumonia. Since the start of the year 390 employees had died of pneumonia. Of those, 375 were black. In October alone, as Roosevelt had been informed, 86 workers had died of pneumonia.

* In an average city in the United States in 1906 the death toll from disease among an equal number of people would have been about thirty.

Malaria, the second worst killer, had taken 186 lives, all but 12 of whom were West Indian Negroes.

The problem was that much of the labor force was particularly vulnerable to viral pneumonia. On Barbados the disease was unknown. And since so many black workers lived where they pleased and as they pleased, often in the jungle, often ignorant of the simplest rules of hygiene, nearly always without the benefit of wire screening, the chances of their contracting almost anything, and malaria in particular, were extremely high.

The ditch-digging, brush-burning, swamp-draining activities carried on by the *Anopheles* brigades, as they were known, had been highly effective within specified areas. Those earlier studies that had shown the *Anopheles* mosquitoes to be susceptible to strong sunshine and wind had produced a calculated program to create as much unshaded, unprotected clear space, as little shade or shelter for the insect, as possible. And thus the new towns along the line stood on open ground, everything neatly clipped and trimmed. *Anopheles* mosquitoes were rarely seen in the immediate vicinity any longer. Roosevelt noted "the extraordinary absence of mosquitoes." He and his party had seen exactly one in three days and it was "not of the dangerous species."

But by no means had every swamp been drained, every breeding ground destroyed. A very large swamp at Miraflores, for example, was especially prolific; the usual catch in a mosquito trap overnight there was about a thousand. The jungle was never much more than a stone's throw from any point along the line and in the jungle the *Anopheles* were as plentiful as always.

Malaria would continue to take more lives, as William Gorgas allowed in his own reports, and the "amount of incapacity" caused by the disease was, as he said, very much greater than that due to all other diseases combined.

Nearly all of the patients Roosevelt saw in the hospitals at Panama had been black men, as he acknowledged. And privately he had been appalled by some of the things he had seen, as we know from his correspondence with Shonts. "The least satisfactory feature of the entire work to my mind was the arrangement for feeding the negroes," he wrote as soon as he reached Washington. "Those cooking sheds with their muddy floors and with the unclean pot which each man had in which he cooked everything, are certainly not what they should be. . . . Moreover, the very large sick rates among the negroes, compared with the whites, seems to me to show that a resolute effort should be made to teach the negro some of the principles of personal hygiene . . ." Could not something be done to provide better housing, better health for these workers? he asked.

Overall, the trip had made him more exuberant than ever on the subject of the canal. Of its ultimate success, he was as "convinced as one can be of any enterprise that is human." His faith in Stevens was implicit throughout the message to Congress, as no one appreciated more than Stevens, who called it an "unqualified endorsement" of his conduct of affairs.

The executive order had been signed at a meeting in the old de Lesseps' Palace at Cristobal on November 17, Roosevelt's last day on the Isthmus. Stevens' authority, therefore, was now firmly fixed.

So it was both puzzling and extremely annoying to Roosevelt when, at the very moment he released his message, Stevens began making trouble. In Washington for a brief visit in December, Stevens was strangely irritable and caustic. He seemed inexplicably resentful of Gorgas and talked of having Gorgas fired. Roosevelt found it "well-nigh impossible to get on with him."

What went sour for Stevens is a mystery that Stevens chose never to explain. With the return of the dry season, the work was rolling ahead as never before. Excavation in Culebra Cut exceeded 500,000 cubic yards in January, more than double the best monthly record of the French. In February the figure was more than 600,000 cubic yards and Stevens' own popularity reached a new high.

In any event, the crisis followed Shonts's resignation on January 22. Shonts was leaving to head the Interborough Rapid Transit Company in New York City, a decision Roosevelt and Stevens knew of in advance and that Roosevelt accepted with none of the fireworks that had attended the Wallace incident. Stevens was formally apprised of the news two days later in a letter from I.C.C. Secretary Joseph Bucklin Bishop.

Then on January 30, at Culebra, Stevens sent a letter to Roosevelt that reached the White House on February 12.

It was six pages in length and as devoid of cant or circumlocution as all his correspondence. It also revealed a very different man from the John Stevens of the previous year, an exhausted and embittered man. He complained of "enemies in the rear" and of the discomforts of being "continually subject to attack by a lot of people . . . that I would not wipe my boots on in the United States." While some "wise lawmakers" might think his salary excessive, he wanted it known that by staying on at Panama he was depriving himself of not less than

$100,000 a year. His home life was disrupted; he was separated from his family much of the year. And at his age he had little enough time left "to enjoy the pleasures and comforts of a civilized life."

He wrote of the tremendous responsibility and strain put upon the man in his position, saying he doubted that he could bear up under them for another eight years. Technical problems were not the issue; it was "the immense amount of detail" one had to keep constantly in mind.

If there was to be glory attached to his role, he was uninterested. Nor in the final analysis did he see any special romance or meaning in the canal itself:

> The "honor" which is continually being held up as an incentive for being connected with this work, appeals to me but slightly. To me the canal is only a big ditch, and its great utility when completed, has never been so apparent to me, as it seems to be to others. Possibly I lack imagination. The work itself . . . on the whole, I do not like. . . . There has never been a day since my connection with this enterprise that I could not have gone back to the United States and occupied positions that to me, were far more satisfactory. Some of them, I would prefer to hold, if you will pardon my candor, than the Presidency of the United States.

This was the passage that settled his fate. The letter was not a formal resignation. He never said specifically that he wanted out, only that he was not "anxious to continue in service." He wanted a rest, and having assured Roosevelt of his high personal regard for him, he asked for his "calm and dispassionate" consideration of the matter.

A reporter who talked to someone who was with Roosevelt at the time Roosevelt received the letter wrote, "To say that the President was amazed at the tone and character of the communication is to describe the feelings mildly." The letter was sent immediately to Taft with a covering note: "Stevens must get out at once." Even if Stevens were to change his mind, it would make no difference "in view of the tone of his letter."

After a brief meeting with Taft, Roosevelt cabled Stevens that his resignation was accepted.

Taft again told Roosevelt that Major Goethals (who was about to become Lieutenant Colonel Goethals) was the best-equipped man for the job, so on the night of February 18 Goethals was summoned to the White House. The change, however, was kept secret until the twenty-sixth, when, with the announcement, Roosevelt issued his widely quoted declaration—a remark made as much for the benefit of the work force on the Isthmus as for the general public—that he would put the canal in the charge of "men who will stay on the job until I get tired of having them there, or till I say they may abandon it. I shall turn it over to the Army."

But in the same breath, according to the New York *Tribune*, he also remarked, "Then if the man in charge suffers from an enlarged cranium or his nerves go to the bad, I can order him north for his health and fill his place without confusion."

Privately Roosevelt was "utterly at sea" over Stevens' behavior. When a friend who was visiting the Isthmus wrote in confidence that Stevens suffered from insomnia, Roosevelt seemed much relieved. "If he were a drinking man or one addicted to the use of drugs, the answer would be simple," he wrote in reply. "As it is, I am inclined to think that it must have been insomnia or something of the kind, due to his tropical surroundings . . ." Then he added: "He has done admirably."

On the Isthmus the announcement had a shattering effect. The *Star & Herald*, standing firmly behind Stevens, declared the top-heavy craniums were all in Washington and that the French must be laughing up their sleeves.

When Stevens' own men appealed to him for some word of explanation, he answered, "Don't talk, dig."

As time passed, numerous theories were put forth. It was said that he had found the Gatun Dam plans to be unsound; that he was angry over a contract that had been agreed to in Washington without his say; that his wife did not like the looks of the contractor; that he had been offered another job; that he was crazy. One editor, exasperated by the absurdity of all that was appearing in print, declared that in fact the problem of green mold on his books was what finally broke the spirit of John Stevens.

The most common and in retrospect the most plausible explanation was that he was overworked and verging on a breakdown, which is what his own letter plainly implied. It was the explanation Taft gave to Congress in the course of later testimony and the conclusion Goethals would reach once he got to the Isthmus. ". . . I think he has broken down with the responsibilities and an evident desire to look after too many details himself," Goethals wrote privately.

"He was not a quitter," Frank Maltby would insist. "He could not have been driven off the canal with a club, if it was a question of fighting for what he thought was the right thing. . . . My own personal opinion . . . is that he disliked notoriety very much."

A more intriguing but wholly unsubstantiated theory was offered some years later by Woodrow Wilson's Secretary of the Navy, Josephus Daniels. According to Daniels, Stevens had inadvertently come upon certain incriminating information concerning the activities of William Nelson Cromwell at the time of the sale of the Compagnie Nouvelle and its franchises. It was information, Daniels wrote, that if revealed "would blow up the Republican Party and disclose the most scandalous piece of corruption in the history of the country."

An explanation that carried great weight on the Isthmus was that Stevens had been merely letting off steam in his letter to Roosevelt and he was as startled as anyone by the reaction it produced. A canal employee who claimed to have been with Stevens in his office when the letter was written said Stevens handed him a carbon, remarking jovially, "I've just been easing my mind to T.R. It's a hot one, isn't it?" When the man told Stevens it was a letter of resignation, Stevens laughed and said Roosevelt would know perfectly well that he did not mean to quit. But if Stevens truly understood Roosevelt, as he claimed, it is inconceivable that he could so misjudge the inevitable effect of belittling remarks concerning the canal, not to mention the decidedly unpleasant edge to his remark about the Presidency.

Stevens had thrived on change his whole career. He left Hill twice because he needed a change. He had accepted the Philippines assignment in 1905 because he was worn out and needed a change. Change, he was to write in an appeal to young men to enter engineering, was for him among the prime attractions of the profession. So possibly a resignation was bound to come sooner or later.

He himself was to assert that all alleged reasons for his sudden departure were alike in one respect: they were all false. "The reasons for the resignation were purely personal . . ." he wrote. "I have never declared these reasons, and probably never will. . . ." He never did.

His work had been outstanding. His railroad scheme in Culebra Cut was, according to George Goethals, beyond the competence of any Army engineer of the day. Others would contend—indeed argue passionately—that in fact it was Stevens who should go down in history as the builder of the canal. Never a modest man, Stevens had his own view about this. He had handed over to the Army engineers, he later said, a "well-planned and well-built machine," which apart perhaps from a squeak or two would run perfectly. His replacement (Goethals) merely "turned the crank," he wrote. "The hardest problems were solved, the Rubicon was crossed, the canal was being built. . . .

Only gross mismanagement or a failure to supply the necessary funds, could militate against its triumphant accomplishment."

But this was manifestly unfair. Closer to the truth was the picture he had implied in the letter to Roosevelt of an immense, complex task, a man-killing responsibility, extending for years to come. Goethals' later tributes to Stevens, that Stevens was one of the greatest engineers who ever lived, that the canal was Stevens' monument, were professional compliments of the highest order offered in all sincerity. But such remarks also say as much about Goethals as they do about Stevens.

Stevens' railroad system would remain the fundamental operating procedure in the Cut until the excavation was finished. But excavation was only beginning in early 1907 and Stevens had not been confronted by major landslides. Surveys were still incomplete. The relocation of the Panama Railroad had not yet begun. The size of the locks had still to be determined. All the complex details of the locks had yet to be designed. Indeed, all the great *construction* work of the canal had still to begin—the building of Gatun Dam, the building of the locks—tasks of unprecedented magnitude requiring technical expertise that Stevens really did not possess.

Stevens' primary tasks—the creation of a well-fed, well-housed, well-equipped, well-organized work force, the conception of a plan of attack—were over by 1907. As a railroad engineer he was inexperienced in the large-scale use of concrete; he knew very little about hydraulics; and these were the specialties of the Army engineers.

Stevens' two-fisted, independent spirit had been exactly what was needed. The critical situation in 1905 had demanded, as he later said, "a kind of politic 'roughneck,' who did not possess too deep a veneration for the vagaries of constituted authority." But ultimately the role called for a larger sense of mission than that.

For a long time now Roosevelt had spoken of building the canal as though it were a mighty battle in which the national honor was at stake, much as Ferdinand de Lesseps had so often spoken. Panama was a tumultuous assault for Progress, the only assault this most bellicose-sounding of American Presidents was ever to launch and lead. At the end of his last day at Cristobal, in an off-the-cuff speech to several hundred Americans, including John Stevens, he had said the canal was a larger, more important endeavor than anyone could as yet realize, and that by bringing it to successful completion they would stand like one of the famous armies of history. It was to be a long, arduous, uphill struggle, he said, one not unlike that of their fathers' in the Civil War.

(His own two-month Cuban war would never have served as an example.)

> When your fathers were in the fighting, they thought a good deal of the fact that the blanket was too heavy by noon and not quite heavy enough by night, that the pork was not as good as it might be . . . and that they were not always satisfied with the way in which the regiments were led. . . . But when the war was done—when they came home, when they looked at what had been accomplished, all those things sank into insignificance, and the great fact remained that they had played their part like men among men; that they had borne themselves so that when people asked what they had done of worth in those great years all they had to say was that they had served decently and faithfully in the great armies. . . . I cannot overstate the intensity of the feeling I have . . . I feel that to each of you has come an opportunity such as is vouchsafed to but few in each generation. . . . Each man must have in him the feeling that, besides getting what he is rightfully entitled to for his work, that aside and above that must come the feeling of triumph at being associated in the work itself, must come the appreciation of what a tremendous work it is, of what a splendid opportunity is offered to any man who takes part in it.

By Roosevelt's lights, Stevens had failed in the most profound and fundamental sense, scarcely less than Wallace had. To Roosevelt the triumph was in the task itself, in taking the dare; the test was in the capacity to keep "pegging away," as he often stressed to his sons. Stevens was not merely giving up; Stevens saw it only as a "job"; there was no commitment of heart, not the slightest apparent sense of duty. To Roosevelt, Stevens was a commander abandoning his army.

He appears to have harbored no bitterness toward Stevens. ("You have done excellent work . . . and I am sorry to lose you," he wrote a few days after receiving Stevens' letter.) It was merely that if Stevens was the sort of man who looked upon the task as something to take or leave at will, then he was someone Roosevelt could quite readily do without and put from mind. In Roosevelt's long essay on the canal in his *Autobiography*, there would be no mention of John Stevens.

III

With the appointment of George Washington Goethals, Roosevelt's worries over the work at Panama came to an end. The canal would now be the "one-man proposition" John Stevens had called for, only the one man was to be an entirely different sort from Stevens.

At forty-eight Goethals was the same age as Roosevelt and of similar ancestry. His Flemish father and mother had arrived in New York with the great wave of immigration in 1848. The second of three children, he had been born in Brooklyn on June 29, 1858, and later, when he was eleven, moved with his family to a house on Avenue D in Manhattan, a block from the East River. But his family had been poor and struggling and unlike Roosevelt he had had to make his way "exclusively by his own exertions." Starting at age fourteen he had worked his way through City College in New York, then went on to West Point, where he was elected president of his class and finished second in his class in 1880, the same year Roosevelt was graduated Phi Beta Kappa from Harvard.

Goethals' career in the Corps of Engineers had been exemplary. In the Department of the Columbia in 1884, William Tecumseh Sherman had singled him out as the finest young officer in his command and predicted a "brilliant future." He had worked on "improvements" in the Ohio River valley (1884–1885); as an instructor of civil and military engineering at West Point (1885–1889); on improvements on the Cumberland and Tennessee rivers and particularly on the Muscle Shoals Canal (1889–1894), where he designed and built a lock with the record lift of twenty-six feet; as assistant to the Chief of Engineers (1894–1898); and harbor works from Block Island to Nantucket (1900–1903). In 1903, the year of Elihu Root's reorganization of the Army, he had been picked to serve on the new General Staff, a corps of forty-four officers who were relieved of all duties in order to assist the new Chief of Staff. And it was thus, as a specialist in coastal defenses, that he had come to Taft's attention.

He was a model officer, but a soldier like many in the Corps of Engineers who had never fought in a war, never fired a shot except on a rifle range, and who seems in fact to have had little affection for conventional "soldiering." Once on a parade ground in Panama, while watching some troops pass in review on a broiling-hot drill field, he would mutter to a civilian companion, "What a hell of a life."

Cool in manner, capable, very correct, he was a man of natural dignity and rigorously high, demanding standards. He had had no experience with notoriety, nor apparently any craving for it. And it would be hard to imagine him losing himself in *Huckleberry Finn* or anything other than his work. Asked years later how "the Colonel" had amused himself, a member of the family would respond, "He did not amuse himself."

A reporter wrote that "above everything he looks alert and fit." Six feet tall, he was in fine physical trim. The salient features were his intent, violet-blue eyes—"rather savage eyes," Alice Roosevelt Longworth would recall—and his close-cut, silvery hair, which he parted in the middle and washed daily. If a bit stiff socially, he was never pompous, largely because he was almost incapable of talking about himself. To pretty young women he could be especially gracious, in a rather fatherly fashion, and they considered him extremely attractive.

He was also a chain smoker and he detested fat people—with the one exception of William Howard Taft. Secretary Taft, Goethals was once heard to remark, was the only *clean* fat man he had ever known.

On the night that he was first summoned to the White House, Goethals and his wife had been entertaining an old friend, Colonel Gustav Fieberger, head of the engineering department at West Point, at their home on S Street. A messenger arrived with a note from William Loeb, Roosevelt's secretary, asking if Goethals would be free to come by the first thing in the morning. Goethals had immediately telephoned Loeb, who told him not to wait until morning but to come over that night at twenty minutes after ten. So Goethals had excused himself from his guest, changed into dress uniform, and left the house having no idea whatever as to why he was being sent for. Nor had he ever met Theodore Roosevelt.

"He entered at once upon the subject of the Canal," Goethals would recall. The canal commission was again to be reorganized and for the final time. Goethals was to be both chairman and chief engineer. Jackson Smith and Dr. Gorgas were to be members of the commission, along with four new men: a former senator from Kentucky named Joseph C. S. Blackburn, Rear Admiral Harry Harwood Rousseau, and Major David Du Bose Gaillard and Major William Sibert, both of the Corps of Engineers. Gaillard was the only one on the list with whom Goethals was personally acquainted—Gaillard, too, had been a member of the first General Staff—but he knew Sibert and Rousseau by reputation and agreed to their appointments.

The critical decision, however, concerned Goethals. "He [Roosevelt] expressed regret that the law required the work to be placed in charge of a commission or executive body of seven men," Goethals remembered, "but . . . his various efforts to work under the law . . . were so unsuccessful that he resolved to assume powers which the law did not give him but which it did not forbid him to exercise."

So while all members of the commission were to be on the Isthmus henceforth, Goethals was to wield supreme authority, an authority that would be backed by another new executive order the following year. Goethals was to be a virtual dictator—"Czar of the Zone"—responsible only to the Secretary of War and the President. In the words of his biographer, Goethals at once became one of the world's absolute despots, who "could command the removal of a mountain from the landscape, or of a man from his dominions, or of a salt-cellar from that man's table."

This was a long way from the spirit of the Spooner Act, but by such means only, Roosevelt insisted, could the task ever be accomplished, a view with which Goethals concurred.

A common misconception later was that the canal was built by the Army, that it was the creation of the Corps of Engineers. It was not. Goethals and the other engineering officers were detached from the Army to serve in Panama. They did not report to the Chief of the Corps of Engineers; they, like the civilian engineers, reported to the canal commission—which was Goethals—and Goethals reported to Taft, exactly as Stevens had according to the previous reorganization.

The critical difference now was that an Army man could not and would not quit. For a West Point graduate to abandon his appointed task in the face of adversity or personal discomfort was all but inconceivable.

In the next several days, Blackburn, Rousseau, Gaillard, and Sibert appeared at the White House one by one to meet with the President and Goethals in the President's office. The same scene was repeated in each instance. Having introduced Goethals, Roosevelt would ask the man to be seated, then would inform him that he was to be appointed to the commission. "It will be a position of ample remuneration and much honor," Roosevelt said. "In appointing you I have only one qualification to make. Colonel Goethals here is to be chairman. He is to have complete authority. If at any time you do not agree with his policies, do not bother to tell me about it—your disagreement with him will constitute your resignation."

Goethals' salary, Roosevelt had decided, would be $15,000 a year, which was substantially more than he had been earning, but only half what Stevens had been paid.

A week or so after his new assignment had been announced in the papers, Goethals wrote in reply to the congratulations of a friend, "It's a case of just plain straight duty. I am ordered down—there was no alternative."

To a whole generation of Americans it was Theodore Roosevelt who built the Panama Canal. It was quite simply his personal creation. Yet the Panama Canal was built under three American Presidents, not one—Roosevelt, Taft, and Wilson—and in fact, of the three, it was really Taft who gave the project the most time and personal attention. Taft made five trips to Panama as Secretary of War and he went twice again during the time he was President. It was Taft who fired Wallace and hired John Stevens, Taft who first spotted Goethals. When Taft replaced Roosevelt in the White House in 1909, the canal was only about half finished.

None of this made much difference, however. Nor ought there ever be any question as to the legitimacy of the Roosevelt stamp on the canal. His own emphatic position was that it would never have been built but for him and it was a position no one tried to dispute. To Goethals, "The real builder of the Panama Canal was Theodore Roosevelt." It could not have been more Roosevelt's triumph, Goethals wrote, "if he had personally lifted every shovelful of earth in its construction. . . ."

The work had not simply begun anew while Roosevelt held office; his leadership had been decisive—in the Hay-Pauncefote Treaty, the choice of the Panama route, the creation of an independent Panama, the defense and support of William Gorgas, the choice of a lock-and-lake plan.

Even with his Panama visit, however brief, he achieved at a stroke something that had never been done before: he made the canal a popular success.

And finally, he had entrusted command of the work to one extremely well-chosen man. "I believe in a strong executive," he once wrote to a correspondent, "I believe in power. . . ."

NATIONAL ARCHIVES

High tea at Culebra (Colonel and Mrs. David D. Gaillard)

Typical housing at Ancon for upper-echelon employees
FROM THE MAKERS OF THE PANAMA CANAL, 1911

Typical dining room in middle-echelon dwelling
NATIONAL ARCHIVES

Momentary pause at a Saturday-night dance at the Tivoli Hotel

Bathers en route to Toro Point

BOTH PHOTOS:
COLLECTION OF J. W. D. COLLINS

BOTH PHOTOS: PANAMA CANAL COMPANY

LEFT ABOVE, bachelor quarters (bottle on the dresser is bay rum)

LEFT BELOW, Culebra Station as it looked in 1911

BELOW, billiard room at one of the Y.M.C.A. clubhouses. Dues were $10 a year.

FROM THE MAKERS OF THE PANAMA CANAL, 1911

ABOVE, West Indian wedding party
BOTH PHOTOS: NATIONAL ARCHIVES

LEFT, the steamer *Ancon* arriving at Cristobal from Barbados with 1,500 laborers
PANAMA CANAL COMPANY

BELOW, typical housing for West Indian laborers

COLLECTION OF D. P. GAILLARD

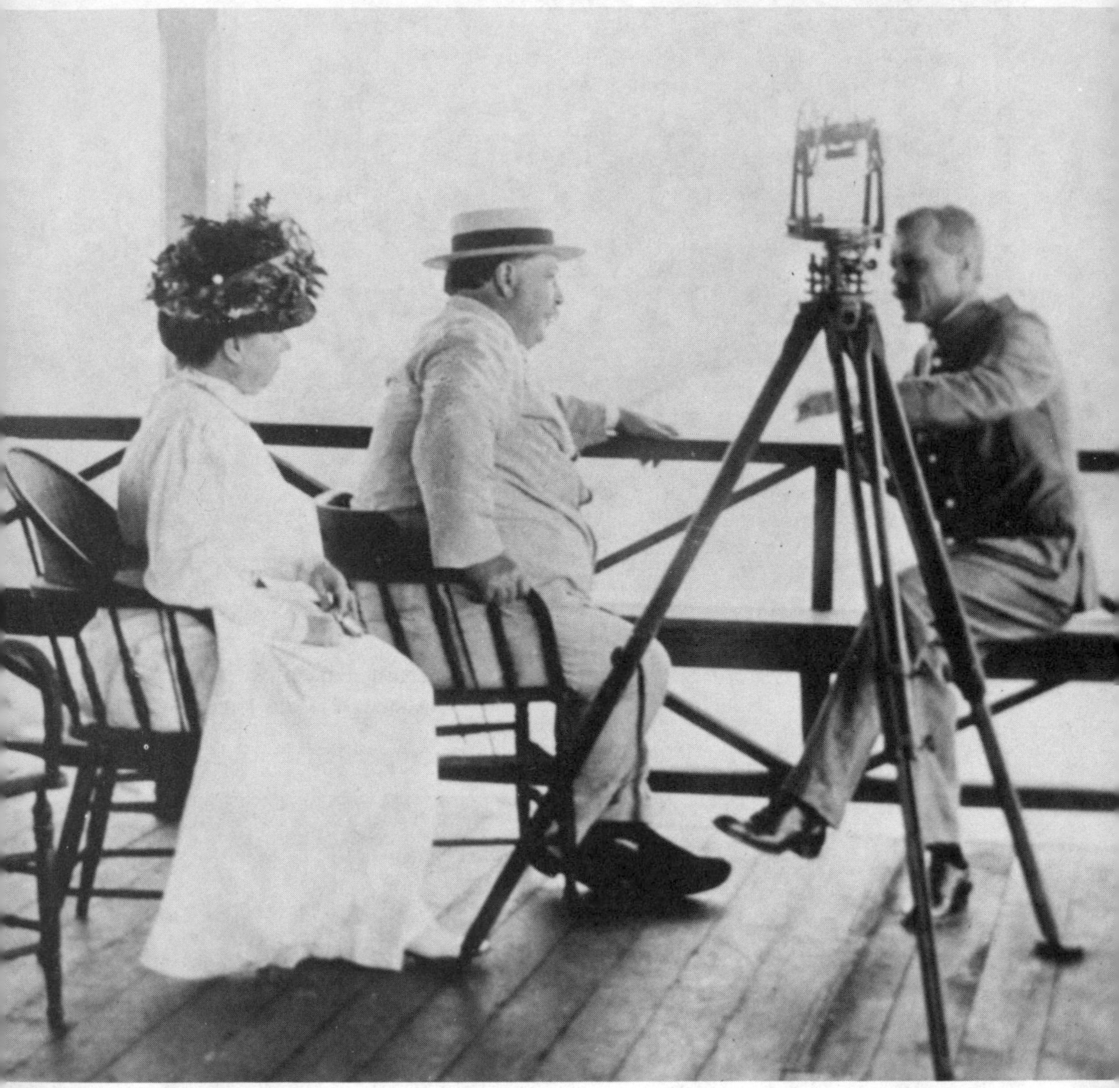

President and Mrs. William Howard Taft with Colonel Gaillard at Culebra

RIGHT, aftermath of a slide in Culebra Cut PANAMA CANAL COMPANY

ABOVE, "Headquarters" at Mount Hope

RIGHT ABOVE, interior of the pay car, which delivered 1,600 pounds of gold, 48,000 pounds of silver coin monthly

RIGHT BELOW, movie still of labor train

ALL PHOTOS: NATIONAL ARCHIVES

PANAMA CANAL COMPANY

ABOVE, *The Approaches to Gatun Locks by* Joseph Pennell

LEFT, the rise of Gatun Locks. Top: Aerial tramway delivers buckets of concrete to steel forms. Center: Giant bull wheel that opens and shuts a lock gate. Bottom: Gate leaves (double gates in foreground, intermediate gates beyond) near completion, 1912.

BOTH PHOTOS: PANAMA CANAL COMPANY

LEFT, shovel No. 222 and shovel No. 230 meet nose to nose on the bottom of the Cut, May 20, 1913.

BELOW, a party of tourists views Culebra Cut and the Cucaracha slide early in 1914, after Goethals had filled the Cut with water and continued the work with dredges.

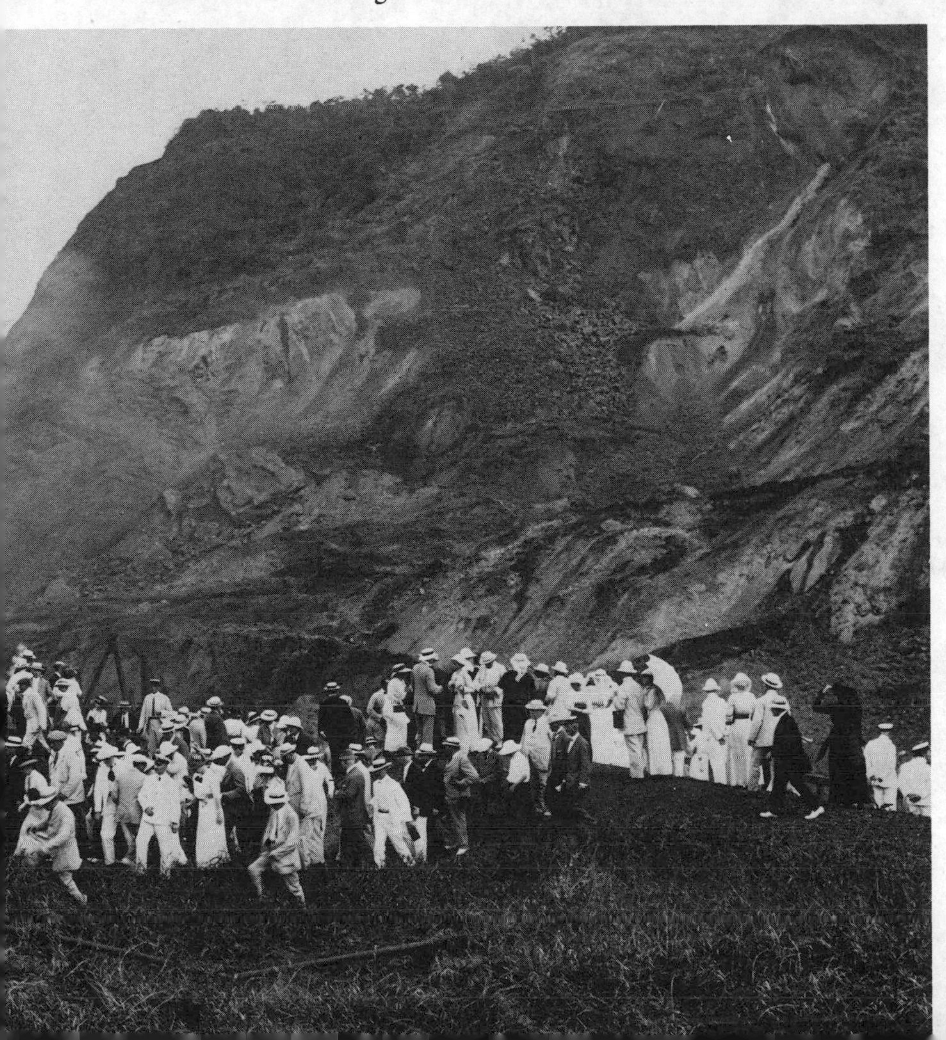

The tug *Gatun* approaches Gatun Locks for the first trial lockage.

Steamer *Ancon* starts into Culebra Cut on the official opening transit of the canal, August 15, 1914.

BOTH PHOTOS: PANAMA CANAL COMPANY

19

The Chief Point of Attack

> *The chief point of attack was, of course, the Culebra Cut, then, as always, the most formidable obstacle to be fought and overcome. How much more formidable it really was than had been suspected was soon to be revealed.*
>
> —Joseph Bucklin Bishop

I

For anyone to picture the volume of earth that had to be removed to build the Panama Canal was an all but hopeless proposition. Statistics were broadcast—15,700,000 cubic yards in 1907, an incredible 37,000,000 cubic yards in 1908—but such figures were really beyond comprehension. What was 1,000,000 cubic yards of dirt? In weight? In volume? In effort?

The illustrative analogies offered by editors and writers were of little help, since they were seldom any less fantastic. The spoil from the canal prism, it was said, would be enough to build a Great Wall of China from San Francisco to New York. If the United States were perfectly flat, the amount of digging required for a canal ten feet deep by fifty-five feet wide from coast to coast would be no greater than what was required at Panama within fifty miles. A train of dirt cars carrying the total excavation at Panama would circle the world four times at the equator. The spoil would be enough to build sixty-three pyramids the size of the Great Pyramid of Cheops. (To help its readers imagine what this might look like, *Scientific American* commissioned an artist to draw Manhattan with giant pyramids lining the length of Broadway from the Battery to Harlem.)

The material taken from Culebra Cut alone, exclaimed one writer toward the completion of the work, would make a pyramid topping the Woolworth Building by 100 feet (the Woolworth, at 792 feet, was then the world's tallest building), while the total spoil excavated in the Canal Zone would form a pyramid 4,200 feet high, or more than seven times the height of the Washington Monument.

If all the material from the canal were placed in one solid shaft with a base the dimension of a city block, it would tower nearly 100,000 feet—nineteen miles—in the air.

But who could imagine such things? Or how many could also take into account the smothering heat of Panama, the rains, the sucking mire of Culebra, none of which was less troublesome or demoralizing than in times past. For however radically systems or equipment were improved upon, however smoothly organized the labor army became, the overriding problem remained Panama itself—the climate, the land, the distance from all sources of supply. At the bottom of Culebra Cut at midday the temperature was seldom less than 100 degrees, more often it was 120 to 130 degrees. As Theodore Shonts once remarked, to have built the same canal in a developed country and a temperate climate would have posed no special difficulties.

More manageable—and more impressive—were the results of, say, a month's or even a day's work in Culebra Cut, or the relative effectiveness of the whole earth-digging, earth-moving system as compared to what the French had achieved at Suez or at Panama, these being the only prior efforts that were really analogous. When the work under Goethals was at its height, the United States was excavating at Panama the equivalent of a Suez Canal every three years. The 37,000,000 cubic yards of earth and rock removed in the one year of 1908 was nearly half as much as two successive French companies had succeeded in digging at Panama in a total of nearly seventeen years, and more than all that portion of excavation by the French that was useful to the present plan.

In any one day there were fifty to sixty steam shovels at work in the Cut, and with the dirt trains running in and out virtually without pause, the efficiency of each shovel was more than double what it had been. Along the entire line about five hundred trainloads a day were being hauled to the dumps. A carload of spoil was being removed every few seconds and the average daily total was considerably more than what the French had been digging in a month's time the year John Wallace arrived on the scene. But all prior effort, American as well as French, was put in the shadows.

Perhaps as extraordinary as anything that can be said is that the work could not have been done any faster or more efficiently in our own day, despite all technological and mechanical advances in the time since, the reason being that no present system could possibly carry the spoil away any faster or more efficiently than the system employed. No motor trucks were used in the digging of the canal; everything ran on rails. And because of the mud and rain, no other method would have worked half so well.

But the canal builders were not merely achieving what others had failed at; they were doing more, much more, than they or anyone had foreseen, for every prior estimate of the size of the task had been woefully inaccurate. The American engineers had been no less naïve in their reckoning of the total mass to be removed than had the French. On November 13, 1904, the day after the first Bucyrus shovel began digging in Culebra Cut there had been a small landslide which put that shovel out of commission for several days. Presently, in Stevens' time, there had been further slides on the order of what the French had experienced. But no one had the remotest conception of what was to occur during the Goethals years. Rather, it was felt that the whole issue of slides had been overemphasized. Professor Burr, of Roosevelt's international advisory board, had testified that there really need be no concern: "All that is necessary to remedy such a condition is simply to excavate the clay or to drain it to keep the water out. It is not a new problem. It is no formidable feature of the work."

The advisory board in its 1906 study—that is, in the minority report for a lock canal—had placed the total volume of excavation still to be accomplished at not quite 54,000,000 cubic yards. But by 1908 that estimate had to be revised to about 78,000,000 cubic yards. In 1910 it was put at 84,000,000; in 1911, at 89,000,000. By 1913 the estimate had reached 100,000,000 cubic yards, or nearly equal to the figure initially given by the advisory board for a canal at sea level!

As in the French time, the more digging that went on, the more digging there was to be done.

For the man who now bore the burden of responsibility for all that occurred, the initial hurdle had been primarily personal and as difficult as anything in his experience.

Goethals' reception upon arrival had been pointedly cool. Plainly, neither he nor the Army was wanted by the rank and file of Americans on the job and everyone seemed eager to make a special point of Stevens' tremendous popularity. Thousands of signatures had been gathered for a petition urging Stevens to withdraw his resignation and stay. No one, it seemed, had anything but the strongest praise for him and all he had done. Never in his career, Goethals remarked, had he seen so much affection displayed for one man.

Stevens and Dr. Gorgas were at the pier the morning Goethals and Major Gaillard landed. No real reception had been arranged; nothing had even been done about a place for Goethals or Gaillard to stay. Stevens still occupied the official residence of the chief engineer, a new six-bedroom house at Culebra that was to be Goethals' once Stevens departed, but since Stevens "didn't seem inclined to take us into his house" (as Goethals wrote to his son George), the two officers had moved in with Gorgas at Ancon, where there was little privacy, not even a desk at which Goethals could work. His letters to his family those first weeks were written on his lap as he sat in a straight-backed chair in one of the bedrooms.

To add to the spirit of gloom, the *Star & Herald* openly deplored the prospect of military rule. Probably no workers would have to wear uniforms, the paper presumed, but neither should anyone be surprised if he had to answer roll call in the morning or salute his new superiors.

That the railroad men around Stevens had scant regard for Army engineers seemed also abundantly plain to Goethals. "Army engineers, as a rule, were said to be, from their very training dictatorial and many of them martinets," he would write, "and it was predicted that if they . . . were placed in charge of actual construction the canal project was doomed to failure." The Army men had only technical training, it was said; they had never "made a success as executive heads of great enterprises."

His own private estimate of the state of the work was entirely favorable. The difference between what he saw now and what he had seen in 1905, during the visit with Taft, was extraordinary. As he wrote to his son, "Mr. Stevens has done an amount of work for which he will never get any credit, or, if he gets any, will not get enough. . . ."

Several days passed before he was granted a more or less official welcome—a Saturday-night "smoker" given as much to entertain a party of visiting congressmen. John Stevens declined to attend and Goethals, at the head table, sat listening without expression as the toastmaster extolled Stevens at length and made several cutting remarks about the military. It was an evening he would never forget. With each mention of Stevens' name there was a resounding cheer, while the few obligatory references to Stevens' successor were met with silence. Goethals was furious at what he regarded as "slurs" on the Army, but kept still until it was his turn. He had come to the affair not in uniform but in a white civilian suit. In fact, he had brought no uniforms to the Isthmus and never in the years to come would he be seen in one.

He was, he told the assembled guests, as appreciative as they of the work Stevens had accomplished and he had no intention of instigating a military regimen. "I am no longer a commander in the United States Army. I now consider that I am commanding the Army of Panama, and that the enemy we are going to combat is the Culebra Cut and the locks and dams at both ends of the Canal, and any man here on the work who does his duty will never have any cause to complain of militarism."

He took over from Stevens officially at midnight, March 31, 1907, and a week later Stevens sailed for home. One of the largest crowds ever seen on the Isthmus jammed the pier at Cristobal to see him off, everyone cheering, waving, and singing "Auld Lang Syne." Stevens was noticeably amazed and touched by the outpouring of affection. This time it was Goethals' turn not to attend.

Having none of Stevens' colorful mannerisms or easy way with people, Goethals impressed many at first as abrupt and arbitrary, a cold fish. The word "*goethals*" in Flemish, it was soon being said, meant "stiff neck."

He hated to have his picture taken. He found the visiting congressmen rude, tiresome, terribly time-consuming. Callers were "an awful nuisance." It was expected that he appear at every dance and social function at the Tivoli or the Culebra Club. He would "brace up" and go "out of a sense of duty" and spend the evening sitting on a porch listening to the music, waiting only for the time when he could politely withdraw.

Stevens' former secretary, having agreed to stay and help with the transition, suddenly resigned. William Bierd, the railroad boss, made a surprise announcement that he was retiring because of his health, but then Goethals learned that Bierd was taking a job with Stevens on the New Haven Railroad. Frank Maltby decided no civilian engineer had a future any longer at Panama and so he too quit. Then the steam-shovel engineers, sensing the time was at last right for a show of strength, threatened to strike unless their demands were met. Goethals refused and they walked off the job. It was the first serious strike since the work had begun. Of sixty-eight shovels, only thirteen were still in operation. He recruited new crews.

Even the newly arrived Major Sibert was proving "cantankerous and hard to hold" in meetings. Mrs. Sibert, Goethals learned, was "disgusted" with the Panama weather.

From surviving letters written to his son George, then in his senior year at the Military Academy, it is apparent that he was also extremely lonely. Mrs. Goethals was still in Washington "doing society at a great rate"; another, younger son, Thomas, was at Harvard. He felt very out of touch, he wrote; there was not time even to read the paper. His sole source of amusement was the French butler, Benoit, who still spoke practically no English but went with the official residence at Culebra, Goethals being his seventh chief engineer.

The day began at first light. At 6:30, with Benoit standing stiffly in attendance, "the Colonel" had his breakfast—one peeled native orange stuck on the end of a fork, two eggs, bacon, one cup of coffee. By seven he had walked down to Culebra Station to catch either the No. 2, northbound, at 7:10, or the No. 3, southbound, at 7:19. The morning was spent inspecting the line. He carried a black umbrella and customarily wore white. Invariably he looked spotless; invariably he was smoking a cigarette.

Back at the house again, immediately upon finishing a light lunch, he would rest for half an hour, then walk to his large, square corner office on the first floor at the Administration Building. There he would receive people until dinner at seven. In the evening, unless otherwise engaged, he would return to the office to concentrate on his paper work until about ten.

To most observers he seemed wholly oblivious of his surroundings, intent only on his work. One employee, relaxing on his own porch one particularly beautiful moonlit evening, witnessed the following scene:

> "There were only a few lights here and there in the Administration Building. One by one they went out, all except that in the old man's office. It was getting on toward ten when his window went dark. . . . A full moon, as big as a dining-room table, was hanging down about a foot and a half above the flagstaff—a gorgeous night. The old man came out and walked across the grass to his house. He didn't stop to look up at the moon; he just pegged along, his head a little forward, still thinking. And he hadn't been in his own house ten minutes before all the lights were out there. He'd turned in, getting ready to catch that early train. . . ."

To his elder son, Goethals wrote that he was better off occupied, since there was nothing else to do. He confessed to working so hard that he would often end the day in a kind of daze. He was not the "clean-desk" man Stevens had been. His "IN" and "OUT" baskets were always jammed. Papers were piled wherever there was room on his desk—correspondence, folded maps, specifications, plans, half a dozen black notebooks, reports in heavy dark-blue bindings. The bit of clear desk surface he managed to maintain directly in front of him was soon peppered with cigarette burns.

He liked things on paper. If during his morning excursions along the line a department head or engineer urged some new approach or improvement, the inevitable response was "Write it down."

It was not in him to court popularity. He wanted loyalty first, not to him but to the work, that above all. He abhorred waste and inefficiency and he was determined to weed out incompetents. Nor was there ever to be any doubt as to his own authority. "What the Colonel said he meant," a steam-shovel engineer remembered. "What he asked for he got. It didn't take us long to find that out." Requests or directives from his office were not to be regarded as subjects for discussion. When the head of the Commissary Department, a popular and influential figure, informed Goethals that he would resign if Goethals persisted in certain changes in the purchasing procedure, Goethals at once informed him that his resignation was accepted and refused to listen when he came to retract the threat. "It will help bring the outfit into line," Goethals noted privately. "I can stand it if they can." He put Lieutenant Wood in as a replacement. ". . . I just put it up to him to make good . . ." he wrote.

"Executive ability," he observed on another occasion, "is nothing more or less than letting the other fellow do the work for you." But to some he gave every appearance of wishing only to dominate everything himself. Marie Gorgas, in particular, found him "grim, self-sufficing." He was much too abrupt for her liking. "His conversation and his manners, like his acts, had no finesse and no spirit of accommodation." She grew to dislike him heartily. Even Robert Wood, who admired his "iron will and terrific energy," found him "stern and unbending—you might say a typical Prussian. . . . I was his assistant for seven years," Wood recalled long afterward, "and I might say that everything in my life since has seemed comparatively easy."

But if the manner was occasionally severe, the standards demanding, he was invariably fair and gave to the job a dignity it had not had before. "I never knew him to be small about anything," recalled an electrical engineer named Richard Whitehead, who joined the force that same summer of 1907. Goethals knew how to pick men. He knew how to instill determination, to get people to want to measure up. He was not loved, not then or later, but he was impressive. And by late summer he had "the outfit in line."

"Another week of observation has confirmed my view . . . that the discontent and uneasiness which followed the departure of Stevens have nearly passed away . . ." wrote Joseph Bucklin Bishop to Theodore Roosevelt in mid-August. Undersized and grouchy-looking, with a little, pointed gray beard and a shiny bald head, Bishop was another new addition. He had been transferred from the Washington office on Roosevelt's orders and was to be at Goethals' side from then on, as secretary of the commission, ghost writer, policy adviser, alter ego. And not incidentally he was to feed confidential reports to the White House on how things were going.

Goethals, reported Bishop, was "worn and tired and says that he has had a veritable 'hell of a time,' but I believe he has won out. When I told him so, he said, 'Well, I don't know.' "

Mrs. Goethals had arrived and departed meantime. So his marriage, characterized years afterward by members of the family as "difficult," became still another topic for local speculation as the lights in his office burned on into the night.

At Bishop's suggestion, Goethals started a weekly newspaper, the *Canal Record*, the first such publication since de Lesseps' *Bulletin du Canal Interocéanique* and very similar in format. Goethals insisted that the paper be neither a rehash of news from the United States nor a means for trumpeting the reputation of anyone on the canal commission. Indeed, quite unlike the de Lesseps' paper, its editorial policy specifically forbade praise of any official. The objective was to provide the American force—as well as Congress—with an accurate, up-to-date picture of the progress being made, something hitherto unavailable in any form, as well as reports on social life within the Zone, ship sailings, sports, any activities "thought to be of general interest."

With Bishop as editor, the first edition appeared September 4, 1907. The style was direct and factual and so it would remain, except for occasional letters from employees. Still it was an amazing morale builder. It did for its readers much what *Stars and Stripes* would do for the A.E.F. in France. It brought the strung-out settlements in closer touch, made the Zone more of a community. In addition, it had an almost instant effect on productivity.

Bishop began publishing weekly excavation statistics for individual steam shovels and dredges, and at once a fierce rivalry resulted, the gain in output becoming apparent almost immediately. "It wasn't so hard before they began printing the *Canal Record*," a steam-shovel man explained to a writer for *The Saturday Evening Post*. "We were going along, doing what we thought was a fair day's work . . . [but then] away we went like a pack of idiots trying to get records for ourselves."

To give employees opportunity to air their grievances, Goethals next established his own court of appeal. Every Sunday morning, from about 7:30 until noon, he was at his desk to receive any and all who had what they believed to be a serious complaint or problem. He saw them personally, individually, on the basis of first come, first served, irrespective of rank, nationality, or color. By late 1907 there were thirty-two thousand people on the payroll, about eight thousand more than when he took over. By 1910 there would be nearly forty thousand. Yet once a week, beginning in the fall of 1907, any of these people—employees or dependents—could "see the Colonel" and speak their minds.

The scene was unique in the American experience, unique and memorable in the eyes of all who saw it. Jules Jusserand, the French ambassador, likened it to the court of justice held by Saint Louis beneath the oak at Vincennes. "One sees the Colonel at his best in these Sunday morning hours," wrote a reporter who had been greatly frustrated by what seemed a congenital inability on Goethals' part to talk about himself. "You see the immensely varied nature of the things and issues which are his concern. Engineering in the technical sense seems almost the least of them."

Some advance screening was done. Bishop saw the English-speaking workers, while the Italians, Spaniards, and other Europeans were seen by a multilingual interpreter, Giuseppe Garibaldi, grandson of the Italian liberator. And often these preliminary interviews were enough to resolve the problem—the mere process of free expression gave the needed relief—but if not, Goethals' door stood open.

On an average Sunday he saw perhaps a hundred people and very few appear to have gone away thinking they had been denied justice. They came to the front of the tall, barnlike Administration Building, entered a broad hallway hung with maps and blueprints and there waited their turn. Their complaints included everything from the serious to the trivial: harsh treatment by a foreman, misunderstandings about pay, failure to get a promotion, dislike of the food or quarters, insufficient furniture. He listened to appeals for special privileges and financial dispensation. One request was for the transfer of a particular steam-shovel engineer to a different division where a particular baseball team needed a pitcher. (The request was granted.) He was given constructive ideas regarding the work and was made party to the private quarrels between husbands and wives or families in adjoining apartments. By all accounts he was a patient listener.

Many complaints could be settled at once with a simple yes or no or by a brief note sent down the line. A serious situation of any complexity was promptly investigated. "He was a combination of father confessor and Day of Judgment," wrote Bishop. The vast majority who came before him were almost excessively respectful. Rarely would anyone challenge his authority and then to no avail. "If you decide against me, Colonel, I shall appeal," one man declared. "To whom?" Goethals asked.

Some of the remaining officials from the Stevens regime had expressed vehement disapproval when these Sunday sessions were first announced. Jackson Smith, of the Labor Department, had been especially exercised, since his own policy in past years had been to tell anyone who had a complaint to feel free to leave on the next ship. And this, apart from Smith's own rude manner, had been considered a perfectly appropriate policy. Stevens had been in full accord. The new approach was in fact wholly unorthodox by the standards of the day. In labor relations Goethals was way in advance of his time, and nothing that he did had so discernible an effect on the morale of the workers or their regard for him: "they were treated like human beings, not like brutes," Bishop recalled, "and they responded by giving the best service within their power."

In Goethals' own estimate, expressed privately many years afterward, it was thus that he won "control of the force," and control of the force was "the big, attractive thing of the job."

When another delegation of congressmen, members of the House Appropriations Committee, arrived in November, they were impressed as much by Goethals as by the strides being made, a point of special satisfaction at the White House. "I was present at all the hearings . . ." Bishop wrote to Roosevelt. "Not only did he [Goethals] show that he knew his business thoroughly, had absolute grasp of the work as a whole, but that he had at his tongue's end more knowledge of details than any of his immediate subordinates."

Before leaving for Washington the chairman of the delegation, Congressman James A. Tawney, told Goethals privately not to worry about appropriations—he could count on whatever he wanted. The committee reported the situation in Panama to be in "excellent shape." And as time went on, Goethals' standing on Capitol Hill was to be a factor of the greatest importance. Money sufficient to do the job correctly was never to become an issue.

"There is only one man who should be heard at Washington on the Canal, and that is Goethals," Bishop stressed to Roosevelt. "He has absolute knowledge, perfect manners, and can talk. . . . He says I am the man who should be spokesman rather than he, but don't let him persuade you into such a belief. He is the man at the helm . . ."

Within less than a year after Goethals took charge, several major changes were made in the basic plan of the canal, and with a sweeping reorganization, beginning in early 1908, he installed his own entirely new regime. The widespread impression was that the plan was firm, that this at last was the canal that was to be built, and that these were the men who would build it. The widespread impression was correct.

The changes, each very important, were as follows:

—The bottom width of the channel through Culebra Cut was to be made half again wider, from two hundred to three hundred feet. Thus it was to be more than four times as broad as the French canal would have been at that point.

—The width of the lock chambers was enlarged, primarily to satisfy the Navy. The locks would be 110 feet wide (rather than 95 feet) to accommodate the largest battleship then on the drawing boards, the *Pennsylvania*, which had a beam of 98 feet. (The largest commercial vessel then being built was the *Titanic*, with a beam of 94 feet.) So each lock chamber was to be 110 feet by 1,000 feet.

—On the Pacific side, where heavy silt-bearing currents threatened to clog the entrance to the canal, the engineers now planned a tremendous breakwater that would reach three miles across the tidal mud flats to Naos Island.

—When trestles began sinking in the mud at the site of the Sosa Dam, a major change had to be made in the placement of the Pacific locks. Previously, there was to have been one lock at the south end of Culebra Cut, at Pedro Miguel, then an intermediate-level lake and another set of two locks close to the Pacific shore, at Sosa Hill. In the new arrangement, the Pedro Miguel complex remained unchanged, but the dam and second set of locks were pulled back from Sosa Hill—back from the Pacific—to a new site at Miraflores. Consequently the terminal lake (called Sosa Lake on the old plan) was greatly reduced in area and the first flight of locks at the Pacific end was now to be as far inland as were the Gatun Locks. From the military viewpoint this was regarded as a far better solution, since the Pacific locks would now be far less vulnerable to bombardment from the sea, a point Goethals had made to Taft as early as 1905, following their tour of the area. The possibility of bombardment from the air had not been considered then, nor was it now late in 1907, since the world had as yet to catch up to the achievements at Kitty Hawk.

With his reorganization Goethals did away with all the old departments first established by Wallace and carried on by Stevens. Under that system the work had been portioned off according to specific types of activity—excavation and dredging, labor and quarters, and so forth. Now everything was simply divided into three geographic units—an Atlantic Division, a Central Divison, and a Pacific Division—each run by one overall chief who was responsible for virtually everything within the district other than sanitary and police activities. It was a scheme very like that used by the French, with the fundamental difference that none of the work was to be done by contract, except for the lock gates. Stevens' contract plan had been dropped at the time Goethals took over.

The Atlantic Division included the four miles of sea-level approach from Limon Bay, Gatun Locks, and Gatun Dam. The Pacific Division included the sea-level entrance at that end, as well as the locks and dams. Everything in between, some thirty-two miles of canal and including Culebra Cut, comprised the Central Division.

The Atlantic side was to be run solely by Army men, with Major Sibert as division head assisted by several other engineering officers. Forty-seven years old, large, headstrong, full of ambition and good humor, William Sibert was cut from much the same pattern as John Stevens, with whom he was one day to collaborate on a book about the canal. Sibert's civilian clothes fit him badly, he chewed on unlit cigars, and he spoke his mind. His relations with Goethals, strained from the start, were to become more and more unpleasant.

Born on a farm in Alabama, Sibert had finished at West Point in 1884, worked on the famous Poe Lock at the Soo Canal and ran a railroad in the Philippines. But for the past six years, assigned to river and harbor work at Pittsburgh, he had built more than a dozen locks and dams on the Allegheny, Monongahela, and Ohio rivers. His experience in such work was second to none, a point neither he nor Goethals would lose sight of.

The Central Division was assigned to Major Gaillard, but his highly competent executive officer was a civilian, a lean, red-haired Bostonian named Louis K. Rourke, who had been running things very well in Culebra Cut for nearly two years.

David Du Bose Gaillard (pronounced Ge-*yard*) was a South Carolinian. He was a year older than William Sibert and a close friend. As cadets at the Military Academy they had been roommates and were known as David and Goliath. Still slim and youthful-looking, Gaillard had had a solid if unspectacular career in the Corps of Engineers and like Goethals had been singled out for the initial General Staff. "Sibert's experience on locks and dams makes his assignment to that work very necessary," Goethals explained to his West Point son, ". . . so Gaillard had to take the Cut."

Like Goethals, these and the other engineering officers who were to serve in Panama considered themselves part of an honored tradition; and this, it should be emphasized, gave to their whole mode of operation a very different tone from that of the previous regime. It was not that they were necessarily superior technicians to the railroad people who preceded them, but that their entire training and experience had been directed toward large construction works in the national interest. They were engineers of the state, no less than those who had come out from France to build the de Lesseps canal. Even their training had been patterned after that of the École Polytechnique, from the time Sylvanus Thayer instituted the sweeping academic reforms at West Point that were to make him "Father of the Military Academy." It was Thayer in the 1820's who, after observing the program of the famous French school, made engineering the heart of the curriculum at West Point and instilled the mission to construct into the academic program. "We must get up early, for we have a large territory," a cadet once explained to a visitor in the 1850's; "we have to cut down the forests, dig canals, and make railroads all over the country." And that had remained the prevailing spirit. Only the top men from each class qualified for the Engineers.

But the Goethals regime did not consist solely of Army people, the common view again notwithstanding. Indeed, the only division head that he personally appointed was a civilian, Sydney B. Williamson, who had been a young assistant at Muscle Shoals when Goethals constructed the high-lift lock. He and Williamson had worked well together then and on several subsequent projects, and their trust in each other was total. Williamson was put at the head of the Pacific Division and all his subordinate engineers were to be civilians. So naturally the lines were drawn: if the Army was to build the Atlantic locks and the civilians the Pacific locks, then it would be a test to see which group was the most resourceful and competent. A sharp rivalry ensued, just as Goethals anticipated.

Meantime, Rear Admiral Harry Harwood Rousseau, who at thirty-eight was the youngest member of the canal commission, was given responsibility for the design and construction of all terminals, wharves, coaling stations, dry docks, machine shops, and warehouses. Lieutenant Frederick Mears, aged twenty-nine, was put in charge of relocating the Panama Railroad, a large and very difficult task. To build the forty-odd miles of the new line would take five years and cost nearly $9,000,000.

Two further resignations were announced, those of Joseph Ripley, who had been Stevens' choice for lock design, and Jackson Smith, whose competence Goethals recognized but whose manner had become more than Goethals was willing to tolerate. As a result Smith's Department of Labor and Quarters was broken up and Major Carroll A. Devol was named Chief Quartermaster of the Zone, with responsibility for labor, quarters, and supplies. The personal choice of Secretary Taft, Devol had been in charge of the Army transport service in San Francisco in 1906 at the time of the earthquake and had managed the distribution of all supplies to the stricken city, an enormous and ably handled operation for which the Army was wholly responsible and for which the Army was to get too little credit.

Up until now all the design work on the locks had been handled in Washington, but with Ripley's departure, Goethals transferred the design staff to the Isthmus and installed still another Army officer, Lieutenant Colonel Harry Foote Hodges, at its head.

Everything considered, Hodges was probably Goethals' most valuable man, as well as the sort journalists and historians could readily overlook. Born in Boston, class of '81 at the Academy, he was small, fussy, humorless, quite unspectacular in manner and appearance. With his sharp little face and large, dark, intense eyes, he looked not unlike a bright mouse. Like Sibert, he had spent several valuable years working with Colonel Poe on the Soo, and like Sydney Williamson, he was Goethals' personal choice. Hodges, henceforth, had overall responsibility for the design and erection of the lock gates, all the tremendous conduits and valves beneath the walls and floors of the locks, every intricate mechanism required. He had, that is, the most difficult technical responsibility in the entire project, upon which depended the canal's success. When Goethals was away from the Isthmus, Hodges would serve as acting chief engineer. According to Goethals, the canal could not have been built without him.

II

The "special wonder of the canal" was Culebra Cut. It was the great focus of attention, regardless of whatever else was happening at Panama. The building of Gatun Dam or the construction of the locks, projects of colossal scale and expense, were always of secondary interest so long as the battle raged in that nine-mile stretch between Bas Obispo and Pedro Miguel. The struggle lasted seven years, from 1907 through 1913, when the rest of the world was still at peace, and in the dry seasons, the tourists came by the hundreds, by the thousands as time went on, to stand and watch from grassy vantage points hundreds of feet above it all. Special trains had to be arranged to bring them out from Colón and Panama City, tour guides provided, and they looked no different from the Sunday crowds on the Boardwalk at Atlantic City. Gentlemen wore white shoes and pale straw hats; ladies stepped along over the grass in ankle-length skirts and carried small, white umbrellas as protection from the sun. A few were celebrities: Alice Roosevelt Longworth, Lord Bryce, President Taft, and William Jennings Bryan (who "evinced more general excitement than anyone since T.R."). "He who did not see the Culebra Cut during the mighty work of excavation," declared an author of the day, "missed one of the great spectacles of the ages—a sight that no other time, or place was, or will be, given to man to see." Lord Bryce called it the greatest liberty ever taken with nature.

A spellbound public read of cracks opening in the ground, of heartbreaking landslides, of the bottom of the canal mysteriously rising. Whole sides of mountains were being brought down with thunderous blasts of dynamite. A visiting reporter engaged in conversation at a tea party felt his chair jump half an inch and spilled a bit of scalding tea on himself.

To Joseph Bucklin Bishop, writing of "The Wonderful Culebra Cut," the most miraculous element was the prevailing sense of organization one felt. "It was organization reduced to a science—the endless-chain system of activity in perfect operation."

On either side were the grim, forbidding, perpendicular walls of rock, and in the steadily widening and deepening chasm between—the first man-made canyon in the world—a swarming mass of men and rushing railway trains, monster-like machines, all working with ceaseless activity, all animated seemingly by human intelligence, without confusion or conflict anywhere. . . . The rock walls gave place here and there to ragged sloping banks of rock and earth left by the great slides, covering many acres and reaching far back into the hills, but the ceaseless human activity prevailed everywhere. Everybody knew what he was to do and was doing it, apparently without verbal orders and without getting in the way of anybody else. . . .

Generally, the more the observer knew of engineering and construction work, the higher and warmer was his appreciation.

Panoramic photographs made at the height of the work gave an idea of how tremendous that canyon had become. But the actual spectacle, of course, was in vibrant color. The columns of coal smoke that towered above the shovels and locomotives—"a veritable Pittsburgh of smoke"—were blue-black turning to warm gray; exposed clays were pale ocher, yellow, bright orange, slate blue, or a crimson like that of the soil of Virginia; and the vibrant green of the near hills was broken by cloud shadow into great patchworks of sea blue and lavender.

The noise level was beyond belief. On a typical day there would be more than three hundred rock drills in use and their racket alone—apart from the steam shovels, the trains, the blasting—could be heard for miles. In the crevice between Gold Hill and Contractors Hill, where the walls were chiefly rock, the uproar, reverberating from wall to wall, was horrible, head-splitting.

For seven years Culebra Cut was never silent, not even for an hour. Labor trains carrying some six thousand men began rolling in shortly after dawn every morning except Sunday. Then promptly at seven the regular work resumed until five. But it was during the midday break and again after five o'clock that the dynamite crews took over and began blasting. At night came the repair crews, men by the hundreds, to tend the shovels, which were now being worked to the limit and taking a heavy beating. Night track crews set off surface charges of dynamite to make way for new spurs for the shovels, while coal trains servicing the shovels rumbled in, their headlights playing steadily and eerily up and down the Cut until dawn. And though it was official I.C.C. policy that the Sabbath be observed as a day of rest, there was always some vital piece of business in the Cut that could not wait until Monday.

Among the most fascinating of the surviving records of the work is a series of Army Signal Corps films made down in the Cut. Watching these rare old motion pictures (now in a collection at the National Archives), seeing the trains cut back and forth across the screen, seeing the dynamite go off and tiny human figures rush about through clouds of dust and smoke, one senses too how extremely dangerous it all was. At one point, when a shovel suddenly swings, Goethals can be seen to jump nimbly out of the way.

Bishop and those others who described the spectacle from the cliffs above had very little to say about such hazards. But year after year hundreds of men were being killed or hideously injured. They were caught beneath the wheels of trains or struck by flying rock, crushed to death, blown to bits by dynamite. "Man die, get blow up, get kill or get drown," recalled one black worker; "during the time someone asked where is Brown? He died last night and bury. Where is Jerry? He dead a little before dinner and buried. So on and so on all the time."

Construction of the canal would consume more than 61,000,000 pounds of dynamite, a greater amount of explosive energy than had been expended in all the nation's wars until that time. A single dynamite ship arriving at Colón carried as much as 1,000,000 pounds—20,000 fifty-pound boxes of dynamite in one shipload—all of which had to be unloaded by hand, put aboard special trains, and moved to large concrete magazines built at various points back from the congested areas.

At least half the labor force was employed in some phase of dynamite work. Those relatively few visitors permitted to walk about down in the Cut saw long lines of black men march by with boxes of dynamite on their heads, gangs of men on the rock drills, more men doing nothing but loading sticks of dynamite into the holes that had been drilled. The aggregate depth of the dynamite holes drilled in an average month in Culebra Cut (another of those statistics that defy the imagination) was 345,223 feet, or more than sixty-five miles.* In the same average month more than 400,000 pounds of dynamite were exploded, which meant that all together more than 800,000 dynamite sticks with their brown paper wrappings, each eight inches long and

* The drills themselves were of two types, a well drill that could bore a hole five inches in diameter to a depth of one hundred feet and a smaller tripod drill that could bore a three-inch hole to a depth of thirty feet. These drills were all powered by compressed air fed into the Cut through some thirty miles of pipe from big compressors at Rio Grande, Empire, and Las Cascadas. The elaborate compressed air system was another of those advances that distinguished the American effort from that of the French.

weighing half a pound, had been placed in those sixty-five miles of drill holes, and again all by hand.

Difficulty was had at first in determining how much dynamite to use in a single shot, depending on the depth of the holes, the spacing of the holes, and the character of the rock, which could be anything from basalt to the softest shale. The foremen responsible for the loading and tamping learned by trial and error. Different grades of powder were tried, different kinds of fuses and methods of firing.

Premature explosions occured all too often as the pace of work increased. "We are having too many accidents with blasts," Goethals noted in June 1907. "One killed 9 men on Thursday at Pedro Miguel. The foreman blown all to pieces." Several fatal accidents were caused when shovels struck the cap of an unexploded charge. Another time a twelve-ton charge went off prematurely when hit by a bolt of lightning, killing seven men. Looking back years later, one West Indian remembered, "The flesh of men flew in the air like birds many days."

The worst single disaster occurred on December 12, 1908, at Bas Obispo. More than fifty holes had been drilled in the solid rock on the west bank of the Cut and these had been loaded with some twenty-two tons of dynamite. The charges had been tamped, the fuses set, but none of the holes had been wired since the blast was not scheduled until the end of the day. As the foreman and one helper were tamping the final charge, the whole blast went off, by what cause no one was ever able to determine. Twenty-three men were killed, forty injured.

As time went on the men became extremely proficient and accidents became comparatively rare considering the volume of explosives being used and the numbers of laborers involved. Still, more men would be killed, and very often, as at Bas Obispo, there would be too little left of them to determine who they were.

The shovels in the Cut set records "never anticipated," as Goethals noted, and in the eyes of most beholders they became something more than mere machines. They had personality and gender–usually feminine, yet they were also likened to Theodore Roosevelt–and accounts of their prodigious feats of strength, as well as their agility, acquired a kind of mythical quality. The *Canal Record*'s full-page reports on their performance were read as avidly as baseball scores.

The peak was in March 1909, when sixty-eight shovels, the largest number ever used at one time in the Cut, removed more than 2,000,000 cubic yards, ten times the volume achieved by the French in their best month. The record for a single shovel was set in March 1910, when a ninety-five-ton Bucyrus (No. 123), working twenty-six days, excavated 70,000 cubic yards. More astonishing is the realization that the vast rift in the earth at Culebra was dug entirely by what, comparatively speaking, was a mere handful of machines. The volume removed from the Cut was 96,000,000 cubic yards. So even allowing for replacements, the average shovel dug well over 1,000,000 cubic yards, despite the worst kind of punishment year in, year out. No machines had ever been subjected to such a test and their record was a tribute to the men who designed and built them.

The shovels were deployed along the entire nine miles of the Cut, but in one section just to the north of Gold Hill they were stacked one above another at seven different levels, while seven parallel tracks carrying the dirt trains were kept constantly busy. "There were any amount of . . . trains, which were going in every direction," noted a young English tourist in her diary; "they must be very well arranged." In fact about 160 trains a day were running in and out of the Cut, and the degree of planning needed to handle such traffic can be further appreciated when it is taken into account that most of the track had to be shifted—removed, replaced, relocated—time and again. There were 76 miles of construction track within the nine-mile canyon, while in the Central Division as a whole there were 209 miles, not counting the Panama Railroad. In any one year well over a thousand miles of track had to be shifted about within that area just to keep the work moving in the Cut. And to complicate the problem further still, the bottom of the Cut, the main work level, kept steadily contracting in width the deeper the Cut became.

No one part of the operation—not the drilling, the blasting, the shoveling, the dirt hauling—could ever be permitted to interfere or disrupt another. So consequently every move was the result of very careful study. All shovels, every mile of track, every one of the hundreds of rock drills in use, were located daily on a map at division headquarters at Empire. Careful estimates were made as to the progress of each individual steam shovel, when it would have to be repositioned, when tracks would have to be shifted, what effect such moves would have on the disposition of drilling and blasting crews. So neatly was everything coordinated, so smooth were communications, that at the close of each day locomotive crews, as an example, had only to check the assignment boards at the roundhouses to see exactly what they were to do the day following.

Traffic in and out of the Cut was directed from towers at either entrance by yardmasters who kept in telephone contact with the various dumps and with a half-dozen small towers strung out along the line of excavation. The yardmasters, who took their orders from the chief dispatcher at Empire, directed the passage of each loaded train to a particular dumping ground and ordered the right of way for the train when it hit the main line of the Panama Railroad. When the empty trains returned, it was the yardmaster again who distributed them to the shovels.

The dumping grounds—the other end of the system—were located anywhere from one to twenty-three miles from the Cut. Sixty-odd locations were used in the course of excavation, and though much of the spoil was simply gotten rid of—that is, put to no useful purpose—a very considerable part of it served to build earth dams, to build embankments on the new line of the railroad, and to create the huge new Naos Island breakwater at the Pacific end. To keep the flooding Chagres from backing up into the Cut as the great trench deepened, an earth dike was thrown across the north end, at Gamboa, seventy-eight feet above sea level.

All the dumps were carefully engineered, with tracks on several terraces. At each dump was another yardmaster who reported the arrivals and departures of trains and his "readiness for spoil," who ordered the distribution of loaded trains to the several dumping tracks, and who, in addition, directed the movements of the Lidgerwood unloaders as well as two additional pieces of equipment that had since come into use: the dirt spreader and the track shifter.

Both devices were of vital importance to the efficiency of the entire system, since the least delay at the dumping end at once decreased progress in the Cut. The dirt spreader was a railroad car with big steel blades mounted on either side, these operated by compressed air. Once a train had been unloaded, its spoil dumped beside the tracks, the spreader came through, pushed by a locomotive, and did the job of several hundred men working with shovels. The track shifter, an even cruder-looking piece of equipment, was the creation of William Bierd, former head of the Panama Railroad, who had built the first one in the shops at Gorgona shortly before Goethals' arrival. It was a huge crane-like contraption that could hoist a whole section of track—rails, ties, and all—and swing it in either direction. And since the tracks at the dumps had to be shifted constantly, to keep pace with the loads being delivered, it was an extremely valuable adjunct. Bierd's own creation could shift track about three feet, but subsequent models, built after he resigned, could reach as much as nine feet. With one such rig, fewer than a dozen men could move a mile of track in a day, a task that would have taken not less than six hundred men working by hand.

The largest of the dumps were at Tabernilla (fourteen miles beyond the north end of the Cut), Gatun Dam (the most distant location), Miraflores, and La Boca, the largest, which had been renamed Balboa. Some of the dumps covered as much as a thousand acres, and in the rainy season they became great seas of mud, with tracks slipping and sinking five or six feet. At Tabernilla, more than 16,000,000 cubic yards of spoil were simply dropped in the jungle. At Balboa, 22,000,000 cubic yards were deposited, with the result that 676 acres were reclaimed from the Pacific as a site for a new town.

By far the most troublesome of the dumps was the Naos breakwater, where, as at Gatun Dam, spoil from the Cut was dumped from a huge trestle, this one being extended slowly across the mud flats of the bay. At first everything went as hoped. But then the soft bottom sediments began to give way beneath the heavier material being poured on top. Overnight whole sections of trestle and track would vanish into mud and everything would have to stop until they were replaced. In some areas the vertical settlement exceeded a hundred feet, while the slippage sideways was three times worse. In time not a single foot of the long trestle remained where it had been to start with. By 1910 well over 1,000,000 cubic yards of spoil had been dumped into the breakwater and still it was a mile short of Naos Island. To reach the island, ultimately, would require 250,000 cubic yards of earth and rock from Culebra, which was ten times what had been originally estimated.

"Culebra Cut was Hell's Gorge," one steam-shovel man would write, recalling the heat and dust and noise. Nor were the rains any less of a problem than in times past. In 1908 and again in 1909, the years of the heaviest work, well over ten feet of rain fell. To check the torrential runoff, to reduce the chance of landslides, Goethals did what the French had done: he had diversion channels dug parallel to the Cut. But he greatly expanded on their plan. The channel on the east side of the Cut, known as the Obispo diversion, ran for a distance of five and a half miles and had a minimum width of fifty feet. To build this ditch, and another similar to it on the opposite side, the so-called Camacho diversion, required another 1,000,000 cubic yards of excavation. And very possibly they were a mistake, as Goethals himself later conceded, since they were dug too close to the Cut and water seeping from them below ground may have been the cause of several of the more disastrous slides.

All technical problems at Panama were small problems compared to the slides in the Cut. The building of the great dam at Gatun, for so long the most worrisome part of the plan, turned out to be one of the least difficult tasks of all. A tremendous man-made embankment simply grew year by year at Gatun, extending a mile and a half across the river valley, a ridge of earth that was to be fifteen times as wide at its base as it was high. At the eastern end were the beginnings of the Gatun Locks; in the center were the beginnings of what was to be the dam's giant concrete spillway. Two big outer walls of "dry" spoil were built first as a base for the embankment. These toes, as they were called, were nearly half a mile apart—the river, meantime, having been turned into an old diversion channel built by the French—and into the space between them was pumped hydraulic, or "wet," fill, a solution of blue clay, which when dry would create a core almost as impervious as concrete. There was no lack of controversy over the project as time went on (much of it stirred up by Philippe Bunau-Varilla, who was convinced that Goethals did not know what he was doing), and once, on November 20, 1908, a section about two hundred feet long slipped sidewise and sank nearly twenty feet at the point where the dam crossed the old French canal. In the face of a storm of criticism and alarm in the newspapers, Goethals insisted that the situation was not serious and as it turned out he was perfectly correct. The damage was repaired; the work went on.

The slides, however, were a wholly different matter. The first occurred early in the fall of 1907, or just as Goethals was beginning to feel he had things under control.

The Cucaracha slide, located on the east bank of the Cut just south of Gold Hill, was the slide that had given the French such grief. On the night of October 4, 1907, after days of unusually heavy rain, Cucaracha "started afresh." Without warning, an avalanche of mud and rock plunged into the bottom of the Cut, destroying two steam shovels, obliterating all track in its path. And for days afterward that same part of the slope, about fifty acres in area, kept moving down and down, slipping anywhere from ten to fifteen feet a day. "It was, in fact, a tropical glacier—of mud instead of ice," Major Gaillard noted in an article for *Scientific American*, "and stakes aligned on its moving surface and checked every 24 hours by triangulation, showed a movement in every respect similar to stakes on moving glaciers in Alaska upon which the writer has made observations in 1896." After ten days, when the slipping stopped, 500,000 cubic yards of mud had been dumped into the canal.

In 1910 Cucaracha let go twice again, burying shovels, track, locomotives, flatcars, and compressed-air lines. The entire south end of the Cut was bottled up for months. Within a year Gaillard reported that the worst of the slides were over, but in fact they were still to come. From 1911 on, as the Cut grew very much deeper, the slides occurred season after season and grew increasingly worse. "No one could say when the sun went down at night what the condition of the Cut would be when the sun arose the next morning," Bishop wrote. "The work of months and years might be blotted out by an avalanche of earth or the toppling over of a small mountain of rock." There were slides at Las Cascadas, La Pita, Empire, Lirio, East Culebra–twenty-two slides all together. Cucaracha was almost never still. It took three months to dig out the rock and mud dumped into the Cut by slides in 1911. In 1912 more than a third of the year, four and a half months, was spent removing slides. On one day more than a hundred trains would roll out of the Cut; the next day there would be none, because a monstrous slide had occurred.

Steam shovels were buried so deep in mud that only the tips of their cranes were left protruding. Hundreds of miles of track disappeared or were twisted into crazy roller-coaster patterns. In one bizarre instance a shovel and track were picked up by a landslide and were deposited unharmed halfway across the floor of the Cut.

On some of the terraced slopes the ground crept ever so slowly, barely inches a day, which was never enough to do any serious damage, but for two years gangs of men had to be kept constantly at hand, day after day, moving the track back to where it belonged.

At another place a slow but relentless slide kept perfect pace with the steam shovel working at its base. The shovel never had to move; as much as it dug, the slide replenished.

For the engineers the problem was not merely the size of the slides. They were also confronted with a type nobody had anticipated. Those slides that had beset the French, like the comparative few experienced by Wallace and Stevens, were normal, or gravity, slides–Cucaracha being the largest and most destructive example. As explained earlier, they nearly always occurred in the rainy season, when a top layer of soft, porous material slid from the sloping plane of underlying rock, "like snow off a roof," as one American said. But the new variety, and much the worst, were what geologists classified as structural break and deformation slides. They were due not to sliding mud, but to unstable rock formations, the height of the slopes, and, in part, to the effects of heavy blasting. As the Cut deepened, the underlying rock formations of the slopes lost their lateral support and were unable to withstand the enormous weight from above. It was as if the flying buttresses had been removed from the wall of a Gothic cathedral: the exposed wall of the Cut simply buckled outward under its own load and fell. Rains and saturation actually had little to do with such slides. In fact, some of the most horrendous happened during the dry season.

The first signs of trouble were huge cracks in the ground running along the rim of the Cut, anywhere from a few feet to a hundred yards back from the edge. The next stage might come weeks or months later, or it might take years. A settling or outward tilt of big blocks, whole sections of the slope, would commence. Then the whole slope would give way, sometimes in an hour or two, sometimes over several days.

The worst of such slides occurred in front of the town of Culebra, on the west bank of the Cut, where huge cracks in the ground began appearing in 1911. By the summer of 1912, "the large and annoying Cucaracha" had put an additional 3,000,000 cubic yards in the path of the canal, but the slide on the west bank at Culebra had deposited more than twice that amount. Thirty buildings in the town of Culebra had to be moved back from the brow of the Cut.

"Now suddenly the people living nearest the Cut were being compelled to move," wrote Rose van Hardeveld, the young wife and mother from Wyoming. "The bank was sliding into the Cut! One after another, the houses were being vacated.

"The neighbors three doors east of us were warned time and again that it was not safe to stay. . . . One morning they awakened to find their back steps well on the way to the bottom of the Cut."

Before long some seventy-five acres of the town broke away and fully half of all the buildings had to be dismantled and removed to save them from being carried over the edge. Ultimately these breaks, all occurring in the dry season, dumped 10,000,000 cubic yards into the Cut, while on the opposite side another 7,000,000 cubic yards fell away, with the result that the top width of the Cut at that point was increased by a quarter of a mile.

The slides "seem to be maneuvered by the hand of some great marshal and sent forth to the fray in every way calculated to put the canal engineers to discomfiture," declared the *National Geographic Magazine*. "Now they are quiescent, attempting to lull the engineers into a false security . . . now they come in the dead of night, spreading chaos and disrupting everything in whatever direction they move . . ." To many of the workers it seemed the task would go on forever. "I personally would say to my fellow men," recalled one Barbadian, "that . . . my children would come and have children, and their children would come and do the same, before you would see water in the Cut, and most all of us agree on the same."

Often wisps of smoke would trail from the moving embankments. Once cracks in the surface below Culebra issued boiling water. When Gaillard arrived to investigate the matter, he took a Manila envelope from his pocket and held it over one of the vents in the earth. In seconds the paper was reduced to ashes. The explanation, according to the geologist who was summoned, was "oxidation of pyrite," but the terrified workers were convinced that they were cutting into the side of a volcano.

The most uncanny of all effects, however, was the rising of the floor of the Cut. Not merely would the walls of the canal come crashing down, but the bottom would rise ten, fifteen, even thirty feet in the air, often quite dramatically. Gaillard on one occasion grew concerned as a steam shovel appeared to be sinking before his eyes, but looking again he realized it was not that the shovel was descending, but that the ground where he stood was steadily rising—about six feet in five minutes, "and so smoothly and with so little jar as to make the movement scarcely appreciable."

This phenomenon, diabolical as it seemed, had a simple explanation. It was caused by the weight of the slipping walls of the Cut acting upon the comparatively soft strata of the exposed canal floor. The effect was exactly that of a hand pressed into a pan of soft dough—the hand being the downward pressure of the slides, the rising dough at the side of the hand being the bottom of the canal.

The slides attracted worldwide attention and inspired all kinds of suggestions as to how the problem might be solved, very few of which were practical. The most popular remedy was to plaster the sides of the Cut with concrete, and this was actually tried in one particularly troublesome area, but without success. The concrete crumpled and fell along with everything else as soon as the slide resumed its downward progress.

To check the deformation slides considerable excavation was also done along the uppermost portions of the slopes in an effort to decrease the pressure on the underlying strata. But by and large there was still only one way to cope with the problem and that was the same as it had been since the time of the French—to work for an angle of repose, to keep cutting back at the slopes, to keep removing whatever came down, until the slides stopped. And no one honestly knew how long that might take. By late 1912 at Cucaracha and at Culebra, the chief trouble spots, the angle of inclination was about one on five (one foot vertical to five horizontal). Still the ground kept moving.

Fifteen thousand tourists came to watch the show in 1911 and in 1912 there were nearly twenty thousand. "You are now overlooking the world-famous Culebra Cut," exclaimed the tour guides at the start of their standard spiel. There was more tonnage per mile moving on the tracks below, the visitors were informed, than on any railroad in the world. But meanwhile a big clubhouse at the town of Culebra was being dismantled and removed ("in order to lighten the weight upon the west bank of the canal at this point"), and on January 19 Cucaracha broke loose once again. It was one of the worst slides on record. It spilled the whole way across the Cut and up the other side. All traffic was blocked at that end; for the sixth or seventh time, the slide had wiped out months of work.

Gaillard was practically in shock, according to one account, and Goethals was hurriedly called to the scene. "What are we to do now?" Gaillard asked. Goethals lit a cigarette. "Hell," he said, "dig it out again."

20

Life and Times

For a while we tramped on in silence, till Umbopa, who was marching in front, broke into a Zulu chant about how brave men, tired of life and the tameness of things, started off into a great wilderness to find new things or die, and how, lo, and behold! when they had got far into the wilderness, they found it was not a wilderness at all, but a beautiful place full of young wives and fat cattle.

—H. Rider Haggard
King Solomon's Mines

And this on the slope of the death-dealing Chagres!

—Charles Francis Adams

I

There were six passenger trains daily on the Panama Railroad, three in each direction, and since the railroad was still the one way to get back and forth, the trains were always crowded and the crowds were always interesting to look at. Especially in the cool of the evening every little station platform would be thronged with people, and to anyone newly arrived on the Isthmus it was astonishing to see American women and children in such numbers and all looking so very healthy, clean, and perfectly at home. They were not merely surviving in such alien—and once deadly—soil, but plainly thriving, and this to many visitors was as impressive, as great a source of patriotic pride, as anything to be seen.

One grew tired of hearing of "the largest dam, the highest locks, the greatest artificial lake, the deepest cut," wrote a correspondent for *The*

Outlook, in an effort to explain the thrill he experienced watching a simple, unimportant scene on the platform at Empire. A man returning from work, grimy and wringing with perspiration, was being met by his wife, a woman of perhaps thirty dressed all in white, who held a baby, and by a still younger woman, apparently her sister, who stood "like a Gibson summer girl," holding the hand of a blond little boy with bare legs and wearing an immaculate white Russian tunic. To the correspondent, watching from the train window, there was something quite miraculous about this "New Jersey group," as he called them, and the way the father, holding his greasy hands stiffly behind, bent forward and kissed his wife and children.

Old Charles Francis Adams—brother of Henry Adams, railroad expert, historian—had a similar experience. In 1911, in his seventy-sixth year, Adams had come to Panama for no other purpose than to see the construction in progress. One evening while waiting for the train at Culebra, his eyes fell on a group of American girls, about ten in number and anywhere from ten to fifteen years of age. Each was nicely dressed in a thin white frock and the sight of them in such a setting affected him profoundly. "A more healthy, well-to-do and companionable group of children could not under similar conditions have been met at any station within twenty miles of Boston," he would report to his fellow members of the Massachusetts Historical Society. Several of the girls had come to take the train, the others to see them off. They were chatting and laughing under the glare of the station lights, oblivious of everything but themselves, without chaperons and wholly without fear—of yellow fever, of malaria, of anything whatever as near as he could surmise. "The material, social and meteorological conditions would in every respect have compared favorably with those to which we are accustomed to during the midsummer season; the single noticeable difference was the more complete absence of insect life. . . . And this on the slope of the death-dealing Chagres!"

Gatun Dam, the locks, Culebra Cut, were all tremendously impressive, he affirmed, but dams and waterways had been built before; the majesty of these was largely a question of degree. But this "vanquishing of pestilence," this clean, prosperous, flourishing Anglo-Saxon civilization in the very heart of the jungle, was, he insisted, unlike anything before in history.

Those resident bystanders who remembered the Isthmus from times past were no less incredulous over the transformation. To any of Adams' own generation, those old enough to remember the hell-roaring days of the gold rush, the change was almost inconceivable. Even the French era seemed part of another time, another world. Tracy Robinson, who had been the sole resident American present at the arrival of *Le Grand Français* in 1880, and who was still a familiar figure in Colón, had lived, he said, to see the confirmation of his lifelong conviction concerning the tropics. The tropics had been "set apart for great things," he had said to de Lesseps. And he said it again now with greater faith in his memoirs, an inscribed copy of which he presented to George Goethals: "It seems to me that Design may be clearly traced in this tendency. The evolution of mankind toward a higher destiny is involved."

The enormous discrepancy between white and black society along this same jungle corridor, the point that the Canal Zone was in actuality a rigid caste society, was barely even implied by such observers (which, of course, was in itself another facet of the life and times). And neither did any but a very few question the kind of white community that had evolved, other than to point out, somewhat apologetically, that it was indeed "a sort of socialism." The statement that the Canal Zone was "a narrow ribbon of standardized buildings and standardized men working at standardized jobs" stands almost alone in all to be found in published or private accounts. To William Franklin Sands, an American diplomat, this new civilization created by his countrymen was simply awful, "a drearily efficient state," "a mechanization of human society." "Every American looked and behaved exactly like every other—to the vast bewilderment of the natives, who had previously thought of us as a race of extreme individualists. . . ." To Sands the view from the train window was cause for despair, as he wrote years later:

> From a railway car one could tell by the type of mission furniture and the color of the hammock swings on the back porches the salaries and social standings of the occupants of all the houses that one passed. In some ways the Canal Zone of the early 1900's was a foretaste of those New Jersey and Long Island suburbs of the 1920's where social ratings were according to the number and cost of one's automobiles. . . .

To those who lived in such houses, however, to the individual men and women like those seen by *The Outlook* correspondent, Panama seems to have been the experience of a lifetime, almost without exception. The work, the way of life, the sense of being part of a creative undertaking so much larger and so much more important than oneself, were like nothing they had known, as they openly and cheerfully expressed then and for the rest of their lives. "It is as if each were individually proud of being one of the chosen people and builders of the greatest work of the modern age," noted a thoughtful young rookie on the Zone police force, an aspiring writer named Harry Franck who had been assigned to take a census. Families returning by ship from home leave invariably spoke of their eagerness to get back. "Not one but was ready and even glad to go back," Charles Francis Adams observed of the group he met during the voyage to Colón; "all looked forward to remaining there for the end–till, as the expression went, they 'saw the thing through.' "

"We felt like pioneers," numbers of them would recall a lifetime afterward. Far from home, they were rolling back the wilderness, serving the cause of progress, serving their country in one of its grandest moments. They were building large and building to last. It was to be a "monument for the world." Every day was a story for the grandchildren.

No one could quite see himself as a soldier under fire any longer, far too much was being provided in the way of creature comforts. There were scarcely even inconveniences any more, let alone real hardships to face. There were never "hard times" in the Zone, as Harry Franck observed, "no hurried, worried faces." Morale was amazing.

Very little if any time was spent worrying over the possible detrimental effects of so structured and paternalistic an order. Their days were too busy, for one thing. Furthermore, there was never any question as to the finite nature of their circumstances. If this was an entirely novel social experiment, it was also to be a brief one, they knew. Vestiges of their way of life would survive–some to the present day–but their own particular era would end abruptly just as soon as the canal was built. The houses they lived in, their schools, offices, whole communities, would disappear with the advance of Gatun Lake. Other towns on higher ground would also be struck like stage sets and carried away, leaving not a trace.

They themselves would depart by the thousands, and for the relative few who would decide to stay on, life would become something quite different. Another era would begin once "Big Job" was no longer the common, galvanizing cause.

The work was everything. "Pride and joy in the work," wrote Bishop, "constituted the magic bond which held the canal colony together . . ." There was no one who was not associated with the work. No one could live within the Zone unless he or she was a worker on the canal or a member of a worker's family. The entire social order existed solely for the work and it rewarded its members according to their importance to the work. Indeed, on a small and limited scale, there existed within the American Zone in Panama between, roughly, 1907 and 1914 something very like what Claude de Rouvroy, the Comte de Saint-Simon, had envisioned for the world a century before. All were caught up in a noble effort that was to benefit humanity: the canal he had also envisioned. Society was controlled by a gifted technician, as he had espoused. His famous maxim, "From each according to his ability, to each according to his work," could well have been the motto of the Isthmian Canal Commission.

Joseph Pennell, an artist who came to do a series of lithographs showing the final heroic stages of construction, called the canal "The Wonder of Work" and said the devotion it inspired, the spirit it engendered among the Americans, were of a quality he had never encountered.

Even the animosity or ill-concealed disdain felt by many Americans toward the Panamanians seems to have stemmed in large measure, if not chiefly, from what the Americans took to be a disgraceful lack of regard for "honest work" on the part of the Panamanians. It was not so much that the Panamanian was lazy, that he had done nothing for centuries, or even that he refused now to take a hand with the canal, but that he appeared to sneer at the fundamental belief that hard work could be good in itself, an ennobling act of faith. To large numbers of young American technicians such contempt was little short of blasphemous.

The fundamental problem with the Panamanian, noted census-taker Harry Franck, was that he could not "rid himself of his racial conviction that a man in an old khaki jacket who is building a canal must be inferior clay to a hotel loafer in a frock coat. . . . Even with seven years of American example about him the Panamanian had not yet grasped the divinity of labor. Perhaps he will eons hence when he has grown nearer true civilization."

The full work force in the last years of construction numbered about 45,000 to 50,000, which was nearly equal to the combined populations of Colón and Panama City. But the total number of white North Americans was only about 6,000, of whom roughly 2,500 were women and children. In 1913 there were 5,362 gold-roll employees and dependents, practically all of whom were Americans. Their average pay was $150 a month. A nurse or teacher received $60 to start; clerks and bookkeepers, $100; a doctor, $150; steam-shovel engineers were by now getting $310. The number of women employed was never more than about 300 and the top salary for a woman was $125 (for a railroad telegraph operator).

A young graduate engineer with two or three years' experience could expect to make $250 to start, which was about $25 more than he could make in the United States. And added to his salary was the host of free benefits and services (housing, hospital care) to which all employees were entitled. His annual vacation was forty-two days with pay (this in a day when two weeks was still the standard) and he was entitled to thirty days' sick leave with pay.

If he was single he generally shared a room with another man in one of the bachelor hotels, as they were called, where the phonograph blared "seven kinds of ragtime" through the night and poker games, strictly forbidden by I.C.C. regulations, were carried on "in much the same spirit as Comanche warfare." The buildings were seldom quiet until 4:30 A.M., when the first alarm clocks began going off.

Meals at this same hotel cost 30 cents each (comparable fare in the United States would have run about 75 cents), and since a man needed only work clothes and one or two light suits, his clothing expenses were minimal. For bowling, billiards, pool, a book or a current magazine and the comforts of a Morris chair, for a game of chess, a chocolate soda, the use of a gymnasium or a quiet place in which to write a letter, he could go to one of the Y.M.C.A. clubhouses, where his dues were $10 a year. (One glowing article about the Y.M.C.A. clubhouses written for home consumption was titled "Uncle Sam's Fight with the Devil.")

"In fact, everything is done to make it as pleasant as possible for the men," one steam-shovel engineer wrote home, after getting settled at Culebra, "and I have not seen a man that was not satisfied. As for myself, I like it very much. This is a pretty town."

For diversions or pleasures not provided at the Y.M.C.A. clubhouses, one could take the train to Colón or to Panama City.

An enormous amount of scrubbing and sanitizing had gone on in the two cities. Yet there remained a sharp division between them and the Zone, on the map and in the mind. On Saturday nights, even on an hour's pass, men would rush over the line like sailors on shore leave, eagerly forsaking the unrelieved wholesomeness of the Zone and very often to do nothing more licentious than stroll about Cathedral Plaza or take in a movie at the Electric Theater.

The cities, observed Harry Franck, "serve as a sort of safety valve, where a man can . . . blow off steam; get rid of the bad eternal vapors that might cause an explosion in a ventless society." There were very few saloons within the Zone itself—half a dozen perhaps—and they were small, rough stand-up bars where, as one writer remarked, "the glitter of mirrors and of cut glass was notably absent," and closing time was eleven sharp. The saloons in the cities, however, were "many, varied, and largely disreputable." According to the *Canal Zone Pilot*, a guidebook that first appeared in 1908, there were 131 saloons in Colón, 40 on Bolívar Street alone. In Panama City there was a total of 220 to choose from. And after dark, things could get pretty rough. Prostitution, if not so gaudy or open as in the French era, was commonplace. Coco Grove, the notorious red-light district in Panama City, was a regular stop for the little horse-drawn coaches, or *carimettas*, that stood waiting at the railroad station as each train pulled in.

Most notable of the brothels was the Navajo, on I Street, run by one of the best known of all Americans on the Isthmus, Mamie Lee Kelly, of New Orleans, who would be remembered vividly by one man more than half a century later as "lusty, large, voluptuous, very profane and very capable." Since all such establishments were known locally as "American houses" and their occupants, irrespective of nationality, were known as "American women," a local ordinance was put through making it unlawful for American women to be on the streets after dark, a rule that not surprisingly gave rise to a number of unfortunate misunderstandings.

If a canal employee were to get married, his entire status changed immediately. If he was earning less than $200 a month, he and his bride moved into a furnished, rent-free, four-room apartment, with a broad screened porch (really a fifth room) and a bath. The apartment would be one of four in what was known as a Type 14 house, the model in which the majority of American families was quartered during the construction years. If the employee was making from $200 to $300 a month, then he was eligible for a Type 17 house (two families) or a small individual cottage. Any American earning $300 to $400 lived in a Type 10: two stories, living room, dining room, kitchen, three bedrooms and bath, porches on both levels, while those in the over $400-a-month bracket were given "large houses of a type distinguished by spaciousness and artistic design."

Whatever the husband's rank, his wife would shop at one of the eighteen I.C.C. commissary stores, each essentially a big department store (and forerunner of the military post exchange), which stocked everything from work pants ("Battleship" brand) to lamb chops to finger bowls and at prices nearly always lower than in the States. (The pants were $1.25; the lamb chops, 24 cents a pound; the finger bowls, 10 cents each.) Moreover, as the purchasing department in Washington grew more proficient, the prices kept going down while prices at home were rising. In early 1909, for example, a porterhouse steak was 29 cents a pound in the commissaries, but a year later it was down to 21 cents, which was less than half what it would have cost in New York.

The I.C.C. bakery produced a different fruit pie fresh daily (Monday, apple; Tuesday, mince; Wednesday, peach; etc.). The garbage was collected, lawns were cut, and a black serving girl could be hired for $10 a month. If she did not work out satisfactorily there was always a dozen eager to take her place.

The whole system was in fact quite intentionally designed to favor the married employee, to provide every inducement for matrimony, to bring *stability* to the skilled white segment of the community.

Books from the best-seller list and recordings of the newest hit tunes were no less current than at home. Indeed, with several ships a week arriving from New York, with thousands of tourists pouring through full of news and wearing the latest fashions, many residents of the Zone felt more in touch and up to date than ever before in their lives. For a very large number of the Americans living and working in Panama, perhaps even the majority, the initial voyage to Colón had been their first experience with salt water. Many had never been away from their hometowns. "Lord, yes, I liked it here," recalled Mrs. Winifred Ewing, a schoolteacher at Empire. "I didn't know anything else but the hills of West Virginia."

The Trail of the Lonesome Pine, by John Fox, Jr., and *The Winning of Barbara Worth*, by Harold Bell Wright, were popular reading along the diggings in the years 1907–1914, as were *When a Man Marries*, by Mary Roberts Rinehart, and Zane Grey's *Riders of the Purple Sage*. In *On the Spanish Main*, by John Masefield, a young American engineer relaxing on his screened porch at the end of the day, the light of an electric lamp falling over his shoulder, could imagine himself accompanying Henry Morgan on the trek up the Chagres to sack Old Panama. (" . . . *They rowed all day, suffering much from mosquitoes, but made little progress. . . . To each side of them were stretches of black, alluvial mud, already springing green with shrubs and water plants. Every now and then as they rowed on, on the dim, sluggish, silent, steaming river, they butted a sleeping alligator as he sunned in the shallows. . . .*")

Whether anyone was reading Frederick Winslow Taylor's *The*

Principles of Scientific Management ("Harmony, not discord. Cooperation, not individualism.") after it appeared in 1911, or Freud's *The Interpretation of Dreams* in 1913, cannot be determined.

In the evenings after dinner, with the soft night air stirring the jungle, couples sat with coffee talking of much the same news being talked of at home—the return of Halley's Comet, the women's suffrage movement, the income-tax amendment. The most-talked-of story in 1912 was the same as everywhere, the sinking of the *Titanic*.

A clerk from the Ancon post office, a Mr. S. C. Russel, walked across the Isthmus in fourteen hours, and that too was news. On April 27, 1913, an aviator from California, Robert G. Fowler, flew a small single-engine hydroplane from the Bay of Panama to Limon Bay, the first trancontinental flight, ocean to ocean in an hour and thirty-five minutes. Four others had attempted the flight before but had failed because of the turbulence of the air. Fowler, who had taken a photographer along for the ride, had banked in a big, slow circle over Culebra Cut to get the first aerial views. Far below, at the bottom of the gaping chasm, men were looking up and wildly waving to him.

The biggest social occasions, year in, year out, were the Saturday-night dances at the Tivoli Hotel, where the band played such favorites of the moment as "Moonlight Bay" and "Wait 'til the Sun Shines, Nellie." Probably the tune danced to more than any other was "Alexander's Ragtime Band," and it was with special zest that everyone sang "Under the Bamboo Tree." A better-looking crowd of young people would be hard to find, wrote the correspondent for *The Outlook*. "Hot water and grit soap had been busy on the men, and the scene, except that some of the men were in white, looked like a college dance."

Besides the dances there were band concerts every Sunday, performed by a "very creditable" thirty-six-piece I.C.C. band. There were lectures, Halloween parties, Thanksgiving and Christmas celebrations. July fourth was the biggest day of the year. Once the Battle of Lexington was staged at Colón in Colonial costume.

The number of clubs and fraternal organizations was simply astonishing. There were camera clubs, bowling clubs, literary clubs, debating clubs, dramatic clubs. The Texans had a club. There were nine women's clubs in nine different towns led by a full-time professional women's club director, Miss Helen Varick Boswell, who was a paid employee of the I.C.C. There was a Strangers Club at Colón and a Century Club in Panama City. There was a club for Spanish-American War veterans, another for college men, and another for college men who were members of national fraternities. The Isthmian Canal Pioneers Association was reserved for those who had been on the work from the start.

Joseph Bucklin Bishop gave the activities of all such organizations full play in a regular column in the *Canal Record* called "Social Life on the Zone." The launching of an Ancon Art Society in 1911 was reported thus:

> The art section of the Ancon Woman's Club, organized under the title of the Ancon Art Society, will hold its first monthly meeting on the evening of January 29, at the residence of Mrs. Herbert G. Squires, the American Legation, Panama, from eight to ten o'clock. In accordance with the regular plan of the society, the evening will consist of music given during the hour's sketching; an exhibit and judgment of work by the critic appointed by the society, and a social half hour during which refreshments will be served. The program of work during the month of January has been figure, landscape, still life, genre and applied design.

"The whole Zone was friendly," remembered Robert Worsley, a stenographer from North Carolina. "People were always willing to help, it was easy to get to know people."

"I just thought I was something. . . . Everybody was so friendly," Winifred Ewing said.

Locomotive engineers, train conductors, and steam-shovel men had their respective "brotherhoods." Fraternal and secret orders abounded. The Masons, the Patriotic Sons of America, the Independent Order of Odd Fellows, the Fraternal Order of Eagles, the Benevolent and Protective Order of Elks, the Improved Order of Red Men, the Modern Woodmen of the World, the Knights of Pythias, and the Knights of Malta had chapters and every chapter was extremely active. The Improved Order of Red Men, "the pioneer of all secret, or fraternal orders and societies in the Canal Zone," had six "tribes," and for special functions, members turned out in full Indian regalia, war bonnets, war paint. But *the* organization—"largest and strongest"—was the Independent Order of Panamanian Kangaroos (motto: *Optimus est qui optima facit*—He is best who does best), founded in 1906 when a number of Americans for their own amusement began staging mock trials, or kangaroo courts, in a boxcar in the yards at Empire. Within a few years, membership had reached a thousand.

Everyone belonged to something. Everyone was someone's "brother" in some fashion, and for large numbers of men their standings in these various organizations were vitally important. In 1911, for example, a large, costly illustrated volume much like a college yearbook, *The Makers of the Panama Canal*, was sold by subscription to American employees, so that by paying the price, any clerk, teacher, or mechanic could have his or her picture, as well as a brief biographical sketch, included along with those of Presidents Roosevelt and Taft, with Goethals, Gorgas, and Panama's prominent political leaders. In nearly all these biographical sketches the orders and brotherhoods figure quite prominently.

Most men listed three or four affiliations. Harvey C. Dew, for example, was an "Assistant Chief Clerk" from Dillon County, South Carolina, who had been on the Isthmus since 1906. He belonged to the Independent Order of Panamanian Kangaroos, the Odd Fellows, and the Knights of Pythias, as well as the University Club of Panama and the Strangers Club of Colón. John L. Davis, a steam-shovel engineer from Indiana, belonged to the Masons, the Fraternal Order of Eagles, the Odd Fellows, and the Junior Order of American Mechanics. Charles Montague, a railroad conductor from Allegan County, Michigan, who worked on a dirt train at Las Cascadas, was "Chancellor Commander of Balboa Lodge, K. of P." (Knights of Pythias), as well as a member of the B.R.T. (Brotherhood of Railroad Trainmen), O.I.T. (Order of Isthmian Trainmen), the Modern Woodmen of the World, and the Kangaroos. The complete entry for Joseph H. Painter reads as follows:

> JOSEPH H. PAINTER
>
> Steam Shovel Engineer, was born in Cincinnati, Missouri. He arrived on the Isthmus in March, 1908. Mr. Painter has taken about all the upright, regular steps in Masonry, having gone up through blue lodge, Royal Arch chapter and Knights Templar commandery on the York Rite side and to the thirty-second degree in the Scottish Rite. He is also a noble of the Mystic Shrine and belongs to the Kangaroos. He is married.

By 1910 there were also thirty-nine churches within the Zone, twenty-six of which, like the Y.M.C.A clubhouses, were built and owned by the I.C.C. Fifteen full-time chaplains were employed–three Catholic, four Episcopal, four Baptist, two Methodist, one Wesleyan, one Presbyterian–their salaries and living expenses being charged off, as someone in Goethals' office decided, to the Sanitary Department.

The cost to the I.C.C. of all these various "privileges and perquisites" was something over $2,500,000 a year.

But neither were the possibilities for unorganized activities limited or ignored. On Sundays, the one day off in the week, hundreds of employees with their wives, girl friends, or families joined the swarms of tourists sight-seeing at Culebra. There were day-long excursions to the beach at Toro Point, across Limon Bay from Colón, and to the old Spanish fort of San Lorenzo, looking down on the mouth of the Chagres.

Taboga Island, in the Bay of Panama, was the most popular of all Sunday destinations. It seldom rained heavily at Taboga, the air was cooler, and while there was also no overabundance of things to do once one arrived, the three-hour boat trip across the glassy bay imparted a sense of getting far away. The one village on the steep little island stood beside a sheltered crescent of bathing beach. Excursionists in their Sunday clothes trailed up and down narrow paths of crushed shells and spread their picnics under the trees near the brown sand. Families went swimming. They took snapshots of themselves grouped in twos and threes beside a lime tree, or in the hard white sunshine in front of a tiny, ancient church at the head of a miniature plaza in the village. In old photograph albums brought back from the Isthmus, they stand arm in arm, their beaming faces half shadowed by hat brims.

They bought pineapples and cold Coca-Cola and a good, crusty bread sold in a pink bakery shop overhung with a filigreed balcony painted an electric blue. On the slope just beyond the village stood the old French sanitarium, the former Hotel Aspinwall, which was still used for convalescents from Ancon Hospital, but took paying guests as well. The view from its long verandas was lovely. For many young couples it seemed very near to paradise.

Numbers of people became keenly interested in Panama's history. Lieutenant Colonel Gaillard, for example, bought and read every book he could find on the subject, and Gorgas' favorite Sunday pastime was to lead day-long excursions on horseback to the ruins of Old Panama. When it was learned that gemstones—sapphires, opals, garnets—as well as fossil shark's teeth could be found in and around the canal diggings, many men spent their Sundays that way, and with considerable success. Mrs. William Sibert raised a dozen or more varieties of orchids on her front porch. Lieutenant Colonel Sibert, among others, sent to the States for a pack of hounds, organized hunting parties in the jungle (for deer mostly), and kept a pet eleven-foot boa constrictor which he fed live possums.

To some visitors it seemed that perhaps everyone was having too good a time, that a little too much was being done at government expense. Others worried more over what the future effect might be of so efficient and apparently so successful a demonstration of socialism.

In this largest of all modern enterprises, reporters were writing, not one man at the top, no one at any level, was working for profit. Visiting bankers and business people went home to report that the government-run Panama Railroad was a "model of efficiency and economy in every department." No railroad in the United States was better equipped with safety devices. No private contractor in the world was feeding laborers so well as the I.C.C. In every phase of employer-employee relations the I.C.C. was more liberal than any private concern of the day, as several publications had already emphasized. The government ran the Tivoli Hotel, very well and at a profit. The steamship line between New York and Colón, also government-run, was earning a profit of some $150,000 a year.

What were to be the consequences when the canal workers, spoiled by such paternalism, came home again?

> When these well paid, lightly worked, well and cheaply fed men return to their native land [warned a New York banker], they will form a powerful addition to the Socialist party. . . . By their votes and the enormous following they can rally to their standard they will force the government to take over the public utilities, if not all the large corporations, of the country. They will force the adoption of government standards of work, wages and cost of living as exemplified in the work on the Canal.

Yet how could it be socialism, some pondered, when those in charge were all technical men and "little interested in political philosophy," as one reporter commented. "The marvel is," wrote this same man, "that even under administrators unfriendly or indifferent to Socialism, these socialistic experiments have succeeded—without exception."

A member of the Socialist Party was found on the payroll, a mechanic who had been on the job almost from the start, and he declared that by no means was it socialism. "First of all, there ain't any democracy down here. It's a Bureaucracy that's got Russia backed off the map. . . . Government ownership don't mean anything to us working men unless we own the Government. We don't here—this is the sort of thing Bismarck dreamed of."

II

To Harry Franck, who was to record his experiences in a book titled *Zone Policeman 88*, one of the most candid and perceptive of published reminiscences, a more fitting analogy was the caste society of India. "The Brahmins," he wrote, "are the gold employees, white American citizens with all the advantages and privileges thereto appertaining."

But this Brahmin caste itself, he emphasized, was divided and subdivided into numerous gradations, each very clearly defined. The ultimate Brahmin was "His Brahmin Highness the Colonel." Immediately below him were the "high priests" of the canal commission—Sibert, Gaillard, Hodges, Gorgas, Rousseau, and the portly former Senator, J. C. S. Blackburn, who was officially Chief of Civil Administration, a job and title that meant almost nothing. (Blackburn's duties were once described unofficially as attending commission meetings, signing cab licenses, and drawing $14,000 a year.)

Down the scale, grade by grade, were the assistant division heads, the highest-paid civil and electrical engineers, the supervisors of construction, the assistant supervisors of construction, heads of machine shops, accountants, paymasters, storekeepers, yardmasters, sanitary inspectors, locomotive engineers, beneath which were the "roughnecks" —steam-shovel men, boilermakers, plumbers, ordinary mechanics, and so forth. (A roughneck, by Franck's description, was a "bull-necked, wholehearted, cast-iron fellow" who was both admirable and likable, but only to a point: "a fine fellow in his way, but you can sometimes wish his way branched off from yours for a few hours.")

There was, however, still one lower level within the white community, that of the regular enlisted man, either Army or Marines, for whom Corporal Jack Fitzgerald, of Boston, may serve as an example. Eighteen years old, single, Corporal Fitzgerald was one of some eight hundred regular Army troops based in the Zone at Camp Otis, near Las Cascadas, beside the western edge of Culebra Cut. Being neither an officer nor an employee of the canal commission, he had no entrée into the social life of the Army elite or among the canal workers. He had no access to the clubhouses, to the mess halls, to anything maintained or put on for the comfort or amusement of the workers. His pay was $18 a month, or considerably less than that of the lowest-paid unskilled laborer. When he went to town in his uniform—and his uniform was all he was permitted to wear—he was an easy target for the Panamanian police, who had no liking for Americans at best, but who could be vicious when dealing with American servicemen.*

* The diplomat William Sands, who spent much of his time settling American-Panamanian differences, wrote that the Panamanian police had "various . . . disconcerting habits, such as carving an American or hammering him unconscious before they arrested him." When fights broke out in a Coco Grove brothel on a night in July 1912, the Panamanian police moved in, shot and killed three unarmed Marines, then rounded up a number of others, who were put in jail, to be beaten and tortured. That Goethals and the American minister refused to order American troops into the city was, according to one journalist present, the cause of "almost universal regret within the American Zone."

It was a very different Panama that Corporal Fitzgerald knew from the one seen, or perhaps even imagined, by his fellow Bostonian Charles Francis Adams, as different nearly as the two Bostons they came from. Later, when the locks were nearing completion, Corporal Fitzgerald would stand guard at Miraflores; but most of his three-year hitch in Panama was taken up with map duty in the jungle and with endless routine chores in camp. Still: "The natives had stills in the jungle and plenty of sugar cane available," he would recall happily. "So a lot of us boys got off to a bad start. No ice, no mixing, just right out of a bottle, ninety proof!" Paydays "you took your turn" at one of the dollar houses in Coco Grove. "There was a few [other] places that were closed to the general run, private like, they were for those high up with the canal, engineers and such."

Corporal Fitzgerald was like those countless others in history who could say in later life that they had been present at some momentous event but who in fact saw almost nothing of it. In his three years he saw little of the excavation going on and took practically no interest in it. He never laid eyes on Goethals, as near as he could recall, but once he *did* see Taft. His only brush with anything like the glamour or historic grandeur that the rest of the world associated with the work at Panama was to stand at attention for nearly three hours one broiling-hot morning at the side of a little spur on the railroad.

> . . . I think it was in 1912, I'm pretty sure it was in the spring. . . . *The President was coming*—Taft! William Howard Taft!—and, gee, they got us out there about nine o'clock in the morning, standing, standing, waiting, waiting, waiting, Christ, the sun kept coming up, and it—*boy!* You know, that place when it rains, it's just like throwing water on a stove, everything steams. And we waited. Finally the damn train came by. And here's Taft with a big white suit on—he weighed about three hundred pounds, a big belly on him—standing on the observation platform. And, well, he just . . . he just *waved*. That was all there was. That was all there was to it.

On another occasion, on a Sunday when he was looking about among the exposed rock down in the Cut, Corporal Fitzgerald saw a fossil print of a fish ("not the meat, just the bones") and still another of a fern, "proving," he would recall, "that millions of years ago it was all in the sea. It all makes you feel very insignificant."

To Charles Francis Adams, as to others, the "innate force" of the entire order was very plainly derived from one man. "The individuality and character of Colonel Goethals today permeate, and permeate visibly, the entire Zone—unconsciously on his part, unconscious on the part of others, his influence is pervasive."

Praise at home for Goethals was boundless. *Collier's* called him "The Solomon of the Isthmus." He received standing ovations when he appeared before congressional committees. Yale, Columbia, and Harvard conferred honorary degrees. Newspapers were full of speculation about his prospects as a dark-horse candidate for the Presidency. "Goethals has created a wonderful, smooth-as-oil, 100 percent efficient machine that is getting results every working day in the year," the Atlanta *Constitution* declared. Everyone returning from the Isthmus spoke of his ability. "Congressmen and senators, civilians, administrators, newspaper and engineering experts, united in the verdict that the digging of the big ditch has developed one of the greatest figures in contemporaneous American life in the person of Colonel Goethals."

To Harry Franck, he was simply "Omnipotent, Omniscient, Omnipresent."

The omnipresence had become legendary almost from the moment he began making his daily tour of the line in an extraordinary self-propelled private car. It was bright yellow, heavily lacquered, shined, spotless, and looked like a dreadful cross between a small locomotive and a stagecoach. Powered with a gasoline engine and run by a uniformed driver, it was known as "The Yellow Peril" and the sight of it rounding a bend in the tracks was an instant warning to look sharp, that the "Old Man" was coming. Alice Roosevelt Longworth, who toured the canal with her husband, Congressman Nicholas Longworth, and who shocked everyone by smoking in public, wrote that "nothing was ever pleasanter than riding along in his track motor, 'The Yellow Peril,' and seeing the big job and hearing him talk about it." "It rattled by at all hours of the day," wrote Rose van Hardeveld of the strange-looking contrivance, "this official car of the I.C.C. . . . The Old Man was so constantly on the job that we never thought of him as being at home or eating or sleeping."

Except for the times he was called to Washington, a few brief vacations, and one official trip to Germany, he was on the Isthmus the entire seven years until the canal was finished.* He seldom entertained, seldom accepted invitations. Only a select coterie appears to have

* The trip to Germany was to visit the new locks on the Kiel Canal. He went in the spring of 1912 and at one point was entertained by the Kaiser, whom he described as "Roosevelt toned down."

known him on anything more than the most formal working basis—Hodges, with whom he seems to have had the most direct day-to-day rapport; Bishop; Williamson; Major Chester Harding, who was one of Sibert's ablest assistants; Dr. W. E. Deeks, a Canadian on Gorgas' staff; and Father Collins, the Catholic priest at Culebra, whose company he appears to have enjoyed above all others.

Mrs. Goethals—Effie Rodman Goethals—tall, vain, a member of an old New Bedford whaling family, was in residence most of the time, but seemed little pleased with the life or ever very comfortable in her role as the first lady of the Zone. It had been arranged also that their older son, George, now a second lieutenant in the Engineers, be assigned to duty on the canal and he brought with him his bride, who, to judge from her photograph, must have been one of the prettiest young women on the Isthmus. Still, to the rank and file Goethals remained a solitary and enigmatic figure and an endless subject for gossip. Stories of supposed romances with the wives of subordinates would persist through all his years in Panama and on into the succeeding generation of canal employees. One especially vivacious young woman named Henrietta Otis, when asked by a close friend and neighbor to confide whether there was any truth to the gossip that she and the Colonel were "very close," affirmed that there was. But according to other stories still told on the Isthmus, the Colonel's lady was someone else entirely.

Intensely partisan factions formed within that small but all-important segment categorized by Harry Franck as the Brahmin high priests. There were the Goethals people, the largest faction, and there were the others, chief of whom were Sibert, Gaillard, Gorgas, and Marie Gorgas. Sibert and Goethals all through the final years of the work were barely speaking to each other.

"Colonel Sibert and my father felt sure he was trying to get rid of them," recalled Gaillard's son, Pierre, then a young M.I.T. graduate who had been employed as a junior engineer on the locks at Pedro Miguel. "He'd give them the silent treatment. Only my father refused to let him get away with that. My father talked to him the same as always. We'd been great friends of the Goethals for years. Our houses at Culebra were right next door to one another. . . . He [Goethals] did a damn good job, but he got delusions of grandeur."

In Washington, reporters quoted an unnamed southern Democrat, a recent visitor to the Isthmus, who said Goethals' "keen dislike" of Gaillard, Sibert, and Gorgas, all three of whom were from the Old

South, had been very apparent to him. The Southerners, he said, were "outraged" because Goethals was attempting to freeze them out of their share of the glory.

Marie Gorgas, the only one of the group ever to say anything in print on such matters, wrote that it was Goethals' "passion for dominating everything and everybody" that made him such a trial. He had become, by her account, a man virtually without feeling, except for power. Power was "the relish and the sweetness of his life." And indeed, one evening while escorting Mrs. Gaillard across the way to her door, Goethals said that he cared very much for the power he exercised. The salary, the title, the prestige, he said, were of but small satisfaction compared to the feeling of such power.*

The Goethals-Gorgas rift came to a head over a difference of views about who should cut the grass. Goethals was concerned with cost efficiency in every department and in the Sanitary Department no less than any other, and, like Taft, his private estimate of Gorgas as an administrator was very low. He wanted the mechanical tasks of cutting grass and clearing brush, a substantial part of Gorgas' operation, put under the quartermaster, where they would normally fall, and he saw no reason why Gorgas should object since such work required no special sanitary knowledge or training. Gorgas did object, however, on the grounds that any change in his campaign against *Anopheles* mosquitoes would jeopardize his success in reducing malaria (and jeopardize thereby the lives of the workers). But they were grounds Goethals refused to accept. So for six months the grass was cut under the direction of the quartermaster, over a larger area than under the old system and, as it turned out, at appreciably less cost. And since the

* This account is Mrs. Gaillard's own, as related by her son to the author. Marie Gorgas, who considered the scene the key to Goethals' character and who, of course, was not present, gave a rather more colorful version:

> One beautiful moonlight night Goethals was walking on a little hill, overlooking the cut, with one of the best-known ladies of the Zone. His companion was much affected by the splendor of the tropical scene.
>
> "Yes, it's a beautiful spot," the Colonel replied to her exclamations, "and I love it! But I love it for other reasons than its beauty or the things I get from it. Above all, I love it for the power."
>
> He was silent for a moment and then went on:
>
> "I remember once visiting a monastery of Jesuit Fathers. I saw the wretched cells they lived in, the little rude cots they slept in, the rough tables at which they had their meals. And then I remembered the vast power that the men who lived like that had once exercised. It was worth living simply in order to have that."
>
> In his enthusiasm he raised his hand.
>
> "That's the only thing in life worth having. Wealth—salaries—these are nothing. It's power, power, power!"

incidence of malaria continued to decline during this same trial period, Goethals felt that he had proved his case and refused to turn the work back to Gorgas.

At this point in the test of wills, according to Marie Gorgas, Goethals exclaimed, "Do you know, Gorgas, that every mosquito you kill costs the United States Government ten dollars?"

"But just think," Gorgas replied, "one of those ten-dollar mosquitoes might bite you, and what a loss that would be to the country."

By Goethals' account, however, no such exchange ever occurred. Like Marie Gorgas' other uncomplimentary remarks and recollections, the story appeared in her biography of her husband in 1924, all of which Goethals chose to ignore. But when Mark Sullivan repeated the story two years later in his popular *Our Times,* Goethals presented his version in a letter to Sullivan. "No further discussion of the matter ever occurred and the sick rate continued to decrease."

What he may have felt privately about Gorgas, Sibert, or Gaillard is less than clear, since he kept most of his feelings very much to himself. Apparently, from remarks he made to the unidentified southern Democrat, he thought seriously of firing Sibert. Sibert was known to regard himself as a better man than Goethals and better qualified for the top job. Goethals sensed this and naturally resented it. Sibert, moreover, at least by Goethals' lights, was a bit too openly ambitious, too political. Primarily the problem seems to have been one of clashing personalities, and though this made the tasks of both no easier, nothing of serious consequence ever came of it. In answer to the charges that he was interested only in the glory his role would bring him, Goethals told one reporter that he intended to resign before the opening of the canal because "I couldn't stand the glamour." But Taft would insist that he stay and in the end Goethals would publicly commend Sibert, Gorgas, all his subordinates, for their professional ability and continued loyalty to the work.

Richard Whitehead, always one of Goethals' staunchest allies among the young engineers, as well as a lifelong friend and admirer afterward, would concede that possibly Goethals "wasn't quite human enough for everyone." But, he added, "None of the others had the responsibilities he did."

The only real scars were those left by the Gaillard tragedy. In the summer of 1913, following the tremendous slides at Culebra, Gaillard appeared suddenly to have cracked under the mental and physical strain of his responsibilities. He left the Isthmus and did not return.

The stories were that he had gone crazy. What actually happened was this:

He was sitting at lunch with his wife and son, looking and acting entirely as usual, when he broke off in the middle of a sentence and began to talk rapidly and incoherently of his childhood at his grandmother's house in South Carolina. From that point on he remained in a state of terrible confusion. "Poor Gaillard . . . went completely to pieces," Goethals wrote privately in August, his diagnosis being the same as that of everyone else. "It is a nervous breakdown. His memory seems to have gone and [Dr.] Deeks doesn't believe he will ever be able to return. He, accompanied by Mrs. Gaillard and Pierre, sailed for the States . . . accompanied by [Dr.] Mason. . . ."

But Gaillard had not broken under stress. He was suffering from a tumor on the brain, "an infiltrating tumor in the brain," according to the records of the case, and an operation performed at Peter Bent Brigham Hospital in Boston and subsequent treatment at Johns Hopkins were to no avail. Gaillard died at Baltimore, December 5, 1913.

His work at Panama had no correlation with the tragedy. The tumor would have killed him had he been serving still at his desk in Washington. However, to his family and numbers of those who had worked with him at Panama, he had been worked to death. At the hospital in Boston, Goethals' son Tom, a medical student at Harvard, had been invited to observe the operation on Gaillard, which was performed by the famous Harvey Cushing, and afterward he had encountered Mrs. Gaillard. "Your father has killed my husband," she told him. Once in Washington, years later, Pierre Gaillard and Colonel (by then General) Goethals passed each other in the street. Each looked directly at the other and neither spoke.

III

In the popular picture of life in the Canal Zone as it emerged in hundreds of magazine and newspaper articles, that vast force of black men and women who were doing the heaviest, most difficult physical labor—some twenty-five thousand to thirty thousand human beings—could be but very faintly seen. As individuals they had no delineation whatsoever. They were there only as part of the workaday landscape. That they too were making a new life in an alien land, that they too were raising families, experiencing homesickness, fear, illness, or exhilaration in the success of the work, was almost never even inferred. In the United States the public had little if any conception of the part played in Panama by "pioneers" who were neither American nor white, *or* how very small numerically the white American force was by contrast. To judge by many published accounts, the whole enormous black underside of the caste system simply did not exist. Cartoons in the newspapers depicted the canal being dug by cheerful white Americans with picks and shovels and many came to Panama expecting to see just that. Harry Franck would write that he had arrived actually believing he could take up a shovel and descend into the canal with other workmen, "that I might someday solemnly raise my hand and boast, 'I helped dig IT.' But that was in the callow days before I . . . learned the awful gulf that separates the sacred white American from the rest of the Canal Zone world."

Official visitors, congressmen on so-called tours of inspection, writers gathering material for books, could not help but be amazed, even astounded, at the degree to which the entire system, not simply the construction, depended on black labor. There were not only thousands of West Indians down amid the turmoil of Culebra Cut or at the lock sites but black waiters in every hotel, black stevedores, teamsters, porters, hospital orderlies, cooks, laundresses, nursemaids, janitors, delivery boys, coachmen, icemen, garbage men, yardmen, mail clerks, police, plumbers, house painters, gravediggers. A black man walking along spraying oil on still water, a metal tank on his back, was one of the most familiar of all sights in the Canal Zone. Whenever a mosquito was seen in a white household, the Sanitary Department was notified and immediately a black man came with chloroform and a glass vial to catch the insect and take it back to a laboratory for analysis.

Yet little official notice would ever be paid to such contributions. In that official journal of Zone life, the *Canal Record*—a reliable, admirable publication in most other respects—the black employee went unrecognized, except in death, and then only in a line or two, his tag number invariably appended, as if he were not quite human. It would be reported that Joshua Steele, of Barbados, Number 23646, was killed in an explosion in the Cut or that Samuel Thomas, of Montserrat, Number 456185, was crushed to death in the pulleys of a mud scow. But no obituaries appeared in the paper, any more than notices of black weddings, social affairs, or the birth of a black child.

As a consequence, the popular mental picture of what life was like in the Canal Zone, and popular pride in the kind of society that had been created there, were founded on a very limited and erroneous view of reality. The measure of Utopia achieved through American know-how and largess was again relative, like the success of the medical crusade.

And as a consequence of such distortion, most all of what would be written in the way of a social history of these years contains but part of the story.

In truth, the color line, of which almost nothing was said in print, cut through every facet of daily life in the Zone, and it was as clearly drawn and as closely observed as anywhere in the Deep South or the most rigid colonial enclaves in Africa. This, some observers later speculated, was the fault of the many Southerners among the skilled workers and among the military officers. Others, including the Southerners, attributed the practices to the upper-class Panamanians, who were notably color conscious, and to the long-established policies of the Panama Railroad. Harry Franck, who as census taker spent the better part of his time among the black workers, wrote acidly, "Even New Englanders grow almost human here among their broader-minded fellow-countrymen. Any northerner can say 'nigger' as glibly as a Carolinian, and growl if one of them steps on his shadow."

The "gold" and "silver" system had become the established practice throughout the Zone; it applied everywhere and nobody misunderstood its purpose. Black West Indians and white North Americans not only stood in different lines when the pay train arrived, but at the post office and the commissary. There were black wards at the hospitals (on the side away from the best views and breezes) and black schools for black children. (Although the number of black children enrolled in Zone schools was twice that of the white children, there were less than half as many black teachers employed.) Black men served on the Zone police force—ninety some out of a force of three hundred were West Indians—but they drew half pay and were not eligible for promotion.

The Y.M.C.A. clubhouses, gold-roll hotels and churches, were all off limits to a West Indian, unless he or she was employed there. "As for the man whose skin is a bit dull," recalled Harry Franck, "he might sit on the steps of an I.C.C. hotel with dollars dribbling out of his pockets until he starved to death—and he would be duly buried in the particular grave to which his color entitled him."

Most conspicuous was the contrast between the black and white living quarters. To much of the white populace—employees, tourists—it was easier not to think about such things, easier to put the black people out of mind except for the services they performed. But in fact the living conditions for the black people were deplorable and the I.C.C. did practically nothing to set things right.

The greatest source of discontent and despair in the early years, for black and white workers alike, had not been the difficulty of the work or the unpleasant climate so much as the prospect of a life almost wholly devoid of women; and just as the I.C.C. had initiated a campaign to bring American women to Panama and thereby establish something approaching normal domestic life within the white community, so it also took steps to bring in large numbers of black women, from all over the West Indies. The first black women to arrive were from Martinique and were listed officially as laundresses; and while some may have been prostitutes, as charged, the accusation that they were all prostitutes, or that they were being shipped in at government expense solely for purposes of prostitution, was absurd and manifestly unfair.* As time went on, as the laborers sent word back to all the islands, several thousand black women came to Panama to join their husbands, a brother, a father, to find a husband, but mainly to find what the men had come for: steady work at better pay than they could ever hope to get at home. "Most of us came from our homelands in search of work and improvements," said John Butcher, of Barbados. "We turned out to be pioneers in a foreign land."

The I.C.C., however, made no provision for housing black women. Only a few crude quarters were provided for black workers who were married. The rest of the quarters available to black workers were the same as they had been at the time of Theodore Roosevelt's visit—the roughest kind of barracks with canvas transport bunks packed in as closely as in steerage. Many hundreds of laborers lived in converted boxcars that were shifted back and forth along the line, according to where the men were most needed. To the single men such accommodations appealed no more than they ever had, while men with wives and families had no choice but to fend for themselves. Furthermore, Gorgas, for several years now, had officially encouraged such dispersal as a way to reduce the spread of pneumonia.

As a result, no fewer than four out of five West Indians paid rent for wretched tenements in Colón or Panama City, where one room usually served an entire family. Or, more often, they settled in the

* The issue had caused a brief sensation at home with the appearance of an article in *The Independent* magazine (March 22, 1906) titled "Our Mismanagement at Panama." The author was Poultney Bigelow, son of John Bigelow, who had spent a few days in Colón and who wrote, "Prostitutes are not needed on the Isthmus—and if they were there is no call to send for them at the expense of the taxpayer." The charge was shown to be based wholly on hearsay, as was virtually all of the article. But to quell the outcry numbers of black women were asked to swear before a duly appointed I.C.C. official that they were leading a moral life and that they were in Panama of their own free choice; and these affidavits were sent on to the appropriate congressional committee.

junglc, building whole villages of dynamite boxes, flattened tin cans, any odd scraps of lumber or corrugated iron that could be scavenged. They lived where they pleased, as best they could, without benefit of screen doors or janitor service, growing small gardens, always a great many chickens pecking about their small shacks, and nobody of official importance cared very much about them one way or the other.

So it was not that the I.C.C. was providing its black labor force—that is, the overwhelming majority of its workers—with substandard quarters; the I.C.C. was providing them with no quarters at all. And in almost nothing that was published for popular consumption was this point ever made explicit; or if touched on at all, it was expressed as another act of generosity on the part of the canal officials, in that they were allowing the Negroes to live the way they were happiest. Yet as one observer did write, no visitor had to search very far to see what a wholly different life these people were subjected to. "The visitor who saw first the trim and really attractive houses and bachelor quarters assigned to the gold employees could hardly avoid a certain revulsion of opinion as to the sweetness and light of Isthmian life when he wandered into the Negro quarters across the railroad in front of the Tivoli Hotel . . . or in some of the back streets of Empire or Gorgona."

A less dramatic but more specific index to the relative inequality of the system was the difference in benefits derived by white and black workers from the $2,500,000 being spent annually on employee entertainment and recreation. To the average skilled (white) worker who was married, this one I.C.C. expenditure meant in plain monetary value about $750 a year. But to the average unskilled (black) worker who was married, it meant $50. In other words, as the system was designed, the white American, who represented about one-fifth of the population, was being treated fifteen times better in the way of free social facilities, sports, and amusements than was the black West Indian, who represented as much as three-quarters of the population.

It could be very naturally assumed that this was all the most blatant kind of racial injustice. And in a very large measure, of course, it was; but not entirely. Simple problems of supply and demand also entered in, that is, experienced technicians (men to run and repair the machines), doctors, and competent clerical people were always in short supply and had to be kept satisfied if the canal was to be built; common unskilled laborers from the impoverished islands of the Caribbean were always available in abundance and expected no better than what they got, which for the most part was better than what they had known at home. And besides, there was the political factor: the labor force was not merely black, it was foreign; these were not United States citizens and in Washington therefore they represented no constituency.

Whether a West Indian working on the Panama Canal was better or worse off than a miner in the coal fields of Kentucky or an immigrant mill hand in Homestead, Pennsylvania, during these same years is debatable. But no coal miner or mill hand of the day received free medical care. As extremely hazardous as the work in Panama was for the black laborer, the safety regulations set down by the I.C.C. were far in advance of those of American industry. No company store could ever compare to the I.C.C. commissaries. Further, it can be said with certainty that no one in Panama went hungry.

The frequent claim was that no labor army in history had ever been so well paid, well fed, well cared for; and this, on balance, was unquestionably true, despite all the obvious inequities of the system, however one-sided or hypocritical other claims made for it may have been. Certainly the fellahin who built the Suez Canal or the West Indians who came to Panama during the French era had been far worse off than any black employee of the I.C.C.

Generally speaking, the West Indian worker on the Panama Canal was soft-spoken, courteous, sober, very religious, as nearly everyone associated with the work came to appreciate. John Stevens once remarked that he never knew such law-abiding people and the records show the crime rate, as well as the incidence of alcoholism and venereal disease, among the black employees to have been abnormally low throughout the construction years.

Approximately 80 percent of the black workers were illiterate. While it was official I.C.C. policy in Washington to hire no one under age twenty, a good many black workers were also little more than children. Joseph Brewster, of Barbados, went to work as a track hand, as he later wrote, at age sixteen; H. B. Clayton, a West Indian born and raised at Gorgona, began as "a young boy," probably fifteen; Jules LeCurrieux, from French Guiana, had just turned seventeen; Alfred Mitchel, who had come from Jamaica with his mother, began as a water boy at fourteen; Jeremiah Waisome, who also had come to Panama with his mother, was twelve or thirteen and proud of his ability to both read and write. His account of the day he first applied for work reads as follows:

> Unknown to my mother one morning instead of going to school, I went to Balboa to look myself a job. . . . I approach a boss . . . I said good morning boss, he retorted good morning boy. At this time he had a big wad of tobacco chewing. I ask him if he needs a water boy, he said yes, he ask what is your name, I told him. I notice that my name did not spell correctly as Jeremiah Waisome, so I said excuse me boss, my name do not spell that way. He gave me a cow look, and spit a big splash, and look back at me and said you little nigger you need a job? I said yes sir, he said you never try to dictate to a white man, take that bucket over there and bring water for those men over there.

Whatever his first job the black worker was not likely to stay with it very long, largely since the steady turnover or some chance turn in his own circumstances seemed always to lead to something more attractive elsewhere along the line. The experience of Clifford St. John was not exceptional. He began at age seventeen working with a steam-shovel gang in the Cut at Gold Hill. Injured by falling rock, he was sent to Ancon Hospital, where he was offered a job he liked better and so stayed there for two years. Next he strung telephone wire on the Panama Railroad, until stricken with typhoid fever. Then after a month of recovery in the hospital at Colón, he returned to the railroad, this time as a watchman. Presently he changed again, to work as a longshoreman at Cristobal, then again, to drive piles at Gatun Dam. When construction of the lock gates began, he was hired to bore holes for the rivets. This made seven different jobs in less than seven years.

Like others who were to reminisce in later life, recalling their "times on the canal," black workers would talk repeatedly of incredible rains and "working all wet," of bugs and mud and the smothering heat. (A waiter at the Tivoli Hotel recalled having to change his suit three times a day, so badly did he perspire.) They remembered the low cost of food and the trains rolling out of the Cut ("It was something marvelous to see"), a particular foreman or engineering officer, or even Goethals ("a great man . . . calm, principled, dignified"). But recurring again and again through all such accounts are memories of tremendous physical exertion and of the constant fear of being killed. "I tell you it was no bed of roses." "It would not be surprising to say those were very rough days." "We had to work very hard." "I worked very hard . . . much danger . . . constant danger." "I had a narrow escape of death." "I had to jump for my life." "I feel blessed to be still alive." "You had to pray every day for God to carry you safe, and bring you back. Those days were horrible days to remember. Those were the times you go to bed at nights and the next day you may be a dead man."

One extremely dangerous task in the last years of the work, for example, was the demolition of the giant trees that stood in what was to be the main channel through Gatun Lake. After the trees were cut down, dynamite crews—hundreds of West Indians—chopped holes in the huge trunks, sometimes as many as fifteen holes in a single tree. Two or three sticks of dynamite were put in each hole, with cap and fuse, then plastered over with mud. The blasting began once the workday had ended and the area was clear, just as dark came on.

"After the 5:15 passenger train pass for Panama, we start lighting," remembered Edgar Simmons, another Barbadian. "Some of us has up to 65 or 72 holes to light and find our way out. So . . . you can judge the situation. . . ." Each man, torches in both hands, dashed from tree to tree, lighting fuses as fast as possible, then ran for cover. "Then it's like Hell. Excuse me of this assertion, but it's a fact . . . it was something to watch and see the pieces of trees flying in the air." Afterward, the pieces were gathered up and piled and burned, a task that went on for months. Gigantic heaps of trees were doused with crude oil, then touched off—and "another Hell roar again."

Sickness among the laborers remained a problem to the very end, the popular impression notwithstanding. Though the incidence of death from malaria and pneumonia was reduced dramatically from the level at the time of Roosevelt's tour, both diseases persisted. Also, typhoid and tuberculosis were on the rise. The much-publicized picture of the Canal Zone as a veritable health resort was genuine so far as it applied to the white community, and medical progress over all in the Zone was far beyond anything ever before achieved in a tropical wilderness. But the hospital records show the situation to have been anything but ideal. And again, as in earlier years, it was the nonwhite, non-American labor force that suffered.

Reports for the fiscal year 1907–1908, the point at which Goethals replaced Stevens, show that 1,273 employees died of all causes. At the end of the construction era, that is, in fiscal year 1913–1914, deaths from all causes totaled 414, a phenomenal reduction. In 1907–1908 there had been 205 deaths from malaria; in the final year of construction, there were only 14. Deaths from pneumonia dropped from 466 to 50. As remarkable as any statistic was the average death rate in the final year among all employees—7.92 per thousand, which was much lower than the general death rate in the United States. Not even in Washington, Montana, or Nebraska, then the healthiest states in the country, was the death rate lower than it was in the Canal Zone.

However, in the 1907–1908 records typhoid fever was not even listed among causes of death, while in 1913–1914 typhoid killed 4 people. Tuberculosis, which in 1907–1908 took 7 lives, took 63 lives in 1913–1914. More astonishing is the fact that during the final year of construction not less than 24,723 employees were treated for illness or accidents, this is to say that *nearly half* of the work force had been in the hospital at one point during those twelve months. In the earlier report the number treated was 11,000. And while only 14 employees died of malaria in 1913–1914, more than 2,200 were hospitalized for the same disease.

Of the total 414 deaths for the final year, 30 were white Americans, 31 were white employees of other nationalities. All the rest, 353, were black. The death rate among all white employees from the United States was actually a mere 2.06 per thousand, an almost unbelievably low figure and deserving all the acclaim that ensued, but the death rate among black workers was 8.23. So, in fact, for all the medical progress that had been made, Panama was still four times more deadly for the black man than it was for the white.

Nor, it must also be emphasized, was the incidence of violent death any less than in years past. In 1907–1908 there were 104 such fatalities; in the final year of construction there were 138. Gorgas, in the earlier report, had called the number of violent fatalities "very excessive" and expressed particular concern that so many were caused by railroad accidents. In 1913–1914 there were 44 people killed in railroad accidents, more than in the earlier year.

Of the 138 who died by violence in the last year, 106 were listed as "colored."

The black laborer who had not spent time in the hospital was the exception. Many were in and out three, four, five times. Nor do the records show the numbers of men who were permanently maimed. "Some of the costs of the canal are here," wrote Harry Franck of the black hospital wards, "sturdy black men in a sort of bed-tick pajamas sitting on the verandas or in wheel chairs, some with one leg gone, some with both. One could not help but wonder how it feels to be hopelessly ruined in body early in life for helping to dig a ditch for a foreign power that, however well it may treat you materially, cares not a whistle-blast more for you than for its old worn-out locomotives rusting away in the jungle."

Very few laborers had ever been inside a hospital until they came to Panama. Most of them had never in their lives been treated by a doctor, let alone a white doctor or a white nurse. So it was usually in a state of abject fear that sick or injured black workers arrived at the hospital the first time, fear, as much as anything, of what might befall them next, inside. To judge by available first-person accounts nothing in their experience made quite so lasting an impression.

James A. Williams, a Jamaican, was taken ill in 1910, when still in his teens. He was then working in a kitchen at one of the labor camps. His own account, written fifty years later, is among the most vivid in the canal archives. It is given here just as he wrote it:

> One morning the Doctor making his usual visit to the kitchen some one reported to him that I am having fever. The Doctor immediately advanced to me and felt my pulse, I could remember he said to me "you are going to be sick boy," go right over to the "Sick Camp" and tell the Clerk to write you up to the Hospital, right away. He further asked me, Are you a God fearing man? I replied yes. He said to me you are going to die. It was near time for the midday train and the Doctor ran over to the Sick Camp and assisted to write up the necessary papers and I was placed on the train to the Ancon hospital. Dr. Beard was the name of the Doctor in that section.
>
> I was placed on a bed on the train to the hospital all the way and when the train arrived in the Panama Station, there were many horse drawn ambulances awaiting to receive the patients to the hospital. We then arrived in the big Ward 30 to be lined up and a very pleasant American nurse was right on the job and started to feel the pulses and assigned each patient to the different bed. As I noticed when she came to me and took my hand, she appeared to be frightened and she called the Orderly and said to him do not put this patient under the shower, give him a bed bath. I wondered to myself as to what is this bed bath. Because I had never been in an Hospital before. However, I was escorted to the Ward by the Orderly Mr. St. Hill and he turned me over to Mr. Norman Piercy right from my home in Partland, Jamaica, W.I. but we at that time did not recognize each other. St. Hill said to Piercy, Give this patient a bed bath. While I kept wondering in my mind what do they mean by this bedbath when I saw this Mr. Piercy placed a heavy waterproof Blanket in the bed and two buckets full of heavy crushed ice and several buckets of water and not even the courtesy as to consult me but stripped me naked and threw me in that cold deadly water. To be truthful, I thought I could not any longer live. However, he gave me a thorough bathing and took me out and dried me with a towel and placed me in a white clean bed. I felt cool for a moment but still fretting over the Iced bath as I had never heard or seen anything of that kind before. . . .
>
> The next shock I had while I felt a little thirsty and when I saw some one coming with a wine glass of water I felt glad as I thought it was some cool water which I felt so much the need and the kindly Nurse

handed it to me and said drink it. So thirsty for a drink of water, I hurried and as it reached my lips it was down my stomach. I tell you, I had never before tasted anything so terribly bitter. I always hearing about Quinine but I thought it was something tasty and nice. And every two hours I was dosed with that bitter liquid night and day and instead of getting rid of the fever it was growing worse.

Then the next thing that happened. I was placed in front of the Nurse's desk and a basin with clean water was placed on the stand beside me, I then thought it was water placed there for me to drink. As I felt thirsty at the time I used my hand and took three hand full three times and swallow and when I heard the Nurse called to me and asked what, that you do drink it? I could not answer her but I saw when she picked up the Telephone I really did not know what happened until I discovered about five Doctors over me and find myself throwing up. And a few hours after I was settled I noticed they drew some blood from my arm. I then noticed from that time there no more of that bitter liquid. The hole night I was not bothered with that stuff.

The next morning two men came with a Stretcher and lifted me from the bed and placed me on the Stretcher and carried me off out of the Ward. I thought they were going to bury me as I was actually given over as dead. However, I was taken to Ward 24 . . . that was the place where typhoid patients were being treated. They found out that the fever I had was typhoid and not malaria.

What I can truthfull say Those American Nurses my own dear mother could not be more kind and tender to me. They did everything lies humanly even to let me take a little nourishment so to keep life in my body. I should right here tell of the incident with the water I drank from the basin on the stand beside my bed. It was poisoned water to kill flies that buzzed around when I thought it was placed there for me to drink as I had never before entered a hospital.

I could never, never in life forget the tender kindness those American Nurses administered to me especially in that Ward 24. I had no desire for nourishment of any kind, my life was ebbing out. But how they plead with me to take some nourishment. Not only that, but they closely watched the colored Orderlies how they handle the patients.

One night, the Nurse on duty came to me, she said to me now, bed 6, if you don't take some nourishment you would never get well and the tone she spoke to me with her hand on my head I forced to swallow a little milk and from that I continued to take little by little and a few days past and she came on duty the night and took my temperature she said to me you are getting better "bed 6." I began to feel a desire for the milk now, very fast. Then they started to give eggnog twice a day also real American Whiskey every day. I was not allowed to raise my head from the pillow even though I am feeling well. I began to feel real hungry but only liquid diet was give me for over three weeks after the fever left me. One mid day at breakfast I was given a toasted Potato. Oh. how I enjoyed it. Even that did not satisfied my starving apetite. I was therefore convinced by the Nurses and Doctors such starvation was for my good.

One morning in the month of May Dr. Connor the Night Doctor came in, that Ward were then run by himself and Dr. Bates. Dr. Connor came to my bed side with the Nurse and took up my Chart, asked me how you feeling James? I replied, ok. doctor. He asked me you hungry? I replied yes doctor. He turned and ordered her to give me light diet and pair of Pajamas.

The morning in question when I was given the Pajamas and was told to get out of bed and tried to walk, every step that tried I had to be supported. That morning I was given a bowl of porridge, two eggs, nice bread and butter, a lovely slice of melon. But they never ceased to give the Whiskey and Eggnog during the days, and a week later I was given "Full Diet." I was then feeling happy and good as when at meal time when we are told that what ever we like having that's no on the table just call.

I was discharged in May and the treatment had me so fat and robus that when I went home to San Pablo Aunt surprise to see how good I was looking.

I took sick in April and was in the Hospital until May 1910.*

Had the black labor force been housed in screened quarters comparable to those provided the white employees, in areas where the sanitary officers exercised control, malaria might possibly have been eradicated, as had yellow fever. As it was, the total loss from disease in the ten years it took to build the canal was less than five thousand people. But Gorgas later declared that if conditions had remained as they had been during the French era, the death toll would have been upward of seventy-eight thousand (figured on a death rate of two hundred out of every one thousand employees). That a man such as James Williams, or the many thousands of others who took sick, survived at all was regarded, by black laborers as much as by anyone, as a miracle of medical progress.

In his ten days of dining at the Tivoli Hotel, Charles Francis Adams had seen exactly three houseflies; it was indeed a wondrous age. The correspondent for *The Outlook*, after a stiff climb in the hills behind Paraíso, was shown a galvanized ash can full of oil placed on a plank over a little stream, so that oil dripped slowly, constantly, onto the water as it flowed toward the canal line. So simple a solution to so large a problem, he wrote, was not the least of the "marvels" of Panama.

There seemed to be but one aspect of progress on the Isthmus over which one might reasonably express some skepticism or concern. Relations between the canal builders and the local populace, uneasy from

* Afterward James Williams worked in one of the machine shops at Gorgona, then as a telephone operator on the railroad. At the time the canal was finished he was a sales clerk at the Corozal commissary. He retired from canal service in 1949.

the beginning, had deteriorated markedly. "In temperament and tradition we are miles away from the Panamanians," noted *The Outlook* correspondent. ". . . The age-old hostility to the 'Gringo' is deep-rooted. Differences in language, customs and religious practices keep the breach wide."

To the average American, Panama was a land of dark, ignorant, undersized people who very obviously disliked him. ("It is hard to like people who have evidently made up their mind to dislike you.") It was said that the whole country had a "chronic case of sulks." The Panamanian—any Panamanian, regardless of position or social status—was a "Spiggotty" or "Spig," terms supposedly derived in earlier years from the erroneous claim of Panama City hackmen that they could "speaks-da-English."

It was thought that the Panamanian showed too little gratitude for all that was being done for him. When Robert Wood declared in a speech years afterward that the United States had created all the wealth in Panama, he was expressing the profound conviction of virtually every American who worked on the canal.

The Panamanian, not surprisingly, resented the Gringo's power, his deprecation of the Panamanian way of life. The Americans were loud, arrogant, impolite, they drank too much. The canal that was to have brought such untold prosperity to everyone appeared to be doing no such thing. Commissaries within the Zone had deprived local merchants of a long-anticipated bonanza; and the Panamanian populace, unable to shop there themselves, resented that goods in such abundance and at such very low prices should be the exclusive privilege of the well-paid canal workers. Even Gorgas' efforts were a source of resentment.

> They hate us because we cleaned their towns and are keeping them clean [one writer surmised], not perhaps because they actually prefer the old filth and fatalities, but because their correction implies that they were not altogether perfect before we came. For the strongest quality of the Panamanian is his pride, and it is precisely that sentiment which we North Americans have either wantonly or necessarily outraged.

Seeing the poverty of those native Panamanians who lived in the old Chagres villages within the Zone, many Americans knew the feelings expressed by Rose van Hardeveld:

> The poor we had literally all around us. I think each one of us found our wash women or fruit vendors always some one or two that were

> more wretchedly miserable than the others, and that seemed to cry personally for help, to us who had so much where they had nothing.
>
> Much neglect and needless cruelty came to our notice every day. It seemed not so much to proceed from wishing to be cruel or neglectful as from the fact that they knew nothing better. . . .
>
> All of our women felt deeply sorry for the sad-eyed children passing by every day, and yet there was little that we could do that would be of any lasting benefit.
>
> If we gave them clothing it would only tend to make them dissatisfied with what they had, and money would usually go for rum or lottery tickets. We came to feel that charity was really not much good as just charity.

One American who had tried very hard to do something had been quickly removed. Rufus Lane, a former seaman from Massachusetts, had arrived looking for a position during the Stevens regime. He had no technical skills, but he had a letter of introduction from Henry Cabot Lodge and he spoke fluent Spanish, so it was decided to put him in charge of "Canal Zone municipalities in the jungle," a wholly meaningless position but one that he took quite seriously. The "jungle Panamanians"—Indians, West Indians—soon began doing as he instructed. "They cleared the jungle around their huts," the diplomat William Sands reported. "They joined their settlements by hard little foottrails . . . they learned how to dispose of disease-fostering refuse and how to set up simple first aid and sanitation centers. They held town meetings on the primitive New England plan . . . Lane's job seemed to me one of the finest things Americans were doing in Panama." But Lane and his work were abolished by a visiting congressional committee, one member of which told Sands, "These people are of no more use than mosquitoes and buzzards; they ought all to be exterminated together."

With the advance of the waters of Gatun Lake, as thousands of villagers were dispossessed of their land and homes and were moved to new sites on higher ground, very few of them felt that they were given fair compensation and bitterly resented the arbitrary fashion in which their new locations were decided for them. "The Americans took awful advantage of the poor people, because they had no one to speak for them," one woman would remark sadly, more than sixty years later, remembering the home her family had been forced to abandon.

Of the officers in charge of the work, only Gorgas seemed to know how not to alienate the Panamanians. Few officials spoke Spanish or made an effort to learn. Goethals spoke none at all. ("Oh, I suppose he knew how to say no," recalled one American disapprovingly.) During his first year on the job, Goethals had written to one of his sons that a state dinner given by President and Señora Amador was "the most trying function that I have had anything to do with." He had been placed between Señora Amador and Señora Obaldía, neither of whom spoke English.

Amador had died in 1909, not long after being succeeded in office by José de Obaldía. What Goethals thought of Obaldía is not known, but he had long before depicted Amador as "no great shakes," and when Obaldía, after taking office, privately revealed some inside facts about the Amador regime, it was as if all the worst American beliefs regarding the high tone of Panamanian politics and politicians had been confirmed. What truth there was to the information is impossible to know; more important was the fact that the American officials believed every word of it.

The charge was that $200,000 to $300,000 had disappeared from the Panamanian treasury, the discovery having been made only when Obaldía took office. Goethals appears to have learned of it first on October 8, 1908, in a confidential letter from Admiral Rousseau written when Goethals was away from the Isthmus on official business. "It is said Amador gets half the loot himself," Rousseau reported. "Not only that Amador charged up to the Panamanian Government all sorts of private entertainment, presents, etc. given President and Mrs. Roosevelt, Secretary Taft, Secretary and Mrs. Root, and also Miss Roosevelt's wedding present. The pieces charged up are 5 to 10 times actual expenditures. Obaldía and his party are 'mad' all the way through and threaten to publish far and wide just what Amador has done." But as Rousseau further explained it had been agreed to keep the story "bottled up," out of respect to Roosevelt and his family.

Nor was any official alarm expressed over the larger animosities between the two peoples. If such feelings prevailed, the consensus seemed to be, they had always prevailed, from the time the first wave of gold prospectors came ashore in 1849. And doubtless there was no quick or ready solution.

To the average American at work on the canal, the aggrieved pride or "smoldering wrath" of the Panamanian (to use the words of one reporter) was of only marginal concern. There would be time enough later to resolve such difficulties. For now the work was going too well, morale was too high, the end was much too plainly in view to think much about anything else.

21
Triumph

Everything is on a colossal scale.
—*Scientific American*
March 18, 1911

I

"It is hard for me to transmit to you the feeling we all possessed toward the work," Robert Wood said. "Rarely can man see his own work, but we saw it physically . . . year by year . . ."

They saw it in the deepening of Culebra Cut; in the rise of docks and warehouses; in the new railroad; in the fortifications being built at Toro Point and Margarita Island in Limon Bay and on the islands of Perico, Flamenco, and Naos in the Bay of Panama. (The giant 16-inch guns being installed were the largest, heaviest weapons in the possession of the United States and had a range of twenty miles.) They saw it in the hydroelectric plant built adjacent to the spillway of Gatun Dam, and in the dam itself, which in its final stage looked as if it had been there always, more like a huge glacial moraine than anything else. It was all measurable progress.

And there was the lake. It had begun its rise with the closing of the West Diversion channel in 1910, when the dam was still incomplete. In the time since, as the water inched steadily up the long, sloped inner face of the dam, as the Chagres gathered and spread inland mile by mile, the realization of what a very different kind of canal it was to be, the conception of a long arm of fresh water suspended in the jungle, began to take hold and with good effect.

Popular interest at home mounted proportionately, the nearer the dream seemed to fulfillment. In the last years of construction, hundreds of articles appeared in magazines and Sunday supplements under such titles as "The Spirit of the Big Job" or "Realizing the Dream of Panama," "Great Work Nobly Done," "The Greatest Engineering Work of All Time," "Our Canal." In 1913, in anticipation of the projected grand opening, close to a dozen different books were published about Panama and the canal.

But it was also in the closing years of the task that the great locks took form for all to see and they were the most interesting and important construction feats of the entire effort. They were the structural triumphs at Panama. In their overall dimensions, mass, weight, in the mechanisms and ingenious control apparatus incorporated in their design, they surpassed any similar structures in the world. They were, as was often said, the mighty portals of the Panama Gateway. Yet they were something much more than monumental; they did not, like a bridge or a cathedral, simply stand there; they *worked*. They were made of concrete and they were made of literally thousands of moving parts. Large essential elements were not built, but were manufactured, made in Pittsburgh, Wheeling, Schenectady, and other cities. In a very real sense they were colossal machines, the largest yet conceived, and in their final, finished form they would function quite as smoothly as a Swiss watch. They were truly one of the engineering triumphs of all time, but for reasons most people failed to comprehend.

To build all the locks took four years, from the time the first concrete was laid in the floor at Gatun, August 24, 1909. Most impressive of all was their size, and especially if seen during the last stages of construction before the water was turned in. Visitors who stood on the dry floor of a single lock chamber when all of it was still open to the light felt as though they had suddenly lost their sense of scale. Each individual chamber was a tremendous concrete basin closed at both ends with steel gates. The walls, one thousand feet long, rose to eighty-one feet, or higher than a six-story building. The impression was of looking down a broad, level street nearly five blocks long with a solid wall of six-story buildings on either side; only here there were no windows or doorways, nothing to give human scale. The gates at the ends, standing partly open to the sky, were like something in a dream.

Greatest of the "ocean leviathans" in the year 1913, a ship larger than the *Titanic* had been, was the proud new 52,000-ton *Imperator*, of the Hamburg-American Line. The *Imperator* accommodated 5,500 people (for whom there were no less than eighty-three lifeboats); she had a "Pompeian Bath," a "social hall" that could accommodate seven hundred passengers traveling first class; and the *Imperator* could have been contained within a lock chamber with ample room to spare (six feet on each side, nearly sixty feet at either end). A single lock if stood on end would have been the tallest structure in the world, taller even than the Eiffel Tower.*

The artist Joseph Pennell, having climbed down to the floor of an empty lock chamber at Pedro Miguel, found the shapes of gates and walls towering above him so "stupendous" that he was almost unable to draw. Walter Bernard, editor of *Scientific American*, returned from a visit to the Isthmus to write an article on "The Mammoth Locks" in which he conceded that it is impossible even to consider the subject "without drifting into the superlative mood." Another visitor would recall "the feeling that follows a service in a great cathedral."

To build the Great Pyramid or the Wall of China or the cathedrals of France, blocks of stone were set one on top of the other in the age-old fashion. But the walls of the Panama locks were poured from overhead, bucket by bucket, into gigantic forms. And within those forms there had to be still other forms to create the different culverts and tunnels, the special chambers and passageways, required inside the walls. Everything had to be created first in the negative, in order to achieve the positive structure wanted.

Moreover, the creation of the building material itself was a "science" requiring specific, controlled measurements and a streamlined system of delivery from mixing plant to construction site. Timing was vital.

Concrete—a combination of sand, gravel, and portland cement (itself a mixture of limestone and clay)—had been known since the time of the Romans, but was used very little as a building material until the late nineteenth century and then mainly for subbasements and floors. Dry docks and breakwaters were built of reinforced (or ferro) concrete—concrete in which metal rods are added—and in the early 1900's several major buildings were built of the same material in Eu-

* If placed upright in present-day Manhattan, a Panama lock would be among the tallest structures on the skyline, surpassed only by the World Trade Center and the Empire State and Chrysler buildings. The difference between the length of a lock and the height of the Empire State Building, for example, is 250 feet.

rope and the United States, as well as silos, some small bridges, and a Montgomery Ward warehouse in Chicago. George Morison drew up plans for his concrete bridge over Rock Creek Park in Washington, and by 1912 a tremendous concrete railroad bridge, the Tunkhannock Viaduct, was under way near Scranton, Pennsylvania. Nothing even approaching the size of the Panama locks had yet been attempted, however, and not until the building of Boulder Dam in the 1930's would any concrete structure equal their total volume. The largest amount of concrete ever poured in a day anywhere else was about 1,700 cubic yards. At Gatun alone the daily average was nearly double that.

"No structure in the world contains as large an amount of material," William Sibert wrote proudly of the great flight of locks at Gatun. With their approach walls, they measured nearly a mile from end to end. The volume of concrete poured was more than 2,000,000 cubic yards—enough, somebody figured, to build a solid wall 8 feet thick, 12 feet high, and 133 miles long. Taken together, the locks at the other end of the canal, at Pedro Miguel and Miraflores, were larger still, with a volume of some 2,400,000 cubic yards.

The lock chambers all had the same dimensions (110 by 1,000 feet) and they were built in pairs, two chambers running side by side in order to accommodate two lanes of traffic. The single flight at Gatun consisted of three such pairs. There was one pair at Pedro Miguel and two at Miraflores, making six pairs (twelve chambers) in all.

The chambers in each pair shared a center wall that was sixty feet wide from bottom to top. The width of the side walls was forty-five to fifty feet at the floor level, but on the outside they were constructed as a series of steps, each step six feet high, starting from a point twenty-four feet from the base level. So at the top, the side walls were only eight feet wide.

The floors of the chambers were solid concrete, anywhere from thirteen to twenty feet thick.

Once completed, the stepped backs of the side walls would be filled in, covered entirely with dirt and rock. And the locks, once they were in use, would never be less than half full of water. So their size would appear nowhere near so overwhelming.

Seen during construction they were a fantasy of huge, raw-looking concrete monoliths, of forms of sheet steel that looked like colossal, blank theatrical flats, of monstrous cranes and soaring cableways—aerial bucket brigades, as somebody said—and of little automatic rail-roads shunting here and there. The swarms of workers at the lock sites appeared lost beside the rising shapes and the incredible array of mechanical contrivances. The noise was shattering.

At Gatun big square buckets of concrete, nearly six tons to a bucket, were swung through the air high above the locks, dropped to position, and dumped, all by means of a spectacular cableway. Eighty-five-foot steel towers stood on either side of the locks (four on each side) and the cables stretched across a span of some eight hundred feet. The towers were on tracks, so they could be moved forward as the work progressed.

Sand and gravel were brought up the old French canal in barges and were stockpiled near a mixing plant. Then sand, gravel, and portland cement were fed into the plant (a battery of eight concrete mixers) by a little automatic railroad, the cars running in and out on a circular track. Another small railroad carried the buckets of wet concrete from plant to cableway, two buckets on two flatcars pushed by one of the French locomotives. At the cableway two empty buckets would descend from overhead, the two full buckets would be snatched up, delivered through the air at a speed of about twenty miles per hour, then returned to repeat the cycle.

The advantage of such an overhead delivery system was that the work area could be kept free of everything except the essential forms within which the concrete was poured. As fast as a bucketload was deposited, men knee-deep in wet concrete would spread it out.

All the locks were constructed in thirty-six-foot sections, each a single monolith that took about a week to build to its full height. The big steel forms, also on tracks, would then be moved ahead to the next position.

At Pedro Miguel and Miraflores, where the terrain was not so open or spacious as at Gatun, division head Williamson and his civilian engineers decided to use cantilever cranes rather than cableways, cranes so enormous in size that they could be seen rising above the jungle from miles distant. Some of these were in the shape of a gigantic T. Others looked like two gigantic T's joined together and were known as "chamber cranes," because they stood within the lock chambers, their long cantilever arms reaching out over the center and side walls. All the cranes moved on tracks and were self-propelling.

The T-shaped variety were the "mixing cranes." One arm of the T hoisted sand and gravel and cement from stockpiles to mixing plants located in the base of the T. The other arm transferred buckets of fresh concrete to the chamber cranes that in turn swung the buckets to the desired position. The complete operation was about as mechanized as it could possibly have been and to the average onlooker a very weird, unearthly sight to behold. The operator of a chamber crane, the man who guided the concrete to its destination, sat alone in a tiny box hanging from the delivery arm of the crane, nearly a hundred feet off the ground.

Five million sacks and barrels of cement were shipped to Panama to build the locks, dams, and spillways, all of it from New York on the *Ancon* and the *Cristobal*, and an idea of what such quantities amounted to is imparted by a single budgetary statistic: an estimated $50,000 was saved in recovered cement after Goethals issued a directive requiring the men to shake each sack after it was emptied.

Gravel and sand for those structures closest to the Atlantic—Gatun Locks, the Gatun spillway—came by water from points twenty to forty miles east of Colón, the gravel from Porto Bello, where a big crushing plant was built, the sand from Nombre de Dios.* On the Pacific side, the rock (basalt, or traprock) was quarried and crushed right at Ancon Hill, while the sand came from Chamé Point, in the Bay of Panama.

By latter-day standards the engineers were novices in the use of concrete. Numerous discoveries had still to be made about the critical water-cement ratio in the "mix design" and the susceptibility of the material to environmental attack. To build anything so large as the concrete locks at Panama was an unprecedented challenge, but what was built had also to hold up in a climate wherein almost everything, concrete included, could go to pieces rapidly. Yet, however comparatively crude the level of theoretical technology may have been regarding the material, the results were extraordinary. After sixty years of service the concrete of the locks and spillways would be in near-perfect condition, which to present-day engineers is among the most exceptional aspects of the entire canal.

The design and engineering of the locks, the results of years of advance planning, can be attributed largely to three men: Lieutenant Colonel Hodges and two exceptionally able civilians, Edward Schildhauer and Henry Goldmark. Schildhauer, slight of build, clean-shaven, very businesslike, was an electrical engineer and still in his thirties.

* In their initial search for sand of the proper quality, the engineers had gone as far as the San Blas Islands, ninety miles east of Colón, and found just what they wanted. But the San Blas Indians declared that the islands—land, water, and sand—were God's gifts to them and that which God had given they would neither sell nor give to the white man. The engineers were permitted only to anchor overnight, and on the condition that they would leave at dawn and never return.

Goldmark, who with his starched collars and thin, well-brushed hair looked like a corporation lawyer, had responsibility for designing the lock gates.

The fundamental element to be reckoned with and utilized in the locks—*the* vital factor in the whole plan and all its structural, mechanical, and electrical components—was water. Water would lift and lower the ships. The buoyancy of water would make the tremendous lock gates, gates two to three times heavier than any ever built before, virtually weightless. The power of falling water at the Gatun spillway would generate the electrical current to run all the motors to operate the system, as well as the towing locomotives or "electric mules." The canal, in other words, would supply its own energy needs.

No force would be required to raise or lower the level of water in the locks (and thus to raise or lower a ship in transit) other than the force of gravity. The water would simply flow into the locks from above—from Gatun Lake or Miraflores Lake—or flow out into the sea-level channels. The water would be admitted or released through giant tunnels, or culverts, running lengthwise within the center and side walls of the locks, culverts eighteen feet in diameter, as large nearly as the Pennsylvania Railroad tubes under the Hudson River. At right angles to these main culverts, built into the floor of each lock chamber, were smaller cross culverts, fourteen to a chamber, these about large enough to admit a two-horse wagon. Every cross culvert had five well-like openings into the floor, which meant there were all together seventy such holes in each chamber, and it was from these that the water would surge or drain, depending on which valves were opened or shut.

The valves in the large culverts were immense sliding steel gates that moved on roller bearings up and down in frames in the manner of a window. There were two gates to each valve and they weighed ten tons apiece. To fill a lock, the valves at the lower end of the chamber would be closed, those at the upper end opened. The water would pour from the lake through the large culverts into the cross culverts and up through the holes in the chamber floor. To release the water from the lock, the valves at the upper end would be shut, those at the lower end opened.

The reason for having as many as seventy wellholes in the chamber floor was to distribute the turbulence of the incoming water evenly over the full area and thereby subject chamber and ships to a minimum of disturbance. It was the engineers' intention to be able to raise or lower a ship in a chamber in about fifteen minutes.

Of all the moving parts in the system, the largest and most conspicuous were, of course, the lock gates, or "miter gates," as they were known, which swung open like double doors and closed in the form of a flattened V. The leaves of the gates weighed many hundreds of tons apiece and were the largest ever erected. Their construction was begun at Gatun in May 1911. As structures they were relatively simple and posed no special challenge, except, again, for their magnitude. A skin of plate steel was riveted to a grid of steel girders in exactly the manner of a steel ship's hull—or of a modern airplane wing, which they much resembled in vastly enlarged form. And being both hollow and watertight, they would actually float, once there was water in the locks, and thus the working load on their hinges would be comparatively little.

The leaves were all a standard sixty-five feet wide and seven feet thick. They varied in height, however, from forty-seven to eighty-two feet, depending on their position. The highest and heaviest (745 tons) were those of the lower locks at Miraflores, because of the extreme variation in the Pacific tides.

During construction, inspectors went down inside the gates through a system of manholes to check every rivet, an extremely uncomfortable task with the sun beating on the outer steel shell. All imperfect rivets were cut out and replaced and the watertightness of the shell was tested by filling the gate leaves with water.

As a safety precaution there were also to be duplicate gates throughout. One set of double doors was backed by another, in the event that the first set failed to function properly or was rammed by a ship. And since each lock chamber (except the lower locks at Miraflores) had its own set of intermediate gates, the complete system consisted of 46 gates (92 leaves), the total tonnage of which (sixty thousand tons) was almost half again greater than that of a ship such as the *Titanic*.

The purpose of the intermediate gates was to conserve water. While the locks were built to accommodate ships as large as the *Titanic* or the *Imperator*, or larger, each lock chamber could be reduced in size, by closing the intermediate gates, if the ship in transit was not one of the giants and could be accommodated by a chamber of six hundred feet or less. And of all the oceangoing ships in the world at that time, approximately 95 percent were less than six hundred feet long.

To lift a great merchant liner, or any ship of more than six hundred feet, to the level of Gatun Lake would require an expenditure from the lake of 26,000,000 gallons of water, the equivalent of a day's water supply for a major city. For a complete lockage through the canal, for one ocean-to-ocean transit, the expenditure would be double that amount, all of it fresh water and all washed out to sea.

The technical challenge of the lock gates was in their mechanical engineering, in the design and manufacture of all the devices needed to make them open, close, shut, and lock. So basic an "accessory" item as a hinge assembly called for specifications unlike any previously prepared for a manufacturer. Flawless, precision hardware had to be cast of special steels in pieces that weighed several thousand pounds and that could withstand a strain of several million pounds. The yoke assembly used to fasten the tops of the gates to the lock walls weighed seven tons and looked not unlike the metal creations of some latter-day sculptors.

The gates were opened and closed by a simple, very powerful mechanism devised by Edward Schildhauer. The leaves of the gates were connected by steel arms, or struts, to enormous horizontal "bull wheels" concealed within the lock walls. These wheels, nearly twenty feet in diameter, were each geared to a big electric motor; and wheel and strut worked like the driving wheel and connecting rod on a locomotive, only here the action was reversed since the power was being delivered *from* the wheel. To open or close a gate, the wheel revolved about 200 degrees.

In the design of such a fundamental piece of apparatus the young engineer had had no established model to go by. Available data "were at variance," as he wrote, and he had to reckon with such forces as mechanical friction, acceleration, wind resistance, and the effect of different water levels on the two sides of a gate. The extreme test would be the opening and closing of the heaviest gates in a dry chamber. But in recalling the first of such "dry lock" tests, Bishop wrote that the gates swung to and fro "as easily and steadily as one would open an ordinary door."

But as resourceful as Schildhauer had been in this and other designs, as notably as he and Goldmark succeeded in everything they undertook, the end results were, above all, a stunning demonstration of how very far industrial technology had advanced. Among the more fascinating facts about the Panama Canal, for example, is that all hardware for the lock gates—the lifting mechanisms for the stem valves, the special bearings, gears, and struts for the gate machines, all ninety-two bull wheels—was made by a single manufacturer in Wheeling, West Virginia. In 1878, only thirty-five years before, the Quaker ironmaster Daniel J. Morrell had marveled at certain relatively simple steel castings displayed by the French at their Universal Exposition in Paris.

Most of what he had seen was quite beyond the most advanced work at Pittsburgh then or at his own mills in Johnstown. Now a comparatively small organization, the Wheeling Mold and Foundry Company, in a comparatively small industrial center, could produce castings in sizes and quantities unimagined in 1878, and of alloy steels formerly used in small quantities only for fine tools and cutlery. Carbon steel, nickel steel, vanadium steel, steels of exceptional strength and high resistance to corrosion, were being developed for naval armament before the turn of the century, but it was the advent of the automobile that spurred their real production. Vanadium steel, for instance, had been adopted by the Ford Motor Company for use in its engines in 1904, and it was of vanadium steel that the most important casting in the lock gates was made, the huge plate upon which the base of each gate leaf turned, a plate that had not only to bear the weight of the gate, but withstand constant immersion in water.

The most obvious and frequently emphasized differences between the French and American efforts at Panama, between failure and success at Panama, were in the application of modern medical science, the methods of financing, and the size of the excavation equipment used. But it should also be understood that the canal that was built was very different from what could have been built by anyone thirty years earlier. It was not only a much larger canal than it would have been (the locks were nearly twice as large as those designed by Eiffel, which measured 59 by 590 feet); it was constructed differently and of different materials. *And* its means of operation and control were altogether different. "Strongly as the Panama Canal appeals to the imagination as the carrying out of an ideal," wrote one astute editor, "it is above all things a practical, mechanical, and industrial achievement."

Nowhere was this more apparent than in the city of Pittsburgh, where some fifty different mills, foundries, machine shops, and specialty fabricators were involved in the canal, making rivets, bolts, nuts (in the millions), steel girders, steel plates, steel forms for the lock walls, special collapsible steel tubes by which the main culverts were formed, steel roller bearings (18,794 steel roller bearings) for the stem valves and spillway gates. The building of the gates themselves had been entrusted to McClintic-Marshall, a Pittsburgh contracting firm that specialized in heavy steel bridge construction.

The giant cranes in use on the Pacific locks traced their structural lineage to the Eiffel Tower. The steel rope—wire cable—used in the cableways and on the cranes, used in fact on every steam shovel and dipper dredge, had its origins in the Brooklyn Bridge, and indeed most of the cable had been manufactured by John A. Roebling Sons.

Cranes, cableways, rock crushers, cement mixers, all ran by electricity. The canal's own motive power, its entire nervous system, was electrical, and an all-electric canal was something quite new under the sun and something that would have been altogether impossible even ten years earlier.

Operation of the locks would depend on no less than 1,500 electric motors. All controls were electrical. The most important part played by any one manufacturer was that of the General Electric Company, which produced approximately half the electrical apparatus needed during construction and virtually all the motors, relays, switches, wiring, and generating equipment that was installed permanently, in addition to the towing locomotives and all the lighting.

Besides the ninety-two motors used to swing the lock gates, there were forty-six small motors to run "miter forcing" mechanisms that locked the gate leaves once they were in the closed position. On top of every gate was a footwalk with a handrail, so attendants could go back and forth from one side of the lock to the other whenever the gates were closed. With the gates open, the handrail would be in the way, so it too was raised or lowered by an electric motor.

There were more than a hundred 40-horsepower motors to operate the big stem valves in the main culverts, while the largest motors installed, motors of 70 and 150 horsepower, similar to those being developed for heavy duty in steel mills, were needed for two "extraordinary precautions" taken to safeguard the lock gates from damage.

As a ship approached the entrance to the locks, its path would be blocked by a tremendous iron "fender" chain stretched between the walls. The chain would be lowered (into a special groove in the channel floor) only if all was proceeding properly—that is, if the ship was in proper position and in control of the towing locomotives. If the ship was out of control and struck the chain, then the chain would be payed out slowly by an automatic release until the ship was brought to a stop, short of the lock gates. (A 10,000-ton ship moving at five knots could be checked within seventy feet.) The length of the chain was more than four hundred feet and its ends were attached to big hydraulic pistons housed in the lock walls. There were pumps to supply water for the pistons and more electric motors to run the pumps.

If by some very remote chance a ship were to smash through the fender chain, the safety gates would still stand in the path, the apex of their leaves pointed toward the ship. To break through the safety gates would take a colossal force, and it was almost inconceivable that the forward motion of any ship could be that great, having just encountered the fender chain. But in the event that this too occurred, there was still one further safeguard.

The most serious threat to the locks would be from a ship out of control as it approached the upper gates, a ship, that is, about to go *down* through the locks and out of the canal. For if the upper gates were destroyed, then the lake would come plunging through the locks.

So on the side walls at the entrance of each upper lock, between the fender chain and the guard gates, stood a big steel apparatus that looked like a cantilever railroad bridge. This was the emergency dam. It was mounted on a pivot and in a crisis it could be swung—turned electrically—across the lock entrance in about two minutes' time. From its underside a series of wicket girders would descend, their ends dropping into iron pockets in the concrete channel floor. The girders would form runways down which huge steel plates would be dropped, one after another, until the channel was sealed off. It was an ungainly contraption, but it worked most effectively.

The likelihood of a ship even hitting the chain was extremely small. The chance of a ship hitting the chain and breaking it was reckoned at perhaps one in ten thousand.*

Under normal procedure a ship would be controlled by the towing locomotives all the way through the locks, with four locomotives to the average-sized ship, two forward pulling, two aft holding the ship steady. At no time in the locks would a ship move under its own power.

Like nearly every detail of the locks, the towing locomotives were the first of a kind. Presently they would become one of the most familiar features of the canal. They were designed by Schildhauer to work back and forth on tracks built into the top of the lock walls and to move a ship from point to point at about two miles an hour or less. But they also had to negotiate the 45-degree incline between the locks.

Built at Schenectady, the early model cost $13,000. The first order was for forty. Each machine was a little more than thirty feet long, weighed forty-three tons, and had identical cabs at either end, duplicate controls and driving engines, so that it could run in either direction without being turned around. The key feature, however, was a big independently powered, center-mounted windlass that handled

* In fact, the emergency dams, like the 16-inch guns, would never be used and eventually they were dismantled and removed.

some eight hundred feet of steel cable. With the windlass the locomotive could control a ship without even moving. Line could be payed out or reeled in at rapid speed and with loads on the line of as much as twenty-five thousand pounds.

For the still young, still comparatively small General Electric Company the successful performance of all such apparatus, indeed the perfect efficiency of the entire electrical system, was of the utmost importance. This was not merely a very large government contract, the company's first large government contract, but one that would attract worldwide attention. It was a chance like none other to display the virtues of electric power, to bring to bear the creative resources of the electrical engineer. The canal, declared one technical journal, would be a "monument to the electrical art." It had been less than a year since the first factory in the United States had been electrified.

In the broader context, the arrangement was also a historic forerunner: a large, novel, technological objective was to be obtained in abnormally little time and according to the most stringent standards through the combined efforts of the federal government and a specialized industry. (It is, to be sure, a very long way from the electrical installations at Panama to the Manhattan Project, but the lineage is plain.) Furthermore, the outstanding success of the arrangement, the most original and important piece of work to come out of the contract, was that for which the spirit of government-industry cooperation was the most pronounced.

The advantages of electrical power were many: it could be transmitted over long distances; in complicated installations each different machine or mechanism could have its own motor drive (exactly as in the locks), instead of the power being transmitted here and there from one central source by an elaborate system of drive shafts, belts, and pulleys (as in a conventional steam-driven factory). The motors themselves were relatively small, compact, watertight; they turned at constant speeds irrespective of the loads put upon them; they required a minimum of attention; they would not blow up.

But the chief virtue of electricity was in the degree of control it afforded. Things could be made to happen—stop, start, open, close—with the mere press of a button or the turning of a few simple switches on a central control board. And so it was to be at Panama, and with one other extremely important feature. In this operation, things could be made to happen only as they were supposed to, in exactly the prescribed sequence.

Though the fundamental principles were much like those developed for railroad switchboards, no comparable control system had been produced heretofore. Again credit for the basic conception belongs to Edward Schildhauer, but otherwise it was a wholly joint effort. "No specifications could have been more exacting or explicit as to the results to be accomplished," wrote one of the engineers at Schenectady, "or have given a wider range as to the method of their accomplishment. . . . It was the single aim of all concerned to produce something better, safer and more reliable than anything before undertaken." A special department was set up at the General Electric works, wherein picked employees concentrated solely on the Panama project. Company engineers were sent to the Isthmus to become thoroughly familiar with all aspects of the problem; Schildhauer and members of his staff came to Schenectady. The result was an unqualified success.

The operation of each flight of locks was to be run from the second floor of a large control house built on the center wall of the uppermost lock. From there, with an unobstructed view of the entire flight, one man at one control board could run every operation in the passage of a ship except the movement of the towing locomotives.

Each control board was a long, flat, waist-high bench, or counter, upon which the locks were represented in miniature—a complete *working* reproduction. The board at Gatun was sixty-four feet in length and about five feet wide. There were little aluminum fender chains that would actually rise in place or sink back out of sight on the board as a switch was turned. The lock chambers were represented by slabs of blue marble. There were aluminum pointers placed in the same relative positions as the lock gates and these opened or closed as the actual lock gates opened and closed. There were upright indicators showing the positions of the rising stem valves, and there were still taller upright indexes showing the level of the water in the chambers to within half an inch.

Everything that happened in the locks—the rise and fall of the fender chains, the opening and closing of the gates—happened on the board in the appropriate place and at precisely the same time. So the situation in the locks could be read in an instant on the board at any stage of the lockage.

In addition, the switches to work the fender chains, lock gates, stem valves, all the switches for every mechanism in the system, were located beside the representation of that device on the board. To lift a 40,000-ton ship twenty-one feet in a lock chamber, one had only to turn a small aluminum handle about like that on an ordinary faucet.

The genius of the system, however, was in the elaborate racks of interlocking bars concealed from view beneath the board. For not only was the operator able to see the entire lockage process in miniature and in operation on the board before him, but the switches were interlocking—mechanically. Each had to be turned in proper sequence, otherwise it would not turn. It was impossible therefore to do anything out of order or to forget to take any crucial step in the necessary order. For example, the switch to lower the fender chain would not operate until the switch to open the lock gates had been thrown into the opening position. Thus no one at the control board could inadvertently lower the chain for a ship to proceed and not have the gates open for the ship to enter the locks. Nor could the same gates be closed once the ship was in the lock without first turning the switch to raise the fender chain again, thus assuring that the chain would always be in the up position to protect the gates whenever the gates were closed.

The gate switch was further interlocked with the switch for the miter-forcing machine (to open the gates the operator had first to unlock the miter-forcing machine). When the stem valves in the culverts were to be opened, to raise the water and lift the ship to the next level, it was possible to open only the correct valves. At Gatun, for example, this would mean that an operator could not possibly flood the lower locks in the flight by opening the valves for the middle and upper chambers at once.

Only with a system run by electricity could the locks have been controlled from a central point. In some instances the distance from an individual motor in the system to the control board was as much as half a mile.

More than half a century later the same control panels would still be in use, functioning exactly as intended, everything as the engineers originally devised. "They were very smart people," a latter-day engineer at Miraflores would remark. "After twenty-one years here I am still amazed at what they did."

Once, just before the canal was completed, the Commission of Fine Arts sent the sculptor Daniel Chester French and the landscape architect Frederick Olmsted, Jr., son of the famous creator of New York's Central Park, to suggest ways in which the appearance of the locks and other components might be dressed up or improved upon. The two men reported:

> The canal itself and all the structures connected with it impress one with a sense of their having been built with a view strictly to their utility.

> There is an entire absence of ornament and no evidence that the aesthetic has been considered except in a few instances. . . . Because of this very fact there is little to find fault with from the artist's point of view. The canal, like the Pyramids or some imposing object in natural scenery, is impressive from its scale and simplicity and directness. One feels that anything done merely for the purpose of beautifying it would not only fail to accomplish that purpose, but would be an impertinence.

Consequently nothing was changed or added. The canal would look as its builders intended, nothing less or more.

II

For all practical purposes the canal was finished when the locks were. And so efficiently had construction of the locks been organized that they were finished nearly a year earlier than anticipated. Had it not been for the slides in the Cut, adding more than 25,000,000 cubic yards to the total amount of excavation, the canal might have opened in 1913.

The locks on the Pacific side were finished first, the single flight at Pedro Miguel in 1911, Miraflores in May 1913. Morale was at an all-time high. Asked by a journalist what the secret of success had been, Goethals answered, "The pride everyone feels in the work."

"Men reported to work early and stayed late, without overtime," Robert Wood remembered. ". . . I really believe that every American employed would have worked that year without pay, if only to see the first ship pass through the completed Canal. That spirit went down to all the laborers."

The last concrete was laid at Gatun on May 31, 1913, eleven days after two steam shovels had met "on the bottom of the canal" in Culebra Cut. Shovel No. 222, driven by Joseph S. Kirk, and shovel No. 230, driven by D. J. MacDonald, had been slowly narrowing the gap all day when they at last stood nose to nose. The Cut was as deep as it would go, forty feet above sea level.

In the second week in June, it would be reported that the newly installed upper guard gates at Gatun had been "swung to a position halfway open; then shut, opened wide, closed and . . . noiselessly, without any jar or vibration, and at all times under perfect control."

On June 27 the last of the spillway gates was closed at Gatun Dam. The lake at Gatun had reached a depth of forty-eight feet; now it would rise to its full height.

Three months later all dry excavation ended. The Cucaracha slide still blocked the path, but Goethals had decided to clear it out with dredges once the Cut was flooded. So on the morning of September 10, photographers carried their gear into the Cut to record the last large rock being lifted by the last steam shovel. Locomotive No. 260 hauled out the last dirt train and the work crews moved in to tear up the last of the track. "The Cut tonight presented an unusual spectacle," cabled a correspondent for *The New York Times*, "hundreds of piles of old ties from the railroad tracks being in flames."

Then on September 26 at Gatun the first trial lockage was made.

A seagoing tug, *Gatun*, used until now for hauling mud barges in the Atlantic entrance, was cleaned up, "decorated with all the flags it owned," and came plowing up from Colón in the early-morning sunshine. By ten o'clock several thousand people were clustered along the rims of the lock walls to witness the historic ascent. There were men on the tops of the closed lock gates, leaning on the handrails. The sky was cloudless, and in midair above the lower gates, a photographer hung suspended from the cableway. He was standing in a cement bucket, his camera on a tripod, waiting for things to begin.

But it was to be a long, hot day. The water was let into the upper chamber shortly after eleven, but because the lake had still to reach its full height, there was a head of only about eight feet and so no thunderous rush ensued when the valves were opened. Indeed, the most fascinating aspect of this phase of the operation, so far as the spectators were concerned, was the quantity of frogs that came swirling in with the muddy water.

With the upper lock filled, however, the head between it and the middle lock was fifty-six feet, and so when the next set of culverts was opened, the water came boiling up from the bottom of the empty chamber in spectacular fashion.

The central control board was still not ready. All valves were being worked by local control and with extreme caution to be sure everything was just so. Nor were any of the towing locomotives in service as yet. Just filling the locks took the whole afternoon. It was nearly five by the time the water in the lowest chamber was even with the surface of the sea-level approach outside and the huge gates split apart and wheeled slowly back into their niches in the walls.

The tug steamed into the lower lock, looking, as one man recalled, "like a chip on a pond." Sibert, Schildhauer, young George Goethals, and their wives were standing on the prow. "The Colonel" and Hodges were on top of the lock wall, walking from point to point, both men in their shirt sleeves, Goethals carrying a furled umbrella,

Hodges wearing glossy puttees and an enormous white hat. The gates had opened in one minute forty-eight seconds, as expected.

The tug proceeded on up through the locks, step by step. The gates to the rear of the first chamber were closed; the water in the chamber was raised until it reached the same height as the water on the other side of the gates ahead. The entire tremendous basin swirled and churned as if being stirred by some powerful, unseen hand and the rise of the water—and of the little boat—was very apparent. Those on board could feel themselves being lifted, as if in a very slow elevator. With the water in the lower chamber equal to that in the middle chamber, the intervening gates were opened and the tug went forward. Again the gates to the stern swung shut; again, with the opening of the huge subterranean culverts, the caramel-colored water came suddenly to life and began its rise to the next level.

It was 6:45 when the last gates were opened in the third and last lock and the tug steamed out onto the surface of Gatun Lake. The day had come and gone, it was very nearly dark, and as the boat turned and pointed to shore, her whistle blowing, the crowd burst into a long cheer. The official time given for this first lockage was one hour fifty-one minutes, or not quite twice as long as would be required once everything was in working order.

That an earthquake should strike just four days later seemed somehow a fitting additional touch, as if that too were essential in any thorough testing-and-proving drill. It lasted more than an hour, one violent shudder following another, and the level of magnitude appears to have been greater than that of the San Francisco quake of 1906. The needles of a seismograph at Ancon were jolted off the scale paper. Walls cracked in buildings in Panama City; there were landslides in the interior; a church fell. But the locks and Gatun Dam were untouched. "There has been no damage whatever to any part of the canal," Goethals notified Washington.

Water was let into Culebra Cut that same week, through six big drain pipes in the earth dike at Gamboa. Then on the afternoon of October 10, President Wilson pressed a button in Washington and the center of the dike was blown sky-high. The idea had been dreamed up by a newspaperman. The signal, relayed by telegraph wire from Washington to New York to Galveston to Panama, was almost instantaneous. Wilson walked from the White House to an office in the Executive Building (as the State, War, and Navy Building had been renamed) and pressed the button at one minute past two. At two minutes past two several hundred charges of dynamite opened a hole more than a hundred feet wide and the Cut, already close to full, at once became an extension of Gatun Lake.

In all the years that the work had been moving ahead in the Cut and on the locks, some twenty dredges of different kinds, assisted by numbers of tugs, barges, and crane boats, had been laboring in the sea-level approaches of the canal and in the two terminal bays, where forty-foot channels had to be dug several miles out to deep water. Much of this was equipment left behind by the French; six dredges in the Atlantic fleet, four in the Pacific fleet, a dozen self-propelled dump barges, two tugs, one drill boat, one crane boat, were all holdovers from that earlier era. Now, to clear the Cut of slides, about half this equipment was brought up through the locks, the first procession from the Pacific side passing through Miraflores and Pedro Miguel on October 25.

The great, awkward dredges took their positions in the Cut; barges shunted in and out, dumping their spoil in designated out-of-the-way corners of Gatun Lake, all in the very fashion that Philippe Bunau-Varilla had for so long championed as the only way to do the job. Floodlights were installed in the Cut and the work went on day and night. On December 10, 1913, an old French ladder dredge, the *Marmot*, made the "pioneer cut" through the Cucaracha slide, thus opening the channel for free passage.

The first complete passage of the canal took place almost incidentally, as part of the new workaday routine, on January 7, when an old crane boat, the *Alexandre La Valley*, which had been brought up from the Atlantic side sometime previously, came down through the Pacific locks without ceremony, without much attention of any kind. That the first boat through the canal was French seemed to everyone altogether appropriate.

The end was approaching faster than anyone had quite anticipated. Thousands of men were being let go; hundreds of buildings were being disassembled or demolished. Job applications were being written to engineering offices in New York and to factories in Detroit, where, according to the latest reports, there was great opportunity in the automobile industry. Families were packing for home. There were farewell parties somewhere along the line almost every night of the week.

William Gorgas resigned from the canal commission to go to South

Africa to help fight an alarming surge of pneumonia among black workers in the gold mines. The understanding was that it would be a brief assignment, after which he was to be made surgeon general of the Army.

Joseph Bucklin Bishop left to resume his literary career in New York.

With the arrival of the new year the Isthmian Canal Commission was disbanded and President Wilson named Goethals the first Governor of the Panama Canal, as the new administrative entity was to be officially known. Goethals' salary as governor was to be $10,000 a year, which was $5,000 less than what he had been paid as chairman of the I.C.C., a decision made in the Senate, which inspired the popular "Mr. Dooley," the syndicated creation of humorist Finley Peter Dunne, to observe:

> "They say republics are ongrateful, but look, will ye, what they've done f'r that fellow that chopped the continent in two at Pannyma. He's a hero, I grant ye, although I'm sorry f'r it, because I can't pronounce his name. . . . What is he goin' to git? says ye? Why, Hinnissy, th' Governmint has already app'inted him Governor iv th' Canal at a greatly rejooced salary."

In Washington after a drawn-out, often acrimonious debate, Congress determined that the clause in the Hay-Pauncefote Treaty stipulating that the canal would be open to the vessels of all nations "on terms of entire equality" meant that American ships could not use the canal toll free, as many had ardently wanted and as much of the press had argued for. American ships would pay the same as the ships of every other nation, 90 cents per cargo ton.

In Washington also, and in San Francisco, plans were being made for tremendous opening celebrations intended to surpass even those at the opening of the Suez Canal. More than a hundred warships, "the greatest international fleet ever gathered in American waters," were to assemble off Hampton Roads on New Year's Day, 1915, then proceed to San Francisco by way of Panama. At San Francisco they would arrive for the opening of the Panama-Pacific International Exposition, a mammoth world's fair in celebration of the canal. The estimate was that it would take four days for the armada to go through the locks.

Schoolchildren in Oregon wrote to President Wilson to urge that the old battleship *Oregon* lead the flotilla through the canal. The idea was taken up by the press and by the Navy Department. The officer who had commanded the ship on her famous "race around the Horn" in 1898, retired Admiral Charles Clark, hale and fit at age seventy, agreed to command her once again and the President was to be his honored guest.

But there was to be no such pageant. The first oceangoing ship to go through the canal was a lowly cement boat, the *Cristobal*, and on August 15 the "grand opening" was performed almost perfunctorily by the *Ancon*. There were no world luminaries on her prow. Goethals again watched from shore, traveling from point to point on the railroad. The only impressive aspect of the event was "the ease and system with which everything worked," as wrote one man on board. "So quietly did she pursue her way that . . . a strange observer coming suddenly upon the scene would have thought that the canal had always been in operation, and that the *Ancon* was only doing what thousands of other vessels must have done before her."

Though the San Francisco exposition went ahead as planned, all but the most modest festivities surrounding the canal itself had been canceled.

For by ironic, tragic coincidence the long effort at Panama and Europe's long reign of peace drew to a close at precisely the same time. It was as if two powerful and related but vastly different impulses, having swung in huge arcs in the forty some years since Sedan, had converged with eerie precision in August 1914. The storm that had been gathering over Europe since June broke on August 3, the same day the *Cristobal* made the first ocean-to-ocean transit. On the evening of the third, the French premier, Viviani, received a telephone call from the American ambassador who, with tears in his voice, warned that the Germans would declare war within the hour. The American ambassador was Myron T. Herrick, who had once been so helpful to Philippe Bunau-Varilla, and at the same moment in Panama, where it was still six hours earlier in the day, Philippe Bunau-Varilla was standing at the rail of the *Cristobal* as she entered the lock at Pedro Miguel, at the start of her descent to the Pacific, he being one of the very few who had come especially for the occasion.

Across Europe and the United States, world war filled the newspapers and everyone's thoughts. The voyage of the *Cristobal*, the *Ancon*'s crossing to the Pacific on August 15, the official declaration that the canal was open to the world, were buried in the back pages.

There were editorials hailing the victory of the canal builders, but the great crescendo of popular interest had passed; a new heroic effort commanded world attention. The triumph at Panama suddenly belonged to another and very different era.

Of the American employees in Panama at the time the canal was opened only about sixty had been there since the beginning in 1904. How many black workers remained from the start of the American effort, or from an earlier time, is not recorded. But one engineer on the staff, a Frenchman named Arthur Raggi, had been first hired by the Compagnie Nouvelle in 1894.

Goethals, Sibert, Hodges, Schildhauer, Goldmark, and the others had been on the job for seven years and the work they performed was of a quality seldom ever known.

Its cost had been enormous. No single construction effort in American history had exacted such a price in dollars or in human life. Dollar expenditures since 1904 totaled $352,000,000 (including the $10,000,000 paid to Panama and the $40,000,000 paid to the French company). By present standards this does not seem a great deal, but it was more than four times what Suez had cost, without even considering the sums spent by the two preceding French companies, and so much more than the cost of anything ever before built by the United States government as to be beyond compare.* Taken together, the French and American expenditures came to about $639,000,000.

The other cost since 1904, according to the hospital records, was 5,609 lives from disease and accidents. No fewer than 4,500 of these had been black employees. The number of white Americans who died was about 350.

If the deaths incurred during the French era are included, the total price in human life may have been as high as twenty-five thousand, or five hundred lives for every mile of the canal.

Yet amazingly, unlike any such project on record, unlike almost any major construction of any kind, the canal designed and built by the American engineers had cost less in dollars than it was supposed to. The final price was actually $23,000,000 below what had been estimated in 1907, and this despite the slides, the change in the width of the canal, and an additional $11,000,000 for fortifications, all factors

* Except for wars, the only remotely comparable federal expenditures up to the year 1914 had been for the acquisition of new territories, and the figure for *all* acquisitions as of that date—for the Louisiana Territory; Florida; California, New Mexico, and other western land acquired from Mexico; the Gadsden Purchase; Alaska; and the Philippines—was $75,000,000, or only about one-fifth of what had been spent on the canal.

not reckoned in the earlier estimate. The volume of additional excavation resulting from slides (something over 25,000,000 cubic yards) was almost equal to all the useful excavation accomplished by the French. The digging of Culebra Cut ultimately cost $90,000,000 (or $10,000,000 a mile). Had such a figure been anticipated at the start, it is questionable whether Congress would have ever approved the plan.

The total volume of excavation accomplished since 1904 was 232,440,945 cubic yards and this added to the approximately 30,000,000 cubic yards of useful excavation by the French gave a grand total, in round numbers, of 262,000,000 cubic yards, or more than four times the volume originally estimated by Ferdinand de Lesseps for a canal at sea level and nearly three times the excavation at Suez.

The canal had also been opened six months ahead of schedule, and this too in the face of all those difficulties and changes unforeseen seven years before.

Without question, the credit for such a record belongs chiefly to George Goethals, whose ability, whose courage and tenacity, were of the highest order.

That so vast and costly an undertaking could also be done without graft, kickbacks, payroll padding, any of the hundred and one forms of corruption endemic to such works, seemed almost inconceivable at the start, nor does it seem any less remarkable in retrospect. Yet the canal was, among so many other things, a clean project. No excessive profits were made by any of the several thousand different firms dealt with by the I.C.C. There had not been the least hint of scandal from the time Goethals was given command, nor has evidence of corruption of any kind come to light in all the years since.

Technically the canal itself was a masterpiece in design and construction. From the time they were first put in use the locks performed perfectly.

Because of the First World War, traffic remained comparatively light until 1918, only four or five ships a day, less than two thousand ships a year on the average. And not until July of 1919 was there a transit of an American armada to the Pacific, that spectacle Theodore Roosevelt had envisioned so long before. Thirty-three ships returning from the war zone, including seven destroyers and nine battleships, were locked through the canal, all but three in just two days.

Ten years after it opened, the canal was handling more than five thousand ships a year; traffic was approximately equal to that of Suez. The British battle cruiser *Hood* and the U.S. carriers *Saratoga* and *Lexington* squeezed through the locks with only feet to spare on their way to the Pacific in the 1920's. By 1939 annual traffic exceeded seven thousand ships.

But in the decades following the Second World War, that figure more than doubled. Channel lighting was installed in 1966 and nighttime transits were inaugurated. Ships were going through the canal at a rate of more than one an hour, twenty-four hours a day, every day of the year. Many of them, moreover—giant container ships, bulk carriers—were of a size never dreamed of when the canal was built: the 845-foot *Melodic*, the 848-foot *Arctic*, the 950-foot *Tokyo Bay*, the largest container ship in the world at the time she made her first transit in 1972. Traffic in the canal by the 1970's was beyond fifteen thousand ships a year, annual tonnage was well beyond the 100,000,000 mark. Tonnage in 1915 had been 5,000,000.

The *Queen Mary*, launched in 1936, was the first ship too large for the locks and others followed—*Queen Elizabeth*, *Normandie*, and, in recent years, supertankers larger even than the *Tokyo Bay*, ships more than 1,000 feet long with beams of more than 150 feet.* As a consequence of this and the steadily mounting traffic in the canal, serious proposals were prepared for a new canal. The President of the United States appointed a commission and all the old routes were surveyed once again—across Tehuantepec, Nicaragua, in the valley of the Atrato, and across the Darien wilderness, at San Blas and at Caledonia Bay.

In 1915 tolls for the year were about $4,000,000. By 1970 they exceeded $100,000,000, even though the rates remained unchanged. In 1973, after sixty years, the Panama Canal Company recorded its first loss, as a result of mounting costs of operation, so in 1974 tolls were raised for the first time, from 90 cents per cargo ton to $1.08, an increase of 20 percent.† Annual revenues from tolls presently exceed $140,000,000.

The largest toll yet paid was for the largest passenger ship ever to pass through the canal, the *Queen Elizabeth II*. She was locked through in March 1975 and paid a record $42,077.88. The average toll per ship (at the present rate) is about $10,000, which is roughly a tenth of the cost of sailing eight thousand miles around Cape Horn.

The lowest toll on record was paid by Richard Halliburton, world

* At this writing there are more than seven hundred "superships" of a size too large to pass through the Panama Canal. But it should be understood that those who built them knew the dimensions of the locks; in other words, use of the canal by such vessels was never intended.

† By law the canal is designed to be self-sustaining and must break even.

traveler, best-selling author, toast of the lecture platform, who in the 1920's swam the length of the canal, doing it by installments one day at a time. He was not the first to swim the canal, but was the first to persuade the authorities to allow him through the locks. So based on his weight, 140 pounds, he was charged a toll of 36 cents.

Changes have been made in the canal as time passed: the Cut was widened to five hundred feet, a storage dam was built across the Chagres about ten miles above Gamboa, and the original towing locomotives were retired and replaced by more powerful models made in Japan. But fundamentally, and for all general appearances, the canal remains the same as the day it opened and its basic plan has been challenged in only one respect. It has been argued that the separation of the two sets of locks at the Pacific end was a blunder, that it would have been a more efficient canal had the Pacific locks been built as a unit at Miraflores, just as at Gatun. But those who have had the most experience with running the canal in recent years do not regard the Pacific arrangement as a limiting factor, and indeed various tests run by the Panama Canal Company through the years indicate that Gatun is actually more of a bottleneck. With certain improvements, the engineers believe the capacity of the present canal could be increased to about twenty-seven thousand ships a year.

The one undeniable misjudgment on Goethals' part was his forecast concerning the slides: he was sure they were over the summer the canal opened. But on a night in October 1914, the side of the Cut at East Culebra gave way and in half an hour the entire channel was blocked. In August the following year the same thing happened again. On September 18, 1915, came the most discouraging break of all in what had been newly renamed Gaillard Cut—an avalanche that closed the canal to traffic for seven months. When the canal reopened, Goethals again insisted that the problem would be "overcome finally and for all time." But that day never arrived. Hundreds of acres of mud and rock slipped into the Cut as the years passed; dredging remained an almost continuous task and a huge expense. And the angle of repose has still to be found. One slide in 1974 dumped an estimated 1,000,000 cubic yards into the Cut.

The creation of a water passage across Panama was one of the supreme human achievements of all time, the culmination of a heroic dream of four hundred years and of more than twenty years of phenomenal effort and sacrifice. The fifty miles between the oceans were among the hardest ever won by human effort and ingenuity, and no statistics on tonnage or tolls can begin to convey the grandeur of what was accomplished. Primarily the canal is an expression of that old and noble desire to bridge the divide, to bring people together. It is a work of civilization.

For millions of people after 1914, the crossing at Panama would be one of life's memorable experiences. The complete transit required about twelve hours, and except for the locks and an occasional community along the shore, the entire route was bordered by the same kind of wilderness that had confronted the first surveyors for the railroad. Goethals had determined that the jungle not merely remain untouched, but that it be allowed to return wherever possible. This was a military rather than an aesthetic decision on his part; the jungle he insisted before a congressional committee was the surest possible defense against ground attack. (Actually he wanted to depopulate the entire Zone, since, as he explained to reporters, "we, as Americans, have no property rights in it.") But for those on board a ship in transit, the effect for the greater part of the journey was of sailing a magnificent lake in undiscovered country. The lake was always more spacious than people expected, Panama far more beautiful. Out on the lake the water was ocean green. The water was very pure, they would learn, and being fresh water, it killed all the barnacles on the ship's bottom.

In the rainy season, storms could be seen long in advance, building in the hills. Sudden bursts of cool wind would send tiny whitecaps chasing over the lake surface. The crossing was no journey down a great trough in the continent, as so many imagined it would be, but a passage among flaming green islands, the tops of hills that protruded still above the surface. For years after the first ships began passing through, much of the shore was lined with half-drowned trees, their dry limbs as white as bones.

The sight of another ship appearing suddenly from around a bend ahead was always startling, so complete was the feeling of being in untraveled waters, so very quiet was everything.

In the Cut the quiet was more powerful, there being little if any wind, and the water was no longer green, but mud-colored, and the sides of what had been the spine of the Cordilleras seemed to press in very close.

Even in the locks there was comparatively little noise. Something so important as the Panama Canal, something so large and vital to world commerce, ought somehow to make a good deal of noise, most people seemed to feel. But it did not. Bells clanged on the towing locomotives now and again and there was the low whine of their engines, but little more than that. There was little shouting back and forth among the men who handled the lines, since each knew exactly what he was to do. The lock gates appeared to swing effortlessly and with no perceptible sound.

Afterword

Among those who were most profoundly stirred by the opening of the canal in August 1914 were Charles de Lesseps and Admirals Alfred Thayer Mahan and Thomas Oliver Selfridge, all three quietly retired, but each still very much alive.

Philippe Bunau-Varilla, having declared it a moment of glory for Goethals (and for "the Genius of the French nation"), rushed home to fight. He lost a leg at Verdun and in later years could be seen "taking his exercise" on the Champs Élysées, a tiny, upright figure marching along on a wooden leg, eyes front, his chauffeur in a limousine following slowly some distance behind. A young American journalist in Paris, Eric Sevareid, who made his acquaintance in 1940, would recall, "I had never encountered such a powerful personality."

Bunau-Varilla died on May 18, 1940, only weeks before Paris was occupied by the German Army.

Theodore Roosevelt never returned to Panama; he never saw the Panama Canal. The passage of the Pacific fleet through the locks in

1919 took place seven months after his death. Nor did he live to see the United States pay Colombia an indemnity of $25,000,000 (in 1921) for the loss of Panama, a move that had been initiated during the Wilson Administration, much to Roosevelt's fury. "One of the rather contemptible features of a number of our worthy compatriots," he wrote privately to Bunau-Varilla, "is that they are eager to take advantage of the deeds of the man of action when action is necessary and then eager to discredit him when the action is once over."

William Gorgas, who headed the Army medical service in the First World War, died of a stroke while in London in 1920. Before his death, Gorgas was visited in the hospital by King George and was knighted for "the great work which you have done for humanity."

George Goethals remained as Governor of the Panama Canal through 1916. During the war he was made quartermaster general in Washington and had charge of procurement, transport, and storage of all supplies for the Army. As a private consulting engineer after the war, with offices on Wall Street, he was extremely active but not a particular success financially, largely because he refused to allow the use of his name "for financial consideration." He died of cancer in 1928 and was buried, as he had requested, at West Point.

John Stevens survived the longest. His work had taken him over much of the country since leaving Panama, and in 1917, at the request of Woodrow Wilson, he went to Russia to reorganize the Trans-Siberian Railway, an assignment that lasted five years. Unlike the others, he made a return trip to Panama, but though tremendously impressed by all that he saw, it was the flight on a Pan American Clipper that gave him the greatest thrill. Vigorous to the end, Stevens died in Pinehurst, North Carolina, in 1943 at the age of ninety.

Once, in a paper addressed "To the Young Engineers Who Must Carry On," Stevens said something with which all of these remarkable men would assuredly have agreed—for all that had happened to the world since Panama.

His faith in the human intellect and its creative capacities remained undaunted, Stevens wrote. The great works had still to come. "I believe that we are but children picking up pebbles on the shore of the boundless ocean. . . ."

Acknowledgments

I wish to express my gratitude to the numbers of persons who over the years have generously helped me with the preparatory work for this book.

First, to my editor, Peter Schwed of Simon and Schuster, who had the idea and who is a steadfast source of creative advice and encouragement; to my literary agents, Paul R. Reynolds and John W. Hawkins, for their counsel; to Captain Miles P. DuVal, Jr., a lifelong student of canal history and author of two fine books on the subject, who offered valuable suggestions at the start of my research; to my sons David and William, who at the ages of fourteen and twelve went with me to Panama and observed things I might otherwise have missed; to my wife, Rosalee, whose confident spirit was never failing, and to whom the book is dedicated. Her part in the work was important beyond measure.

Of enormous value, year by year, were my conversations with numbers of actual participants in the American effort in Panama prior to

1914 and with the descendants and friends of many of the central figures in the book—American, French, Panamanian, and West Indian. The names of twenty-nine of these persons, several since deceased, are listed in the Sources.

Many interviews lasted two to three hours; the majority were taped in part or whole, depending on the wishes of the individual being interviewed. And though all who participated helped tremendously, recalling what life and work on the Isthmus meant in human terms, several supplied explanations, recollections of events, and people, of the kind that no amount of conventional research could ever produce.

Especially in this respect do I acknowledge my indebtedness to Mme. Hervé Alphand, who by a former marriage was the daughter-in-law of Philippe Bunau-Varilla; Alice Anderson, who recalled her childhood beside the canal "diggings" and what the coming of the Americans meant to her family and others of Panama's West Indian community; Crede Calhoun of Panama City; Arthur H. Dean, who as a young man was a protégé of William Nelson Cromwell's; Katharine Harding Deeble, whose father, Chester Harding, served on Goethal's staff and afterward became Governor of the Canal Zone; John Fitzgerald; David St. Pierre Gaillard; Mrs. Thomas Goethals; Alice Roosevelt Longworth; Aminta Meléndez of Colón.

Hubert de Lesseps, whom I interviewed in his office on the Boulevard Haussmann, had not only a fund of sparkling stories about his famous grandfather, Ferdinand, but a clear memory of his Uncle Charles. Moreover, with his physical appearance and bearing, his exceeding charm, M. de Lesseps provided the author with a living example of the legendary family personality.

Professor Elting E. Morison of M.I.T., historian, author, authority on Theodore Roosevelt, guided me in my research into the career of his great-uncle, George Shattuck Morison, and read and commented on portions of the manuscript. The late Richard H. Whitehead of Laconia, New Hampshire, with whom I spent days and with whom I corresponded over several years, was then the last surviving member of Goethals' staff. Aileen Gorgas Wrightson, not long before her death, reminisced at length about her father, William Crawford Gorgas.

In the course of my reading, more than four hundred books were consulted (the listing in Sources is a "select" bibliography) and nearly a hundred different newspapers, magazines, and technical journals. Many months—well over a year in total—were spent combing through collections of unpublished correspondence, diaries, field journals and notebooks, company reports, bulletins, contracts, meteorological records, maps, surveys, boxes of press clippings, scrapbooks, photograph albums—often in out-of-the-way places. The background material has been pieced together from libraries and archives and private collections in twelve different states, in Paris, in Bogotá, in Panama and the Canal Zone, and in Washington, where the major portion of the surviving record of both the French and the American efforts is on file at the Library of Congress and the National Archives. (A rough idea of the volume of material in Washington alone can be conveyed with a few statistics. The general records of the two French canal companies take up 111 feet of shelf space at the National Archives; the general correspondence of the Isthmian Canal Commission from 1905–1914 occupies 186 feet. The papers of Philippe Bunau-Varilla at the Library of Congress include approximately ten thousand items; the Goethals' papers, fifteen thousand items.)

Yet a remarkable amount of rare material has been within arm's reach on my own office shelves, thanks to the interest and generosity of Elinor T. Douglas of Santa Barbara, who let me borrow the pick of what was once General Goethals' own collection of Panama history—books long since out of print, some extremely rare, all hard to come by; Mrs. James B. Moore, Jr., of Plandome, New York, who loaned letters, private memoranda, and technical reports passed down from her grandfather, the eminent civil engineer Alfred Noble (this collection has since been given to the University of Michigan at Ann Arbor); and again to Richard H. Whitehead, who contributed volumes of material collected over a lifetime (as well as a suitcase to carry it all off in).

For their assistance with translation I wish to thank Harold Bell (in Paris), Annie Geohegan, Constance C. Jewett, and Alfred E. Street. Keith L. Oberg worked with newspaper files in Bogotá. Anne Rauffet of the Paris office of *The Reader's Digest* helped with introductions and interviews at the Société de Géographie and the École Polytechnique. Kate Lewin (Paris) and Judith Harkison (Washington) helped track down old photographs. I am a strong believer in the research value of photographs and literally thousands have been examined.

The help and cooperation received from the Canal Zone Library at Balboa and from the staff of the Panama Canal Information Office were unstinting. Especially am I indebted to Librarian Emily J. Price and the late Ruth C. Stuhl, and to Mrs. Nan S. Chong, who is the Panama Collection Librarian; to Frank A. Baldwin, Panama Canal Information Officer, and his able associates Victor G. Canel and Annie R. Rathgeber. (They answered my many queries as time went on. They arranged interviews, provided statistics, photographs, and a complete transcript of the written reminiscences of 112 canal laborers, most of them West Indians and all veterans of construction years—an invaluable collection assembled in 1963 by the Isthmian Historical Society.)

Others in Panama who gave of their time and knowledge and hospitality were Major General David S. Parker, then Governor of the Canal Zone; Ruth Rickarby; and Frank H. Robinson. Dr. Patricia A.

Webb of the Gorgas Memorial Institute of Tropical and Preventive Medicine kindly guided me in my reading on yellow fever.

To the following I am also indebted in many different ways: James C. Andrews, Director of Libraries, Rensselaer Polytechnic Institute; Thomas Alton Ashley; the Baker Library, Dartmouth College; Clarence A. Barnes, Jr.; Doreen Kane Barnes; Peter McC. Barnes; Samuel E. Barnes; Mrs. Ira Barrows; Roy P. Basler, John C. Broderick, Carolyn H. Sung, and the other staff members of the Manuscript Division of the Library of Congress; Dr. William B. Bean of the University of Iowa; A. L. Bentley, Jr.; the Bibliothèque Nationale, Paris; Stephen Birmingham; the Boston Public Library; Jean-François Burgelin, Secrétaire Général de la Première Présidence de la Cour d'Appel de Paris; Roger Butterfield; the Carnegie Library, Pittsburgh; Charleton Coulter, III; Mrs. Gerrit Duys; the staff of the Da Rosa Corporation; Maria Ealand; the Eastern Massachusetts Regional Public Library System and in particular Ann Haddad of the Falmouth Public Library; Gerald Feuille; Mrs. Harry A. Franck and her daughter, Patricia Sheffield; Valarie Franco of the Huntington Library; François Geoffroy of *The Reader's Digest*, Paris; George W. Goethals, II; Peter Goethals; Herbert R. Hands of the American Society of Civil Engineers; Mrs. John U. Hawks; Henry B. Hough; Paula R. Hymes of the American Geographical Society; Kathleen Moore Knight; Mrs. Bronislaw Lesnikowski, Librarian, Martha's Vineyard Regional High School; the staff of the newspaper files, Library of Congress; Dennis Longwell of the Museum of Modern Art; Maria Look; H. H. McClintic, Jr.; John McCullough; the late John McKenna, Director of the Middlebury College Library; Leonard Martin; W. V. Graham Matthews, whose own determined explorations into the career of Philippe Bunau-Varilla have been an inspiration; Burroughs Mitchell; Hazel M. Murdock of the Washington office of the Panama Canal Company; the National Geographic Society; the staff of the New York Historical Society Library; the staff of the Science and Technology Room, New York Public Library; Norah Nicholls; Royall and Sally O'Brien; Charleton and Vera Ogburn; Professor Aimé Perpillou of the Société de Géographie; Eulalie Morris Regan of the Vineyard Haven Public Library and her staff, Margaret Cunningham and Carol McCulloch; Robert L. Reynolds; Mme. Giselle Bunau-Varilla Rocco, daughter of Philippe Bunau-Varilla, who corresponded from her home in Kenya; Soeur Lucie Rogé, Supérieure Générale de la Compagnie des Filles de la Charité de Saint Vincent de Paul, Paris; Cornelius Van S. Roosevelt; Betty Ross; John B. Rothrock; Colonel Charles H. Schilling and Marie T. Capps of the United States Military Academy; Cecil O. Smith, Jr., of Drexel University; John F. Stevens, Jr.; Carolyn T. Stewart, Director, Gorgas Home and Library, University of Alabama; Margot Barnes Street; the University of North Carolina Library; the University of Vermont Library; Robert Vogel of the Smithsonian Institution; James D. Walker and Joel Barker of the National Archives; Richard H. Whitehead, Jr.; Nancy Whiting of the West Tisbury Library; the Widener Library, Harvard University; the Yale University Library.

Finally, I wish to thank my mother and father, Mr. and Mrs. C. Hax McCullough of Pittsburgh, for their abiding interest and encouragement; Audre Proctor, who typed the manuscript; Pat Miller, the copy editor; Frank Metz; Sophie Sorkin; Rafael D. Palacios; Wendell Minor; Edith Fowler, the designer; my daughter Melissa, who did a variety of chores; my son Geoffrey; and Dorie McCullough, who at age seven cannot remember when her father was not working on a book about the Panama Canal.

Notes

Authors and/or book titles are given here in the most convenient abbreviated form; full details on all works cited may be found in the Sources.

BOOK ONE

1. Threshold

The official account of the first Darien Expedition—*Reports of Explorations and Surveys to Ascertain the Practicability of a Ship-Canal between the Atlantic and Pacific Oceans by way of the Isthmus of Darien*, by Thomas Oliver Selfridge—is not only exhaustive but a delight to read and is accompanied by magnificent maps. This and Selfridge's own handwritten journal (Library of Congress) have been the primary sources here. From the large and exceedingly colorful body of material on the gold rush and the Panama Railroad, I have relied primarily on Kemble, *The Panama Route;* Otis, *History of the Panama Railroad;* Taylor, *Eldorado;* and Tomes, *Panama in 1855.*

PAGE

19 "There is a charm": *N.Y. Times*, Jan. 24, 1870.

22–24 Strain expedition: Cullen, *Over Darien;* "Darien Exploring Expedition," *Harper's New Monthly*, Mar., Apr., May, 1855; Kirkpatrick,

PAGE

"Strain's Panama Expedition," *Naval Proceedings*, Aug. 1935; Strain, letter to the Sec. of the Navy, Oct. 1854.

23 "In nearly all, the intellect": Cullen, *Over Darien*, 27–28.

23 "It is to the isthmus of Darien": Isthmian Canal Commission, *Report, 1899–1901*, 38.

25 "One of the most formidable enemies": Beatty, 256.

26 "To Europeans the benefits": *North American Review*, Feb. 1881.

27 "Sufficient is it to add": Selfridge, *Reports*, 6. "It may be the future": Shufeldt, 20.

27 Views of Francisco López de Gómara: Mack, 43.

31 Views of Charles Biddle: *ibid.*, 126.

31 Stephens' cost estimate: *Incidents*, Vol. 1, 413.

32 "At home, this volcano": *ibid.*, Vol. 2, 13.

32 Article XXXV: I.C.C., *Report, 1899–1901*, 451.

34 "I have no time to give reasons": Howe, 26.

34 "overwhelmed with the thought": Taylor, *Eldorado*, 15.

34 Young man from Bennington: letter from Charles G. Lincoln, May 12, 1852, Park-McCullough Papers.

34 Footnote: *Harper's New Monthly*, Jan. 1859.

36 Opinion of Matthew Fontaine Maury: DuVal, 3.

37 Blacks on payroll: P.R.R. *Stockholders' Report*, 1853.

37 Trade in pickled dead: Schott, 68.

37 "well-picked skeletons and bones": Tomes, 207.

38 Grant's ordeal: Richardson, 139–145.

39 Childs survey: I.C.C., *Report, 1899–1901*, 75–76.

39 Kelley calculations: *Scribner's*, June 1879.

39 Trautwine explorations and conclusions: *Journal of the Franklin Institute*, May 1854.

41 "When you give an order": *N.Y. Times*, Apr. 10, 1870.

41 "I am at the front": *ibid.*

"The entire column": *ibid.*

42–43 Diary entries: Selfridge, *Reports*, 20.

44 Top of the ridge: *ibid.*, 28.

44 Concluding views: *ibid.*, 6–7.

2. The Hero

Strange to say, there is no adequate modern biography of Ferdinand de Lesseps. The best of those in French is that by G. Edgar-Bonnet (1951). Of those in English my preference is for the contemporary work by G. Barnett Smith. The biography by Beatty (1956) devotes very little to Panama and is marred by inaccuracies. De Lesseps' own *Recollections* is characteristically high-spirited and one-sided and contains little more than glowing forecasts for Panama, since it was written in the late 1880's.

PAGE

45 "He is loved with true affection": Edgar-Bonnet, 116.

45 "Ferdinand is so good": *ibid.*, 118.

PAGE

Love of life, wives of others: *ibid.*, 119.

46 "lovers of progress": Bertrand, 9.

47 "These healthful occupations": Smith, 8.

48 "Madame de Lesseps received": Beatty, 47.

48–49 Saint-Simon and Enfantin: Dondo.

50 New life at La Chesnaye: Edgar-Bonnet, 125–127.

51 "My dear de Lesseps": *Bulletin du Canal Interocéanique*, Nov. 15, 1879; also Mack, 300.

52 Luxury in the desert: de Lesseps, *Suez Canal*, 9.

52 Description of rainbow: de Lesseps, *Lettres*, Vol. 1, 16.

53 "I am going to accomplish": de Lesseps to Mme. Delamalle, Jan. 22, 1855, quoted in Beatty, 114.

54 "He persevered, you see": Hubert de Lesseps, conversations with the author.

54 Verne on de Lesseps: *Twenty Thousand Leagues Under the Sea.* "I wait with patience": Edgar-Bonnet, 117.

54 "There was a real Egyptian sky": Aronson, 30.

56 "We have had a lot of other men": Hubert de Lesseps, conversations with the author.

56–57 "a small man": Robinson, 139.

57 "He bears his years": N.Y. *Herald*, Feb. 25, 1880.

57–58 "Her form": quoted in Panama *Star & Herald*, Apr. 6, 1880.

58 "We had no difficulty": Hyndman, 119.

58 "Astonish the world": Siegfried, 236.

58 Great geographical movement: Murphy, 1–35.

58–59 Louvre exhibition: *N.Y. Times*, Aug. 1, 1875.

59 "Inevitably the whirlpool": Beatty, 283.

59 "insidious influences": *N.Y. Times*, May 31, 1879.

60 "Either I am the head": *N.Y. Tribune*, Mar. 4, 1880.

60 Syndicate launched: Edgar-Bonnet, 86.

61 Wyse background: Bonaparte-Wyse.

62–66 Wyse describes his two isthmian expeditions and the resulting journey to Bogotá, San Francisco, and Washington in his *Canal Interocéanique* and *Le Canal.* Also Rodrigues, 43–49; Panama *Star & Herald*, Jan. 21, 24, 26, 1878.

62 "I told Messrs. Wyse and Réclus": N.Y. *World*, Feb. 25, 1880.

64 Lack of interest in Wyse by Bogotá press: *El Relator* and others, Mar. 1878.

66 Turndown by Menocal and Lull: Menocal, *North American Review*, Sept. 1879; also Wyse letter to Ammen, July 26, 1876, quoted in *Canal Interocéanique*, 277.

66 Wyse Concession: English translation, Isthmian Canal Commission, *Report, 1899–1901*, 473–478.

67 Sole dissenting view: Gerster, 20.

67 Wyse-Park meeting: Wyse, *Le Canal*, 268–269; Wyse interview, N.Y. *World*, Jan. 29, 1880.

68 Authorized delegates: *Instructions*, 3.

68 "What do you wish to find at Panama?": Courau, 135–136. (Also

PAGE

quoted in somewhat different versions in Beatty, Siegfried, and Mack.)

3. Consensus of One

The complete proceedings of the Paris congress are contained in the *Compte Rendu des Séances* (Paris, 1879) and the event was well covered by the daily press. (I have used *Le Temps, The New York Times,* the New York *Tribune,* and reports by the Paris correspondent for the Panama *Star & Herald,* May 29, June 14, 1879.) The observations by Dr. Johnston are from the *Journal of the American Geographical Society,* Vol. 11, 1879, 172–180. Menocal, in addition to his official report to Secretary Evarts, wrote of the affair in the *North American Review,* Sept. 1879.

PAGE

70 "Great blunders": Bartlett, 497.
72 De Lesseps' opening remarks: *Compte Rendu,* 28.
73 Missing trunk: Menocal to Evarts, June 21; *Instructions,* 12.
75 Wyse speech: *Compte Rendu,* 223.
77 "The surprise and painful emotion": Ammen to Evarts, June 21; *Instructions,* 6.
77 "a modern American political boss": Bishop, *Gateway,* 69–70.
78 "threw off the mantle of indifference": Cristano Medina (from Guatemala), *North American Review,* Sept. 1903.
79–81 Proposal of Godin de Lépinay *Compte Rendu,* 293–299; personal background: Forot.
82 Menocal's disgust: Menocal to Evarts, June 21; *Instructions,* 20.
82 Sea-level canal at Panama recommended: *Compte Rendu,* 454.
83 "The hall was densely crowded": Ammen to Evarts, June 21; *Instructions,* 10.
85–86 "The Congress believes": *Compte Rendu,* 646.
86 Menocal's analysis: *North American Review,* Sept. 1879.

4. Distant Shores

The Panama *Star & Herald* reported the de Lesseps visit to the Isthmus in great detail, as well as the U.S. tour, beginning with the issue of Jan. 1, 1880. De Lesseps' *Bulletin du Canal Interocéanique* is full of descriptive material; the New York *Tribune* carried eyewitness accounts by "an occasional correspondent"; the Rodridgues dispatches ran in the New York *World,* Jan. 11, 22, 23, 24, 29.

PAGE

101 "Panama will be easier": *Recollections,* 200.
102 "M. de Lesseps is convinced": *N.Y. Times,* July 25, 1879.
102 Rumors on the Bourse: Edgar-Bonnet, 123–124.
103 Scene with Wyse: *ibid.,* 129.
104 Welcoming ceremonies: Robinson, 139.
104 "The canal will be made": *ibid.,* 140.

PAGE

107 "M. de Lesseps himself rode": N.Y. *World*, Jan. 22, 1880.
108 "*Le plus belle région*": *Bulletin du Canal Interocéanique*, Feb. 1.
109–110 Crossing at Barbacoas: N.Y. *World*, Jan. 22.
110 "a railroad like this": *ibid.*
112 "Panama is a very miserable old town": Pomfret, 222–223.
113 Mass cleanup: N.Y. *Tribune*, Jan. 22.
113 "14,000 inhabitants": N.Y. *World*, Jan. 22.
114 Striking the first blow: *Bulletin du Canal Interocéanique*, Feb. 1; also Rodridgues, N.Y. *World*, Jan. 22; Rodrigues, *Panama Canal*, 65–66.
115 Group portrait: Archives, Panama Canal Company.
116 "And now, gentlemen, you see": N.Y. *World*, Feb. 25.
116 "he left us entirely to ourselves": *ibid.*
116 "rather bright" sun: *Bulletin du Canal Interocéanique*, Feb. 1.
116 Wright's conclusion: Edgar-Bonnet, 135.
116 Blanchet wedding: Panama *Star & Herald*, Feb. 7.
116 "Her form was voluptuous": Robinson, 143.
117 "one of those men who know how to please": N.Y. *World*, Jan. 22.
117 "Nothing . . . to dampen the ardor": Robinson, 146.
117 Emily Crawford interview: N.Y. *Tribune*, Jan. 13, 1889.
118 "cannot understand why they hesitated so long": Siegfried, 246–247; also *Bulletin du Canal Interocéanique*, Feb. 1, 1880.
118 Shipboard revisions: Bishop, *Gateway*, 75.
119 Windsor Hotel interview: N.Y. *Tribune*, Feb. 25.
119–120 Delmonico's banquet: New York papers, Mar. 2; also *Addresses.*
120 De Lesseps on Capitol Hill: *Testimony*, 1880.
120 Eads project: *Address . . . Before the House Select Committee*, Mar. 9, 1880.
121 President Hayes's message: N.Y. *Tribune*, Mar. 9, 1880; also *Sen. Exec. Doc. No. 112*, 46th Cong., 2nd Sess., 1–2.
121 "The message of President Hayes": New York papers, Mar. 10; also *Bulletin du Canal Interocéanique*, Mar. 15.
121 "What the President said": Rodrigues, *Panama Canal*, 69.
121–122 Cross-country tour: Panama *Star & Herald*, Apr. 20, 21.
122 "It is France alone": Edgar-Bonnet, 127.
122 Arrival in Paris: Panama *Star & Herald*, Apr. 29.
122–123 Lectures, exuberance: *ibid.*

5. The Incredible Task

Appendix B (pages 197–213) of the Isthmian Canal Commission, *Report, 1899–1901* gives a clear, brief history of the Universal Interoceanic Canal Company from 1880 to 1889, with pertinent dates and statistics to document both the financial and engineering sides of the story.

PAGE

125 Spirit of venture capitalism: Siegfried, 240.
125 Dire forecast by Lévy-Crémieux man: Emily Crawford in the N.Y. *Tribune*, Jan. 13, 1889.

PAGE

125–127 Launching of the Compagnie Universelle: *Bulletin du Canal Interocéanique;* also Isthmian Canal Commission, *Report, 1899–1901,* 202–203; also Mack, 310–311.

125–126 Press response: Mack, 312.

126 Crédit Foncier and Rothschilds: N.Y. *Tribune*, Jan. 1, 1881.

127 Comparison to Paris-Lyon-Mediterranée railway: Rodrigues, 81.

127 First stockholders' meeting: *N.Y. Times,* Jan. 31, Feb. 1, 1881.

127 Salaries: Mack, 314.

127 Grant to Ammen: quoted in *N.Y. Times*, Feb. 15, 1888.

128 Comité Américain, expectations and performance: *Sen. Doc. 429*, 59th Cong., 1st Sess.; also Rodrigues, 112–113.

129 French civil engineers: Artz.

129 Henry Barnard view: *Scientific Schools.*

130 "never a more complicated problem": Bigelow, *Panama Canal*, 6–7.

131 "*Travail commencé*": *Bulletin du Canal Interocéanique*, Feb. 15, 1881.

133–134 "We must make certain": Charles to Réclus, July 15, 1881, quoted in Edgar-Bonnet, 171.

134 "Everything you can do": Charles to Richier, May 17, 1882, quoted in Edgar-Bonnet, 172.

134 "The canal hospitals": Nelson, 236.

134 "a very much better institution": Gorgas, *Sanitation*, 224.

135 "overrun with Yankees": Bidwell.

135 "I am persuaded": Réclus to Charles, Mar. 30, 1881, quoted in Edgar-Bonnet, 173.

135 "It is necessary at any price": Chambre des Députés *Rapport*, Vol. 3, 205; also Simon, 48.

135–136 Purchase of the Panama Railroad: Robinson, 159–163; also Mack, 315.

137 "Perhaps no other man": *Illustrated London News,* Nov. 27, 1869.

138 Bionne's death: Bishop, *Gateway*, 93–94.

138 "The truth is": *Bulletin du Canal Interocéanique*, Sept. 1, 1881.

138 Mallet story: Bishop, *Gateway*, 92–93.

138–139 De Lesseps' claim at Vienna: Panama *Star & Herald*, Oct. 2, 1881.

139–145 My material on yellow fever and malaria has been drawn from a variety of sources, chief of which were Gorgas and Hendrick, Dr. Gorgas' own writings, Heiser, Nelson, Theiler and Downs, and Warshaw.

144 Mosquitoes breeding in gardens and hospital wards: Gorgas, *Sanitation*, 232.

145 "Many foreigners": Robinson, 239.

145 "the people who best resist": Nelson, 17.

145 "Certainly his moral character": Bishop, *Gateway*, 94.

146 "genuine bacchanalian orgy": *ibid.*, 88.

146 Lines from Froude: *English in the West Indies*, 177.

146 "Vice flourished": Robinson, 142.

146 Stabbing: Panama *Star & Herald,* May 27, 1881.

PAGE

148 "ten thousand snow shovels": Haskin, 209.
149–150 Earthquake: Panama *Star & Herald*, Sept. 14, 1882; also Nelson, 170–178.
150 "With $30,000,000 already invested": N.Y. *Tribune*, Sept. 28, 1882.
151 "The truth is": Chambre des Députés *Rapport*, Vol. 1, 451.
152 "With your good judgment": Ferdinand to Charles, Feb. 23, 1883, quoted in Edgar-Bonnet, 165.

6. Soldiers Under Fire

While the recollections of Philippe Bunau-Varilla (*Panama: The Creation, Destruction, and Resurrection*) and the files of the Panama *Star & Herald* provide much of the color and human interest to be found in the surviving record of the French years, the most balanced views of the project are the intelligence reports of Navy lieutenants Kimball, Rodgers, and Rogers.

PAGE

153 "soldiers under fire": Bunau-Varilla, *Panama*, 52.
154 "only drunkards and the dissipated": Haskins, 194.
154 "The purge continues": Dingler to Charles, Apr. 1, 1884, quoted in Edgar-Bonnet, 182.
155 "It was put into [their heads]": Dingler to Charles, Oct. 5, 1884, *ibid.*, 183.
155 "quadruple our efforts": Dingler to Charles, Apr. 29, 1884, *ibid.*, 183.
155 Dingler's estimate: Isthmian Canal Commission, *Report, 1899–1901*, 205.
156–158 Slaven operations: Robinson, 150–158.
160 Photograph of Dingler's daughter: Bibliothèque Nationale.
160 "My poor husband": Edgar-Bonnet, 184.
160–161 "I cannot thank you enough": *ibid.*
161 "love for the great task undertaken": Bunau-Varilla, *Panama*, 36.
162 Illegitimate birth of Bunau-Varilla: Archives of the École Polytechnique, Paris.
163 "As an officer runs to it": Bunau-Varilla, *Panama*, 44–45.
163–164 Costly expenditures: Nelson, 235.
163 "a kind of miniature Bois de Boulogne": Chambre des Députés, *Rapport*, Vol. 1, 451; also Mack, 344.
163–164 Search for tribe of giants: Mack, 344.
164–165 "no one can appreciate more than these men": Rogers, 57.
165–166 Recollections of S. W. Plume: *Hearings, H.R. 3110.*
166–167 Geological conditions: Donald F. MacDonald, "Outline of Canal Zone Geology," in Goethals, ed., *Panama Canal*, Vol. 1, 67–83.
168–169 Method of excavation: Goethals, "The Dry Excavation of the Panama Canal," *ibid.*, 335–338; also Bunau-Varilla, *Panama*, 67–70.
170 "this mountain is full of gold": quoted in Whitehead, *Our Faith*, 36.
170 "There is enough bureaucratic work": Nelson, 235.

PAGE

171 "the gallant employees": Pim, *Remarks*, quoted in DuVal, *Mountains*, 82.
171 Dingler kills horses: Bennett, 196.
172 Death for two, perhaps three, out of four French workers: Bishop, *Gateway*, 91.
172 Career of Sister Marie Rouleau: Archives, Compagnie des Filles de la Charité de Saint Vincent de Paul, Paris.
172 "one of those rare women": N.Y. *Tribune*, Aug. 22, 1886.
173 Chief physician locked in his cabin: Mimande.
173 Bodies rolled down the embankment: N.Y. *Tribune*, Aug. 5, 1886.
173 "Sitting on your veranda": *ibid.*, Aug. 29, 1886.
173 Doctors advise staying out of the sun: *ibid.*, Oct. 1, 1886.
174 Gauguin's experiences: Mack, 341–342; also Perruchot, 136–140.
175–179 Prestan and Aizpuru uprisings: Bennett, 84–85; also N.Y. *Times*, Apr. 4, 18, 19, 22, Sept. 4, 1885; *Papers on Naval Operations, 1885; Report of Commander McCalla.*
180 "Men's energies are spontaneously influenced": Bunau-Varilla, *Panama*, 48.

7. Downfall

Events leading to the demise of the canal company were covered in detail by the principal papers on both sides of the Atlantic. I have relied primarily on *The New York Times*, the New York *Tribune*, and *The Times* of London.

PAGE

182 De Lesseps at the Académie: Smith, 211–227.
183 "the most terrible financial disaster": Rodrigues, iv–v.
184 "The whole thing is a humbug": *ibid.*, 124.
184 Storm hits Colón: Bunau-Varilla, *Panama*, 53–56; also Panama *Star & Herald*, Dec. 4, 7, 8, 1885.
186 "You are for us": Panama *Star & Herald*, Feb. 20, 1886.
186 "With hearts and minds like yours": *ibid.*
186 "always indefatigable": Bishop, *Gateway*, 83.
187 Charles "very clear headed": Bigelow, *Diary*.
187 "Any homage paid": Bunau-Varilla, *Panama*, 59.
187–188 "Till the money is secured": Bigelow, *Panama Canal.*
188 "incontestable advantage": *ibid.*
188 Ten times more difficult: Ernest Lambert, "Panama: The Story of a Colossal Bubble," *Forum*, Mar. 1893.
189 Crawford interview: N.Y. *Tribune*, Mar. 30, 1886.
189 Rousseau Report: *Bulletin du Canal Interocéanique*, July 1, 1886; also Mack, 329–330.
189–190 Jacquet and Boyer views: Isthmian Canal Commission, *Report, 1899–1901*, 206.
190 "I am postponed": *The Times*, July 13, 1886; also Siegfried, 265.
191 Baïhaut's reputation and personal life: Brogan, 271; also Simon, 64–65.

PAGE

191–192 Emily Crawford: obituary, *Contemporary Review* (London), Feb. 1916.

192 Salesmen converse with de Lesseps: Panama *Star & Herald*, Feb. 27, 1888.

192 "Soon, gentlemen, we shall meet": N.Y. *Tribune*, Nov. 29, 1886.

193–194 Bunau-Varilla plan: Bunau-Varilla, *Panama*, 79–85.

194 Eiffel engaged: *The Times*, Nov. 16, 1887.

194–195 New "temporary" plan: I.C.C., *Report, 1899–1901*, 207–208.

196 "From all information received": Smith, 287.

196 "The ruin is getting on fine!": Siegfried, 268.

197 "The prudence, the maturity": Bunau-Varilla, *Panama*, 88.

198 False telegram: *ibid.;* also Chambre des Députés, *Rapport*, Vol. 1, 2073.

199 "M. Lesseps soon showed he was not dead": Smith, 293. "unscrupulous audacity": N.Y. *Tribune*, July 1, 1888.

199 "All France, it may be said": Chambre des Députés, *Rapport*, Vol. 3, 101–102.

200 Threatens to reveal "every step": Panama *Star & Herald*, Nov. 26, 1888.

200 "I appeal to all Frenchmen": *Bulletin du Canal Interocéanique*, Dec. 2, 1888; also Mack, 363.

200–201 Last day of the sale: *Pall Mall Gazette*, London *Daily News*, N.Y. *Tribune;* also Simon, 79–80.

8. The Secrets of Panama

There is an abundance of material on the Panama scandal and the resulting trials in Mack, *The Land Divided;* Simon, *The Panama Affair;* and D. W. Brogan's *France Under the Republic;* in Dansette, *Les Affaires de Panamá*, and in both the Smith and Courau biographies of Ferdinand de Lesseps. Again I have relied heavily on the voluminous coverage of the story in the New York and London papers. The most vivid, if something less than an objective, portrayal of the political side of the story is to be found in *Leurs Figures*, the novel by Maurice Barrès.

PAGE

205–206 Drumont and his impact: Byrnes; also Tuchman, 183–184.

206 "This evil doer": Drumont, 325–326.

206 Drumont on Dingler: *ibid.*, 362.

208 Charles describes father's visit: Mack, 378–379.

208 "I imagined I was summoned": Smith, 310.

209 Arthur Meyer's motivation: Vassili, 262.

211 Barrès view: Barrès, "The Panama Scandal," *Cosmopolitan*, June 1894.

211 "I would stake here my honor": Smalley, N.Y. *Tribune*, Nov. 22, 1892.

212 "we can with difficulty": *ibid.*, Nov. 23.

213–215 Career of Cornelius Herz: Simon, 183–197; also N.Y. *Herald*,

PAGE

Dec. 4, 1892; N.Y. *World,* Dec. 22, 23; Vizetelly, *Republican France*, 357–358.

215 Herz and Clemenceau: Bruun, 47–48.

215 "Everything is ranged against me": Bertaut, 117.

217 "That fatal word": *ibid.*, 121.

223 Anti-Semitic rally: *N.Y. Times*, Jan. 7, 1893.

223 "The real France": N.Y. *Tribune*, Jan. 1, 1893.

224 "Oh, I hesitated": Harriss, 157.

225 "With as much intention": *N.Y. Times*, Jan. 13, 1893.

225 Barboux plea: Cour d'Appel, *Barboux.*

226 Sentences: *ibid.*, Feb. 10, 1893.

226 Charles's visit to his father: Smith, 401–402.

227 Charles a changed man: Dansette, 202; also Smith, 411.

228–229 Crawford interview with Herz: N.Y. *Tribune*, Feb. 12, 1893.

229 "It smells bad": Courau, 236–237.

229 "For fifteen years": *ibid.*, 225.

229 "His intelligence, his ability": Siegfried, 275.

230–231 Charles's testimony: *N.Y. Times*, Mar. 9, 1893.

234 "He would not talk about it": Hubert de Lesseps, conversations with the author.

234–235 Funeral: *Times* (London), Dec. 17, 1894.

235 "dancing and pirouetting": Bishop, *Gateway*, 65.

236 "I will not protest": Beatty, 309.

238 "Beautiful illusions!": Cour d'Appel, *Barboux.*

239–240 "In the end one almost believed": Siegfried, 277.

240 Money "as clean gone": *Times*, Dec. 8, 1894.

240–241 Richard Harding Davis reflections: *Harper's Weekly*, Jan. 11, 1896.

BOOK TWO

9. Theodore the Spinner

A biography of the extraordinary John Tyler Morgan has as yet to be written. I have relied chiefly on the numerous newspaper clippings contained among his own papers and among those of Philippe Bunau-Varilla; on obituaries, the published recollections of some of his colleagues in the Senate (Cullom especially), and upon his own speeches and correspondence.

PAGE

247 "Now look! That damned cowboy": quoted in Sullivan, *Our Times*, Vol. II, 380.

247 "400 percent bigger": quoted in Lord, 1.

247 "I did not care a rap": Roosevelt, *Autobiography*, 357.

247 "He strode triumphant": Steffens, 503.

248 "a stream of fresh, pure": Sullivan, *Our Times*, Vol. II, 399.

248 "It is a dreadful thing": *ibid.*, 393.

248 "We need not tell our readers": *ibid.*, 404.

PAGE

248 "As usual Theodore absorbed": Adams, *Letters*, 365.
249 "His walk": quoted in Keller, ed., 18.
249 "No single great material work": *Messages*, 6663.
251 "drifting on the lines": Mahan, *Recollections*, 274.
unable to do knots: Puleston, 24.
251 "It is sea power": Tuchman, 135.
252 "very much the clearest": Morison, *Letters*, Vol. I, 221.
252–253 Mahan on the canal: Mahan, *Influence*, 33–34.
253 "I curled up on the seat opposite": quoted in Thayer, Vol. II, 333.
253 Build the Nicaragua canal "at once": Roosevelt to Mahan, May 3, 1897, quoted in Pringle, *Roosevelt*, 171.
253 "Gradually, a slight change": Roosevelt: quoted in *Autiobiography*, 207.
254 "We cannot sit huddled": quoted in Hill, 1.
254 "The Race of the Oregon": quoted from Sullivan, *Our Times*, Vol. I, 456.
255 "I saw first a mast": Joshua Slocum, *Sailing Alone Around the World*, 1900 (Dover reprint, 1956), 264.
255 "By that experience": Sullivan, *ibid.*, 456.
255 "For after all": Mahan, *Recollections*, 200.
255 "I wish to see": quoted in Beale, 81.
256 "great achievement": quoted in Leech, 508.
256 "no one out of a madhouse": quoted in Beale, 104.
257 "You can imagine": *ibid.*, 102–103.
257 "I do not see why": Bishop, *Roosevelt*, Vol. I, 144.
257 "fraught with very great mischief": Roosevelt to Hay, Feb. 18, 1900, quoted in Thayer, Vol. II, 339–340.
257 "If that canal": *ibid.*
257 "filthy newspaper abuse": *ibid.*, 229.
258 "We must bear the atmosphere of the hour": *ibid.*, 228.
258 "I have hideous forebodings": Hay to Adams, July 11, 1901, quoted in Thayer, Vol. II, 263.
258 "young fellow of infinite dash": Hay to Lady Jeune, Sept. 14, 1901, *ibid.*, 266.
258 "modest withal": Jusserand, 262.
258 "Oh, *dear* little Mr. Hay": Alice Roosevelt Longworth, conversation with the author.
258 "the most delightful man to talk to": Roosevelt to Lodge, Jan. 28, 1909, quoted in Pringle, *Roosevelt*, 243.
259 Signing of the treaty: *N.Y. Times*, Nov. 19, 1901.
259 Panama never mentioned in prior speeches: Pringle, *Roosevelt*, 302.
259 "You know the high regard": Roosevelt to Morgan, Oct. 5, 1901, Morison, *Letters*, Vol. III, 161.
260 "Senator Morgan was an extraordinary man": Cullom, 348.
262 "a job which disgusted France": Pringle, *Roosevelt*, 303.
264 Walker delivers the report: Panama *Star & Herald*, Dec. 3, 1901.
264 Hearst releases report: N.Y. *Journal*, Nov. 21, 22.

PAGE

265 "gumshoe campaign": *N.Y. Times,* Dec. 8, 1901.

265 "I haven't heard a brush crack": *ibid.*

265–266 Paris stockholders' meeting: *ibid.,* Dec. 22.

266 Commission's appraisal of French holdings: Isthmian Canal Commission, *Report, 1899–1901,* 171–175.

266 "It put things": *Hearings, H.R. 3110.*

267 Oval Office meeting: *Story of Panama,* 166.

267 Morgan goes to the White House: *N.Y. Times,* Jan. 17, 1902.

267 Inventory of French property: I.C.C., *Report, 1899–1901,* "Supplementary Report," 673–681.

268 "All the objections shown": N.Y. *Herald,* Jan. 30, 1902.

268 *Courier-Journal* response: Jan. 28, clipping, Morgan papers.

268 "Talk about buying a lawsuit": reprinted, Panama *Star & Herald,* Feb. 22, 1902.

269 "no mean nor grasping spirit": *N.Y. Times,* Jan. 20, 1902.

10. The Lobby

My chief sources here have been the Bunau-Varilla papers. Bunau-Varilla's own published accounts (particularly his *Panama: The Creation, Destruction, and Resurrection*), and the voluminous transcript of hearings before the House Committee on Foreign Affairs, the document known as *The Story of Panama.*

PAGE

270 "In the course of a very active": *Story of Panama,* 207.

270 "The first bugle-note": Bunau-Varilla, *Panama,* 174–175.

271 eyes "as clear as a baby's"; complexion: N.Y. *World,* Oct. 4, 1908.

271 the look of a clever drama student: oil portrait in the Wall Street offices of Sullivan & Cromwell.

271 delight in mysterious reputation: Arthur H. Dean (Sullivan & Cromwell), conversation with the author.

271 "the most dangerous man": *Story of Panama,* 61–62.

271 "No life insurance agent could beat him": N.Y. *World,* Oct. 4, 1908.

271–272 "Accidents don't happen": Arthur H. Dean, conversation with the author.

272 Cromwell background: Dean, *Cromwell.*

272 "skills unusual to lawyers": Dean, *Cromwell.*

273 $800,000 fee rendered: *Story of Panama,* 141.

274 Influence on McKinley concerning I.C.C. appointments: *ibid.,* 152, 227.

274 Cromwell "ready at all times to assist": Isthmian Canal Commission, *Report, 1899–1901,* 14; also *Story of Panama,* 228; personal *Diary, 1899,* George S. Morison.

275 Materials assembled for the commission: a complete set of all the maps, reports, and other documents is included among the George S. Morison papers.

PAGE

275 The elegant Pavillon Paillard, now called the Pavillon Élysée, is still in business and little changed.
275 personalized menus: one survives among the Alfred Noble papers.
275 "a very fine lunch": *Diary, 1899*, Morison papers.
276 "kept in constant . . . communication": *Story of Panama*, 231.
277 "At every turn of my steps": Bunau-Varilla, *Panama*, 177.
278 "When I came to know him": Mitchell, 343.
278 Morison view of Bunau-Varilla: *Bulletin of the American Geographical Society*, Feb. 1903.
278 a mistake to underrate this man: Noble to Morgan, Apr. 8, 1901, Morgan papers.
278 Bunau-Varilla's English: Alice Roosevelt Longworth, conversation with the author.
278 "He didn't just come into a room": *ibid.*
278 "Every phase of the canal question": Mitchell, 345.
279 "a sort of resourceful energy": N.Y. *Sun*, Mar. 19, 1903.
279 "father-son relationship": Clapp, 307.
279 Bunau-Varilla's Russian escapade: *Panama*, 145–151; also "Russia and the Panama Canal," *Review of Reviews*, June 1904.
280 Bigelow letter: Nov. 9, 1898, Bunau-Varilla papers.
280 "Our conferences were long and frequent": Bunau-Varilla, *Panama*, 166.
280 Pavey recollection: *Story of Panama*, 5.
281 Involvement in the Dreyfus case: Dreyfus, 112.
281 "When my three eminent new friends": Bunau-Varilla, *Panama*, 166.
281 "The fight to a finish": *ibid.*, 174.
281 "We have not forgotten": William Watts Taylor to Bunau-Varilla, Dec. 13, 1900, Bunau-Varilla papers.
282 "Every time I was in need": Bunau-Varilla, *Panama*, 177.
282 "Everything has been done for Philippe": A. C. Baker to Pavey, Feb. 7, 1901, Bunau-Varilla papers.
282 Cincinnati speech: Bunau-Varilla, *Panama*, 179.
282 Note and drawing from daughter: Mar. 1902, Bunau-Varilla papers.
282 "intensity of conviction": William Watts Taylor to Bunau-Varilla, no date, Bunau-Varilla papers.
284 "Never did a more propitious occasion": Bunau-Varilla, *Panama*, 181.
284 "This French engineer": Boston *Herald*, Jan. 26, 1901.
284 "He lectured before 250": Baker to Pavey, Feb. 7, 1901, Bunau-Varilla papers.
285 Expenditures: Bunau-Varilla papers.
285 "Open any dictionary": Bunau-Varilla, *Nicaragua*, 31.
286 Replies from Hay and Walker: Bunau-Varilla papers.
286 "Towards midnight": Bunau-Varilla, *Panama*, 184–185.
286 "Monsieur Bunau-Varilla, you have convinced me": *ibid.*, 186–187.
287 Hanna's opinion was McKinley's "own": *ibid.*, 187.

PAGE

287 Bunau-Varilla's account of the scene: *ibid.*, 187–188.
288 Method of organizing the Compagnie Nouvelle: I.C.C., *Report, 1899–1901*, 213; also Mack, 406–408.
289 Bunau-Varilla the creation of the Seligmans: This is the theory offered by Stephen Birmingham in "*Our Crowd*." When I wrote to ask Mr. Birmingham what his sources were, he replied that he could not remember.
289–290 Footnote: *Life*, June 19, 1939.
290 "servants to wait on the servants": Mme. Hervé Alphand, conversation with the author.
290 Origins of family fortune remain a mystery: Mme. Hervé Alphand and Philippe Bunau-Varilla, II, conversations with the author.
290 "An active go-between": *Story of Panama*, 13.
291 $60,000 donation from Cromwell: *ibid.*, 70–71.
291 A strong case for why $60,000 could never have "fixed" Hanna is made by Miner, 102–103.
291–292 Walker letter: Bunau-Varilla, *Panama*, 205–206.
292 Walker sees Cromwell: Bunau-Varilla, *Panama*, 240–241.
292 Bunau-Varilla returns: Bunau-Varilla, *Panama*, 207.
292–293 Wellman message: Bunau-Varilla papers.
293 Bunau-Varilla's answer: *ibid.*
293 Bunau-Varilla meets with Bô: Bunau-Varilla, *Panama*, 209.
293 Cables sent Jan. 3: Bunau-Varilla papers.
294 Cromwell reinstated: *Story of Panama*, 244–246.
294 Terms of reinstatement: *ibid.*, 245.

11. Against All Odds

The testimony of Admiral Walker and others is quoted directly from the published transcript, *Hearings before the Senate Committee on Interoceanic Canals on H.R. 3110.* The speeches of Morgan and Hanna on the Senate floor are from the *Congressional Record* (57th Cong., 1st Sess.).

PAGE

315 Reaction at Brooklyn dinner: *N.Y. Times*, Mar. 16, 1902.
316 Pelée disaster: *ibid.*, May 9, 10, 12, 19, 28; also Lately Thomas, "Prelude to Doomsday," *American Heritage*, Aug. 1961.
316 "What an unexpected turn": Bunau-Varilla, *Panama*, 228.
317 "He greatly prefers": Hay to Morgan, May 12, 1902, quoted in Thayer, Vol. II, 302.
320 "Ladies and diplomats": Beer, 265.
320 "He has a mass of material": telegram, June 3, 1902, Bunau-Varilla papers.
321 "This plain old person": Beer, 265.
322 "*Mais, il est formidable!*": *ibid.*
323 "It was absolutely necessary": Bunau-Varilla, *Panama*, 246.
324 Hanna claims forty-five votes: *N.Y. Times*, June 11.
324 Outcome: *ibid.*, June 20.

PAGE

326 Morison letter of Dec. 10: Morison papers.

326–327 Career and character of Morison: Morison papers; also Morison, *George Shattuck Morison;* Elting E. Morison, conversations with the author.

326 "There is a kind of man": Morison, *Morison*, 19.

327 "I hate to eat my lunch": Fullerton L. Waldo, "An Engineer's Life in the Field on the Isthmus," *Engineering Magazine*, Dec. 1905.

328 "Make way for the canal!": June 27, 1902.

12. Adventure by Trigonometry

The primary source for Bunau-Varilla's activities is Bunau-Varilla; still, it should be stressed that there is little or nothing in the rest of the surviving record to lead one to dispute or doubt his version of the story to any serious degree. Nor did others involved (such as Cromwell or Loomis) issue denials or contrary accounts following publication of Bunau-Varilla's *Panama: The Creation, Destruction, and Resurrection* in 1920.

The finest overall study of the struggle between the Colombian ministers and the State Department, and of the Marroquín regime, is Dwight C. Miner's *The Fight for the Panama Route*.

PAGE

329 Concha in a strait jacket: Dennis, 314.

329–330 "new impelling force": Thayer, Vol. II, 297.

331 "The desire to make themselves appear": *Story of Panama*, 192.

331 Concha "subject to great nervous excitement": Hart to Hay, Nov. 3, *State Department Dispatches from Colombia*, Vol. 50; also Miner, 182.

331 Concha's "conscience": quoted in Miner, 184.

332 Hay ultimatum: *ibid.*, 195.

332 "waking from a horrible nightmare": quoted in DuVal, *Cadiz*, 207.

332 Morgan proposes sixty amendments: *Story of Panama*, 273.

333 Raúl Perez article: "A Colombian View of the Panama Question," *North American Review*, July 1903.

333 "Without question": *State Department Dispatches from Colombia*, Vol. 59; also Miner, 249.

333 "the whole document is favorable": quoted in Pringle, *Roosevelt*, 310–311.

333 "urbane, dignified manners": *Dictionary of American Biography*.

334 June 9 message to Beaupré: Hill, 53.

334 "an aggressiveness rarely found": *ibid.*, 48.

336 "An impartial investigation": quoted in Freehoff, 43.

337 "You want to be very careful, Theodore": quoted in Sullivan, *Our Times*, Vol. II, 319.

339 Gap in Cromwell files: Arthur H. Dean, conversation with the author.

339 Cullom interview: N.Y. *Herald*, Aug. 15, 1903.

PAGE

340 "Those contemptible little creatures": Roosevelt to Hay, July 14, 1903, quoted in Pringle, *Roosevelt*, 311.

340 "I would come at once": Dennis, 342–343.

340 "We may have to give a lesson": Roosevelt to Hay, Aug. 17, quoted in Pringle, *Roosevelt*, 311.

340 Adee cautions: Adee to Hay, Aug. 18, Hay papers.

340 "We are very sorry": Adee to Hay, Aug. 19, Hay papers.

340 "It seems that the great bulk": Roosevelt to Hay, Aug. 19, Morison, *Letters*, Vol. III, 567.

341 "It will, doubtless, be a surprise": N.Y. *Herald*, Aug. 29.

343 Cromwell ready to "go the limit": *Story of Panama*, 349.

343 "the responsible person": *ibid.*, 348.

343–344 Amador's code: *ibid.*, 358–359.

345 Poker en route: *ibid.*, 359.

345 Hay gives Duque promise of American assistance: *ibid.*, 360.

346 Cromwell made "a thousand offers": *ibid.*, 362.

346 "I was to go to Washington": Bunau-Varilla, *Panama*, 291.

347 "Disappointed": *Story of Panama*, 364.

347 "We can never know too much": Allan Nevins, in the introduction to Nicholas Roosevelt's *Theodore Roosevelt*, vii.

347 "The warning I gave": quoted in Miner, 349.

347 Corinne Robinson recollection: Hagedorn, 177.

348 "It is for you to decide": quoted in Miner, 351.
Roosevelt's reply: *ibid.*

348 "No one can tell": Roosevelt to Taft, Sept. 15, Morison, *Letters*, Vol. III, 598.

348 "When I make up my mind": quoted in Roosevelt, *Theodore Roosevelt*, 64–65.

349 "I naturally took advantage": Bunau-Varilla, *Panama*, 289.

349 "With your imprudence": *ibid.*, 291.

350 "Tell me what are your hopes": *ibid.*, 292.

350 "Had I the moral right": *ibid.*, 302–303.

350–351 Meeting and exchange with Roosevelt: *ibid.*, 310–312.

351 "Nothing was said": Loomis to Roosevelt, Jan. 5, 1904, quoted in DuVal, *Cadiz*, 299.

351 "Of course I have no idea": Roosevelt to Bigelow, Jan. 6, 1904, Morison, *Letters*, Vol. III, 689.

352 "Bunau-Varilla was up over Sunday": quoted in Clapp, 313.

353 "Never before was this problem": Cromwell to Roosevelt, Oct. 14, 1903, quoted in DuVal, *Cadiz*, 303.

354 "But we shall not be caught napping": Bunau-Varilla, *Panama*, 318.

354 "'There is not a fruit nor a grain'": Davis, *Captain Macklin*, 197–198.

355 Bunau-Varilla's reaction to the interview: Bunau-Varilla, *Panama*, 318–319.

355–356 Mission of Captain Humphrey and Lieutenant Murphy; meeting

PAGE

with Roosevelt: Miner, 353–354; also *Story of Panama*, 367–368; Pringle, *Roosevelt*, 321.

356 Bunau-Varilla demands that he represent the new republic: Bunau-Varilla, *Panama*, 320–321.

357 Telegram provided by Bunau-Varilla: *ibid.*, 324.

358 "The plan seems to me good": *Story of Panama*, 71.

358 Telegram from "Smith": Bunau-Varilla, *Panama*, 328.

358 "It was not information": *ibid.*, 329.

359 "If I succeeded": *ibid.*

359 "The words I had heard": *ibid.*, 331.

359 "My only reply to such critics": *ibid.*, 333.

360 "The United States gunboat": *N.Y. Times*, Nov. 1, 1903.

13. Remarkable Revolution

The Story of Panama: Hearings on the Rainey Resolution before the Committee on Foreign Affairs of the House of Representatives is *the* source for what transpired at Panama during the course of the revolution, almost minute by minute. It includes Cromwell's plea for fees, a lengthy statement by Bunau-Varilla, the log kept by Commander Hubbard, and, most important, the testimony of Henry N. Hall, of the New York *World*. Hall had been assigned by the *World* to investigate every facet of the revolution–who did what, when, both in Panama and in Washington and New York–in preparation for the subsequent Roosevelt-*World* lawsuit. *The Story of Panama* runs to 736 pages.

PAGE

361 "It was a remarkable revolution": *Fifty Years*, 383.

362 "You are an old man": *Story of Panama*, 379.

363 $35,000 payoff for Varón: *ibid.*, 382, 445.

364 "Have just wired you": *ibid.*, 386.

365 Stratagem and Señora Amador: *ibid.*, 387.

365 Personality and background of Shaler: N.Y. *Tribune*, Jan. 2, 1904.

366 "I pointed out to him": *Story of Panama*, 388.

366–367 Hubbard's orders: *Diplomatic History*, 362.

367 Hubbard's return cable: *ibid.*, 365.

367–368 Mission of Aminta Meléndez: Aminta Meléndez, conversation with the author; also *Story of Panama*, 389–390.

368 "If you will aid us": *ibid.*, 390.

368–369 Payoffs to soldiers, $65,000 for Huertas: *ibid.*, 382, 447, 459.

369 "There was nothing that did not show": *ibid.*, 390.

370 Footnote: *ibid.*, 391.

370 "Generals, you are my prisoners": *ibid.*, 394.

371 Ehrman's cable: *ibid.*, 395.

372 Meléndez and Torres at the hotel bar: *ibid.*, 441.

373 "The world is astounded": *ibid.*, 446–447.

374 "We have the money! We are free!": *ibid.*

374 "You must understand": *ibid.*, 450–451.

PAGE

374 "I answered Señor Amador": *ibid.*
375 Arrival of *Dixie;* Torres payoff, *ibid.*, 456–457.
376 Marines land: *ibid.*, 458; also Lejeune, 155.
376 Checks on Brandon bank: *Story of Panama*, 462.
377 Reply from Washington: *ibid.*, 463–464.
377 Cleveland on sovereignty: Message to Congress, Dec. 1885, quoted in Freehoff, 158.
377 Seward statement: Miner, 168.
377–378 Fish statement: Freehoff, 234–235.
378 Two cables ordering Hubbard to take the railroad: *Story of Panama*, 440.
378 "Uprising on Isthmus reported": *ibid.*, 393.
379 Taft view: Pringle, *Taft*, Vol. I, 281.
379–380 "I did not consult Hay": Thayer, Vol. II, 328.
380 Civilization to the "waste places": quoted in Beale, 76.
380 "We have no choice": *ibid.*, 159.
380 "covenant running with the land": memorandum written by Oscar Straus, Nov. 6, 1903, quoted in Morison, *Letters*, Vol. III, 648–649, fn.
381 "It is reported we have made a revolution": quoted in Jusserand, 253.
381 "act of sordid conquest": *N.Y. Times*, Nov. 5, 1903.
381 "oppression habitual . . . our Government was bound": Bishop, *Roosevelt*, Vol. 1, 294–295.
381–382 Message to Congress: *House Doc. No. 1* (58th Cong., 2nd Sess.).
382 "We did our duty": *Outlook*, Oct. 7, 1911.
Colombia not a "responsible" power: Roosevelt to W. R. Thayer, July 2, 1915, quoted in Thayer, Vol. II, 327.
382 "the exercise of intelligent forethought": Roosevelt, *Autobiography*, 508.
382 "most important action": *ibid.*, 512.
382 "our course was straight-forward": *ibid.*, 524.
382–383 "Some of our greatest scholars": Hay to Professor George P. Fisher (Yale), Jan. 20, 1904, quoted in Thayer, Vol. II, 323.
383 Knox remark: Jessup, Vol. I, 404. Root view: *ibid.*, 404–405.
Root view: *ibid.*, 404–405.
383–384 Berkeley speech: San Francisco *Examiner*, Mar. 24, 1911; also DuVal, *Cadiz*, 438.
384 "because Bunau-Varilla brought it . . . on a silver platter": Bunau-Varilla, *Great Adventure*, 34.
386 "By refusing to allow Colombia": Hill, 68.

14. Envoy Extraordinary

PAGE

387 "I had fulfilled my mission": Bunau-Varilla, *Panama*, 429.
388 "the same fatal germs": *ibid.*, 357.

PAGE

388 "Against my work": *ibid.*, 352–353, 358.
388 "So long as I am here, Mr. Secretary": *ibid.*, 358.
388 "I am officially informed": *ibid.*, 359.
388 "to proceed in everything strictly in accord with them": *ibid.*, 360.
390 "What do you think, Mr. Minister": *ibid.*, 366.
390 "I think, Mr. President": *ibid.*
390 "Doubtless M. Bunau-Varilla": *N.Y. Times*, Nov. 14, 1903.
391 Pavey "corrected": Bunau-Varilla, *Panama*, 369.
391 "It was with anxiety": *ibid.*, 372.
391 "I cannot refrain": *ibid.*, 373.
392 "As for your poor old dad": Thayer, Vol. II, 318.
392 Davis letter: Nov. 15, 1903, Hay papers.
392 "very satisfactory, vastly advantageous": Hay to Spooner, quoted in Dennis, 341.
393 wording of Article III: Hay–Bunau-Varilla Treaty, National Archives; also quoted in full in DuVal, *Cadiz*, 476–486.
394 "You see that from a practical standpoint": Bunau-Varilla, *Panama*, 376.
394 "I had not a long time to think it over": *ibid.*, 377.
394 "We separated not without emotion": *ibid.* Bunau-Varilla's greeting at the station: *ibid.*, 378.
395 Boyd said to strike Bunau-Varilla: Aminta Meléndez, conversation with the author.
397 "This time I hit the mark": Bunau-Varilla, *Panama*, 384.
397 Senator Money's speech: *ibid.*, 427.
397 "The debates will be long and heated": quoted in DuVal, *Cadiz*, 407.
397 Opposition "pretty well over": Roosevelt to Theodore, Jr., Feb. 10, 1904, Morison, *Letters*, Vol. IV, 724.
397–398 Herrán departs "with crushed spirits": quoted in DuVal, *Cadiz*, 418.
401 Bunau-Varilla's rush of private thoughts: Bunau-Varilla, *Panama*, 429.
401 Departing words to Hay: *ibid.*

BOOK THREE

15. The Imperturbable Dr. Gorgas

Material on the family background of Dr. Gorgas and on his early career has been drawn almost entirely from the biography *William Crawford Gorgas*, written by his wife Marie D. Gorgas and her collaborator, Burton J. Hendrick. Accounts of Gorgas' work in Cuba can also be found in the first volume of Mark Sullivan's *Our Times* and in Joseph Bucklin Bishop's *The Panama Gateway*. The excellent sketches of Walter Reed, Surgeon General Sternberg, Carlos Finlay, and Henry Rose Carter in the *Dictionary of American Biography* have been of particular value.

PAGE

405 "The world requires": *Memoirs.*
405 "It is all unspeakably loathsome": N.Y. *Tribune,* Feb. 4, 1904.
405 "There is nothing in the nature of the work": *ibid.,* Jan. 23, 1904.
406 "As you know": Roosevelt to Walker, Feb. 24, 1904, Morison, *Letters,* Vol. IV, 738.
406 seven members required: Spooner Act, Sec. 7; also Miner, 411.
407 "We passed through a room": quoted in Sullivan, *Our Times,* Vol. I, 458.
408 "What this nation will insist upon": Instructions to the Isthmian Canal Commission, Morison, *Letters,* Vol. IV, 746.
408 "Tell them that I am going to make the dirt fly": *Nation,* Nov. 23, 1905.
408 Footnote: Roosevelt to Taft, May 9, 1904, Morison, *Letters,* Vol. IV, 786.
410 "The door is unlocked": Ross, *Memoirs,* 301.
411 "not . . . a life to look forward to": Gorgas and Hendrick, 53.
411 "I am not much of a doctor": Martin, 61.
412 "more than an education in medicine": Gorgas and Hendrick, 60.
415 "For the first time since English occupation": quoted in Ross, *Memoirs,* 453.
416 "It is hardly an exaggeration to say": Gorgas and Hendrick, 151.
416 Bessie Murdock's recollection: "Ancon Hospital in 1904 and 1905," Society of the Chagres, *Yearbook, 1913.*
417 "Men who achieve greatness": Martin, 62.
419 "If we can control malaria": Gorgas, article prepared for *Engineering Record,* May 1904, Gorgas papers.
419 "It was not known how many different species": Le Prince, 43.
419 "We had no means": *ibid.,* 44.
420 "The condition is very much the same": Gorgas, article for *Engineering Record,* Gorgas papers.
420 Study of mosquitoes in the wards: Le Prince, 21–22.
421 Gorgas told to use the mails: Gorgas and Hendrick, 151–152.
422 Walker's views on mosquito theory: *ibid.,* 162–164.
423 ". . . whether we build the canal": Pepperman, 54.
423 Davis response: Gorgas and Hendrick, 164–165.
423 "That persistence": *ibid.,* 152.
424 ". . . Dr. Carter was on hand to greet us": *ibid.,* 154.
424 "There is an alluring something": *ibid.,* 155–156.
424 "We were on a high point": *ibid.,* 156–159.
425 "Old rusted French beds": Bessie Murdock, "Ancon Hospital in 1904 and 1905," Society of the Chagres, *Yearbook, 1913;* also quoted in Pepperman, 228–230.
425 "One straight-backed chair": *Canal Record,* Sept. 11, 1907.
425 "He bowed to the man, shook his hand": Dr. Victor Heiser, conversation with the author.
426 "He loved especially the adventure stories": Mrs. Aileen Gorgas Wrightson, conversation with the author.
426 Ross meets Gorgas, goes to Panama: Ross, *Memoirs,* 492–493.

16. Panic

PAGE

439 Commission approval of employees at $1,800 or more: Pepperman, 48.
439 Six vouchers for a handcart: *ibid.*, 50.
439 Carpenters forbidden to cut ten-foot boards: John Foster Carr, "The Chief Engineer and His Work," *Outlook*, June 2, 1906.
439 Paper work for payroll: Pepperman, 49.
439 Incident of the hinges: Bates, 30; also Cameron, 120.
440 Abandoned French equipment and parts: F. L. Waldo, "The Present Status of the Panama Canal," *Engineering* (London), Mar. 15, 1907.
440 "One cannot spend much time": Lindsay Denison, "Making Good at Panama," *Everybody's*, May 1906.
440 "They showed skill": John Foster Carr, "The Chief Engineer and His Work," *Outlook*, June 2, 1906.
441 "One appreciated more and more": Eugene P. Lyle, Jr., "The Real Conditions at Panama," *World's Work*, Nov. 1905.
442 Wallace claim to have had a "regular system" in mind: Wallace testimony, Mar. 20, 1906, *Hearings . . . an Investigation.*
443 "fighting becomes a righteous duty": Stevens, "The Panama Canal," A.S.C.E., *Transactions*, XCI, 949. Wallace and handwritten reports: DuVal, *Mountains*, 148.
443 "One young man came down": Karner, 24.
444 "The beginnings of the force": Wood, 17.
444 "Nothing was said as to how I came to be there": Frank B. Maltby, "In at the Start at Panama," *Civil Engineering*, June 1945.
445 "the cost of explosives, cost of loosening": Wallace, "Building the Foundations," in Bennett, 190–191.
445 U.S. has no imperialist designs: Pringle ,*Taft*, Vol. I, 280–281.
446 Taft wants to revise the I.C.C., wants sea-level canal: *ibid.*, 282.
446 Taft impressed by Wallace: Taft testimony, Apr. 19, 1906, *Hearings . . . an Investigation.*
446 Wallace has chair reinforced for Taft: Karner, 75.
447 "very superior man": Davis to Taft, Jan. 6, 1905, quoted in DuVal, *Mountains*, 151.
447 Wallace and the atmosphere of the old Dingler house: Wallace, "Building the Foundations," in Bennett, 195–196; also Karner, 13–14.
447 First case of yellow fever: Le Prince, 275. Cases in December, January, on the *Boston: ibid.*, 276.
448 Death of Philip G. Eastwick: Panama *Star & Herald*, Jan. 30, 1905.
449 Davis calls press stories "cruelly exaggerated": *ibid.*, Feb. 20, 1905.
450 Remarks by Will Schaefer: *ibid.*, Feb. 13, 1905.
451 Wallace in *Harper's Weekly: ibid.*, May 29, 1905.
451 "ill-paid, over-worked, ill-housed": Magoon to Shonts, June 13, 1905, I.C.C. Records.
451 Death of architect Johnson: Panama *Star & Herald*, May 15, 1905.

PAGE

451 "ending of many a bright young man": quoted in DuVal, *Mountains*, 176.

451 Embittered nurse's views: N.Y. *Tribune*, July 6, 1905.

451 "A white man's a fool": Panama *Star & Herald*, July 17, 1905.

451 "A feeling of alarm": Isthmian Canal Commission, *Annual Report for 1905*, 30.

452 Men feel "doomed": Gorgas, *Sanitation*, 154.

452 Gorgas calls for old newspapers: Karner, 69–70.

454 "To say the least": Panama *Star & Herald*, June 26, 1905.

454 Outbreak of plague: I.C.C., *Annual Report for 1905*, 35–38.

455 "complicated business": Wallace to Taft, June 5, 1905, quoted in Pepperman, 109.

455 Magoon to Taft: June 11, 1905, *ibid.*, 111–14.

455 Second Magoon letter: June 13, 1905, *ibid.*, 115.

456 "This action, of course": quoted in Johnson, *Four Centuries*, 312.

456 "Mr. Wallace, I am inexpressibly disappointed": quoted in Pepperman, 118.

457 "For mere lucre": *ibid.*, 121–123.

457 Press reaction: Panama *Star & Herald*, July 31, 1905.

457–458 Wallace claims to have had yellow fever: "Building the Foundations," in Bennett, 199.

458 Wallace dislike of Cromwell, claim he stepped aside for others better qualified: Wallace testimony, Feb. 5, 1906, *Hearings . . . an Investigation.*

458 "thorough case of fright": Stevens to Taft, Mar. 22, 1906, I.C.C. Records.

17. John Stevens

No one who studies the American effort at Panama can fail to admire John Stevens, of whom a full-scale biography has still to be written. In addition to the sources cited, I have been aided by conversations and/or correspondence with William Russell and Richard H. Whitehead, who knew Stevens personally, by his son, John F. Stevens, Jr., and his granddaughter, Mrs. John U. Hawks. The remarks and observations attributed to Frank Maltby are from a series of four articles that he wrote for *Civil Engineering* (June, July, August, and September 1945) under the title "In at the Start at Panama."

PAGE

459 "pose and draw a salary": Pringle, *Roosevelt*, 262.

459 "Mr. Hill told the President": John F. Stevens, Jr., Speech before the Panama Canal Society, Washington, May 19, 1951.

460 Wallace refuses Stevens' invitation to talk: Stevens to Taft, Mar. 22, 1906, I.C.C. Records.

460 Stevens confers with wife: John F. Stevens, Jr., Speech.

461 "I became tough and hard": Marion T. Colley, "Stevens Has Blasted His Way Across America," *American Magazine*, Feb. 1926.

PAGE

461 "With respect to supermen": Stevens, *Engineer's Recollections,* 44.
462 "He is always in the right place": John Foster Carr, "The Chief Engineer and His Work," *Outlook,* June 2, 1906.
462 "diplomacy . . . unfit to exercise": Stevens, *Engineer's Recollections,* 36.
462 Things in a "devil of a mess": Stevens, "Panama Canal," A.S.C.E., *Transactions,* XCI.
462 "buttle like hell": *ibid.*
462 Shonts's press conference: N.Y. *Tribune,* July 15, 1905.
462 Taft cable to Magoon: June 30, 1905, I.C.C. Records.
422–463 "about as discouraging a proposition": Stevens, "The Truth of History," in Bennett, 210.
463 "scared out of their boots": Stevens testimony, Jan. 16, 1906, *Hearings . . . an Investigation.*
463 "I found no organization": Stevens, *Engineer's Recollections,* 44.
463 Exchange of views on Magoon's veranda: Pepperman, 132–136.
464 "There are three diseases in Panama": Bishop, *Goethals,* 133.
464 "A collision has its good points": Isthmian Canal Commission, *Annual Report for 1905,* 121.
465 "The digging is the least thing": quoted in Eugene P. Lyle, "The Real Conditions at Panama," *World's Work,* Nov. 1905.
465 "regardless of clamor or criticism": Stevens to Taft, Mar. 22, 1906, I.C.C. Records.
466 $90,000 for screening: Stevens, *Engineer's Recollections,* 45–46.
466 Supplies ordered: *Minutes of Meetings,* 1906.
466–467 ". . . a man was reported ill": I.C.C., *Annual Report of 1905,* 32.
467 "Like probably many others": Stevens, "Panama Canal," A.S.C.E., *Transactions,* XCI.
467 Roosevelt's reply to Welch: Gorgas and Hendrick, 198.
467–468 Advice of Dr. Lambert: *ibid.,* 198–202; also Hagedorn, 240–242.
468 Gorgas' tribute to Shonts: Pepperman, 6–7.
468 Gorgas' tribute to Stevens: Gorgas, *Sanitation,* 155.
468 "The fact is that you are the only one": Gorgas to Stevens, Apr. 17, 1914, quoted in Stevens, "Panama Canal," A.S.C.E., *Transactions,* XCI.
469 Hill dictum: *Dictionary of American Biography.*
469 "Personally, I have always felt grateful": Wood, 19–20.
470 "Now I would liken that [French] plant": Stevens, *Report of the Board of Consulting Engineers,* 1906.
473 Different specialties wanted: *Minutes of Meetings,* 1906.
474 *Outlook* writer's view: John Foster Carr, "The Silver Men," *Outlook,* May 19, 1906.
475 Correspondent's account of recruiting in Barbados: Edwards, 29.
476 Martinique workers forced ashore: I.C.C., *Annual Report for 1905,* 56.
476 "I load cement": Albert Banister in Reminiscences.

PAGE

477 "and never forget our ice cream": George H. Martin, *ibid.*

477–478 "The West Indian, while slow": R. E. Wood, "The Working Force of the Panama Canal," in Goethals, ed., *Panama Canal*, Vol. I, 199.

478 Goethals sees only chaos: Panama *Star & Herald*, Mar. 19, 1907.

478 Funds for baseball fields: William Russell, conversation with the author.

479 "There is no element of mystery": I.C.C., *Annual Report for 1905*, 120.

480 Stevens observations to Shonts: Dec. 19, 1905, I.C.C. Records.

483 Views of Chairman Davis: *Report of the Board of Consulting Engineers*, 1906.

484 "Danger to ships in a canal": *ibid.*

485 Stevens' refusal to endorse any plan: *ibid.*

485–486 Stevens' argument for lock plan: quoted in *Engineer's Recollections*, 41.

486 "Yes, if it is absolutely safe": Stevens' testimony, Jan. 23, 1906, *Hearings . . . an Investigation.*

486 "If we had [had] you on our side": Stevens, *Engineer's Recollections*, 42.

487 "a peculiar personal interest": *Report of the Board of Consulting Engineers*, 1906.

487 "The one great problem": DuVal, *Mountains*, 206.

488 "I never regard difficulties": Hart, 287.

18. The Man with the Sun in His Eyes

PAGE

490 "And never did a President": *The Future in America*, 250.

490 Stevens and *Huckleberry Finn:* Roosevelt to George Otto Trevelyan, Jan. 22, 1906, Morison, *Letters*, Vol. V, 137.

490–491 "You all know damn well": quoted in Alfred D. Chandler, Jr., "Theodore Roosevelt and the Panama Canal," Morison, *Letters*, Vol. VI, Appendix, 1547.

491 Stevens' Aug. 5 letter to Roosevelt: quoted in Stevens, "The Truth of History," in Bennett, 222.

493 "Mother and I walk briskly": Roosevelt to Kermit, Nov. 11, 1906, Morison, *Letters*, Vol. V, 495.

493 "It would have been impossible": *Special Message*, 4.

494 Panama declaration: quoted in DuVal, *Mountains*, 232–233.

494–495 Gorgas and Roosevelt disappear: Gorgas and Hendrick, 206–207.

495 Speeches of Amador and Roosevelt: Panama *Star & Herald*, Nov. 19, 1906.

495 "He seemed obsessed": Maltby, "In at the Start at Panama," *Civil Engineering*, Sept. 1945.

496 "I have blisters on both feet": *ibid.*

498 Roosevelt's description of Culebra Cut: Roosevelt to Kermit, Nov. 20, 1906, Morison, *Letters*, Vol. V, 497–498.

498–499 Wells's response to Niagara dynamos: *Future in America*, 54–55.

PAGE

499 Wells's exchange with Roosevelt: *ibid.*

499–500 "We saw him . . . on the end of the train": Hardeveld, 78.

500 "It is a stupendous work": Roosevelt, *Special Message*, 28.

500 "No man can see": *ibid.*, 15.

500 Message "a corker": Gorgas and Hendrick, 210.

500–501 Health figures: all from Roosevelt, *Special Message*, Appendix II.

502 "The least satisfactory feature": Roosevelt to Shonts, Nov. 27, 1906, Morison, *Letters*, Vol. V, 504.

503 Roosevelt finds Stevens "impossible to get on with": Roosevelt to Richard Rogers Bowker, *ibid.*, 629.

503–504 Stevens' letter of Jan. 30: I.C.C. Records.

505 "If he were a drinking man": Roosevelt to Richard Rogers Bowker, Mar. 22, 1907, Morison, *Letters*, Vol. V, 630.

505 "Don't talk, dig": Panama *Star & Herald*, Mar. 1, 1907.

505 ". . . I think he has broken down": quoted in Bishop, *Goethals*, 151.

505 "He was not a quitter": Maltby, "In at the Start at Panama," *Civil Engineering*, Sept. 1945.

505–506 Josephus Daniels' theory: *Wilson Era*, 214.

506 "I've just been easing my mind": Edwards, 495.

506 "The reasons for the resignation": Stevens, "Panama Canal," A.S.C.E., *Transactions*, XCI.

506 "well-planned and well-built machine": Stevens, *Engineer's Recollections*, 52.

506 "The hardest problems were solved": Stevens, "Panama Canal."

508 Roosevelt's speech at Colón: quoted in *Special Message*, Appendix I.

509 "What a hell of a life": R. H. Whitehead, conversation with the author.

509 "He did not amuse himself": Mrs. Thomas Goethals, conversation with the author.

510 "savage eyes": Longworth, *Crowded Hours*, 182–183.

510 Taft, only *clean* fat man: Mrs. Thomas Goethals, conversation with the author.

510 Summons from the White House: Bishop, *Goethals*, 141.

510 "He entered at once": *ibid.*, 143.

510 "He [Roosevelt] expressed regret": *ibid.*, 144.

511 "command the removal of a mountain": *ibid.*, 239.

511 "It will be a position of ample remuneration": quoted in Sullivan, *Our Times*, Vol. I, 466.

511 "a case of just plain straight duty": Bishop, *Goethals*, 149.

512 "The real builder": *ibid.*, 144.

512 "I believe in a strong executive": quoted in Morison, *Letters*, Vol. V, xvii.

19. The Chief Point of Attack

Were it not for Goethals' letters to his son George, included among his papers in the Library of Congress, one might mistakenly assume that he was as devoid of warmth and human emotion as his critics insisted. The letters are a window thrown open upon a sensitive and appealing inner man; for the author they were a revelation and one of the high points of the research.

PAGE

529 "The chief point of attack was": Bishop, *Gateway*, 184. Yardage figures: *ibid.*, 198, 200.

530 *Scientific American* illustration: issue of Nov. 9, 1912. A pyramid to top the Woolworth Building: Abbot, 135.

530 Temperature in Culebra Cut: *Canal Record*, Jan. 20, 1909.

530 A carload every few seconds: *ibid.*, Apr. 1, 1908.

531 System could not be improved upon today: R. H. Whitehead, conversation with the author.

531 Burr statement: *Report of the Board of Consulting Engineers;* also quoted in Bishop, *Gateway*, 186.

531 Estimates revised: Bishop, *Gateway*, 185.

532 Never such affection displayed: quoted in DuVal, *Mountains*, 266–267.

532 Moves in with Gorgas: Goethals to G. R. Goethals, Mar. 10, 1907, Goethals papers.

532 "Army engineers, as a rule": Goethals, "The Building of the Panama Canal," *Scribner's*, Mar. 1915.

532 "Mr. Stevens has done an amount of work": Goethals to G. R. Goethals, Mar. 17, 1907, Goethals papers.

532–533 Suffers through "smoker": Goethals' recollections quoted in Bishop, *Goethals*, 155.

533 ". . . I am commanding the Army of Panama": Panama *Star & Herald*, Mar. 19, 1907.

534 Sibert "cantankerous and hard to hold": Goethals to G. R. Goethals, June 13, 1907, Goethals papers.

534 Morning routine: Bishop, *Goethals*, 227.

534 "There were only a few lights": Edwards, 497.

535 "What the Colonel said he meant": Edgar Young, "The Colonel Passes," N.Y. *Herald-Tribune*, Feb. 5, 1928.

535 "It will help bring the outfit into line": Goethals to G. R. Goethals, May 1, 1907, Goethals papers.

535 "Executive ability": *ibid.*, June 24, 1907.

535–536 "His conversation and his manners": Gorgas and Hendrick, 218.

535 Robert Wood's assessment: *Monument*, 34.

536 Whitehead view: conversation with the author.

536 Bishop report to Roosevelt: quoted in *Goethals*, 181.

536 Goethals has had "a veritable 'hell of a time' ": *ibid.*, 180.

537 "It wasn't so hard": Samuel G. Blythe, "Life in Spigotty Land," *Saturday Evening Post*, Mar. 21, 1908.

537 Goethals' Sunday-morning sessions were described by any number of reporters and visitors; among the best accounts are: Peter C. Macfarlane, "The Solomon of the Ishtmus," *Collier's*, Dec. 9, 1911; Edwards, 504.

PAGE

538 "If you decide against me": Bishop, *Gateway*, 295.
538 "treated like human beings": *ibid.*, 294.
538 "I was present at all the hearings": Bishop, *Goethals*, 190.
539 "He has absolute knowledge": *ibid.*, 190–191.
541 "We must get up early": *West Point, 1975–76 Catalog*, 5.
543 "He who did not see the Culebra Cut": Abbot, 210.
543 Lord Bryce remark: *South America*, 36.
543–544 Bishop's description: *Gateway*, 185–186.
544 Night activities: "Repairing Steam Shovels," *Canal Record*, Apr. 13, 1910.
545 "Man die, get blow up": Albert Banister in Reminiscences.
545 Greater amount of explosive energy than in all previous wars: Dutton, 200.
545 Million pounds of dynamite in one shipment: *Canal Record*, Oct. 21, 1908.
545 Aggregate depth of dynamite holes: Goethals, "Dry Excavation," in Goethals, ed., *Panama Canal*, Vol. I, 350.
546 "We are having too many accidents": Goethals to G. R. Goethals, June 13, 1907, Goethals papers.
546 "The flesh of men flew in the air": Berisford G. Mitchell in Reminiscences.
546 Worst single disaster: *Canal Record*, Dec. 16, 1908.
546 sixty-eight shovels: Goethals, "Dry Excavation," in Goethals, ed., *Panama Canal*, Vol. I, 351.
547 Record for single shovel: *ibid.*, 354.
547 96,000,000 cubic yards: *ibid.*, 384.
547 Miles of track within the Cut: *ibid.*, 359.
548 Handling of traffic: *ibid.*, 357–359.
548 Dumping grounds: "Utilizing Spoil," *Canal Record*, Apr. 7, 1909.
549 Volume deposited at Balboa: Goethals, "Dry Excavation," in Goethals, ed., *Panama Canal*, Vol. I, 369.
549–550 Diversion channels a mistake: *ibid.*, 339.
550 Cucaracha "started afresh": Bishop, *Gateway*, 187.
550 "It was, in fact, a tropical glacier": Gaillard, "Culebra Cut and the Problem of the Slides," *Scientific American*, Nov. 9, 1912.
551 "No one could say when": Bishop, *Gateway*, 193–194.
551 twenty-two slides altogether: Goethals, "Dry Excavation," in Goethals, ed., *Panama Canal*, Vol. I, 377.
552 Rose van Hardeveld's recollection: *Make the Dirt Fly*, 135.
552–553 *National Geographic* quote: "Battling the Panama Slides," Feb. 1914.
553 "I personally would say to my fellow men": George H. Martin in Reminiscences.
553 Sees shovel rise: Gaillard, "Culebra Cut and the Problem of the Slides," *Scientific American*, Nov. 9, 1912.
554 Angle of inclination in 1912: Bishop, *Gateway*, 190.
554 "What are we to do now?": Bishop, *Goethals*, 208–209.

20. Life and Times

Because the American community in Panama was a government creation from start to finish, there are endless records of every facet of daily life—food, housing, varieties of entertainment, health, job efficiency. The body of documentary material on this one subject alone would be enough for several scholarly theses. Moreover, a vast photographic record was made of that now vanished community, a record from which I have drawn again and again for this account.

PAGE

555–556 Observations by the *Outlook* reporter: Edwards, 48–50.

556 "A more healthy, well-to-do and companionable group": Adams, *Panama Canal Zone*, 6–7.

557 "standardized buildings and standardized men": Sands, 25.

557 "Every American looked and behaved": *ibid.*

557 Sands' train-window view: *ibid.*, 27.

558 "It is as if each were individually proud": Franck, 225.

559–560 Pay scale: *Manual of Information*, 1909; *Canal Record*, May 27, 1908.

560 "In fact, everything is done": *Canal Record*, Sept. 18, 1907.

561 Remembrance of Mamie Lee Kelly: Crede Calhoun, conversation with the author.

562 Winifred Ewing recollection: conversation with the author.

563 Fowler flight: John O. Collins, "The Year 1913 in Canal History," in Society of the Chagres, *Yearbook, 1913*.

563 "Hot water and grit soap": Edwards, 52.

564 Robert Worsley recollection: conversation with the author.

565 $2,500,000 expenditure: Bishop, *Gateway*, 279.

566 Sapphires and shark's teeth: John Fitzgerald, conversation with the author.

566 orchids and hounds: General Edwin L. Sibert, conversation with the author.

567 New York banker on socialism and consequences: Abbot, 326–327.

567 "The marvel is": Edwards, 571.

567 "First of all, there ain't any democracy": *ibid.*

567–568 Harry Franck on the caste system: *Zone Policeman 88*, 219.

568 "bull-necked, wholehearted, cast-iron fellow": *ibid.*, 86.

568–569 Corporal Fitzgerald's experiences: John Fitzgerald, conversation with the author.

568 Footnote: Sands, 5; also Abbot, 235–237.

569–570 "The individuality and character": Adams, *Panama Canal Zone*, 16.

570 "Goethals has created": Atlanta *Constitution*, Jan. 14, 1912.

570 "Congressmen and senators": *ibid.*, Jan. 13, 1912.

570 Rose van Hardeveld's recollection: *Make the Dirt Fly*, 100.

571 Goethals' romantic life: R. H. Whitehead, Mrs. R. H. Whitehead, Winter Collins, conversations with the author.

PAGE

571 Pro- and anti-Goethals' factions: General Edwin L. Sibert, Katharine Harding Deeble, R. H. Whitehead, D. St. P. Gaillard, conversations with the author.

571 D. St. P. Gaillard recollection: conversation with the author.

571 Goethals' dislike of others reported in Washington: miscellaneous newspaper clippings, Goethals papers.

572 Footnote: Gorgas and Hendrick, 217.

572 Goethals' estimate of Gorgas as an administrator: R. H. Whitehead, conversation with the author.

573 Goethals-Gorgas exchange: Gorgas and Hendrick, 222.

573 Whitehead opinion: conversation with the author.

574 Gaillard stricken at lunch: D. St. P. Gaillard, conversation with the author.

574 "Poor Gaillard": Bishop, *Goethals*, 207.

574 "Your father has killed my husband": Mrs. Thomas Goethals, conversation with the author.

574 Goethals and D. St. P. Gaillard do not speak: D. St. P. Gaillard, conversation with the author.

575 "that I might someday solemnly raise my hand": Franck. 12.

576 "Even New Englanders": *ibid.*, 225.

576 Black teacher-pupil ratio: *Canal Record*, July 7, 1909.

576 "As for the man whose skin": Franck, 221.

577 "Most of us came from our homelands": John Butcher in Reminiscences.

578 "The visitor who saw first": Abbot, 344–345.

578 Difference in benefits derived by black and white workers: Bishop, *Gateway*, 279.

579 Stevens' view: Stevens' testimony, Jan. 16, 1906, *Hearings . . . an Investigation.*

579 Amount of illiteracy: Franck, 129.

579–580 Account by Jeremiah Waisome: Reminiscences.

580 Experience of Clifford St. John: *ibid.*

581 Account by Edgar Simmons: *ibid.*

582 Figures of illness and fatalities: Isthmian Canal Commission, *Annual Report, 1914*, 381–391.

582 Accident figures: *ibid.*

582 "Some of the costs of the canal are here": Franck, 85.

583–585 Account by James A. Williams: Reminiscences.

585 78,000 would have died: Gorgas, *Sanitation*, 283.

586 "In temperament and tradition": Edwards, 81–82.

586 "It is hard to like people who": *ibid.*, 93.

586 Robert Wood view: *Monument*, 43.

586 "They hate us because": Abbot, 235.

586–587 Rose van Hardeveld feelings: *Make the Dirt Fly*, 116–117.

587 "They joined their settlements": Sands, 30–31.

587 "These people are of no more use": *ibid.*, 31.

587 "The Americans took awful advantage": Alice Anderson, conversation with the author.

PAGE

588 "It is said Amador gets half the loot": Rousseau to Goethals, Oct. 8, 1908, Goethals papers.

21. Triumph

The collection of superb papers in Vol. II of *The Panama Canal: An Engineering Treatise,* edited by Goethals, has been the main source for this account of the locks and how they were built. The authors of the papers were the principal builders themselves: Hodges, Goldmark, Whitehead, and Schildhauer, among others. Colonel Sibert's essay, "Construction of Gatun Locks, Dam, and Spillway," is contained in Vol. I.

The various figures pertaining to the canal after 1914 are from materials provided by the Panama Canal Information Office.

PAGE

589 "It is hard for me to transmit to you": Wood, 44.

597 Gates swing "easily and steadily": Bishop, *Gateway,* 372.

598 "Strongly as the Panama Canal appeals": Ira E. Bennett, in Bennett, 322.

598 Pittsburgh participation: "Industrial Roll of Honor," *ibid.,* 436–460.

602 "No specifications could have been more exacting": J. W. Upp, "An Instance of Co-operation," *General Electric Review,* Jan. 1914.

603 "They were very smart people": Roy Wallace, conversation with the author.

603–604 French-Olmsted report: quoted in Mack, 512.

604 "The pride everyone feels": *Scientific American,* Mar. 9, 1912.

604 "Men reported to work early": Wood, 36.

604 Upper gates operated: *Scientific American,* June 22, 1912.

605 "The Cut tonight": *N.Y. Times,* Sept. 10, 1913.

605–606 *Gatun* makes first passage through the locks: John O. Collins, "The Year 1913 in Canal History," in Society of the Chagres, *Yearbook, 1913;* photographs, National Archives; *N.Y. Times,* Sept. 27, 1913.

606 President Wilson sets off Gamboa blast: *N.Y. Times,* Oct. 11, 1913.

608 "Mr. Dooley" on Goethals: Bishop, *Goethals,* 255.

609 "So quietly did she pursue her way": John Barrett, "The Opening of the Panama Canal," *Bulletin of the Pan-American Union,* Sept. 1914.

613 Engineers do not regard Pacific arrangement as a limiting factor: Maj. Gen. David S. Parker, correspondence with the author.

613 1974 slide: *Panama Canal Spillway,* Oct. 18, 1974.

614 "we . . . have no property rights": quoted in W. Va. *Gazette,* undated clipping, Goethals papers.

AFTERWORD

616 "I had never encountered such a powerful personality": Eric Sevareid, "The Man Who Invented Panama," *American Heritage,* Aug. 1963.

PAGE

617 "One of the rather contemptible features": Roosevelt to Bunau-Varilla, July 7, 1914, quoted in Wagenknecht, 272.

617 Goethals refuses use of name: R. H. Whitehead, conversation with the author.

617 "I believe that we are but children picking up pebbles": Stevens, *Engineer's Recollections*, 69.

Sources

I. Manuscript and Archival Materials

Library of Congress
- Philippe Bunau-Varilla Papers
- George Goethals Papers
- William Gorgas Papers
- John Hay Papers
- A. T. Mahan Papers
- Theodore Roosevelt Papers
- Thomas Oliver Selfridge Papers
- John F. Stevens Papers
- William H. Taft Papers
- John G. Walker Papers

National Archives
- Hay–Bunau-Varilla Treaty
- Records of the Compagnie Nouvelle du Canal de Panamá
- Records of the Compagnie Universelle du Canal Interocéanique
- Records of the Isthmian Canal Commission

John Bigelow Papers, New York Public Library
George S. Morison Papers, Peterborough, N.H.
Alfred Noble Papers, University of Michigan, Ann Arbor

Park-McCullough Papers, North Bennington, Vermont

General Francis E. Pinto Diary (1848), New York Public Library

Richard H. Whitehead Papers, Laconia, N.H.

Reminiscences of life and work during the construction of the Panama Canal (letters in response to a competition in 1963), Isthmian Historical Society, Balboa Heights, C.Z.

II. Official and Semiofficial Publications

Canal Record. Ancon, C.Z. 1907–1914.

Chambre des Députés. *5e Législature, Session de 1893, Rapport Général Fait au Nom de la Commission d'Enquête Chargée de Faire la Lumière sur les Allégations Portées à la Tribune à l'Occasion des Affaires de Panamá.* 3 vols. Paris, 1893.

Congressional Record. Washington: Government Printing Office.

Cour d'Appel de Paris, 1re Chambre. *Plaidoirie de Me. Henri Barboux pour MM. Ferdinand et Charles de Lesseps.* Paris, 1893.

Davis, Rear Admiral Charles H. *Report on Interoceanic Canals and Railroads between the Atlantic and Pacific Oceans.* Washington: Government Printing Office, 1867.

Diplomatic History of the Panama Canal (*Sen. Doc. 474*, 63rd Cong., 2nd Sess.). Washington: Government Printing Office, 1914.

Goethals, George W., ed. *The Panama Canal: An Engineering Treatise.* 2 vols. New York: McGraw-Hill, 1916

Hearings before the Senate Committee on Interoceanic Canals on H.R. 3110 (*Sen. Doc. 253*, 57th Cong., 1st Sess.) Washington: Government Printing Office, 1902.

Hearings before the Senate Committee on Interoceanic Canals on the Senate Resolution Providing for an Investigation of Matters Relating to the Panama Canal (*Sen. Doc. 401*, 59th Cong. 2nd Sess.). Washington: Government Printing Office, 1907.

Howard, L. O. *The Yellow Fever Mosquito* (U.S. Dept. of Agriculture Farmer's Bulletin 547). Washington: Government Printing Office, 1913.

Instructions to Rear Admiral Daniel Ammen and Civil Engineer A. G. Menocal, U.S. Navy, Delegates on the Part of the United States to the Interoceanic Canal Congress, Held at Paris May, 1879, and Reports of the Proceedings of the Congress. Washington: Government Printing Office, 1879.

Isthmian Canal Commission. *Annual Reports.* 1904–1914.

Kimball, Lieutenant William W. *Special Intelligence Report on the Progress of the Work on the Panama Canal during the Year 1885* (*House Misc. Doc. 395*, 49th Cong., 1st Sess.). Washington: Government Printing Office, 1886.

List of Books and of Articles in Periodicals Relating to Interoceanic Canal and Railway Routes (*Sen. Doc. No. 59*, 56th Cong., 1st Sess.). Compiled by Hugh A. Morrison, Jr., of the Library of Congress. Washington: Government Printing Office, 1900.

Lull, Edward P. *Reports of Explorations and Surveys for the Location of Interoceanic Ship-Canals through the Isthmus of Panama, and by the Valley of the River Napipi, by U.S. Naval Expeditions, 1875* (*Sen. Exec. Doc.*

75, 45th Cong., 3rd Sess.). Washington: Government Printing Office, 1879.

——— *Reports of Explorations and Surveys for the Location of a Ship-Canal between the Atlantic and Pacific Oceans through Nicaragua, 1872–1873* (*Sen. Exec. Doc.* 57, 43rd Cong., 1st Sess.). Washington: Government Printing Office, 1874.

Manual of Information Concerning Employments for Service on the Isthmus of Panama. Washington: Government Printing Office, 1909.

Minutes of Meetings of the Isthmian Canal Commission. 1905–1914.

The Panama Canal, 25th Anniversary. Canal Zone Publication, 1939.

Papers on Naval Operations during the Year Ending 1885. Navy Department, Bureau of Navigation, Office of Naval Intelligence. Washington, 1885.

Report of the Board of Consulting Engineers for the Panama Canal. Washington; Government Printing Office, 1906.

Report of the Isthmian Canal Commission, 1889–1901 (*Sen. Doc.* 222, 58th Cong., 2nd Sess.). Washington: Government Printing Office, 1904.

Reports of the United States Commissioners to the Paris Universal Exposition, 1878. Washington: Government Printing Office, 1878.

Rodgers, Lieutenant Raymond P. *Progress of Work on Panama Ship-Canal* (*Sen. Doc. 123,* 48th Cong., 1st Sess.). Washington: Government Printing Office, 1884.

Rogers, Lieutenant Charles C. *Intelligence Report of the Panama Canal, March 30, 1887* (*House Misc. Doc. 599,* 50th Cong., 1st Sess.). Washington: Government Printing Office, 1889.

Roosevelt, Theodore. *Special Message of the President of the United States Concerning the Panama Canal.* Washington, 1906.

Selfridge, Commander Thomas Oliver. *Reports of Explorations and Surveys to Ascertain the Practicability of a Ship-Canal between the Atlantic and Pacific Oceans by way of the Isthmus of Darien.* Washington: Government Printing Office, 1874.

Shufeldt, Captain Robert W. *Reports of Explorations and Surveys to Ascertain the Practicability of a Ship-Canal between the Atlantic and Pacific Oceans by way of the Isthmus of Tehuantepec* (*Sen. Exec. Doc 6,* 42nd Cong., 2nd Sess.). Washington: Government Printing Office, 1872.

Sullivan, Lieutenant John T. *Report of Historical and Technical Information relating to the Problem of Interoceanic Communication by way of the American Isthmus* (*House Exec. Doc.* 107, 47th Cong., 2nd Sess.). Washington: Government Printing Office, 1883.

The Story of Panama: Hearings on the Rainey Resolution before the Committee on Foreign Affairs of the House of Representatives. Washington: Government Printing Office, 1913.

Testimony Taken Before the Select Committee on the Interoceanic Ship Canal. Washington, 1880.

III. Other Published Primary Sources

Abbot, Willis John. *Panama and the Canal in Picture and Prose.* New York: Syndicate Publishing Co., 1913.

Adams, Charles Francis. *The Panama Canal Zone*. Boston: Massachusetts Historical Society, 1911.

Adams, Frederick Upham. *Conquest of the Tropics: The Story of the Creative Enterprise Conducted by the United Fruit Company*. New York: Doubleday, Page & Co., 1914.

Adams, Henry. *The Education of Henry Adams*. Boston: Houghton Mifflin Co., 1918. (Sentry Edition, 1961.)

——— *Letters*, ed. Worthington Chauncey Ford. 2 vols. Boston: Houghton Mifflin, 1930–1938.

Addresses at the de Lesseps Banquet Given at Delmonico's, March 1, 1880. New York, 1880.

Amicis, Edmondo de. *Studies of Paris*. New York: G. P. Putnam's Sons, 1882.

Ammen, Daniel. *American Isthmian Canal Routes*. Philadelphia,1889.

——— *The Old Navy and the New*. Philadelphia: J. B. Lippincott Co., 1891.

Arango, José Augustín. *Datos para la historia de la independencia del istmo*. Panama, 1922.

Avery, Ralph Emmett. *America's Triumph at Panama*. Chicago: L. W. Walter Co., 1913.

Baedeker, Karl. *Paris and Environs*. London, 1900.

Bancroft, George. *History of the United States, from the Discovery of the Continent*. New York: D. Appleton and Co., 1884.

Bancroft, Hubert Howe. *History of California*. San Francisco: A. L. Bancroft and Co., 1884.

——— *The New Pacific*. San Francisco Bancroft Co., 1912.

Barnard, Henry. *Scientific Schools, Part 1, France. The Polytechnic School at Paris*. 1862.

Barrès, Maurice. *Leurs Figures*. Paris: F. Juven, 1911.

Batbedat, Th. *DeLesseps Intime*. Paris, 1899.

Bates, Lindon Wallace. *Retrieval at Panama*. New York, 1907.

Bennett, Ira E., ed. *History of the Panama Canal: Its Construction and Builders*. Washington: Historical Publishing Co., 1915.

Bertrand, Alphonse, and E. Ferrier. *Ferdinand de Lesseps*. Paris: G. Charpentier et Cie., 1887.

Bidwell, Charles. *The Isthmus of Panama*. London: Chapman & Hall, 1865.

Bigelow, John. *The Panama Canal. Report of the Hon. John Bigelow, Delegated by the Chamber of Commerce of New York to Assist at the Inspection of the Panama Canal in February, 1886*. New York: Press of the Chamber of Commerce, 1886.

Bishop, Farnham. *Panama, Past and Present*. New York: D. Appleton-Century Co., 1913.

Bishop, Joseph Bucklin. *The Panama Gateway*. New York: Charles Scribner's Sons, 1913.

——— *Theodore Roosevelt and His Time*. 2 vols. New York: Charles Scribner's Sons, 1920.

Bishop, Joseph Bucklin and Farnham. *Goethals: Genius of the Panama Canal. A Biography*. New York: Harper and Bros., 1930.

Bryce, James. *South America. Observations and Impressions.* New York: Macmillan Co., 1912.

Bunau-Varilla, Philippe. *From Panama to Verdun: My Fight for France.* Philadelphia: Dorrance and Co., 1940.

——— *The Great Adventure of Panama.* New York: Doubleday, Page & Co., 1920.

——— *Panama: The Creation, Destruction, and Resurrection.* New York: Robert M. McBride, 1920.

——— *Panama or Nicaragua?* (pamphlet). New York, 1901.

Butler, Benjamin F. *Butler's Book.* Boston: A. M. Thayer & Co., 1892.

Cermoise, Henri. *Deux Ans à Panamá. Notes et Récits d'un Ingénieur au Canal.* Paris: C. Marpon et E. Flammarion, 1886.

Chatfield, Mary A. *Light on Dark Places at Panama. By an Isthmian Stenographer.* New York: Broadway Publishing Co., 1908.

Collins, John Owen. *The Panama Guide.* Panama: Vibert and Dixon, 1912.

Colquhoun, Archibald Ross. *The Key of the Pacific: The Nicaragua Canal.* London: Archibald Constable and Co., 1895.

Communication of the Board of Directors of the Panama Railroad Company to the Stockholders. New York, 1853.

Congrès International d'Études du Canal Interocéanique. *Compte Rendu des Séances.* Paris, 1879.

Conrad, Joseph. *Nostromo.* New York: Dell Publishing Co., 1960.

Cornish, Vaughan. *The Panama Canal and Its Makers.* London: T. Fisher Unwin, and Boston: Little, Brown & Co., 1909.

Cullen, Edward. *The Isthmus of Darien Ship Canal.* London: Effingham Wilson, 1852.

Cullen, Edward, and others. *Over Darien by a Ship Canal. Reports of the Mismanaged Darien Expedition of 1854, with Suggestions for a Survey by Competent Engineers, and an Exploration by Parties with Compasses.* London: Effingham Wilson, 1856.

Cullom, Shelby Moore. *Fifty Years of Public Service.* Chicago: A. C. McClurg and Co., 1911.

Davis, Richard Harding. *Captain Macklin.* New York: Charles Scribner's Sons, 1906.

Dean, Arthur H. *William Nelson Cromwell.* Privately published, New York, 1957.

Dreyfus, Alfred and Pierre. *The Dreyfus Case.* New Haven: Yale University Press, 1937.

Drumont, Édouard. *La Dernière Bataille.* Paris: E. Dentu, 1890.

Eads, James Buchanan. *Address . . . Before the House Select Committee on Inter-oceanic Canals, 9th of March, 1880, in Reply to Count de Lesseps.* (No pl., no d.)

——— *Inter-oceanic Ship Railway. Address . . . Delivered Before the San Francisco Chamber of Commerce, August 11, 1880.* (No pl., no d.)

Edwards, Albert. *Panama: The Canal, the Country and the People.* New York: Macmillan Co., 1913.

Fabens, Joseph. *A Story of Life on the Isthmus.* New York: G. P. Putnam & Co., 1853.

Forbes-Lindsay, Charles H. A. *Panama. The Isthmus and the Canal.* Philadelphia: J. C. Winston Co., 1906.
Forot, Victor. *L'Ingénieur Godin de Lépinay.* Paris, 1910.
Franck, Harry A. *Zone Policeman 88. A Close Range Study of the Panama Canal and Its Workers.* New York: Century Co., 1913.
Fraser, John Foster. *Panama and What It Means.* London: Cassell and Co., 1913.
Freehoff, Joseph C. *America and the Canal Title.* New York, 1916.
Froude, James Anthony. *The English in the West Indies.* New York: Charles Scribner's Sons, 1888.
Gerster, Arpad. *Recollections of a New York Surgeon,* New York, 1917.
Gilbert, James Stanley. *Panama Patchwork.* Panama: Star & Herald Co., 1905.
Gorgas, Marie D., and Burton J. Hendrick. *William Crawford Gorgas: His Life and Work.* New York: Doubleday, Page & Co., 1924.
Gorgas, General William Crawford. *Sanitation in Panama.* New York: D. Appleton and Co., 1915.
Griswold, C. D. *The Isthmus of Panama, and What I Saw There.* New York, 1852.
Haggard, H. Rider. *King Solomon's Mines.* New York: Airmont Publishing Co., 1967.
Hamley, W. G. *A New Sea and an Old Land.* London: William Blackwood and Sons, 1871.
Handy Guide to the City of Washington. Chicago: Rand McNally & Co., 1899.
Hanotaux, Gabriel. *Contemporary France,* Vol. IV (1877–1882), New York: G. P. Putnam's Sons, 1909.
Hardeveld, Rose van. *Make the Dirt Fly.* Hollywood, Calif.: Pan Press, 1956.
Haskin, Frederic J. *The Panama Canal.* New York: Doubleday, Page & Co., 1914.
Haskins, William C., ed. *Canal Zone Pilot.* Panama: Star & Herald Co., 1908.
Heiser, Victor. *An American Doctor's Odyssey.* New York: W. W. Norton & Co., 1936.
Howard, Leland O. *The Insect Book.* New York: Doubleday, Page & Co., 1902.
Hughes, Lt. Col. George W. *Letter in answer to the Hon. John M. Clayton, Secretary of State, on Intermarine Communications.* Washington, 1850.
Humboldt, Alexander von. *Political Essay on the Kingdom of New Spain.* London: Longman, 1811.
Humboldt, Alexander von, and Aimé Bonpland. *Personal Narrative of Travels to the Equinoctial Regions of America, During the Years 1799–1804.* London: George Bell and Sons, 1881.
Huntington, C. P. *The Nicaragua Canal* (pamphlet). 1900.
Hyndman, H. M. *Clemenceau, the Man and His Times.* New York: Frederick A. Stokes Co., 1919.
The Isthmus of Panama Inter-Oceanic Canal M. le Comte de Lesseps at Liverpool. Exeter, England, 1880.

Johnson, Theodore T. *Sights in the Gold Region, and Scenes by the Way*. New York, 1849.
Johnson, Willis Fletcher. *Four Centuries of the Panama Canal*. New York: Henry Holt & Co., 1906.
Jusserand, J. J. *What Me Befell*. Boston: Houghton Mifflin Co., 1933.
Karner, William J. *More Recollections*. Boston, 1921.
Kelley, Frederick M. *The Union of the Oceans by Ship-canal Without Locks, via the Atrato Valley*. New York: Harper and Bros., 1859.
Lawton, Frederick. *The French Third Republic*. Philadelphia: J. B. Lippincott Co., 1909.
Lejeune, Major General John A. *The Reminiscences of a Marine*. Philadelphia: Dorrance and Co., 1930.
Le Prince, Joseph A., A. J. Orenstein, and L. O. Howard. *Mosquito Control in Panama*. New York: G. P. Putnam's Sons, 1916.
Lesseps, Ferdinand de. *Lettres, Journal et Documents*. 5 vols. Paris: Didier et Cie., 1875–1881.
——— *Recollections of Forty Years*. New York: D. Appleton and Co., 1888.
——— *The Suez Canal, Letters and Documents Descriptive of Its Rise and Progress in 1854–1856*. London: Henry S. King and Co., 1876.
Lonergan, W. F. *Forty Years of Paris*. New York: Brentano's, 1907.
McCarty, Mary L. *Glimpses of Panama and the Canal*. Kansas City, 1913.
Mahan, Captain A. T. *From Sail to Steam, Recollections of a Naval Life*. New York: Harper and Bros., 1907.
——— *The Influence of Sea Power upon History*. Boston: Little, Brown & Co., 1890.
Maréchal, Henri. *Voyage d'un Actionnaire à Panamá*. Paris: E. Dentu, 1885.
Meyer, Arthur. *Forty Years of Parisian Society*. London, 1912.
Mimande, Paul. *Souvenirs d'un Échappé de Panamá*. Paris: Perrin et Cie., 1893.
Mitchell, Edward P. *Memoirs of an Editor*. New York: Charles Scribner's Sons, 1924.
Moore, Charles, ed. *The St. Mary's Canal*. Detroit, 1907.
Morison, Elting E., ed. *The Letters of Theodore Roosevelt*. Vols. I–V. Cambridge, Mass.: Harvard University Press, 1952.
Morison, George S. *The Isthmian Canal* (a lecture delivered before the Contemporary Club, Bridgeport, Conn.), 1902.
Munchow, Mrs. Ernst Ulrich von, ed. *The American Woman on the Panama Canal, 1904–1916*. Panama: Star & Herald Co., 1916.
Nelson, Wolfred. *Five Years at Panama: The Trans-isthmian Canal*. New York: Belford Co., 1889.
Otis, Fessenden Nott. *History of the Panama Railroad; and of the Pacific Mail Steamship Company. Together with a Traveller's Guide and Business Man's Hand-book for the Panama Railroad*. New York: Harper and Bros., 1867.
Pennell, Joseph. *Joseph Pennell's Pictures of the Panama Canal*. Philadelphia: J. B. Lippincott Co., 1912.
Pepperman, Walter Leon. *Who Built the Panama Canal?* New York: E. P. Dutton & Co., 1915.

Pim, Bedford Clapperton. *The Gate of the Pacific*. London: L. Reeve & Co., 1863.

Ponsolle, Paul. *Le Tombeau des Milliards*. Paris, 1893.

Rainey, Congressman Henry T. *The Story of a Trip to Panama* (pamphlet). Washington, 1907.

Richardson, Albert D. *Personal History of Ulysses S. Grant*. Hartford: American Publishing Co., 1885

Robinson, Tracy. *Fifty Years at Panama, 1861–1911*. New York: Trow Press, 1911.

Rodrigues, José Carlos. *The Panama Canal. Its History, Its Political Aspects, and Financial Difficulties*. New York: Charles Scribner's Sons, 1885.

Roosevelt, Theodore. *An Autobiography*. New York: Charles Scribner's Sons, 1920.

Ross, Sir Ronald. *Memoirs*. New York: E. P. Dutton & Co., 1923.

——— *Mosquito Brigades*. New York: Longmans, Green & Co., 1902.

Sala, George Augustus. *Paris Herself Again* (dispatches from the London *Daily Telegraph*). London: Golden Gallery Press, 1948.

Sands, William Franklin. *Our Jungle Diplomacy*. Chapel Hill: University of North Carolina Press, 1944.

Schierbrand, Wolf von. *America, Asia and the Pacific*. New York: Henry Holt & Co., 1904.

Scrapbook kept by W. W. Wheildon of Concord, Mass., 1879–1884, Boston Public Library.

Shaw, Albert. *A Cartoon History of Roosevelt's Career*. New York: *Review of Reviews*, 1910.

Sibert, William L., and John F. Stevens. *The Construction of the Panama Canal*. New York: D. Appleton and Co., 1915.

Siegfried, André. *Suez and Panama*. New York: Harcourt, Brace & Co., 1940.

Society of the Chagres. *Yearbook, 1913*. Culebra, C.Z.

Sonderegger, C. *L'Achèvement du Canal de Panamá*. Paris: Veuve C. Dunod, 1902.

Stark, James H. *Stark's History and Guide to Barbados and the Caribbean Islands*. Boston, 1903.

Steffens, Lincoln. *The Autobiography of Lincoln Steffens*. New York: Harcourt, Brace & Co., 1931.

Stephens, John Lloyd. *Incidents of Travel in Central America, Chiapas, and Yucatán*. New York: Harper and Bros., 1841.

Stevens, John Frank. *An Engineer's Recollections*. New York: McGraw-Hill, 1936.

Sullivan, Mark. *Our Times*, Vols. I and II. New York: Scribner's Sons, 1928.

Taylor, Bayard. *Eldorado, or, Adventures in the Path of Empire*. New York: G. P. Putnam, and London: Richard Bentley, 1850.

Tomes, Robert. *Panama in 1855: An Account of the Panama Railroad, of the Cities of Panama and Aspinwall, with Sketches of Life and Character on the Isthmus*. New York: Harper and Bros., 1855.

Trollope, Anthony. *The West Indies and the Spanish Main*. New York: Harper and Bros., 1860.

Tyson, James L., M.D. *Diary of a Physician in California*. New York, 1850.

Vassili, Count Paul. *France from Behind the Veil*. London: Cassell & Co., 1914.

Verne, Jules. *Around the World in Eighty Days*, 1873.

——— *Twenty Thousand Leagues Under the Sea*, 1873.

Vizetelly, Ernest A. *Court Life of the Second French Empire*. New York: Charles Scribner's Sons, 1907.

——— *Paris and Her People Under the Third Republic*. New York: Frederick A. Stokes Co., 1919.

——— *Republican France, 1870–1912*. Boston: Small, Maynard & Co., 1912.

Washburne, E. B. *Recollections of a Minister to France, 1869–1877*, Vol. 1. New York: Charles Scribner's Sons, 1887.

Wells, H. G. *The Future in America*. New York: Harper and Bros., 1906.

Whitehead, Richard H. *Our Faith Moved Mountains*. Newcomen Society, 1944.

Winthrop, Theodore. *The Canoe and the Saddle. Adventures Among the Northwestern Rivers and Forests; and Isthmiana*. Boston: James R. Osgood and Co., 1871.

Wood, Robert E. *Monument for the World*. Chicago: Encyclopaedia Britannica, Inc., 1963.

Wyse, Lucien Napoleon-Bonaparte. *Canal Interocéanique 1876–77. Rapport Sur les Études de la Commission Internationale d'Exploration de l'Isthme du Darien*. Paris: A. Chaix et Cie., 1877.

——— *Le Canal de Panamá, l'Isthme Américain. Explorations, Comparison des Tracés Études; Négociations; État des Travaux. . . .* Paris: Hachette et Cie., 1886.

IV. Secondary Sources

Allott, Kenneth. *Jules Verne*. London: Cresset Press, 1940.

Anderson, C. L. G. *Old Panama and Castilla del Oro*. Washington: Sudworth Co., 1911.

Aronson, Theo. *The Fall of the Third Napoleon*. Indianapolis: Bobbs-Merrill Co., 1970.

Artz, Frederick B. *The Development of Technical Education in France, 1500–1850*. Cambridge: M.I.T. Press, 1966.

Auchmuty, James Johnston. *Sir Thomas Wyse, 1791–1862. The Life and Career of an Educator and Diplomat*. London: P. S. King and Son, 1939.

Beale, Howard K. *Theodore Roosevelt and the Rise of America to World Power*. Baltimore: Johns Hopkins Press, 1956. (Paperback edition, Collier Books, New York, 1970.)

Beatty, Charles. *De Lesseps of Suez: The Man and His Times*. New York: Harper and Bros., 1956.

Beer, Thomas. *Hanna*. New York: Alfred A. Knopf, 1929.

Bertaut, Jules. *Paris, 1870–1935*. New York: Appleton-Century, 1936.

Biesanz, John and Mavis. *The People of Panama*. New York: Columbia University Press, 1955.

Birmingham, Stephen. *"Our Crowd": The Great Jewish Families of New York*. New York: Harper and Row, 1967.

Bonaparte-Wyse, Olga. *The Spurious Brood.* London: Victor Gollancz, 1969.
Brogan, D. W. *France Under the Republic: The Development of Modern France (1870–1939).* New York: Harper, 1940.
Bruun, Geoffrey. *Clemenceau.* Cambridge: Harvard University Press, 1943.
Burchill, S. C. *Building the Suez Canal.* New York American Heritage Publishing Co., 1966.
Burnett, Robert. *The Life of Paul Gauguin.* New York: Oxford University Press, 1937.
Byrnes, Howard Fertig. *Antisemitism in Modern France.* New York: Rutgers University Press, 1969.
Cameron, Ian. *The Impossible Dream: The Building of the Panama Canal.* New York: William Morrow & Co., 1972.
Carles, Ruben D. *The Centennial City of Colón.* Panama, 1952.
Chapman, Frank M. *My Tropical Air Castle, Nature Studies in Panama.* New York: D. Appleton and Co., 1929.
Chapman, Guy. *The Third Republic of France.* New York: St. Martin's Press, 1962.
Clapp, Margaret. *Forgotten First Citizen: John Bigelow.* Boston: Little, Brown & Co., 1947.
Courau, Robert. *Ferdinand de Lesseps. De l'Apothéose de Suez au Scandale de Panamá.* Paris: Bernard Grasset, 1932.
Daniels, Josephus. *The Wilson Era, Years of Peace, 1910–1917.* Chapel Hill: University of North Carolina Press, 1944.
Dansette, Adrien. *Les Affaires de Panamá.* Paris: Perrin, 1934.
Dennis, A. L. P. *Adventures in American Diplomacy.* New York: E. P. Dutton & Co., 1927.
Dondo, Mathurin. *The French Faust, Henri de Saint-Simon.* New York: Philosophical Library, 1955.
Dutton, William S. *DuPont, One Hundred and Forty Years.* New York: Charles Scribner's Sons, 1942.
DuVal, Captain Miles P., Jr. *Cadiz to Cathay: The Story of the Long Struggle for a Waterway Across the American Isthmus.* Stanford University: Stanford University Press, and London: Oxford University Press, 1940.
——— *And the Mountains Will Move.* Stanford University: Stanford University Press, 1947.
Edgar-Bonnet, G. *Ferdinand de Lesseps.* Paris: Plon, 1951.
Evans, I. O. *Jules Verne and His Work.* London: Arco Publications, 1965.
Freeman, T. W. *A Hundred Years of Geography.* Chicago: Aldine Publishing Co., 1961.
Giedion, Sigfried. *Space, Time and Architecture.* Cambridge: Harvard University Press, 1971.
Hagedorn, Hermann. *The Roosevelt Family of Sagamore Hill.* New York: Macmillan Co., 1954.
Harriss, Joseph. *The Tallest Tower, Eiffel and the Belle Epoque.* Boston: Houghton Mifflin, 1975.
Hart, Albert Bushnell, ed. *Theodore Roosevelt Cyclopedia.* New York: Roosevelt Memorial Association, 1941.

Herring, Hubert. *A History of Latin America.* New York: Alfred A. Knopf, 1968.

Hill, Howard Copeland. *Roosevelt and the Caribbean.* Chicago: University of Chicago Press, 1927.

Howarth, David. *Panama, Four Hundred Years of Dreams and Cruelty.* New York: McGraw-Hill, 1966.

Howe, Octavius Thorndike. *Argonauts of '49.* Cambridge: Harvard University Press, 1923.

Jessup, Philip C. *Elihu Root,* Vol. I. New York: Dodd, Mead & Co., 1938.

Keller, Morton, ed. *Theodore Roosevelt.* New York: Hill and Wang, 1967.

Kemble, John Haskell. *The Panama Route, 1848–1869.* Berkeley: University of California Press, 1943.

Lee, W. Storrs. *The Strength to Move a Mountain.* New York: G. P. Putnam's Sons, 1958.

Leech, Margaret. *In the Days of McKinley.* New York: Harper and Bros., 1959.

Lewis, Oscar. *Sea Routes to the Gold Fields.* New York: Alfred A. Knopf, 1949.

Lofts, Norah, and Margery Weiner. *Eternal France, A History of France, 1789–1944.* Garden City: Doubleday & Co., 1968.

Lorant, Stefan. *The Life and Times of Theodore Roosevelt.* Garden City: Doubleday & Co., 1959.

Lord, Walter. *The Good Years.* New York: Harper and Bros., 1960.

Mack, Gerstle. *The Land Divided.* New York: Alfred A. Knopf, 1944. (An excellent overall history of the canal based on careful scholarship.)

Manuel, Frank E. *The New World of Henri Saint-Simon.* Cambridge: Harvard University Press, 1956.

Marlowe, John. *World Ditch.* New York: Macmillan Co., 1964.

Martin, Franklin. *Major General William Crawford Gorgas.* Chicago: Gorgas Memorial Institute. (No d.)

Miner, Dwight Carroll. *The Fight for the Panama Route.* New York: Columbia University Press, 1940.

Minter, John Easter. *The Chagres.* New York: Rinehart & Co., 1948.

Morison, George Abbot. *George Shattuck Morison, 1842–1903.* Peterborough, N.H.: Peterborough Historical Society, 1932.

Morison, Samuel Eliot. *Admiral of the Ocean Sea.* Boston: Little, Brown & Co., 1942.

——— *The European Discovery of America, The Southern Voyages, 1492–1616.* New York: Oxford University Press, 1974.

Murphy, Agnes. *The Ideology of French Imperialism, 1871–1881.* Washington: Catholic University of America Press, 1948.

Padelford, Norman J. *The Panama Canal in Peace and War.* New York: Macmillan Co., 1942.

Perruchot, Henri. *Gauguin.* New York: World Publishing Co., 1913.

Pomfret, John E., ed. *California Gold Rush Voyages, 1848–1849.* San Marino, Calif.: Huntington Library, 1954.

Pringle, Henry F. *The Life and Times of William Howard Taft.* 2 vols.

New York: Holt, Rinehart, and Winston, 1939. (Archon Books reprint, 1964.)

——— *Theodore Roosevelt: A Biography*. New York: Harcourt, Brace & Co., 1931.

Puleston, W. D. *Mahan: The Life and Work of Captain Alfred Thayer Mahan*. New Haven: Yale University Press, 1939.

Roosevelt, Nicholas. *The Restless Pacific*. New York: Charles Scribner's Sons, 1928.

——— *Theodore Roosevelt, The Man As I Knew Him*. New York: Dodd, Mead & Co., 1967.

Schonfield, Hugh J. *Ferdinand de Lesseps*. London: Herbert Joseph, 1937.

Schott, Joseph L. *Rails Across Panama*. Indianapolis: Bobbs-Merrill Co., 1967.

Simon, Maron J. *The Panama Affair*. New York: Charles Scribner's Sons, 1971.

Smith, G. Barnett. *The Life and Enterprises of Ferdinand de Lesseps*. London: W. H. Allen, 1895.

Taylor, Charles Carlisle. *The Life of Admiral Mahan*. London: John Murray, 1920.

Thayer, William Roscoe. *The Life and Letters of John Hay*. Boston: Houghton Mifflin Co., 1915.

Theiler, Max, and W. G. Downs. *Arthropod-Borne Viruses of Vertebrates*. New Haven: Yale University Press, 1973.

Tuchman, Barbara W. *The Proud Tower, A Portrait of the World Before the War, 1890–1914*. New York: Macmillan Co., 1966.

Wagenknecht, Edward. *The Seven Worlds of Theodore Roosevelt*. New York: Longmans, Green & Co., 1960.

Warshaw, Leon J. *Malaria, The Biography of a Killer*. New York: Rinehart & Co., 1949.

Washington, D.C., A Guide to the Nation's Capital. American Guide Series. New York: Hastings House, 1942.

Weigley, Russell F. *History of the United States Army*. New York: Macmillan Co., 1967.

Whitehead, Richard H. *Our Faith Moved Mountains*. New York: Newcomen Society, 1944.

Williams, Wythe. *The Tiger of France*. New York: Duell, Sloan, & Pearce, 1949.

Williamson, Harold F., and Kenneth H. Myers, II. *Designed for Digging*. Evanston: Northwestern University Press, 1955.

V. Interviews

Mme. Hervé Alphand, Paris
Alice Anderson, Vineyard Haven, Mass.
Philippe Bunau-Varilla, II, New York City
Crede Calhoun, Panama City
Winter Collins, Balboa Heights, C.Z.
Lavinia Dahlhoff, St. Petersburg, Fla.
Arthur H. Dean, New York City

Katharine Harding Deeble, West Tisbury, Mass.
Winifred Ewing, Panama City
John Fitzgerald, Amherst, Mass.
David St. Pierre Gaillard, Washington, D.C.
Mrs. Thomas Goethals, Vineyard Haven, Mass.
Dr. Victor Heiser, New York City
Walter L. Hersh, St. Petersburg, Fla.
Keith E. Kelley, St. Petersburg, Fla.
Hubert de Lesseps, Paris
Tauni de Lesseps, New York City
Prisca Bunau-Varilla Lionelli, Paris
Alice Roosevelt Longworth, Washington, D.C.
Aminta Meléndez, Colón
Elting E. Morison, Peterborough, N.H.
William E. Russell, New York City
General Edwin L. Sibert, West Tisbury, Mass.
Maurice Thatcher, Washington, D.C.
Mary Weller, St. Petersburg, Fla.
Richard H. Whitehead, Laconia, N.H.
Mrs. Richard H. Whitehead, Laconia, N.H.
Robert Worsley, Panama City
Aileen Gorgas Wrightson, Washington, D.C.

VI. Newspapers, Magazines, and Technical Journals

American Geographical Society Bulletin
American Heritage
American Magazine
American Society of Civil Engineers, *Proceedings*
American Society of Civil Engineers, *Transactions*
Atlanta *Constitution*
Atlantic Monthly
Boston *Evening Transcript*
Boston *Herald*
British Medical Journal
Brooklyn *Eagle*
Bulletin de la Société de Géographie
Bulletin du Canal Interocéanique
Bulletin of the Pan-American Union
California Historical Society Quarterly
Catholic World
The Century
Chicago *Tribune*
Civil Engineering
Collier's
Contemporary Review (London)
El Derecho (Bogotá)
Engineering (London)
Engineering Magazine

Engineering News
Engineering News-Record
Everybody's Magazine
Le Figaro (Paris)
Forum
General Electric Review
Harper's New Monthly Magazine
Harper's Weekly
Illustrated London News
Independent
Journal of the American Geographical Society
Journal of the Franklin Institute
Journal of Inter-American Studies and World Affairs
Journal of the Society of Arts
The Leisure Hour
Leslie's Newspaper
La Libre Parole (Paris)
Life
Lippincott's Magazine
Literary Digest
London *Daily News*
Louisville *Courier-Journal*
Martha's Vineyard *Gazette*
Le Matin (Paris)
Medical Journal
Montana History
Munsey's Magazine
Nation
National Geographic Magazine
New York *Daily Commercial Bulletin*
New York *Economist*
New York *Evening Post*
New York *Herald*
New York *Journal*
New York *Journal of Commerce*
New York *Sun*
The New York Times
New York *Tribune*
New York *World*
Newcomen Society, *Transactions*
North American Review
The Outlook
Pacific Historical Review
Pall Mall Gazette
Panama Canal Spillway
Panama *Star & Herald*
El Relator (Bogotá)
Review of Reviews

San Francisco *Chronicle*
San Francisco *Examiner*
The Saturday Evening Post
Scientific American
Scribner's Magazine
Tacoma *Evening News*
Le Temps
The Times (London)
United States Naval Proceedings
The University Magazine
Washington *Post*
Washington *Star*
World's Work

VII. Pictorial Sources

American Geographical Society
American Museum of Natural History
Clarence A. Barnes, Jr.
Bibliothèque Nationale
Canal Zone Library-Museum
Winter Collins
Culver Picture Services
Mrs. Katharine Deeble
École Polytechnique
David St. Pierre Gaillard
Mrs. Thomas Goethals
Library of Congress
Museum of Modern Art
National Archives
New York Public Library
Smithsonian Institution
United States Military Academy Archives

VIII. Reference Works

The American Heritage Dictionary of the English Language. Boston: Houghton Mifflin Co., 1969.

Bartlett, John. *Bartlett's Familiar Quotations.* Boston: Little, Brown & Co., 1955.

Dictionary of American Biography. New York: Charles Scribner's Sons, 1936.

Dictionary of American History. New York: Charles Scribner's Sons, 1942.

Dictionary of National Biography. London: Oxford University Press, 1921.

Encyclopaedia Britannica. 15th ed. Chicago: Encyclopaedia Britannica, 1974.

Morison, Samuel Eliot, and Henry Steele Commager. *The Growth of the American Republic.* New York: Oxford University Press, 1960.

National Geographic Atlas of the World. Rev. 3rd ed. Washington: National Geographic Society, 1970.

Webster's Geographical Dictionary. Springfield: G. & C. Merriam Co., 1949.

Index

Index